Advances in Neurobiology

Volume 16

Series Editor
Arne Schousboe

More information about this series at http://www.springer.com/series/8787

Arturo Ortega • Arne Schousboe
Editors

Glial Amino Acid Transporters

Springer

Editors
Arturo Ortega
Department of Toxicology
Cinvestav-IPN México City
México, Mexico

Arne Schousboe
Department of Drug Design
and Pharmacology
Faculty of Health and Medical Sciences
University of Copenhagen
Copenhagen, Denmark

ISSN 2190-5215 ISSN 2190-5223 (electronic)
Advances in Neurobiology
ISBN 978-3-319-55767-0 ISBN 978-3-319-55769-4 (eBook)
DOI 10.1007/978-3-319-55769-4

Library of Congress Control Number: 2017946841

Printed on acid-free paper

This Springer imprint is published by Springer Nature
The registered company is Springer International Publishing AG
The registered company address is: Gewerbestrasse 11, 6330 Cham, Switzerland

Contents

Contributors

Penélope Aguilera Laboratorio de Patología Vascular Cerebral, Instituto Nacional de Neurología y Neurocirugía "Manuel Velasco Suárez", Ciudad de México, México

Anas Al-Khawaja Department of Drug Design and Pharmacology, Faculty of Health and Medical Sciences, University of Copenhagen, Copenhagen, Denmark

Ivan Alquisiras-Burgos Laboratorio de Patología Vascular Cerebral, Instituto Nacional de Neurología y Neurocirugía "Manuel Velasco Suárez", Ciudad de México, México

Cynthia M. Anderson Biological Sciences Department, California State Polytechnic University, Pomona, CA, USA

Carmen Aragón Departamento de Biología Molecular, Centro de Biología Molecular "Severo Ochoa", Universidad Autónoma de Madrid-Consejo Superior de Investigaciones Científicas, Madrid, Spain

IdiPAZ-Hospital Universitario La Paz, Universidad Autónoma de Madrid, Madrid, Spain

Centro de Investigación Biomédica en Red sobre Enfermedades Raras, ISCIII, Madrid, Spain

Marina Arribas-Blázquez Centro de Biología Molecular Severo Ochoa, Facultad de Ciencias, Consejo Superior de Investigaciones Científicas, Universidad Autónoma de Madrid, Madrid, Spain

Centro de Investigación Biomédica en Red de Enfermedades Raras and IdiPAZ, Instituto de Salud Carlos III, Madrid, Spain

Michael Aschner Department of Molecular Pharmacology, Albert Einstein College of Medicine, Bronx, NY, USA

David Bartolomé-Martín Centro de Biología Molecular Severo Ochoa, Facultad de Ciencias, Consejo Superior de Investigaciones Científicas, Universidad Autónoma de Madrid, Madrid, Spain

Centro de Investigación Biomédica en Red de Enfermedades Raras and IdiPAZ, Instituto de Salud Carlos III, Madrid, Spain

Cristina Benito-Muñoz Departamento de Biología Molecular, Centro de Biología Molecular "Severo Ochoa", Universidad Autónoma de Madrid-Consejo Superior de Investigaciones Científicas, Madrid, Spain

Ole Jannik Bjerrum Department of Drug Design and Pharmacology, Faculty of Health and Medical Sciences, University of Copenhagen, Copenhagen, Denmark

Birger Brodin Department of Pharmacy, The Faculty of Health and Medical Sciences, University of Copenhagen, Copenhagen, Denmark

Donají Chi-Castañeda Laboratorio de Neurotoxicología, Departamento de Toxicología, Centro de Investigación y de Estudios Avanzados del Instituto Politécnico Nacional, Ciudad de México, México

Soluciones para un México Verde S.A. de C.V., Ciudad de México, México

Andrew N. Clarkson Department of Anatomy, Brain Health Research Centre and Brain Research New Zealand, University of Otago, Dunedin, New Zealand

Faculty of Pharmacy, The University of Sydney, Sydney, NSW, Australia

Rasmus Prætorius Clausen Department of Drug Design and Pharmacology, Faculty of Health and Medical Sciences, University of Copenhagen, Copenhagen, Denmark

Maria Damgaard Department of Drug Design and Pharmacology, Faculty of Health and Medical Sciences, University of Copenhagen, Copenhagen, Denmark

Gerhard F. Ecker Department of Pharmaceutical Chemistry, University of Vienna, Vienna, Austria

Roberto Elizondo-Vega Departamento de Biología Celular, Universidad de Concepción, Concepción, Chile

Sepehr Eskandari Biological Sciences Department, California State Polytechnic University, Pomona, CA, USA

Mónica Espinoza-Rojo Laboratorio de Biología Molecular y Genómica, Universidad Autónoma de Guerrero, Chilpancingo, Guerrero, México

Bente Frølund Department of Drug Design and Pharmacology, Faculty of Health and Medical Sciences, University of Copenhagen, Copenhagen, Denmark

María Angeles García-Robles Departamento de Biología Celular, Universidad de Concepción, Concepción, Chile

Georgi Gegelashvili Institute of Chemical Biology, Ilia State University, Tbilisi, Georgia

Department of Drug Design and Pharmacology, Faculty of Health and Medical Sciences, University of Copenhagen, Copenhagen, Denmark

Cecilio Giménez Centro de Biología Molecular Severo Ochoa, Facultad de Ciencias, Consejo Superior de Investigaciones Científicas, Universidad Autónoma de Madrid, Madrid, Spain

Centro de Investigación Biomédica en Red de Enfermedades Raras and IdiPAZ, Instituto de Salud Carlos III, Madrid, Spain

Germán Fernando Gutiérrez Aguilar Laboratorio de Patología Vascular Cerebral, Instituto Nacional de Neurología y Neurocirugía "Manuel Velasco Suárez", Ciudad de México, México

Anne Stæhr Haugaard Department of Drug Design and Pharmacology, Faculty of Health and Medical Sciences, University of Copenhagen, Copenhagen, Denmark

Hans Christian Cederberg Helms Department of Pharmacy, The Faculty of Health and Medical Sciences, University of Copenhagen, Copenhagen, Denmark

Peter Hong Department of Physiology, Meharry Medical College, Nashville, TN, USA

Ignacio Ibáñez Centro de Biología Molecular Severo Ochoa, Facultad de Ciencias, Consejo Superior de Investigaciones Científicas, Universidad Autónoma de Madrid, Madrid, Spain

Centro de Investigación Biomédica en Red de Enfermedades Raras and IdiPAZ, Instituto de Salud Carlos III, Madrid, Spain

James Johnson Jr Department of Physiology, Meharry Medical College, Nashville, TN, USA

Pratap Karki Department of Pharmaceutical Sciences, Florida A&M University, Tallahassee, FL, USA

Stefanie Kickinger Department of Pharmaceutical Chemistry, University of Vienna, Vienna, Austria

Povl Krogsgaard-Larsen Department of Drug Design and Pharmacology, Faculty of Health and Medical Sciences, University of Copenhagen, Copenhagen, Denmark

Beatriz López-Corcuera Departamento de Biología Molecular, Centro de Biología Molecular "Severo Ochoa", Universidad Autónoma de Madrid-Consejo Superior de Investigaciones Científicas, Madrid, Spain

IdiPAZ-Hospital Universitario La Paz, Universidad Autónoma de Madrid, Madrid, Spain

Centro de Investigación Biomédica en Red sobre Enfermedades Raras, ISCIII, Madrid, Spain

Eunsook Lee Department of Pharmaceutical Sciences, Florida A&M University, Tallahassee, FL, USA

Maria E.K. Lie Department of Drug Design and Pharmacology, Faculty of Health and Medical Sciences, University of Copenhagen, Copenhagen, Denmark

Carsten Uhd Nielsen Department of Physics, Chemistry and Pharmacy, University of Southern Denmark, Odense, Denmark

Arturo Ortega Departamento de Toxicología, Cinvestav-IPN, México, México

Herminia Pasantes-Morales División de Neurociencias, Instituto de Fisiología Celular, Universidad Nacional Autónoma de México, Mexico City, Mexico

Dolores Piniella Centro de Biología Molecular Severo Ochoa, Facultad de Ciencias, Consejo Superior de Investigaciones Científicas, Universidad Autónoma de Madrid, Madrid, Spain

Centro de Investigación Biomédica en Red de Enfermedades Raras and IdiPAZ, Instituto de Salud Carlos III, Madrid, Spain

Angelina Rodríguez Facultad de Ciencias Naturales, Universidad Autónoma de Querétaro, Querétaro, México

Lauren Taylor Rosenblum Jefferson Weinberg ALS Center, Vickie and Jack Farber Institute for Neuroscience, Department of Neuroscience, Thomas Jefferson University, Philadelphia, PA, USA

Arne Schousboe Department of Drug Design and Pharmacology, Faculty of Health and Medical Sciences, University of Copenhagen, Copenhagen, Denmark

Edna Suárez-Pozos Laboratorio de Neurotoxicología, Departamento de Toxicología, Centro de Investigación y de Estudios Avanzados del Instituto Politécnico Nacional, Ciudad de México, México

Davide Trotti Jefferson Weinberg ALS Center, Vickie and Jack Farber Institute for Neuroscience, Department of Neuroscience, Thomas Jefferson University, Philadelphia, PA, USA

Helle Sønderby Waagepetersen Department of Drug Design and Pharmacology, The Faculty of Health and Medical Sciences, University of Copenhagen, Copenhagen, Denmark

Petrine Wellendorph Department of Drug Design and Pharmacology, Faculty of Health and Medical Sciences, University of Copenhagen, Copenhagen, Denmark

Samantha L. Willford Biological Sciences Department, California State Polytechnic University, Pomona, CA, USA

Francisco Zafra Centro de Biología Molecular Severo Ochoa, Facultad de Ciencias, Consejo Superior de Investigaciones Científicas, Universidad Autónoma de Madrid, Madrid, Spain

Centro de Investigación Biomédica en Red de Enfermedades Raras and IdiPAZ, Instituto de Salud Carlos III, Madrid, Spain

Editors Biographies

Arne Schousboe obtained his M.Sc. in biochemistry from the University of Copenhagen in 1968 at a time where an M.Sc. was considered equivalent to a Ph.D. He subsequently (1978) earned his doctor of science (D.Sc.) degree also from the University of Copenhagen. After a postdoc period with Dr. Eugene Roberts at the Department of Neuroscience, City of Hope National Medical Center in Los Angeles (1972–1973), he came back to the University of Copenhagen to resume a tenured position as associate professor at the Medical Faculty. In 1990, he moved to the Royal Danish School of Pharmacy, Copenhagen, as a full professor of biochemistry and has remained in this position also when this institution changed to become an independent university and subsequently the Faculty of Pharmaceutical Sciences which lately was incorporated in the Faculty of Health and Medical Sciences still at the University of Copenhagen. He has served as the department chair from 2005 to 2010 and is now part-time professor of neuropharmacology in the Department of Drug Design and Pharmacology at the University of Copenhagen. He has been working on astrocyte function focusing on amino acid neurotransmission during the past more than 40 years and is currently engaged in studies of glutamate and GABA homeostasis and metabolism. He has published over 550 papers on these and related topics. He has served on the editorial board of numerous neuroscience journals over the years and is currently the editor in chief of *Neurochemical Research* and editor of the book series *Advances in Neurobiology*.

Arturo Ortega holds a B.Sc. degree in chemistry from the National Autonomous University of Mexico (1985); an M.Sc. in physiology from Cinvestav, Mexico City (1985); and a Ph.D. in neurobiology (1991) from the Weizmann Institute of Science in Rehovot, Israel, under the supervision of the late Vivian I. Teichberg, with whom he started to work in glutamate-mediated signaling in glial cells. After spending 2 years in the Department of Biochemistry in Cinvestav as an assistant professor, he moved to the Department of Genetics and Molecular Biology where he was promoted to associate professor in 1997 and to the rank of professor in 2004. More recently (2013), he moved to the Department of Toxicology where he established a neurotoxicology lab. He has been visiting professor in the Institute of Neuroscience of the Autonomous University of Barcelona (2008) and in the Catholic University of Louvain (2014–2015). He served as the department chair of the Department of Genetics and Molecular Biology. He has been working on glutamate receptors and transporters focusing on signaling and gene expression regulation at the transcriptional and translational levels for more than 30 years. He has published over 100 research articles and mentored 33 Ph.D. and 67 M.Sc. students. He is currently member of the editorial board of two major neurochemistry journals and of the book series *Advances in Neurobiology*.

Manganese Control of Glutamate Transporters' Gene Expression

Eunsook Lee, Pratap Karki, James Johnson Jr, Peter Hong, and Michael Aschner

Abstract Manganese (Mn) is an essential trace element, serving as a cofactor for several enzymes involved in various cellular and biochemical reactions in human body. However, chronic overexposure to Mn from occupational or environmental sources induces a neurological disorder, characterized by psychiatric, cognitive, and motor abnormalities, referred to as manganism. Mn-induced neurotoxicity is known to target astrocytes since these cells preferentially accumulate Mn. Astrocytes are the most abundant non-neuronal glial cells in the brain, and they play a critical role in maintaining the optimal glutamate levels to prevent excitotoxic death. The fine regulation of glutamate in the brain is accomplished by two major glutamate transporters – glutamate transporter-1 (GLT-1) and glutamate aspartate transporter (GLAST) that are predominantly expressed in astrocytes. Excitotoxic neuronal injury has been demonstrated as a critical mechanism involved in Mn neurotoxicity and implicated in the pathological signs of multiple neurodegenerative diseases including Alzheimer's disease, Parkinson's disease, and amyotrophic lateral sclerosis. Recent evidences also establish that Mn directly deregulates the expression and function of both astrocytic glutamate transporters by decreasing mRNA and protein levels of GLT-1 and GLAST. Herein, we will review the mechanisms of Mn-induced gene regulation of glutamate transporters at the transcriptional level and their role in Mn toxicity.

Keywords Manganese • Astrocytes • Glutamate transporters • GLT-1 • GLAST • Yin Yang 1

E. Lee (✉) • P. Karki
Department of Pharmaceutical Sciences, Florida A&M University, Tallahassee, FL 32307, USA
e-mail: eunsook.lee@famu.edu

J. Johnson Jr • P. Hong
Department of Physiology, Meharry Medical College, Nashville, TN 37208, USA

M. Aschner
Department of Molecular Pharmacology, Albert Einstein College of Medicine, Bronx, NY 10461, USA

A. Ortega, A. Schousboe (eds.), *Glial Amino Acid Transporters*, Advances in Neurobiology 16, DOI 10.1007/978-3-319-55769-4_1

1 Introduction

Manganese (Mn) is an abundantly available trace element that is required for normal functioning and development of the central nervous system (CNS) (Takeda 2003). Mn acts as a cofactor for many crucial enzymes such as arginase, pyruvate decarboxylase, superoxide dismutase, and glutamine synthetase (Bentle and Lardy 1976; Stallings et al. 1991; Wedler and Denman 1984; Diez et al. 1992). But, excessive CNS accumulation of Mn may cause toxicity, resembling Parkinson's disease (PD), and is referred to as manganism (Chen et al. 2015; Kwakye et al. 2015). The occupational and environmental sources of Mn exposure include welding, mining, and ferroalloy industries as well as Mn-contaminated drinking water and also from the use of gasoline additive methylcyclopentadienyl manganese tricarbonyl (MMT) and pesticide maneb (Bast-Pettersen et al. 2004; Bowler et al. 2007; Montes et al. 2008; Williams et al. 2012). Mn is transported into the CNS via multiple transporters including transferrin, divalent metal transporter-1 (DMT-1), *N*-methyl-D-aspartate (NMDA) receptor channel, and the divalent metal/bicarbonate ion symporters ZIP8 and ZIP14 (Aschner and Gannon 1994; Au et al. 2008; Fujishiro et al. 2012; Itoh et al. 2008). Once Mn enters into the brain, astrocytes appear to be more vulnerable to Mn toxicity compared to other cell types since they preferentially accumulate Mn (Morello et al. 2008). One of the critical functions of astrocytes in the CNS is to maintain optimal glutamate levels to prevent the excitotoxic neuronal death (Danbolt 2001). Astrocytes express two glutamate transporters – glutamate transporter-1 (GLT-1) and glutamate aspartate transporter (GLAST), also known as excitatory amino acid transporter (EAAT) 1 and 2 in humans, respectively, which are responsible for uptaking more than 80% of extracellular glutamate. Since among the five subtypes of glutamate transporters, GLT-1/EAAT2 and GLAST/EAAT1 carry out most of the glutamate uptake in the CNS, and these astrocytic isoforms are the primary target of Mn toxicity; herein we will focus on the effects of Mn on these two transporters. Mn is known to interfere with the astrocytic glutamate regulation by inhibiting the gene expression of glutamate transporters (Lee et al. 2009, 2012) which will be discussed in the next sections.

2 Astrocytes and Mn Neurotoxicity

Astrocytes are the principal reservoir for Mn accumulation in the brain with the presence of efficient Mn transport system. Astrocytes contain 50–60-fold higher Mn concentration than their neuron counterparts (Morello et al. 2008; Aschner et al. 1992). Further, the preferential sequestration of Mn in mitochondria makes this energy-producing organelle more prone to Mn toxicity by Mn-induced mitochondrial dysfunction and oxidative stress (Erikson et al. 2004; Chen and Liao 2002; Gavin et al. 1999). Mn directly inhibits the enzymes involved in ATP-generating pathways and also activates mitochondrial apoptotic pathway to exert cytotoxic effects (Gavin et al. 1992; Gonzalez et al. 2008). Furthermore, Mn also induces oxidative stress by inhibiting glutathione synthetase, an astrocyte-specific enzyme that is critical for the synthesis of antioxidant glutathione (Erikson et al. 2004, 2006). More importantly, Mn also interferes with the glutamate-glutamine cycle that leads to the imbalance of neurotransmitters, a common trigger for various neurodegenerative disorders (Sidoryk-Wegrzynowicz and Aschner 2013).

2.1 Glutamate Excitotoxicity in Mn Neurotoxicity

Glutamate is the major excitatory neurotransmitter in the CNS, and it plays an important role in various essential brain functions including cognition, learning, and memory (Danbolt 2001). However, the increased extracellular levels of glutamate, followed by the overstimulation of glutamate receptors, induce excitotoxic neuronal injury. The survival and proper functioning of neurons is regulated by astrocytes given that astrocytes not only provide structural, metabolic, and trophic support for neurons but also produce and supply neuronal growth factors and antioxidants (Seifert et al. 2006). Mn-elicited excitotoxicity could result from the interference with the astrocyte function of glutamate uptake or through the activation of glutamate receptors. The study by Brouillet et al. first established that Mn produces excitotoxic lesions in rat striatum by impairing the ATP generation, and treatment with NMDA receptor antagonist MK-801 ameliorates these injuries (Brouillet et al. 1993). These observations were further confirmed in a later study, which showed that MK-801 prevents Mn-induced neurotoxicity (Xu et al. 2010a, b). The same group also showed that Mn causes neurotoxicity in rats by increasing extracellular glutamate, secondary to the altered expression of NMDA receptors (Xu et al. 2010c). Similarly, the role of glutamate receptor activation in Mn neurotoxicity was evident in Mn-caused neuronal loss in globus pallidus where Mn increased the sensitivity of postsynaptic glutamate receptors to glutamate (Spadoni et al. 2000). However, more severe effects of Mn toxicity may be mediated by impairment of astrocyte function caused by reduced expression and function of astrocytic glutamate transporters.

2.2 Mn Inhibition of Glutamate Transporters' Gene Expression

The reduced expression and function of astrocytic glutamate transporters is linked to the pathogenesis of a myriad of neurological disorders including PD, Alzheimer's disease (AD), amyotrophic lateral sclerosis (ALS), stroke, HIV-associated dementia, and glaucoma (Potter et al. 2013; Rao et al. 2001; Robelet et al. 2004; Rothstein et al. 1995; Yanagisawa et al. 2015). Since many of these diseases are also associated with Mn toxicity, it prompted researchers to investigate the effects of Mn on glutamate transporters. The studies revealed that Mn decreases glutamate uptake in astrocytes (Hazell and Norenberg 1997). Mn inhibition of glutamate uptake was further confirmed by another study demonstrating that Mn decreases glutamate uptake in astrocytes by reducing GLAST expression (Erikson and Aschner 2002). Consistently, another study showed that Mn decreases both GLAST and GLT-1-mediated glutamate uptake (Mutkus et al. 2005). A decrease in the expression of GLT-1 and GLAST was also noted in nonhuman primates exposed to Mn although the reduction in expression was dependent on brain areas and exposure duration (Erikson et al. 2007, 2008). Later studies from our group demonstrated that Mn decreases glutamate uptake activity of GLAST by reducing its protein expression and membrane trafficking (Lee et al. 2009). We also showed that Mn decreases the promoter activity, mRNA/protein levels, and activity of GLT-1 in astrocytes (Lee et al. 2012). These studies illustrated that Mn-induced reduction in the expression of transforming growth factor (TGF)-α and -β mediates Mn inhibition of glutamate transporters' expression and function.

3 Mn Induces Glutamate Transporters' Gene Dysregulation

Since Mn reduces the promoter activity as well as mRNA and protein levels of glutamate transporters, it is apparent that Mn acts at the transcription level to exert its repressive effects. However, the mechanism of Mn-induced transcriptional repression of glutamate transporters is not completely known. Multiple intracellular signaling pathways and transcription factors are suggested to mediate the Mn's inhibitory action on glutamate transporters.

3.1 Intracellular Signaling Pathways

Mn is known to activate some intracellular signaling pathways that mediate its effects on glutamate transporters. Among these, protein kinase C (PKC) appears to be one of the major pathways involved in Mn-induced regulation of glutamate transporters. Mn activates PKCα and PKCδ to decrease glutamate uptake, and inhibition of either PKC isoforms reverses Mn-induced reduction of glutamate uptake in astrocytes (Sidoryk-Wegrzynowicz et al. 2011, 2012). Furthermore, inhibition of the PKC pathway also attenuated Mn-induced decrease in protein expression levels of GLT-1 and GLAST (Sidoryk-Wegrzynowicz et al. 2012). These findings established a major role of the PKC pathway in Mn-induced repression of glutamate transporters. The same study also showed that Mn enhances the interaction between GLT-1 and PKCδ and knockdown of PKCδ alleviates the Mn-induced decrease in glutamate uptake (Sidoryk-Wegrzynowicz et al. 2012). The caspase-3-dependent cleavage of PKCδ is also implicated in Mn-induced neurotoxicity (Kitazawa et al. 2005;

Latchoumycandane et al. 2005). Corroborating with these findings, inhibition of caspase-3 with Z-Ala-Glu (OMe)-Val-Asp (OMe)-fluoromethyl+ ketone (Z-VAD-FMK) abrogated Mn-induced decrease in GLT-1 and GLAST protein expression as well as glutamate uptake (Sidoryk-Wegrzynowicz et al. 2012). Moreover, caspase-3-mediated cleavage of GLT-1 results in inactivation of the GLT-1 transporter, suggesting that apoptotic signaling also modulates the glutamate transporters' function (Boston-Howes et al. 2006). Mn activation of PKCs might also result in reduced membrane trafficking of glutamate transporters given that phorbol ester-induced PKC activation has been shown to decrease the cell surface expression of GLT-1 (Kalandadze et al. 2002). A similar role of PKC-induced phosphorylation of GLAST leading to its decreased glutamate uptake activity has been reported (Conradt and Stoffel 1997). In addition to PKCs, several in vitro and in vivo studies have shown that Mn activates other signaling kinases such as extracellular signal-regulated kinase (ERK), c-Jun N-terminal kinase (JNK), p38 mitogen-activated protein kinase, and Akt, but the definitive role of these pathways in Mn-induced downregulation of glutamate transporters remains to be elucidated (Cordova et al. 2012; Ito et al. 2006; Peres et al. 2013; Yin et al. 2008).

3.2 Transcriptional Regulation

Mn-induced inhibition of glutamate transporters starts from promoter levels, so there must be some crucial transcription factors that mediate the repressive effects of Mn on the gene expression of transporters. However, the role of transcription factors in regulating the gene expression of glutamate transporters during Mn toxicity had not been investigated. We recently demonstrated that a transcription factor Yin Yang 1 (YY1) mediates Mn-induced repression of GLT-1 and GLAST (Karki et al. 2014a, 2015). These studies established that Mn activates YY1 to inhibit the expression and function of astrocytic glutamate transporters. Both GLT-1 and GLAST promoters contain consensus-binding sites for YY1, and Mn increased the binding of YY1 to these sites in the promoters. Previous studies have noted the role of YY1 in repressing glutamate transporters, and our findings illustrated that Mn inhibition of glutamate transporters is mediated by YY1 (Lee et al. 2011; Rosas et al. 2007). Multiple studies have shown that various positive modulators of glutamate transporters such as soluble neuronal factors, ceftriaxone, epidermal growth factor, estrogen, and selective estrogen receptor modulators (SERMs) all activate nuclear factor-κB (NF-κB) to upregulate glutamate transporters (Karki et al. 2013, 2014b, 2015; Ghosh et al. 2011; Lee et al. 2008). We demonstrated that Mn-activated YY1 can completely suppress NF-κB-mediated stimulatory effects on glutamate transporters, indicating that the repressive effects of YY1 can easily surpass the positive regulatory pathways (Karki et al. 2014a, 2015). Our studies also showed that tumor necrosis factor-α (TNF-α) facilitates Mn-induced YY1 activation given that Mn treatment increases TNF-α secretion in astrocytes and TNF-α decreases YY1 expression (Karki et al. 2014a). Earlier studies have established that TNF-α is a repressor of glutamate transporters and Mn increases TNF-α expression (Kim et al. 2003; Sitcheran et al. 2005; Su et al. 2003; Zhao et al. 2009). Furthermore, TNF-α increases YY1 expression as well as its DNA-binding activity (Huerta-Yepez et al. 2006). Accordingly, it appears that Mn-TNFα-YY1 activation cascade is responsible for the transcriptional repression of astrocytic glutamate transporters. Further studies are required to investigate if other repressive transcription factors of glutamate transporters such as nuclear factor of activated T cells (NFAT) and N-myc are also involved in Mn-induced repression of glutamate transporters (Sitcheran et al. 2005; Abdul et al. 2009).

3.3 Epigenetic Regulation

Methylation and acetylation represent two major epigenetic regulatory pathways that modulate the expression of glutamate transporters. For example, methylation of the EAAT2 promoter reduces its activity, and inhibition of DNA methyltransferases increases EAAT2 mRNA levels (Zschocke et al. 2007). The increased expression and activity of various histone deacetylases (HDACs) is linked to neurological disorders, and accordingly several HDAC inhibitors have been shown to be neuroprotective against a wide range of neurotoxic insults including glutamate excitotoxicity (Baltan et al. 2011; Bardai and D'Mello 2011; Janssen et al. 2010; Leng et al. 2010). The epigenetic regulation of glutamate transporters was previously demonstrated by a study where valproic acid, a HDAC inhibitor, increases acetylated histone H4 levels in the GLT-1 promoter (Perisic et al. 2010). Direct evidence for the role of HDACs in repressing glutamate transporters was established by our recent studies where overexpression of various HDAC isoforms resulted in decreased glutamate transporters' promoter activities (Karki et al. 2014a, 2015). Furthermore, HDACs were recruited as corepressors by YY1 to inhibit glutamate transporters, and activation of HDACs suppressed stimulatory effects of NF-κB. Given that recruitment of repressor proteins is one of the mechanisms involved in YY1-mediated gene repressions (Shi et al. 1997), Mn-induced inhibition of glutamate transporters occurs with the formation of YY1-HDAC repressor complex that also sequesters NF-κB rendering it inactive. This was further supported by the findings that Mn increases interactions between HDACs, YY1, and p65, suggesting that Mn exerts its inhibitory actions on glutamate transporters by inducing the formation of a transcriptional repressor comprised of YY1, HDACs, and NF-κB. Moreover, the involvement of HDACs in negatively regulating glutamate transporters is further corroborated by findings that a wide range of HDAC inhibitors increase the expression and function of glutamate transporters and attenuate Mn-induced impairment of the transporters (Karki et al. 2014a, 2015).

3.4 *Attenuation of Mn-Induced Glutamate Transporters' Repression*

Mn toxicity is associated with a plethora of neurodegenerative disorders, including AD, PD, HD, and ALS, and current knowledge suggests that Mn-induced impairment of astrocytic glutamate transporters might play a crucial role in triggering the pathogenesis of these diseases (Bowman et al. 2011). The pharmacological compounds that can reverse Mn-induced repression of astrocytic glutamate transporters could be developed as potential therapeutics against the diseases elicited by Mn neurotoxicity and the dysregulation of glutamate transporters. In this regard, the studies from our group have established that estrogen and SERMs could be promising therapeutic candidates to combat Mn toxicity (Lee et al. 2009, 2012; Karki et al. 2014b). The protective effects of estrogen and SERMs might be via production of TGF-α that stimulates transporters expression by activating NF-κB and cAMP response element-binding protein (CREB) pathways (Karki et al. 2013, 2014b). Likewise, activation of the ERK and Akt pathways facilitates the stimulatory effects of estrogenic compounds on glutamate transporters (Lee et al. 2009). The findings that SERMs upregulate glutamate transporters and reverse Mn inhibitory actions have an important clinical significance since these SERMs are already in clinic utilities. For instance, tamoxifen and raloxifene are US Food and Drug Administration (FDA)-approved drugs for breast cancer and osteoporosis, respectively. Given their clinical safety record and ability to attenuate Mn-induced repression of glutamate transporters, the efficacy of SERMs in treating Mn-induced neurological disorders merits further evaluation. It has been shown that riluzole, the only drug for ALS in clinics, exerts protective effects against Mn-induced disruption of expression and function of astrocytic glutamate transporters (Deng et al. 2012). Various HDAC inhibitors are also known to enhance glutamate transporters' expression, and our studies demonstrated that these compounds can attenuate Mn-induced repression of glutamate transporters (Karki et al. 2014a, 2015). As discussed above, the neuroprotective roles of HDAC inhibitors are well appreciated, and with these new findings that they can also offer protection against Mn-caused impairment of glutamate transporters, at least some of these HDAC inhibitors could offer a plausible alternative therapeutics to be developed against glutamate excitotoxicity and Mn toxicity. At the mechanistic level, the protective actions of HDAC inhibitors on Mn toxicity might be due to their ability to interfere with the YY1 pathway. This notion is supported by observations that Mn activates YY1 to repress glutamate transporters and valproic acid, a HDAC inhibitor, decreases YY1 binding to the GLAST promoter, relieving the repressive effects of YY1 on GLAST (Aguirre et al. 2008).

4 Summary

The dysregulation of astrocytic glutamate transporters and ensuing excitotoxicity appears to be one of the major mechanisms involved in Mn neurotoxicity. The accumulating evidences suggest that Mn acts at the transcription level to downregulate glutamate transporters and epigenetic regulation, especially HDACs, which also play a crucial role in this process. At the cellular level, the increased expression of TNF-α with the subsequent activation of the YY1 pathway mediates Mn-induced impairment of astrocytic glutamate transporters. Pharmacological compounds that effectively attenuate Mn inhibition of glutamate transporters could be potential therapeutics against both Mn neurotoxicity and excitotoxicity. To this end, estrogen, SERMs, riluzole, and HDAC inhibitors might be considered as promising therapeutic candidates against the neurological disorders elicited by Mn toxicity-mediated dysfunction of astrocytic glutamate transporters. Future studies could be profitable directed to provide more precise information on the mechanisms by which Mn regulates glutamate transporters' gene expression, paving the way for exploring critical cellular pathways and novel pharmacological compounds with an ultimate goal of developing effective therapeutics against Mn-caused excitotoxicity.

Acknowledgments MA was supported in part by NIH grants, R01 ES 010563, R01 ES 003771, and R01 ES 020852. EL was supported in part by NIH grants, NIGMS SC1089630 and R01 ES024756.

Conflict of Interest The author declares no conflicts of interest.

References

Abdul HM, Sama MA, Furman JL, Mathis DM, Beckett TL, Weidner AM, Patel ES, Baig I, Murphy MP, LeVine H 3rd, Kraner SD, Norris CM. Cognitive decline in Alzheimer's disease is associated with selective changes in calcineurin/NFAT signaling. J Neurosci Off J Soc Neurosci. 2009;29:12957–69.

Aguirre G, Rosas S, Lopez-Bayghen E, Ortega A. Valproate-dependent transcriptional regulation of GLAST/EAAT1 expression: involvement of Ying-Yang 1. Neurochem Int. 2008;52:1322–31.

Aschner M, Gannon M. Manganese (Mn) transport across the rat blood-brain barrier: saturable and transferrin-dependent transport mechanisms. Brain Res Bull. 1994;33:345–9.

Aschner M, Gannon M, Kimelberg HK. Manganese uptake and efflux in cultured rat astrocytes. J Neurochem. 1992;58:730–5.

Au C, Benedetto A, Aschner M. Manganese transport in eukaryotes: the role of DMT1. Neurotoxicology. 2008;29:569–76.

Baltan S, Murphy SP, Danilov CA, Bachleda A, Morrison RS. Histone deacetylase inhibitors preserve white matter structure and function during ischemia by conserving ATP and reducing excitotoxicity. J Neurosci. 2011;31:3990–9.

Bardai FH, D'Mello SR. Selective toxicity by HDAC3 in neurons: regulation by Akt and GSK3beta. J Neurosci. 2011;31:1746–51.

Bast-Pettersen R, Ellingsen DG, Hetland SM, Thomassen Y. Neuropsychological function in manganese alloy plant workers. Int Arch Occup Environ Health. 2004;**77**:277–87.

Bentle LA, Lardy HA. Interaction of anions and divalent metal ions with phosphoenolpyruvate carboxykinase. J Biol Chem. 1976;251:2916–21.
Boston-Howes W, Gibb SL, Williams EO, Pasinelli P, Brown RH Jr, Trotti D. Caspase-3 cleaves and inactivates the glutamate transporter EAAT2. J Biol Chem. 2006;281:14076–84.
Bowler RM, Roels HA, Nakagawa S, Drezgic M, Diamond E, Park R, Koller W, Bowler RP, Mergler D, Bouchard M, Smith D, Gwiazda R, Doty RL. Dose-effect relationships between manganese exposure and neurological, neuropsychological and pulmonary function in confined space bridge welders. Occup Environ Med. 2007;64:167–77.
Bowman AB, Kwakye GF, Herrero Hernandez E, Aschner M. Role of manganese in neurodegenerative diseases. J Trace Elem Med Biol. 2011;25:191–203.
Brouillet EP, Shinobu L, McGarvey U, Hochberg F, Beal MF. Manganese injection into the rat striatum produces excitotoxic lesions by impairing energy metabolism. Exp Neurol. 1993;120:89–94.
Chen CJ, Liao SL. Oxidative stress involves in astrocytic alterations induced by manganese. Exp Neurol. 2002;175:216–25.
Chen P, Chakraborty S, Peres TV, Bowman AB, Aschner M. Manganese-induced neurotoxicity: from to humans. Toxicol Res. 2015;4:191–202.
Conradt M, Stoffel W. Inhibition of the high-affinity brain glutamate transporter GLAST-1 via direct phosphorylation. J Neurochem. 1997;68:1244–51.
Cordova FM, Aguiar AS Jr, Peres TV, Lopes MW, Goncalves FM, Remor AP, Lopes SC, Pilati C, Latini AS, Prediger RD, Erikson KM, Aschner M, Leal RB. In vivo manganese exposure modulates Erk, Akt and Darpp-32 in the striatum of developing rats, and impairs their motor function. PLoS One. 2012;7:e33057.
Danbolt NC. Glutamate uptake. Prog Neurobiol. 2001;65:1–105.
Deng Y, Xu Z, Xu B, Xu D, Tian Y, Feng W. The protective effects of riluzole on manganese-induced disruption of glutamate transporters and glutamine synthetase in the cultured astrocytes. Biol Trace Elem Res. 2012;148:242–9.
Diez AM, Campo ML, Soler G. Trypsin digestion of arginase: evidence for a stable conformation manganese directed. Int J Biochem. 1992;24:1925–32.
Erikson K, Aschner M. Manganese causes differential regulation of glutamate transporter (GLAST) taurine transporter and metallothionein in cultured rat astrocytes. Neurotoxicology. 2002;23:595–602.
Erikson KM, Dobson AW, Dorman DC, Aschner M. Manganese exposure and induced oxidative stress in the rat brain. Sci Total Environ. 2004;334–335:409–16.
Erikson KM, Dorman DC, Fitsanakis V, Lash LH, Aschner M. Alterations of oxidative stress biomarkers due to in utero and neonatal exposures of airborne manganese. Biol Trace Elem Res. 2006;111:199–215.
Erikson KM, Dorman DC, Lash LH, Aschner M. Manganese inhalation by rhesus monkeys is associated with brain regional changes in biomarkers of neurotoxicity. Toxicol Sci. 2007;97:459–66.
Erikson KM, Dorman DC, Lash LH, Aschner M. Duration of airborne-manganese exposure in rhesus monkeys is associated with brain regional changes in biomarkers of neurotoxicity. Neurotoxicology. 2008;29:377–85.
Fujishiro H, Yano Y, Takada Y, Tanihara M, Himeno S. Roles of ZIP8, ZIP14, and DMT1 in transport of cadmium and manganese in mouse kidney proximal tubule cells. Metallomics. 2012;4:700–8.
Gavin CE, Gunter KK, Gunter TE. Mn2+ sequestration by mitochondria and inhibition of oxidative phosphorylation. Toxicol Appl Pharmacol. 1992;115:1–5.
Gavin CE, Gunter KK, Gunter TE. Manganese and calcium transport in mitochondria: implications for manganese toxicity. Neurotoxicology. 1999;20:445–53.
Ghosh M, Yang Y, Rothstein JD, Robinson MB. Nuclear factor-kappaB contributes to neuron-dependent induction of glutamate transporter-1 expression in astrocytes. J Neurosci. 2011;31:9159–69.

Gonzalez LE, Juknat AA, Venosa AJ, Verrengia N, Kotler ML. Manganese activates the mitochondrial apoptotic pathway in rat astrocytes by modulating the expression of proteins of the Bcl-2 family. Neurochem Int. 2008;53:408–15.

Hazell AS, Norenberg MD. Manganese decreases glutamate uptake in cultured astrocytes. Neurochem Res. 1997;22:1443–7.

Huerta-Yepez S, Vega M, Garban H, Bonavida B. Involvement of the TNF-alpha autocrine-paracrine loop, via NF-kappaB and YY1, in the regulation of tumor cell resistance to Fas-induced apoptosis. Clin Immunol. 2006;120:297–309.

Ito Y, Oh-Hashi K, Kiuchi K, Hirata Y. p44/42 MAP kinase and c-Jun N-terminal kinase contribute to the up-regulation of caspase-3 in manganese-induced apoptosis in PC12 cells. Brain Res. 2006;1099:1–7.

Itoh K, Sakata M, Watanabe M, Aikawa Y, Fujii H. The entry of manganese ions into the brain is accelerated by the activation of N-methyl-D-aspartate receptors. Neuroscience. 2008;154:732–40.

Janssen C, Schmalbach S, Boeselt S, Sarlette A, Dengler R, Petri S. Differential histone deacetylase mRNA expression patterns in amyotrophic lateral sclerosis. J Neuropathol Exp Neurol. 2010;69:573–81.

Kalandadze A, Wu Y, Robinson MB. Protein kinase C activation decreases cell surface expression of the GLT-1 subtype of glutamate transporter. Requirement of a carboxyl-terminal domain and partial dependence on serine 486. J Biol Chem. 2002;277:45741–50.

Karki P, Webb A, Smith K, Lee K, Son DS, Aschner M, Lee E. cAMP response element-binding protein (CREB) and nuclear factor kappaB mediate the tamoxifen-induced up-regulation of glutamate transporter 1 (GLT-1) in rat astrocytes. J Biol Chem. 2013;288:28975–86.

Karki P, Webb A, Smith K, Johnson J Jr, Lee K, Son DS, Aschner M, Lee E. Yin Yang 1 is a repressor of glutamate transporter EAAT2, and it mediates manganese-induced decrease of EAAT2 expression in astrocytes. Mol Cell Biol. 2014a;34:1280–9.

Karki P, Webb A, Zerguine A, Choi J, Son DS, Lee E. Mechanism of raloxifene-induced upregulation of glutamate transporters in rat primary astrocytes. Glia. 2014b;62:1270–83.

Karki P, Kim C, Smith K, Son DS, Aschner M, Lee E. Transcriptional regulation of the astrocytic excitatory amino acid transporter 1 (EAAT1) via NF-kappaB and Yin Yang 1 (YY1). J Biol Chem. 2015;290:23725–37.

Kim SY, Choi SY, Chao W, Volsky DJ. Transcriptional regulation of human excitatory amino acid transporter 1 (EAAT1): cloning of the EAAT1 promoter and characterization of its basal and inducible activity in human astrocytes. J Neurochem. 2003;87:1485–98.

Kitazawa M, Anantharam V, Yang Y, Hirata Y, Kanthasamy A, Kanthasamy AG. Activation of protein kinase C delta by proteolytic cleavage contributes to manganese-induced apoptosis in dopaminergic cells: protective role of Bcl-2. Biochem Pharmacol. 2005;69:133–46.

Kwakye GF, Paoliello MM, Mukhopadhyay S, Bowman AB, Aschner M. Manganese-induced parkinsonism and Parkinson's disease: shared and distinguishable features. Int J Environ Res Public Health. 2015;12:7519–40.

Latchoumycandane C, Anantharam V, Kitazawa M, Yang Y, Kanthasamy A, Kanthasamy AG. Protein kinase Cdelta is a key downstream mediator of manganese-induced apoptosis in dopaminergic neuronal cells. J Pharmacol Exp Ther. 2005;313:46–55.

Lee SG, Su ZZ, Emdad L, Gupta P, Sarkar D, Borjabad A, Volsky DJ, Fisher PB. Mechanism of ceftriaxone induction of excitatory amino acid transporter-2 expression and glutamate uptake in primary human astrocytes. J Biol Chem. 2008;283:13116–23.

Lee ES, Sidoryk M, Jiang H, Yin Z, Aschner M. Estrogen and tamoxifen reverse manganese-induced glutamate transporter impairment in astrocytes. J Neurochem. 2009;110:530–44.

Lee SG, Kim K, Kegelman TP, Dash R, Das SK, Choi JK, Emdad L, Howlett EL, Jeon HY, Su ZZ, Yoo BK, Sarkar D, Kim SH, Kang DC, Fisher PB. Oncogene AEG-1 promotes glioma-induced neurodegeneration by increasing glutamate excitotoxicity. Cancer Res. 2011;71:6514–23.

Lee E, Sidoryk-Wegrzynowicz M, Yin Z, Webb A, Son DS, Aschner M. Transforming growth factor-alpha mediates estrogen-induced upregulation of glutamate transporter GLT-1 in rat primary astrocytes. Glia. 2012;60:1024–36.

Leng Y, Marinova Z, Reis-Fernandes MA, Nau H, Chuang DM. Potent neuroprotective effects of novel structural derivatives of valproic acid: potential roles of HDAC inhibition and HSP70 induction. Neurosci Lett. 2010;476:127–32.

Montes S, Riojas-Rodriguez H, Sabido-Pedraza E, Rios C. Biomarkers of manganese exposure in a population living close to a mine and mineral processing plant in Mexico. Environ Res. 2008;106:89–95.

Morello M, Canini A, Mattioli P, Sorge RP, Alimonti A, Bocca B, Forte G, Martorana A, Bernardi G, Sancesario G. Sub-cellular localization of manganese in the basal ganglia of normal and manganese-treated rats: an electron spectroscopy imaging and electron energy-loss spectroscopy study. Neurotoxicology. 2008;29:60–72.

Mutkus L, Aschner JL, Fitsanakis V, Aschner M. The in vitro uptake of glutamate in GLAST and GLT-1 transfected mutant CHO-K1 cells is inhibited by manganese. Biol Trace Elem Res. 2005;107:221–30.

Peres TV, Pedro DZ, de Cordova FM, Lopes MW, Goncalves FM, Mendes-de-Aguiar CB, Walz R, Farina M, Aschner M, Leal RB. In vitro manganese exposure disrupts MAPK signaling pathways in striatal and hippocampal slices from immature rats. Biomed Res Int. 2013;2013:769295.

Perisic T, Zimmermann N, Kirmeier T, Asmus M, Tuorto F, Uhr M, Holsboer F, Rein T, Zschocke J. Valproate and amitriptyline exert common and divergent influences on global and gene promoter-specific chromatin modifications in rat primary astrocytes. Neuropsychopharmacol. 2010;35:792–805.

Potter MC, Figuera-Losada M, Rojas C, Slusher BS. Targeting the glutamatergic system for the treatment of HIV-associated neurocognitive disorders. J Neuroimmune Pharmacol. 2013;8:594–607.

Rao VL, Dogan A, Todd KG, Bowen KK, Kim BT, Rothstein JD, Dempsey RJ. Antisense knockdown of the glial glutamate transporter GLT-1, but not the neuronal glutamate transporter EAAC1, exacerbates transient focal cerebral ischemia-induced neuronal damage in rat brain. J Neurosci. 2001;21:1876–83.

Robelet S, Melon C, Guillet B, Salin P, Kerkerian-Le Goff L. Chronic L-DOPA treatment increases extracellular glutamate levels and GLT1 expression in the basal ganglia in a rat model of Parkinson's disease. Eur J Neurosci. 2004;20:1255–66.

Rosas S, Vargas MA, Lopez-Bayghen E, Ortega A. Glutamate-dependent transcriptional regulation of GLAST/EAAT1: a role for YY1. J Neurochem. 2007;101:1134–44.

Rothstein JD, Van Kammen M, Levey AI, Martin LJ, Kuncl RW. Selective loss of glial glutamate transporter GLT-1 in amyotrophic lateral sclerosis. Ann Neurol. 1995;38:73–84.

Seifert G, Schilling K, Steinhauser C. Astrocyte dysfunction in neurological disorders: a molecular perspective. Nat Rev. 2006;7:194–206.

Shi Y, Lee JS, Galvin KM. Everything you have ever wanted to know about Yin Yang 1. Biochim Biophys Acta. 1997;1332:F49–66.

Sidoryk-Wegrzynowicz M, Aschner M. Manganese toxicity in the central nervous system: the glutamine/glutamate-gamma-aminobutyric acid cycle. J Intern Med. 2013;273:466–77.

Sidoryk-Wegrzynowicz M, Lee E, Mingwei N, Aschner M. Disruption of astrocytic glutamine turnover by manganese is mediated by the protein kinase C pathway. Glia. 2011;59:1732–43.

Sidoryk-Wegrzynowicz M, Lee E, Aschner M. Mechanism of Mn(II)-mediated dysregulation of glutamine-glutamate cycle: focus on glutamate turnover. J Neurochem. 2012;122:856–67.

Sitcheran R, Gupta P, Fisher PB, Baldwin AS. Positive and negative regulation of EAAT2 by NF-kappaB: a role for N-myc in TNFalpha-controlled repression. EMBO J. 2005;24:510–20.

Spadoni F, Stefani A, Morello M, Lavaroni F, Giacomini P, Sancesario G. Selective vulnerability of pallidal neurons in the early phases of manganese intoxication. Exp Brain Res. 2000;135:544–51.

Stallings WC, Metzger AL, Pattridge KA, Fee JA, Ludwig ML. Structure-function relationships in iron and manganese superoxide dismutases. Free Radic Res Commun. 1991;12–13(Pt 1):259–68.

Su ZZ, Leszczyniecka M, Kang DC, Sarkar D, Chao W, Volsky DJ, Fisher PB. Insights into glutamate transport regulation in human astrocytes: cloning of the promoter for excitatory amino acid transporter 2 (EAAT2). Proc Natl Acad Sci U S A. 2003;100:1955–60.

Takeda A. Manganese action in brain function. Brain Res. 2003;41:79–87.

Wedler FC, Denman RB. Glutamine synthetase: the major Mn(II) enzyme in mammalian brain. Curr Top Cell Regul. 1984;24:153–69.

Williams M, Todd GD, Roney N, Crawford J, Coles C, McClure PR, Garey JD, Zaccaria K, Citra M. Agency for Toxic Substances and Disease Registry (ATSDR) toxicological profiles. In: Toxicological profile for manganese. Atlanta: Agency for Toxic Substances and Disease Registry (US); 2012.

Xu B, Xu ZF, Deng Y. Protective effects of MK-801 on manganese-induced glutamate metabolism disorder in rat striatum. Exp Toxicol Pathol. 2010a;62:381–90.

Xu Z, Jia K, Xu B, He A, Li J, Deng Y, Zhang F. Effects of MK-801, taurine and dextromethorphan on neurotoxicity caused by manganese in rats. Toxicol Ind Health. 2010b;26:55–60.

Xu B, Xu ZF, Deng Y. Manganese exposure alters the expression of N-methyl-D-aspartate receptor subunit mRNAs and proteins in rat striatum. J Biochem Mol Toxicol. 2010c;24:1–9.

Yanagisawa M, Aida T, Takeda T, Namekata K, Harada T, Shinagawa R, Tanaka K. Arundic acid attenuates retinal ganglion cell death by increasing glutamate/aspartate transporter expression in a model of normal tension glaucoma. Cell Death Dis. 2015;6:e1693.

Yin Z, Aschner JL, dos Santos AP, Aschner M. Mitochondrial-dependent manganese neurotoxicity in rat primary astrocyte cultures. Brain Res. 2008;1203:1–11.

Zhao F, Cai T, Liu M, Zheng G, Luo W, Chen J. Manganese induces dopaminergic neurodegeneration via microglial activation in a rat model of manganism. Toxicol Sci. 2009;107:156–64.

Zschocke J, Allritz C, Engele J, Rein T. DNA methylation dependent silencing of the human glutamate transporter EAAT2 gene in glial cells. Glia. 2007;55:663–74.

Glycine Transporters in Glia Cells: Structural Studies

Beatriz López-Corcuera, Cristina Benito-Muñoz, and Carmen Aragón

Abstract Glycine, besides exerting essential metabolic functions, is an important inhibitory neurotransmitter in caudal areas of the central nervous system and also a positive neuromodulator at excitatory glutamate-mediated synapses. Glial cells provide metabolic support to neurons and modulate synaptic activity. Six transporters belonging to three solute carrier families (SLC6, SLC38, and SLC7) are capable of transporting glycine across the glial plasma membrane. The unique glial glycine-selective transporter GlyT1 (SLC6) is the main regulator of synaptic glycine concentrations, assisted by the neuronal GlyT2. The five additional glycine transporters $ATB^{0,+}$, SNAT1, SNAT2, SNAT5, and LAT2 display broad amino acid specificity and have differential contributions to glial glycine transport. Glial glycine transporters are divergent in sequence but share a similar architecture displaying the 5 + 5 inverted fold originally characterized in the leucine transporter LeuT. The availability of protein crystals solved at high resolution for prokaryotic and, more recently, eukaryotic homologues of this superfamily has advanced significantly our understanding of the mechanism of glycine transport.

Keywords Glycine • Transporter • Structure • Glia • Astrocyte • LeuT

B. López-Corcuera (✉) • C. Aragón
Departamento de Biología Molecular, Centro de Biología Molecular "Severo Ochoa", Universidad Autónoma de Madrid-Consejo Superior de Investigaciones Científicas, C/ Nicolás Cabrera 1, 28049 Madrid, Spain

IdiPAZ-Hospital Universitario La Paz, Universidad Autónoma de Madrid, Madrid, Spain

Centro de Investigación Biomédica en Red sobre Enfermedades Raras, ISCIII, Madrid, Spain
e-mail: blopez@cbm.csic.es; blopez@cbm.uam.es

C. Benito-Muñoz
Departamento de Biología Molecular, Centro de Biología Molecular "Severo Ochoa", Universidad Autónoma de Madrid-Consejo Superior de Investigaciones Científicas, C/ Nicolás Cabrera 1, 28049 Madrid, Spain

A. Ortega, A. Schousboe (eds.), *Glial Amino Acid Transporters*, Advances in Neurobiology 16, DOI 10.1007/978-3-319-55769-4_2

List of Abbreviations

APCS	Amino acid organocation superfamily
ASCT	Alanine, serine, threonine transporter
$ATB^{0,+}$	Neutral and cationic amino acid transporter
DAT	Dopamine transporter
EAAT	Glutamate transporters
EL	External loop
GAT1	GABA transporter 1
GlyT	Glycine transporter
LAT	Leucine-preferring amino acid transporter
MeAIB	*N*-methyl-aminoisobutyric acid
mTOR	Mechanistic target of rapamycin
NMDA	*N*-methyl-D-aspartate
NSS	Neurotransmitter sodium symporters
SERT	Serotonin transporter
SLC	Solute carrier
SNAT	Sodium-coupled neutral amino acid transporter
TM	Transmembrane domain

1 Introduction

Glycine, the smallest amino acid, exerts important metabolic and signaling functions in the mammalian central nervous system (CNS). Although it is a nonessential compound, glycine is involved not only in protein synthesis but also as a precursor for biosynthetic pathways of creatine, glutathione, and nucleotides and has a key role in the metabolism of one-carbon fragments (Pai et al. 2015). As a signaling molecule, glycine serves as both neurotransmitter and neuromodulator. Glycine neurotransmitter fulfills the fast inhibitory transmission in caudal regions of the adult CNS predominantly in neuronal circuits of the central auditory pathway, receptive fields in the retina, and spinal cord motor and sensitive pathways. Glycine released from glycinergic inhibitory interneurons by calcium-dependent exocytosis of glycine-containing synaptic vesicles binds to the strychnine-sensitive glycine-A-binding site on glycine receptors. These ligand-gated anion channels promote chloride ion fluxes into postsynaptic neurons resulting in the generation of inhibitory postsynaptic potentials (Legendre 2001; Dutertre et al. 2012). To terminate signaling, synaptic glycine is cleared by sodium-driven uptake to the presynaptic terminals and surrounding glial cells (Zafra and Gimenez 2008). Two specific plasma membrane glycine transporters GlyT1 and GlyT2, with different cellular locations, are in charge of glycine clearance and maintain synaptic glycine concentrations in the low micromolar range in resting conditions.

In addition to its inhibitory action, glycine exerts a modulatory function in fast excitatory neurotransmission by acting as an essential co-agonist of the main excitatory neurotransmitter in the CNS, glutamate, on *N*-methyl-D-aspartate (NMDA) receptors (Johnson and Ascher 1987). Binding of glycine to a glycine-B NMDA receptor site is necessary both for ion channel opening and receptor internalization (Nong et al. 2003). This ligand-gated cation channel promotes neuronal excitation. Therefore, glycine functions as a positive allosteric modulator of NMDA receptors and thereby may contribute to neural plasticity (Harvey and Yee 2013; Zhang et al. 2008). Moreover, glycine alone is required for activation of a purely excitatory NMDA receptor subtype in the absence of glutamate (Chatterton et al. 2002). Neurotransmitter clearance from excitatory synaptic clefts mostly relies on the glycine transporter GlyT1 located in neighboring astrocytes, although lower levels of the transporter are present in glutamatergic terminals and in the postsynaptic membranes of glutamatergic synapses (Cubelos et al. 2005a). GlyT1 is a potential target for the treatment of several nervous system disorders in which the increase of synaptic glycine concentrations could be beneficial such as schizophrenia, alcohol dependence, pain, epilepsy, and breathing disorders (Harvey and Yee 2013; Singer et al. 2015; Nunez et al. 2000).

Neurotransmitter clearance from the synaptic cleft is a major function of astrocytes to sustain the rapid kinetics of chemical neurotransmission. In addition, astrocytes play a pivotal role in brain homeostasis providing structural, trophic, and metabolic support to neurons and modulating synaptic activity. Glycine, as a small neutral amino acid, is transported by several glial plasma membrane transporters that have overlapping substrate reactivity. Six transporters that accept glycine as a substrate have been identified at the molecular level in the astrocyte plasma membrane. Only one is selective for glycine, and five share specificity for the amino acid but show differences in structure and transport modes (Table 1). According to the Human Genome Organization nomenclature, these carriers belong to three gene families: SLC6, SLC38, and SLC7 (solute carriers 6, 38, and 7). In addition, a very minor contribution to glycine transport is made by two SLC1 transporter variants that are farther from the scope of this review.

2 Glial Glycine Transporters

Glial GlyT1 is the only specific glycine transporter present in astroglial cells. It is exclusively selective for glycine and its *N*-methyl analog sarcosine (*N*-methyl glycine). GlyT1 is the most important glycine transporter in astrocytes and is critical for the regulation of glycine levels at inhibitory synapses during early postnatal life (Eulenburg et al. 2010; Gomeza et al. 2003a). GlyT1 protein is present in astrocytic processes associated to both glycinergic and glutamatergic pathways of the brain stem, spinal cord, cerebellum, and also neocortex, hippocampus, and thalamus (Aragon and Lopez-Corcuera 2005; Zafra et al. 1995). Its activity has been measured in several astrocytic types including retinal Müller cells, cerebellar Bergmann

Table 1 Glial glycine transporters

Gene name	Protein name	Alias	Mechanism	Substrate specificity	Functional classification system	Expression profile
SLC6A9	GlyT1		S:$2Na^+$:$1Cl^-$	G, Sarcosine	Gly	Brain stem, spinal cord, neocortex, hippocampus, thalamus, retina (kidney, pancreas, lung, placenta, liver)
SLC6A14	$ATB^{0,+}$		S:$2Na^+$:$1Cl^-$	Neutral and cationic amino acids	$B^{0,+}$	Lung, trachea, salivary gland, mammary gland, stomach, pituitary, colon, uterus, prostate, testis, brain[a]
SLC38A1	SNAT1	GlnT, SAT1, ATA1, SA2, NAT2	S:$1Na^+$	G, A, S, C, N, Q, H, M	A	Cerebral cortex, placenta, heart, lung. Skeletal muscle, spleen, stomach, testis
SLC38A2	SNAT2	SAT2, ATA2, SA1	S:$1Na^+$	G, A, S, P, C, N, Q, H, M	A	Ubiquitous (widely expressed in astrocytes and neurons in central nervous system)
SLC38A5	SNAT5	SN2	S:$1Na^+$/A:$1H^+$	Q, N, H, A, S, G	N	Stomach, brain, liver, lung, intestinal tract
SLC7A8	LAT2		A	Neutral amino acids, except P	L	Kidney, placenta, brain (liver, prostate, testis, ovary, lymph node, thymus, spleen, skeletal muscle, heart)

Amino acids are given in one-letter codes
S symport, *A* antiport
[a]Distribution not fully characterized

glia, cerebral cortex cultured astrocytes (Werdehausen et al. 2012), and mature protoplasmic astrocytes, cells that fulfill all requirements necessary for stabilization of synaptic transmission (Szoke et al. 2006). Its contribution to the whole glycine uptake has been estimated in about 60% in several glial cells such as Müller cells (Hosoya et al. 2010), lateral superior olive astrocytes (Stephan and Friauf 2014), or Bergmann glia (Lopez et al. 2005), about 70% in cultured astrocytes

(Werdehausen et al. 2012) and more than 85% in C6 glioma cells (Zafra and Gimenez 1989). Most of the high-affinity glycine transport in astrocytes and in the brain is fulfilled by GlyT1. Genetic ablation of mouse GlyT1 provokes a 70–80% reduction in glycine transport (Gomeza et al. 2003a). As its neuronal counterpart GlyT2, GlyT1 belongs to the SLC6 transporter family that includes Na^+- and Cl^--dependent co-transporters for neurotransmitters, osmolytes, and amino acids (Broer and Gether 2012; Broer and Palacin 2011; Broer 2013; Rudnick et al. 2014; Kanner and Zomot 2008). These carriers perform concentrative high-affinity uptake with a Km value in the μM concentration range. As most of the members of the SLC6 family, GlyT1 (*SLC6A9*) depicts a $2Na^+/1Cl^-$/glycine co-transport stoichiometry, and it is an electrogenic transporter moving one neat charge per transport cycle (Roux and Supplisson 2000; Lopez-Corcuera et al. 1998). Glial GlyT1 is close to its reversal potential at the astrocyte resting membrane potential, what permits uptake or release of substrate depending on the electrochemical conditions. By contrast, the neuronal Glyt2 (*SLC6A5*) with a $3Na^+/1Cl^-$/glycine stoichiometry generates about hundred times larger driving force for glycine uptake and establishes a high (10–40 mM) concentration of glycine in the presynaptic terminal (Supplisson and Roux 2002; Kirischuk et al. 2015). This concentration, about one order of magnitude higher than that estimated in glia (2 mM), is crucial for supplying glycine for synaptic vesicle refilling by the vesicular inhibitory amino acid transporter (VIAAT), a low affinity transporter (McIntire et al. 1997; Sagne et al. 1997; Apostolides and Trussell 2013). For this reason, the absence of GlyT2 activity provokes a deficient glycinergic transmission and may cause the human neurological disorder of hyperekplexia (Rees et al. 2006; Arribas-González et al. 2015; Gomeza et al. 2003b; Gimenez et al. 2008). GlyT2 is exclusively localized in the presynaptic terminals of glycinergic neurons in the spinal cord and the brainstem. Very low levels occur in the diencephalon, and only in the cerebellum it is expressed both in terminal boutons and in glial elements (Zafra et al. 1995). GlyT2 is absent from glial cells in the forebrain (Werdehausen et al. 2012; Zeilhofer et al. 2005), from Müller cells (Hosoya et al. 2010), and from lateral superior olive astrocytes (Stephan and Friauf 2014). However, an atypical GlyT2-like activity has been recently reported in astrocytes, which does not fulfill high-affinity GlyT2 glycine transport features (Aroeira et al. 2014).

Another SLC6 transporter, whose presence in astrocytes has been elusive, is $ATB^{0,+}$ (*SLC6A14*). This is the only very broad substrate specificity transporter of the SLC6 family, and its expression is upregulated in cancer (Broer 2013; Karunakaran et al. 2008). $ATB^{0,+}$ transports all neutral and cationic amino acids including glycine with high affinity (Km about 100 μM) but does not transport anionic amino acids. Therefore, it transports 18 of the 20 proteinogenic amino acids (Broer and Gether 2012; Broer and Palacin 2011). In addition, it carries several interesting compounds such as carnitine (Nakanishi et al. 2001), β-alanine (Sloan and Mager 1999), or several D-amino acids including D-serine (Karunakaran et al. 2008; Hatanaka et al. 2002). It co-transports glycine together with $2Na^+$ and $1Cl^-$ (Sloan and Mager 1999). $ATB^{0,+}$ is mainly expressed in lung, colon, pituitary, and mammary gland (Sloan et al. 2003; Bhutia and Ganapathy 2016), but its presence in astrocytes has been only lately reported (Samluk et al. 2010, 2012). Unfortunately, a complete characterization of the $ATB^{0,+}$ transport activity in astrocytes is missing. It would be interesting to examine the effect of α-methyl-DL-tryptophan, a selective $ATB^{0,+}$ inhibitor, on glycine transport by astrocytes since the small inhibition of Na^+-and Cl^--dependent glial glycine transport by D-Alanine (Zafra and Gimenez 1989) or β-alanine (Lopez et al. 2005) might be due to a $ATB^{0,+}$contribution. However, $ATB^{0,+}$ mRNA is not present in Müller cells (Hosoya et al. 2010). The deletion of the *SLC6A14* gene suppresses tumor growth in mouse models of breast cancer, but there is no information on the brain phenotype of these mice (Babu et al. 2015). Molecular tools such as quantitative real-time PCR (qPCR) to determine *SLC6A14* brain expression would help to clarify this issue.

Three of the characterized members of the SLC38 family, the SNATs (sodium-coupled neutral amino acid transporters), contribute to glycine transport in astrocytes. SNATs perform medium- to low-affinity ($K_{1/2}$ mM range) Na^+-dependent net uptake and efflux of small neutral amino acids. SNAT1 (*SLC38A1*) and SNAT2 (*SLC38A2*) are the molecular entities that carry out system A transport (alanine-preferring) (Christensen et al. 1965). They transport Ala, Ser, Cys, Asn, Gln, His, and also Met and Gly and are inhibited by *N*-methyl-aminoisobutyric acid (MeAIB) (Schiöth et al. 2013; Broer 2014). The third SLC38 glial glycine transporter is SNAT5 (*SLC38A5*). It performs system N transport, MeAIB-insensitive, with narrower substrate selectivity (Gln, Asn, His, Ala, Ser, Gly) and preference for glutamine. SNATs can also operate in channel mode displaying uncoupled currents of H^+ (system N) or anion currents (system A) (Jenstad and Chaudhry 2013). Coupled transport by system A transporters is electrogenic with 1:1 Na^+-amino acid co-transport stoichiometry and pH-sensitive. System N SNAT5 co-transports the neutral substrate with Na^+ and simultaneously antiports H^+, resulting in an electroneutral and bidirectional movement of substrate (Broer 2014; Mackenzie and Erickson 2004). The uncoupled substrate-gated conductance of protons through SNAT5 favors the reverse flow through the plasma membrane. The magnitude of the uncoupled currents differs among the different substrates and makes operation of the transporter sensitive to membrane potential (Hamdani el et al. 2012). The release of glutamine and glycine from astrocytes through SNAT5 has been associated with important aspects of brain physiology such as the glutamate-glutamine cycle and the NMDA receptor regulation (Cubelos et al. 2005a; Schiöth et al. 2013; Broer 2014). Tolerance of Li^+ instead of Na^+ seems to be a feature of some members of this family, mainly system N SNATs (Chaudhry et al. 2001; Kilberg et al. 1980; Mackenzie et al. 2003). SNAT1 and SNAT2 are present in neurons and in glial cells all over the CNS, with remarkable expression in perivascular glial profiles and lower expression in astrocytes of the cerebral parenchyma (Melone et al. 2004; Gonzalez-Gonzalez et al. 2005). This supports a role in controlling the amino acid permeability at the blood-brain barrier. The contribution of system A glycine transport to total glycine uptake has been estimated in about 30–40% in Müller cells (Hosoya et al. 2010) and Bergmann glia (Lopez et al. 2005) and 15% in C6 glioma cells (Zafra and Gimenez 1989). Moreover, SNAT5 is exclusively expressed in astrocytes, and it is undetectable in neuronal elements showing the highest presence in the striatum, spinal cord, and neocortex (Cubelos et al. 2005a). The SLC38 transporters display a distinctive subcellular distribution showing a reduced basal surface expression, and only in certain conditions such as amino acid deprivation (system A), plasma membrane expression is increased (Hamdani el et al. 2012). In C6 glioma cells, after 19 h incubation in amino acid-free medium, the low-affinity transport of glycine was increased three-fold above the values in cells incubated in regular medium (Zafra and Gimenez 1989).

Na^+-independent glycine transport in glial cells is brought about by one variant of the general system L, LAT2, leucine-preferring amino acid transporter 2 (Pineda et al. 1999; Segawa et al. 1999). LAT2 is the catalytic subunit of a disulfide-linked heterodimer of a chaperone-like subunit 4F2hc (*SLC3A2*) and LAT2 (*SLC7A8*). LATs belong to the subfamily of heteromeric amino acid transporters (Broer and Palacin 2011). LAT2/4F2hc (*SLC7A8/SLC3A2*) accepts all neutral amino acids except proline including small amino acids such as glycine. LATs can be functionally identified by response to the non-metabolizable amino acid analog 2-aminobicyclo-(2.2.1)-heptane-2-carboxylic acid (BCH). LAT2 performs sodium-independent, obligatory exchange (1:1 stoichiometry) with high to medium affinity showing some asymmetry due to μM affinity from the extracellular but mM for the intracellular amino acid (Fotiadis et al. 2013). LAT2/4F2hc is present in many tissues and is abundant in neurons and astrocytes, where a role in L-alanine transport has been recognized (Broer et al. 2007). Deletion of the LAT2 gene in mice (*slc7a8*) revealed a slight impairment in movement coordination (Braun et al. 2011) and a role in thyroid hormone transport in neurons (Nunez et al. 2014). Only few studies have evaluated the contribution of LAT2 to the total uptake of glycine in glial cells. In the absence of Na^+, glycine uptake is reduced to about 25% in Bergmann glia (Lopez et al. 2005), to about 3% in Müller cells (Hosoya et al. 2010), and to a negligible proportion in C6 glioma cells (Zafra and Giménez 1989). This latter figure is in agreement with the absence of expression of LAT2 in C6 cells revealed through a RT-PCR analysis and functional evaluation (Kim et al. 2004). Although glycine is not a preferred substrate for LAT2, and it displays low mM affinity for external or internal glycine, the amino acid is able to induce important *trans*-stimulation of neutral amino acid efflux by the equilibrative transporter. Characterization of LAT2 expressed in *Xenopus laevis* oocytes disclosed that the antiport function is controlled by the intracellular amino acid concentration (Meier et al. 2002). Since the concentration of glycine in astrocytes is maintained by SLC6 and SLC38 cumulative transporters at around 2 mM, glycine transport by LAT2 may provide a mechanism to import essential amino acids and in general small and large neutral amino acids such as alanine, serine, threonine, cysteine, phenylalanine, tyrosine, leucine, arginine, and tryptophan. This function has been called tertiary-active transport (Rudnick et al. 2014; Verrey 2003). This may have remarkable consequences besides contributing to link different amino acid concentration gradients. Leucine and some neutral amino acids are potent activators of mTOR signaling, and mTOR-controlled pathways regulate many physiological and pathological aspects in the CNS (Milkereit et al. 2015; Bockaert and Marin 2015).

Two isoforms of the SLC1 family ASCT1 (*SLC1A4*) and ASCT2 (*SLC1A5*) are ACS-type transporters because of their preference for alanine, serine, and cysteine. They mediate high-affinity Na^+-dependent obligatory exchange of small neutral amino acids including glycine, though glycine is a very minor non-preferred substrate (Kanai et al. 2013; Broer et al. 1999; Utsunomiya-Tate et al. 1996). ASCT1 is the main transport system of L-serine in astrocytes (Sakai et al. 2003), and its function is distinctively sensitive to L-trans-4 hydroxyproline (Pinilla-Tenas et al. 2003). Recently, the exchange of D-serine by both isoforms has been proven (Foster et al. 2016). ASCT2 also accepts glutamine and asparagine as high-affinity substrates and might contribute to the glutamine homeostasis in neuronal and glial cells (Broer et al. 1999; Broer and Brookes 2001). An ASCT2 variant has been shown to be significantly upregulated in cancer tissues (Broer et al. 2016). ASCTs mediate electroneutral one-for-one amino acid homo- or heteroexchange and negligible net amino acid flux. They are hence equilibrative transporters, harmonizers (Broer et al. 2016), that also balance different pools of neutral amino acids. Although the transport cycle requires one extracellular Na^+ ion (Scopelliti et al. 2014; Broer et al. 2000), the sodium gradient is not the driving force, contrasting with the homologous glutamate transporters (EAAT), also members of this family (Danbolt 2001; Grewer et al. 2014). In addition, the binding of Na^+ and substrate to the transporters activates a chloride conductance that is thermodynamically uncoupled from amino acid flux (Zerangue and Kavanaugh 1996). Therefore, ASCTs function as both transporters and chloride channels, a feature shared by SLC1 transporters (Kanai et al. 2013; Wadiche et al. 1995). ASCT1 is present in the CNS in co-localization with the astrocytic marker GFAP and is low or negative in neuronal elements. Its distribution is parallel to that of the first enzyme needed for de novo synthesis of serine, 3-phosphoglycerate dehydrogenase, mainly present in astrocytes (Sakai et al. 2003). The expression of ASCT2 in astrocytes has also been verified (Broer et al. 1999; Sakai et al. 2003), and the activity of both isoforms has been identified in cultured rat hippocampal astrocytes by functional assays (Foster et al. 2016). The contribution of ASCTs to the transport of glycine in glia has not been directly evaluated, but in contrast to old reports based in transporter functional classification, a very minor role of these ASC-type transporters in glycine transport is consistent with recent information (Broer et al. 2007; Foster et al. 2016).

3 Structure of the Glycine Transporters

The three families encompassing glycine transporters from glial cells (SLC6, SLC38, and SLC7) are divergent in sequence (less than 15% identity) (Schlessinger et al. 2010) but are characterized by a similar structural fold. Hence, they are comprised into the amino acid-polyamine-organocation superfamily (APCS), which includes secondary carriers with different coupling modes and nature of the substrates (Wong et al. 2012; Schweikhard and Ziegler 2012). The sodium-and-chloride-dependent symporters of the SLC6 family and the sodium-independent exchanger (LAT2) of the SLC7 family are 12 transmembrane (TM) domain proteins with cytoplasmic N and C termini (Palacin et al. 2016). However, SLC38 members, which are sodium symporters (SNAT1 and 2) or sodium symporter/proton antiporters (SNAT5), share the first 11 TMs and are predicted to have an extracellular C terminus (Broer 2014). The common fold shared by the APCS transporters was first described in a prokaryotic SLC6 homologue LeuTAa (leucine transporter) from *Aquifex aeolicus*, which has been presently crystalized in three different conformations in the presence of several inhibitors and substrates (Yamashita et al. 2005; Krishnamurthy and Gouaux 2012). The LeuT fold, also called the 5 + 5 inverted repeat fold, involves 10 TMs with two structural repeats of five membrane helices: the arrangement of TMs 1–5 is repeated in the organization of TMs 6–10 except that each one is topologically inverted. In this way, the protein has a pseudo twofold symmetry axis parallel to the plane of the membrane (Rudnick et al. 2014; Shi 2013). The two repeats intertwine to form two four-helix bundles with equal contribution of the two: a scaffold bundle (hash) including TMs 3, 4, 8, and 9 and a core bundle including TMs 1, 2, 6, and 7. TMs 5 and 10 are linkers between the two bundles. TM1 and TM6 are partially unwound in the central portion of the membrane and are involved in sodium and substrate binding. According to the alternating access hypothesis proven for the physical explanation of transport, the transporter protein with a central substrate-binding site has to open to one side of the membrane for substrate/s upload and then to the opposite side for substrate/s release, undergoing conformational changes during this transformation (Jardetzky 1966). The mechanistic predictions based on the available crystal structures in the superfamily suggest a rocking bundle-type alternating-access mechanism in which the core bundle moves relative to the scaffold bundle (Rudnick et al. 2014; Shi 2013; Drew and Boudker 2016; Forrest and Rudnick 2009; Forrest et al. 2011). In LeuT, which displays a 2:1 Na^+/substrate stoichiometry, one sodium-binding site (Na2) is located between core and scaffold (contributed by TM1 and TM8), and its binding is predicted to stabilize the outward-facing conformation (Claxton et al. 2010). This site is conserved in all the sodium-dependent transporters from the APCS (Drew and Boudker 2016; Zhang et al. 2009; Krishnamurthy et al. 2009). Another, sodium site (Na1) is located within the core bundle, and it is exclusive of LeuT homologues of the neurotransmitter sodium symporters, NSS. In amino acid transporters, the carboxyl group of the substrate contributes to the coordination in Na1 (Yamashita et al. 2005; Krishnamurthy and Gouaux 2012; Zdravkovic et al. 2012; Perez-Siles et al. 2012). In Na^+-independent transporters of APCS, it has been proposed that the amine group of a lysine residue (Lys-158 in LAT1), located in a position equivalent to the Na2 site of LeuT, serves the same functional role of the sodium at Na2 (Broer 2014; Shi 2013; Shaffer et al. 2009). The general location of the substrate-binding sites appears conserved and involves TMs 1, 6, 3, and 8, though the composition of the residues coordinating the substrates varies significantly (Forrest et al. 2011). Many of the SLC6 transporters symport one chloride ion and have a chloride-binding site formed by residues of TM2, TM6, and TM7 near the Na1 site in the core bundle (Ben-Yona et al. 2011). A serine residue in TM6 (i.e., Ser336 in SERT) is involved in the coordination of both the chloride site and Na1 site, promoting a stringent coupling (Zomot et al. 2007; Penmatsa et al. 2013; Kantcheva et al. 2013). This serine replaces a glutamate (Glu290) in TM7 of LeuTAa, and the presence of the acidic amino acid is responsible for chloride independence of the bacterial transporter (Zomot et al. 2007; Penmatsa and Gouaux 2014). Two crystals from eukaryotic SLC6 transporters bound to antidepressant inhibitors have been analyzed to high resolution: the *Drosophila melanogaster* dopamine transporter (*d*DAT) and the human serotonin transporter (*h*SERT) (Penmatsa et al. 2013; Coleman et al. 2016). Their outward-facing overall structures confirmed previous LeuT architecture including ion and substrate-binding sites and inhibition mechanism but added relevant new information. *d*DAT and *h*SERT have a pronounced kink in transmembrane helix 12 halfway across the membrane bilayer. Since TMs 11 and 12 are presumably involved in oligomer formation (Broer and Gether 2012; Gether et al. 2006), this suggests a unique eukaryotic quaternary structure. For the first time, part of the carboxyl-terminal helix could be visualized showing a latch-like structure that caps the cytoplasmic gate near the internal loop 1 (Penmatsa et al. 2013). Very relevant to transport function is the presence of cholesterol molecules associated to both transporter proteins although in different positions (Penmatsa et al. 2013; Coleman et al. 2016). In the *h*SERT, an allosteric site in the extracellular vestibule of the transporter has been allocated. The occupancy of this site hinders ligand unbinding from the central site, explaining how allosteric ligands such as (*S*)-citalopram slow the off-rate of inhibitors bound to the central site. These data may boost the design of new ligands targeting the central and allosteric-binding sites (Coleman et al. 2016).

4 GlyT1 Transporter: Structural Aspects

The architecture of the glycine-selective transporter GlyT1 has been investigated by homology modeling of its membrane-embedded three-dimensional structure using LeuTAa (PDB 2A65) outward-facing substrate-bound conformation as a template (Perez-Siles et al. 2011). As all eukaryotic SLC6 transporters, the homology of GlyT1 with LeuTAa is low (about 20–25% sequence identity), but the residues involved in the binding and permeation of substrates, located in the transmembrane segments, show adequate degree of conservation. In addition, the molecular modeling of the highly homologous neuronal GlyT2 (50% sequence identity) allowed making structural comparisons between the two glycine-selective transporters (Perez-Siles et al. 2012). Molecular dynamic simulations on the generated GlyT1 and GlyT2 homology models predict the conservation of the substrate and the two sodium sites (Na1 and Na2) present in the template, although the location of the additional sodium site in GlyT2 remains to be confirmed (Subramanian et al. 2016). In addition, although not present in LeuT, the location of the GlyTs chloride-binding site seems to be conserved as compared to Cl^--dependent SLC6 transporters. The mutation of Ser339 in GlyT1 and Ser515 in GlyT2 provokes chloride-independent transport in both GlyTs, indicating the conservation of the

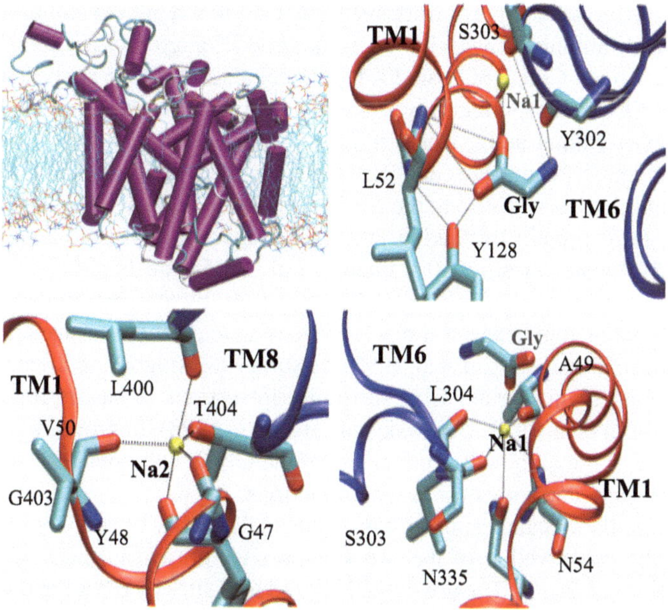

Fig. 1 GlyT1 structural model obtained by homology modeling using LeuTAa (PDB code 2A65) as the template. (**a**) Lateral view of modelled GlyT1. (**b**) Glycine-binding site. (**c** and **d**) Sodium-binding sites

chloride site (Pérez-Siles et al. unpublished results). The validity of the working models has been confirmed by mutagenesis and functional analysis (Fig. 1). As already noted, the residues that form the binding pocket for the leucine side chain in LeuT are substituted by more voluminous ones in GlyTs, defining a reduced pocket that accommodates the smaller glycine (Yamashita et al. 2005). Interestingly, the different sensitivity to sarcosine of the two GlyTs is due to a single amino acid substitution flanking this region. GlyT1 contains a small TM6 glycine residue (Gly305) that permits the access of sarcosine to the substrate-binding site, but the substitution in GlyT2 by a bigger amino acid (Ser481) prevents the transport of the N-methyl analog (Vandenberg et al. 2007). Since sarcosine is able to reduce the action of a selective inhibitor on GlyT2 transport, we believe that Ser481 does not prevent binding but the translocation of sarcosine (Pérez-Siles, unpublished results). This could be due to a competitive action on the primary substrate site, although the discovery of the allosteric site in the extracellular vestibule of *h*SERT raises the question of whether an analogous site is also present in the GlyTs and could contribute to the observed effect (Coleman et al. 2016). The external vestibule is lined by several external loops, including the long EL2 between TMs 3 and 4 that, in GlyT1, contains four N-linked glycans (Martinez-Maza et al. 2001; Olivares et al. 1995). The resolution of *d*DAT and *h*SERT structures has revealed for the first time the arrangement of EL2 and its extensive interactions with EL4. This may allow widening the modeled GlyTs structures to include this region essential for glycine transport and inhibitor sensitivity (Ju et al. 2004; Edington et al. 2009; Gimenez et al. 2012; Lopez-Corcuera et al. 2001; Roux et al. 2001). The analysis of the two eukaryotic SLC6 transporters has also evidenced interesting structural aspects of the carboxyl-terminal helix, which interacts with the internal loop 1 capping the cytoplasmic gate (Penmatsa et al. 2013; Coleman et al. 2016). N and C termini are regions of high diversity among the NSS transporters that have shown to display a relevant role in protein trafficking and variant specification (Fernandez-Sanchez et al. 2008; Poyatos et al. 2000; Cubelos et al. 2005b; de Juan-Sanz et al. 2011, 2013). Five variants of GlyT1 (GlyT1a-e), built by combinations of N- and C-terminal exons that occur as a result of alternative promoter usage and/or splicing, have been identified (Liu et al. 1992; Nelson 1998; Aragon and Lopez-Corcuera 2003). Among them, variant GlyT1b (transcript variant 1) has a preferred brain expression (Liu et al. 1993; Borowsky and Hoffman 1998). Improved knowledge of the structure of the carboxyl-terminal end of the GlyTs will open research avenues on their relevant, and somehow elusive, functions.

The molecular dynamic simulations on the modeled GlyT1 and GlyT2 structures predicted the conservation of the two sodium sites present in the LeuT template (Perez-Siles et al. 2011). The use of the Na1 and Na2 sites by GlyT1 and GlyT2 was confirmed through site-directed mutagenesis of the proposed coordinating residues and through the analysis of the sensitivity to lithium ion. Lithium is a non-transport-driving ion for the GlyTs (Lopez-Corcuera et al. 1998), but it can replace the transport-driving ion (sodium) at Na2 site but not at Na1 site, in agreement with previous data on the GABA transporter GAT1 (Zhou et al. 2006). Therefore, the mutation of residues involved in Na2 coordination (but not Na1) prevents lithium binding, confirming that Na1 and Na2 sites have dissimilar contributions to the interaction with lithium. This supports that Na1 is a high-affinity/high-selectivity sodium-binding site, and Na2 is a more promiscuous cation-binding site where lithium can bind (Zhou et al. 2006; Noskov and Roux 2008). Moreover, lithium binding to GlyT1 and GlyT2 Na2 sites has profoundly different consequences on glycine transport as it inhibits GlyT1 but stimulates GlyT2, probably due to the different conformational changes that may promote in the two transporters (Zhao and Noskov 2013). Kinetic analysis of lithium responses by the GlyTs revealed two lithium-binding events in GlyT2 but only one in GlyT1, what boosts the speculation on the capability of lithium binding by the unknown Na3 site. Nevertheless, the nature of the additional lithium site in GlyT2 remains to be established (Perez-Siles et al. 2011). In the search for the additional Na^+ site in GlyT2 whose nature is presently obscure, comparative molecular dynamics simulations in the presence of Na^+ were performed on GlyT1 and GlyT2 models (Perez-Siles et al. 2012). Calculated molecular interaction potential (MIP) maps and in silico mutagenesis revealed a conserved region in the GlyT2 external vestibule that displayed Na^+-binding energy values sufficiently stable to accommodate the sodium ion with higher probability than in GlyT1. Three aspartate residues within this region (Asp471, Asp542, and Asp635) exhibited the highest MIP for Na^+ ions, but only Asp471 tolerated some amino acid substitutions, suggesting a crucial role of the cluster in glycine transport. By mutagenesis and biochemical and electrophysiological analysis, Asp471 was revealed as a residue involved in cation affinity, cooperativity, and coupling of transport that fulfills the features of a cation-sensitive and conformationally active residue involved in Na^+-and Li^+-induced conformational changes. These properties contrast with those of the homologous aspartate in GlyT1 (Asp295) and sustain interesting functional differences in the external vestibule of GlyT1 and GlyT2 (Perez-Siles et al. 2012). The location of the aspartate cluster in GlyT2 seems to partially overlap with the recently discovered transient Na^+-binding site Na1' in LeuTAa, located more toward extracellular side compared with Na1. By using microsecond simulations, a role in the reposition of Na^+ ion into Na1 site has been suggested (Zomot et al. 2015). The proposal that Na1' is probably necessary for the binding of substrate and the transition from the outward-open to occluded state strongly coincides with the properties exhibited by the aspartate cluster in GlyT2.

SLC6 transporters are oligomeric proteins although the role of the oligomerization in the establishment of a catalytic-competent transporter is unclear (Farhan et al. 2006; Chiba et al. 2014). Most of the crystalized transporters appear as dimers. In LeuT the dimer interface is formed by TM9 and TM12 (Yamashita et al. 2005). However, recent solved structures from eukaryotic members show divergence in this structural aspect. In *d*DAT, the differential arrangement of TM12 showing a bend in the center at Pro572 that causes the second half of the helix to turn away from the transporter indicates that the dimerization interface is not conserved as compared to LeuT (Penmatsa et al. 2013). Furthermore, the crystal structure of *h*SERT does not fulfill the necessary protomer alignment features to validate the structure as oligomeric (Coleman et al. 2016). It seems luckily that these proteins need some additional molecules or the membrane bilayer to adopt the quaternary structure since *d*DAT and *h*SERT remain monomeric in detergent micelles despite its oligomeric structure has been proven (Anderluh et al. 2014; Zhen et al. 2015; Sitte et al. 2015). This behavior could be the cause of the reported monomeric state for the glycine transporters in detergent micelles (Horiuchi et al. 2001; Lopez-Corcuera et al. 1993). Up to date, only GlyT oligomers obtained in the presence of cross linkers have been described to exist in the cell surface (Fernandez-Sanchez et al. 2008; Bartholomaus et al. 2008). However, for other SLC6 members, a variety of oligomerization states and even cooperativity between protomers has been proposed (Anderluh et al. 2014; Zhen et al. 2015). For glycine transporters, oligomerization has been clearly proven as a requirement for the exit from the endoplasmic reticulum both for GlyT1 (Fernandez-Sanchez et al. 2008) and for GlyT2 (Arribas-González et al. 2015). The presence of GlyT1 (Fernandez-Sanchez et al. 2008) and for GlyT2 (Gimenez et al. 2012) oligomers at the plasma membrane has also been proven. In this later case, the first in vivo evidence of GlyT2 oligomers at the plasma membrane came from a dominant mutation found in hyperekplexia patients, for which the dominance relies in the existence of heteroligomers carrying wild type and mutant protomers (Gimenez et al. 2012). The structural domains involved in the oligomerization of GlyTs remain to be defined. GlyTs oligomerization models could be a valuable aid for this purpose.

5 Concluding Remarks

Glycine, as a small neutral amino acid, is transported by several glial plasma membrane transporters that have overlapping substrate reactivity. The signaling functions of glycine as a inhibitory neurotransmitter and a neuromodulator of glutamate transmission are terminated by the glycine selective transporters, predominantly the glial GlyT1. The contribution of the five additional broad-specificity transporters to the global glycine transport is starting to be recognized after the understanding of its molecular structure. Three transporter families encompass carriers able to transport glycine with no sequence homology: the SLC6, SLC38, and SLC7 families. All share a common structural pattern, the LeuT fold, displayed by many crystallized transport proteins that is widening our understanding of the alternate access transport mechanism. The resolution of the three-dimensional structure of new members of this superfamily provides a solid platform for homology modeling of glycine transporters.

Acknowledgments This work was supported by grants from the Spanish Dirección General de Investigación Científica y Técnica (SAF2014-58045-R), the Fondo de Investigaciones Sanitarias (CIBER of rare diseases), and by an institutional grant from the Fundación Ramón Areces. C.B.M. is a recipient of a grant BES-2015-072392 from the Spanish Ministerio de Economia y Competitividad.

Conflict of Interest The author declares no conflicts of interest.

References

Anderluh A, Klotzsch E, Reismann AW, Brameshuber M, Kudlacek O, Newman AH, et al. Single molecule analysis reveals coexistence of stable serotonin transporter monomers and oligomers in the live cell plasma membrane. J Biol Chem. 2014;289(7):4387–94.

Apostolides PF, Trussell LO. Rapid, activity-independent turnover of vesicular transmitter content at a mixed glycine/GABA synapse. J Neurosci. 2013;33(11):4768–81.

Aragon C, Lopez-Corcuera B. Structure, function and regulation of glycine neurotransporters. Eur J Pharmacol. 2003;479(1–3):249–62.

Aragon C, Lopez-Corcuera B. Glycine transporters: crucial roles of pharmacological interest revealed by gene deletion. Trends Pharmacol Sci. 2005;26(6):283–6.

Aroeira RI, Sebastiao AM, Valente CA. GlyT1 and GlyT2 in brain astrocytes: expression, distribution and function. Brain Struct Funct. 2014;219(3):817–30.

Arribas-González E, de Juan-Sanz J, Aragón C, López-Corcuera B. Molecular basis of the dominant negative effect of a glycine transporter 2 mutation associated with hyperekplexia. J Biol Chem. 2015;290(4):2150–65.

Babu E, Bhutia YD, Ramachandran S, Gnanaprakasam JP, Prasad PD, Thangaraju M, et al. Deletion of the amino acid transporter Slc6a14 suppresses tumour growth in spontaneous mouse models of breast cancer. Biochem J. 2015;469(1):17–23.

Bartholomaus I, Milan-Lobo L, Nicke A, Dutertre S, Hastrup H, Jha A, et al. Glycine transporter dimers: evidence for occurrence in the plasma membrane. J Biol Chem. 2008;283(16):10978–91.

Ben-Yona A, Bendahan A, Kanner BI. A glutamine residue conserved in the neurotransmitter:sodium: symporters is essential for the interaction of chloride with the GABA transporter GAT-1. J Biol Chem. 2011;286(4):2826–33.

Bhutia YD, Ganapathy V. Glutamine transporters in mammalian cells and their functions in physiology and cancer. Biochim Biophys Acta Mol Cell Res. 2016;1863(10):2531–9.
Bockaert J, Marin P. mTOR in brain physiology and pathologies. Physiol Rev. 2015;95(4):1157–87.
Borowsky B, Hoffman BJ. Analysis of a gene encoding two glycine transporter variants reveals alternative promoter usage and a novel gene structure. J Biol Chem. 1998;273(44):29077–85.
Braun D, Kinne A, Brauer AU, Sapin R, Klein MO, Kohrle J, et al. Developmental and cell type-specific expression of thyroid hormone transporters in the mouse brain and in primary brain cells. Glia. 2011;59(3):463–71.
Broer S. Diseases associated with general amino acid transporters of the solute carrier 6 family (SLC6). Curr Mol Pharmacol. 2013;6(2):74–87.
Broer S. The SLC38 family of sodium-amino acid co-transporters. Pflugers Arch. 2014;466(1): 155–72.
Broer S, Brookes N. Transfer of glutamine between astrocytes and neurons. J Neurochem. 2001;77(3):705–19.
Broer S, Gether U. The solute carrier 6 family of transporters. Br J Pharmacol. 2012;167(2):256–78.
Broer S, Palacin M. The role of amino acid transporters in inherited and acquired diseases. Biochem J. 2011;436(2):193–211.
Broer A, Brookes N, Ganapathy V, Dimmer KS, Wagner CA, Lang F, et al. The astroglial ASCT2 amino acid transporter as a mediator of glutamine efflux. J Neurochem. 1999;73(5):2184–94.
Broer A, Wagner C, Lang F, Broer S. Neutral amino acid transporter ASCT2 displays substrate-induced Na+ exchange and a substrate-gated anion conductance. Biochem J. 2000;346(Pt 3): 705–10.
Broer S, Broer A, Hansen JT, Bubb WA, Balcar VJ, Nasrallah FA, et al. Alanine metabolism, transport, and cycling in the brain. J Neurochem. 2007;102(6):1758–70.
Broer A, Rahimi F, Broer S. Deletion of amino acid transporter ASCT2 (SLC1A5) reveals an essential role for transporters SNAT1 (SLC38A1) and SNAT2 (SLC38A2) to sustain Glutaminolysis in cancer cells. J Biol Chem. 2016;291(26):13194–205.
Chatterton JE, Awobuluyi M, Premkumar LS, Takahashi H, Talantova M, Shin Y, et al. Excitatory glycine receptors containing the NR3 family of NMDA receptor subunits. Nature. 2002;415(6873):793–8.
Chaudhry FA, Krizaj D, Larsson P, Reimer RJ, Wreden C, Storm-Mathisen J, et al. Coupled and uncoupled proton movement by amino acid transport system N. EMBO J. 2001;20(24):7041–51.
Chiba P, Freissmuth M, Stockner T. Defining the blanks – pharmacochaperoning of SLC6 transporters and ABC transporters. Pharmacol Res. 2014;83:63–73.
Christensen HN, Oxender DL, Liang M, Vatz KA. The use of N-methylation to direct route of mediated transport of amino acids. J Biol Chem. 1965;240(9):3609–16.
Claxton DP, Quick M, Shi L, de Carvalho FD, Weinstein H, Javitch JA, et al. Ion/substrate-dependent conformational dynamics of a bacterial homolog of neurotransmitter:sodium symporters. Nat Struct Mol Biol. 2010;17(7):822–9.
Coleman JA, Green EM, Gouaux E. X-ray structures and mechanism of the human serotonin transporter. Nature. 2016;532(7599):334–9.
Cubelos B, Gimenez C, Zafra F. Localization of the GLYT1 glycine transporter at glutamatergic synapses in the rat brain. Cereb Cortex. 2005a;15(4):448–59.
Cubelos B, Gimenez C, Zafra F. The glycine transporter GLYT1 interacts with Sec3, a component of the exocyst complex. Neuropharmacology. 2005b;49(6):935–44.
Danbolt NC. Glutamate uptake. Prog Neurobiol. 2001;65(1):1–105.
de Juan-Sanz J, Zafra F, Lopez-Corcuera B, Aragon C. Endocytosis of the neuronal glycine transporter GLYT2: role of membrane rafts and protein kinase C-dependent ubiquitination. Traffic. 2011;12(12):1850–67.
de Juan-Sanz J, Nunez E, Lopez-Corcuera B, Aragon C. Constitutive endocytosis and turnover of the neuronal glycine transporter GlyT2 is dependent on ubiquitination of a C-terminal lysine cluster. PLoS One. 2013;8(3):e58863.
Drew D, Boudker O. Shared molecular mechanisms of membrane transporters. Annu Rev Biochem. 2016;85:543–72.

Dutertre S, Becker CM, Betz H. Inhibitory glycine receptors: an update. J Biol Chem. 2012;287(48): 40216–23.

Edington AR, McKinzie AA, Reynolds AJ, Kassiou M, Ryan RM, Vandenberg RJ. Extracellular loops 2 and 4 of GLYT2 are required for N-arachidonylglycine inhibition of glycine transport. J Biol Chem. 2009;284(52):36424–30.

Eulenburg V, Retiounskaia M, Papadopoulos T, Gomeza J, Betz H. Glial glycine transporter 1 function is essential for early postnatal survival but dispensable in adult mice. Glia. 2010;58(9):1066–73.

Farhan H, Freissmuth M, Sitte HH. Oligomerization of neurotransmitter transporters: a ticket from the endoplasmic reticulum to the plasma membrane. Handb Exp Pharmacol. 2006;175:233–49.

Fernandez-Sanchez E, Diez-Guerra FJ, Cubelos B, Gimenez C, Zafra F. Mechanisms of endoplasmic-reticulum export of glycine transporter-1 (GLYT1). Biochem J. 2008;409(3):669–81.

Forrest LR, Rudnick G. The rocking bundle: a mechanism for ion-coupled solute flux by symmetrical transporters. Physiology (Bethesda). 2009;24:377–86.

Forrest LR, Kramer R, Ziegler C. The structural basis of secondary active transport mechanisms. Biochim Biophys Acta. 2011;1807(2):167–88.

Foster AC, Farnsworth J, Lind GE, Li Y-X, Yang J-Y, Dang V, et al. D-serine is a substrate for neutral amino acid transporters ASCT1/SLC1A4 and ASCT2/SLC1A5, and is transported by both subtypes in rat hippocampal astrocyte cultures. PLoS One. 2016;11(6):e0156551.

Fotiadis D, Kanai Y, Palacin M. The SLC3 and SLC7 families of amino acid transporters. Mol Asp Med. 2013;34(2–3):139–58.

Gether U, Andersen PH, Larsson OM, Schousboe A. Neurotransmitter transporters: molecular function of important drug targets. Trends Pharmacol Sci. 2006;27(7):375–83.

Gimenez C, Zafra F, Lopez-Corcuera B, Aragon C. Molecular bases of hereditary hyperekplexia. Rev Neurol. 2008;47(12):648–52.

Gimenez C, Perez-Siles G, Martinez-Villarreal J, Arribas-Gonzalez E, Jimenez E, Nunez E, et al. A novel dominant hyperekplexia mutation Y705C alters trafficking and biochemical properties of the presynaptic glycine transporter GlyT2. J Biol Chem. 2012;287(34):28986–9002.

Gomeza J, Hulsmann S, Ohno K, Eulenburg V, Szoke K, Richter D, et al. Inactivation of the glycine transporter 1 gene discloses vital role of glial glycine uptake in glycinergic inhibition. Neuron. 2003a;40(4):785–96.

Gomeza J, Ohno K, Hulsmann S, Armsen W, Eulenburg V, Richter DW, et al. Deletion of the mouse glycine transporter 2 results in a hyperekplexia phenotype and postnatal lethality. Neuron. 2003b;40(4):797–806.

Gonzalez-Gonzalez IM, Cubelos B, Gimenez C, Zafra F. Immunohistochemical localization of the amino acid transporter SNAT2 in the rat brain. Neuroscience. 2005;130(1):61–73.

Grewer C, Gameiro A, Rauen T. SLC1 glutamate transporters. Pflugers Arch. 2014;466(1):3–24.

Hamdani el H, Gudbrandsen M, Bjorkmo M, Chaudhry FA. The system N transporter SN2 doubles as a transmitter precursor furnisher and a potential regulator of NMDA receptors. Glia. 2012;60(11):1671–83.

Harvey RJ, Yee BK. Glycine transporters as novel therapeutic targets in schizophrenia, alcohol dependence and pain. Nat Rev Drug Discov. 2013;12(11):866–85.

Hatanaka T, Huang W, Nakanishi T, Bridges CC, Smith SB, Prasad PD, et al. Transport of D-serine via the amino acid transporter ATB(0,+) expressed in the colon. Biochem Biophys Res Commun. 2002;291(2):291–5.

Horiuchi M, Nicke A, Gomeza J, Aschrafi A, Schmalzing G, Betz H. Surface-localized glycine transporters 1 and 2 function as monomeric proteins in Xenopus oocytes. Proc Natl Acad Sci U S A. 2001;98(4):1448–53.

Hosoya K, Ichikawa T, Akanuma S, Hirose S, Tachikawa M. Glycine and L-arginine transport in cultured Muller glial cells (TR-MUL). Neurochem Int. 2010;57(3):262–8.

Jardetzky O. Simple allosteric model for membrane pumps. Nature. 1966;211(5052):969–70. doi:10.1038/211969a0.

Jenstad M, Chaudhry FA. The amino acid transporters of the glutamate/GABA-glutamine cycle and their impact on insulin and glucagon secretion. Front Endocrinol (Lausanne). 2013;4:199.

Johnson JW, Ascher P. Glycine potentiates the NMDA response in cultured mouse brain neurons. Nature. 1987;325(6104):529–31.

Ju P, Aubrey KR, Vandenberg RJ. Zn2+ inhibits glycine transport by glycine transporter subtype 1b. J Biol Chem. 2004;279(22):22983–91.

Kanai Y, Clemencon B, Simonin A, Leuenberger M, Lochner M, Weisstanner M, et al. The SLC1 high-affinity glutamate and neutral amino acid transporter family. Mol Asp Med. 2013;34(2–3):108–20.

Kanner BI, Zomot E. Sodium-coupled neurotransmitter transporters. Chem Rev. 2008;108(5): 1654–68.

Kantcheva AK, Quick M, Shi L, Winther AM, Stolzenberg S, Weinstein H, et al. Chloride binding site of neurotransmitter sodium symporters. Proc Natl Acad Sci U S A. 2013;110(21):8489–94.

Karunakaran S, Umapathy NS, Thangaraju M, Hatanaka T, Itagaki S, Munn DH, et al. Interaction of tryptophan derivatives with SLC6A14 (ATB0,+) reveals the potential of the transporter as a drug target for cancer chemotherapy. Biochem J. 2008;414(3):343–55.

Kilberg MS, Handlogten ME, Christensen HN. Characteristics of an amino acid transport system in rat liver for glutamine, asparagine, histidine, and closely related analogs. J Biol Chem. 1980;255(9):4011–9.

Kim DK, Kim IJ, Hwang S, Kook JH, Lee MC, Shin BA, et al. System L-amino acid transporters are differently expressed in rat astrocyte and C6 glioma cells. Neurosci Res. 2004;50(4):437–46.

Kirischuk S, Heja L, Kardos J, Billups B. Astrocyte sodium signaling and the regulation of neurotransmission. Glia. 2015;64:1655–66.

Krishnamurthy H, Gouaux E. X-ray structures of LeuT in substrate-free outward-open and apo inward-open states. Nature. 2012;481(7382):469–74.

Krishnamurthy H, Piscitelli CL, Gouaux E. Unlocking the molecular secrets of sodium-coupled transporters. Nature. 2009;459(7245):347–55.

Legendre P. The glycinergic inhibitory synapse. Cell Mol Life Sci. 2001;58(5–6):760–93.

Liu QR, Nelson H, Mandiyan S, Lopez-Corcuera B, Nelson N. Cloning and expression of a glycine transporter from mouse brain. FEBS Lett. 1992;305(2):110–4.

Liu QR, Mandiyan S, Lopez-Corcuera B, Nelson H, Nelson N. A rat brain cDNA encoding the neurotransmitter transporter with an unusual structure. FEBS Lett. 1993;315(2):114–8.

Lopez E, Lee-Rivera I, Lopez-Colome AM. Characteristics and regulation of glycine transport in Bergmann glia. Neurochem Res. 2005;30(12):1567–77.

Lopez-Corcuera B, Alcantara R, Vazquez J, Aragon C. Hydrodynamic properties and immunological identification of the sodium- and chloride-coupled glycine transporter. J Biol Chem. 1993;268(3):2239–43.

Lopez-Corcuera B, Martinez-Maza R, Nunez E, Roux M, Supplisson S, Aragon C. Differential properties of two stably expressed brain-specific glycine transporters. J Neurochem. 1998;71(5): 2211–9.

Lopez-Corcuera B, Nunez E, Martinez-Maza R, Geerlings A, Aragon C. Substrate-induced conformational changes of extracellular loop 1 in the glycine transporter GLYT2. J Biol Chem. 2001;276(46):43463–70.

Mackenzie B, Erickson JD. Sodium-coupled neutral amino acid (System N/A) transporters of the SLC38 gene family. Pflugers Arch. 2004;447(5):784–95.

Mackenzie B, Schafer MK, Erickson JD, Hediger MA, Weihe E, Varoqui H. Functional properties and cellular distribution of the system A glutamine transporter SNAT1 support specialized roles in central neurons. J Biol Chem. 2003;278(26):23720–30.

Martinez-Maza R, Poyatos I, Lopez-Corcuera B, Nunez E, Gimenez C, Zafra F, et al. The role of N-glycosylation in transport to the plasma membrane and sorting of the neuronal glycine transporter GLYT2. J Biol Chem. 2001;276(3):2168–73.

McIntire SL, Reimer RJ, Schuske K, Edwards RH, Jorgensen EM. Identification and characterization of the vesicular GABA transporter. Nature. 1997;389(6653):870–6.

Meier C, Ristic Z, Klauser S, Verrey F. Activation of system L heterodimeric amino acid exchangers by intracellular substrates. EMBO J. 2002;21(4):580–9.

Melone M, Quagliano F, Barbaresi P, Varoqui H, Erickson JD, Conti F. Localization of the glutamine transporter SNAT1 in rat cerebral cortex and neighboring structures, with a note on its localization in human cortex. Cereb Cortex. 2004;14(5):562–74.

Milkereit R, Persaud A, Vanoaica L, Guetg A, Verrey F, Rotin D. LAPTM4b recruits the LAT1-4F2hc Leu transporter to lysosomes and promotes mTORC1 activation. Nat Commun. 2015;6:7250.

Nakanishi T, Hatanaka T, Huang W, Prasad PD, Leibach FH, Ganapathy ME, et al. Na+− and Cl −coupled active transport of carnitine by the amino acid transporter ATB(0,+) from mouse colon expressed in HRPE cells and Xenopus oocytes. J Physiol. 2001;532(Pt 2):297–304.

Nelson N. The family of Na+/Cl- neurotransmitter transporters. J Neurochem. 1998;71(5):1785–803.

Nong Y, Huang YQ, Ju W, Kalia LV, Ahmadian G, Wang YT, et al. Glycine binding primes NMDA receptor internalization. Nature. 2003;422(6929):302–7.

Noskov SY, Roux B. Control of ion selectivity in LeuT: two Na+ binding sites with two different mechanisms. J Mol Biol. 2008;377(3):804–18.

Nunez E, Lopez-Corcuera B, Martinez-Maza R, Aragon C. Differential effects of ethanol on glycine uptake mediated by the recombinant GLYT1 and GLYT2 glycine transporters. Br J Pharmacol. 2000;129(4):802–10.

Nunez B, Martinez de Mena R, Obregon MJ, Font-Llitjos M, Nunes V, Palacin M, et al. Cerebral cortex hyperthyroidism of newborn mct8-deficient mice transiently suppressed by lat2 inactivation. PLoS One. 2014;9(5):e96915.

Olivares L, Aragon C, Gimenez C, Zafra F. The role of N-glycosylation in the targeting and activity of the GLYT1 glycine transporter. J Biol Chem. 1995;270(16):9437–42.

Pai YJ, Leung KY, Savery D, Hutchin T, Prunty H, Heales S, et al. Glycine decarboxylase deficiency causes neural tube defects and features of non-ketotic hyperglycinemia in mice. Nat Commun. 2015;6:6388.

Palacin M, Errasti-Murugarren E, Rosell A. Heteromeric amino acid transporters. In search of the molecular bases of transport cycle mechanisms. Biochem Soc Trans. 2016;44(3):745–52.

Penmatsa A, Gouaux E. How LeuT shapes our understanding of the mechanisms of sodium-coupled neurotransmitter transporters. J Physiol. 2014;592(5):863–9.

Penmatsa A, Wang KH, Gouaux E. X-ray structure of dopamine transporter elucidates antidepressant mechanism. Nature. 2013;503(7474):85–90.

Perez-Siles G, Morreale A, Leo-Macias A, Pita G, Ortiz AR, Aragon C, et al. Molecular basis of the differential interaction with lithium of glycine transporters GLYT1 and GLYT2. J Neurochem. 2011;118(2):195–204.

Perez-Siles G, Nunez E, Morreale A, Jimenez E, Leo-Macias A, Pita G, et al. An aspartate residue in the external vestibule of GLYT2 (glycine transporter 2) controls cation access and transport coupling. Biochem J. 2012;442(2):323–34.

Pineda M, Fernandez E, Torrents D, Estevez R, Lopez C, Camps M, et al. Identification of a membrane protein, LAT-2, that co-expresses with 4F2 heavy chain, an L-type amino acid transport activity with broad specificity for small and large zwitterionic amino acids. J Biol Chem. 1999;274(28):19738–44.

Pinilla-Tenas J, Barber A, Lostao PM. Transport of proline and hydroxyproline by the neutral amino-acid exchanger ASCT1. J Membr Biol. 2003;195(1):27–32.

Poyatos I, Ruberti F, Martinez-Maza R, Gimenez C, Dotti CG, Zafra F. Polarized distribution of glycine transporter isoforms in epithelial and neuronal cells. Mol Cell Neurosci. 2000;15(1):99–111.

Rees MI, Harvey K, Pearce BR, Chung SK, Duguid IC, Thomas P, et al. Mutations in the gene encoding GlyT2 (SLC6A5) define a presynaptic component of human startle disease. Nat Genet. 2006;38(7):801–6.

Roux MJ, Supplisson S. Neuronal and glial glycine transporters have different stoichiometries. Neuron. 2000;25(2):373–83.

Roux MJ, Martinez-Maza R, Le Goff A, Lopez-Corcuera B, Aragon C, Supplisson S. The glial and the neuronal glycine transporters differ in their reactivity to sulfhydryl reagents. J Biol Chem. 2001;276(21):17699–705.

Rudnick G, Kramer R, Blakely RD, Murphy DL, Verrey F. The SLC6 transporters: perspectives on structure, functions, regulation, and models for transporter dysfunction. Pflugers Arch. 2014;466(1):25–42.

Sagne C, El Mestikawy S, Isambert MF, Hamon M, Henry JP, Giros B, et al. Cloning of a functional vesicular GABA and glycine transporter by screening of genome databases. FEBS Lett. 1997;417(2):177–83.

Sakai K, Shimizu H, Koike T, Furuya S, Watanabe M. Neutral amino acid transporter ASCT1 is preferentially expressed in L-Ser-synthetic/storing glial cells in the mouse brain with transient expression in developing capillaries. J Neurosci. 2003;23(2):550–60.

Samluk L, Czeredys M, Nalecz KA. Regulation of amino acid/carnitine transporter B 0,+ (ATB 0,+) in astrocytes by protein kinase C: independent effects on raft and non-raft transporter subpopulations. J Neurochem. 2010;115(6):1386–97.

Samluk L, Czeredys M, Skowronek K, Nalecz KA. Protein kinase C regulates amino acid transporter ATB(0,+). Biochem Biophys Res Commun. 2012;422(1):64–9.

Schiöth HB, Roshanbin S, Hägglund MGA, Fredriksson R. Evolutionary origin of amino acid transporter families SLC32, SLC36 and SLC38 and physiological, pathological and therapeutic aspects. Mol Asp Med. 2013;34(2–3):571–85.

Schlessinger A, Matsson P, Shima JE, Pieper U, Yee SW, Kelly L, et al. Comparison of human solute carriers. Protein Sci. 2010;19(3):412–28.

Schweikhard ES, Ziegler CM. Amino acid secondary transporters: toward a common transport mechanism. Curr Top Membr. 2012;70:1–28.

Scopelliti AJ, Heinzelmann G, Kuyucak S, Ryan RM, Vandenberg RJ. Na+ interactions with the neutral amino acid transporter ASCT1. J Biol Chem. 2014;289(25):17468–79.

Segawa H, Fukasawa Y, Miyamoto K, Takeda E, Endou H, Kanai Y. Identification and functional characterization of a Na+–independent neutral amino acid transporter with broad substrate selectivity. J Biol Chem. 1999;274(28):19745–51.

Shaffer PL, Goehring A, Shankaranarayanan A, Gouaux E. Structure and mechanism of a Na+–independent amino acid transporter. Science. 2009;325(5943):1010–4.

Shi Y. Common folds and transport mechanisms of secondary active transporters. Annu Rev Biophys. 2013;42:51–72.

Singer P, Dubroqua S, Yee BK. Inhibition of glycine transporter 1: the yellow brick road to new schizophrenia therapy? Curr Pharm Des. 2015;21(26):3771–87.

Sitte HH, Schutz GJ, Freissmuth M. Cooperativity between individual transporter protomers: new data fuelling old complexes. J Neurochem. 2015;133(2):163–6.

Sloan JL, Mager S. Cloning and functional expression of a human Na(+) and Cl(−)-dependent neutral and cationic amino acid transporter B(0+). J Biol Chem. 1999;274(34):23740–5.

Sloan JL, Grubb BR, Mager S. Expression of the amino acid transporter ATB 0+ in lung: possible role in luminal protein removal. Am J Physiol Lung Cell Mol Physiol. 2003;284(1):L39–49.

Stephan J, Friauf E. Functional analysis of the inhibitory neurotransmitter transporters GlyT1, GAT-1, and GAT-3 in astrocytes of the lateral superior olive. Glia. 2014;62(12):1992–2003.

Subramanian N, Scopelitti AJ, Carland JE, Ryan RM, O'Mara ML, Vandenberg RJ. Identification of a 3rd Na+ binding site of the glycine transporter, GlyT2. PLoS One. 2016;11(6):e0157583.

Supplisson S, Roux MJ. Why glycine transporters have different stoichiometries. FEBS Lett. 2002;529(1):93–101.

Szoke K, Hartel K, Grass D, Hirrlinger PG, Hirrlinger J, Hulsmann S. Glycine transporter 1 expression in the ventral respiratory group is restricted to protoplasmic astrocytes. Brain Res. 2006;1119(1):182–9.

Utsunomiya-Tate N, Endou H, Kanai Y. Cloning and functional characterization of a system ASC-like Na+–dependent neutral amino acid transporter. J Biol Chem. 1996;271(25):14883–90.

Vandenberg RJ, Shaddick K, Ju P. Molecular basis for substrate discrimination by glycine transporters. J Biol Chem. 2007;282(19):14447–53.

Verrey F. System L: heteromeric exchangers of large, neutral amino acids involved in directional transport. Pflugers Arch. 2003;445(5):529–33.

Wadiche JI, Amara SG, Kavanaugh MP. Ion fluxes associated with excitatory amino acid transport. Neuron. 1995;15(3):721–8.

Werdehausen R, Kremer D, Brandenburger T, Schlosser L, Jadasz J, Kury P, et al. Lidocaine metabolites inhibit glycine transporter 1: a novel mechanism for the analgesic action of systemic lidocaine? Anesthesiology. 2012;116(1):147–58.

Wong FH, Chen JS, Reddy V, Day JL, Shlykov MA, Wakabayashi ST, et al. The amino acid-polyamine-organocation superfamily. J Mol Microbiol Biotechnol. 2012;22(2):105–13.

Yamashita A, Singh SK, Kawate T, Jin Y, Gouaux E. Crystal structure of a bacterial homologue of Na+/Cl – dependent neurotransmitter transporters. Nature. 2005;437(7056):215–23.

Zafra F, Gimenez C. Characteristics and adaptive regulation of glycine transport in cultured glial cells. Biochem J. 1989;258(2):403–8.

Zafra F, Gimenez C. Glycine transporters and synaptic function. IUBMB Life. 2008;60(12):810–7.

Zafra F, Aragon C, Olivares L, Danbolt NC, Gimenez C, Storm-Mathisen J. Glycine transporters are differentially expressed among CNS cells. J Neurosci. 1995;15(5 Pt 2):3952–69.

Zdravkovic I, Zhao C, Lev B, Cuervo JE, Noskov SY. Atomistic models of ion and solute transport by the sodium-dependent secondary active transporters. Biochim Biophys Acta. 2012;1818(2):337–47.

Zeilhofer HU, Studler B, Arabadzisz D, Schweizer C, Ahmadi S, Layh B, et al. Glycinergic neurons expressing enhanced green fluorescent protein in bacterial artificial chromosome transgenic mice. J Comp Neurol. 2005;482(2):123–41.

Zerangue N, Kavanaugh MP. ASCT-1 is a neutral amino acid exchanger with chloride channel activity. J Biol Chem. 1996;271(45):27991–4.

Zhang LH, Gong N, Fei D, Xu L, Xu TL. Glycine uptake regulates hippocampal network activity via glycine receptor-mediated tonic inhibition. Neuropsychopharmacology. 2008;33(3):701–11.

Zhang Z, Albers T, Fiumera HL, Gameiro A, Grewer C. A conserved Na(+) binding site of the sodium-coupled neutral amino acid transporter 2 (SNAT2). J Biol Chem. 2009;284(37):25314–23.

Zhao C, Noskov SY. The molecular mechanism of ion-dependent gating in secondary transporters. PLoS Comput Biol. 2013;9(10):e1003296.

Zhen J, Antonio T, Cheng SY, Ali S, Jones KT, Reith ME. Dopamine transporter oligomerization: impact of combining protomers with differential cocaine analog binding affinities. J Neurochem. 2015;133(2):167–73.

Zhou Y, Zomot E, Kanner BI. Identification of a lithium interaction site in the gamma-aminobutyric acid (GABA) transporter GAT-1. J Biol Chem. 2006;281(31):22092–9.

Zomot E, Bendahan A, Quick M, Zhao Y, Javitch JA, Kanner BI. Mechanism of chloride interaction with neurotransmitter: sodium symporters. Nature. 2007;449(7163):726–30.

Zomot E, Gur M, Bahar I. Microseconds simulations reveal a new sodium-binding site and the mechanism of sodium-coupled substrate uptake by LeuT. J Biol Chem. 2015;290(1):544–55.

Taurine Homeostasis and Volume Control

Herminia Pasantes-Morales

Abstract Taurine content is high (mM) in mammalian brain. By its major role as an osmolyte, taurine contributes to the cell volume control, which is particularly critical in the brain. Taurine participates in osmotic adjustments required to maintain the organization and size of intracellular compartments. It counteracts volume fluctuations in unbalanced transmembrane fluxes of ions and neurotransmitters, preserving the functional synaptic contacts. Taurine has a key role in the long-term adaptation to chronic hyponatremia as well as in other pathologies leading to brain edema. Together with other osmolytes, taurine corrects cell shrinkage, preventing mysfunction of organelles and apoptosis. Swelling corrective taurine efflux occurs through a leak pathway, likely formed by LCRR8 protein isoforms. Shrinkage-activated influx comes largely by the increased activity of the Na^+/Cl^--dependent transporter. The brain taurine pool results from the equilibrium between (*i*) dietary intake and active transport into the cell, (*ii*) synthesis in the brain itself or import of that synthesized elsewhere, and (*iii*) leak and posterior excretion. The interplay between these elements preserves brain taurine homeostasis in physiological conditions and permits the proper adjustments upon deviations of normal in the internal/external environment.

Keywords Volume regulation • Hyponatremia • Taurine deficiency • Brain edema

List of Abbreviations

CDO	Cysteine dioxygenase
CSAD	Cysteine sulfinic acid decarboxylase
KO	Knockout
LCRR8	Leucine-rich repeat containing 8
TAUT	Taurine transporter
VRAC	Volume-regulated anion channel

H. Pasantes-Morales (✉)
División de Neurociencias, Instituto de Fisiología Celular, Universidad Nacional Autónoma de México, Ciudad Universitaria, Coyoacán, 04500 Ciudad de México, Mexico
e-mail: hpasante@ifc.unam.mx

A. Ortega, A. Schousboe (eds.), *Glial Amino Acid Transporters*,
Advances in Neurobiology 16, DOI 10.1007/978-3-319-55769-4_3

1 Introduction

Taurine is a β-aminosulfonic acid, widely distributed and present in high concentration in animal cells. It is largely free in the cytosol, it is not incorporated into proteins, and, excluding the conjugation with bile acids, it participates in a limited number of metabolic reactions. Taurine has been proposed to act as membrane stabilizer, as regulator of immune system and of Ca^{2+} homeostasis (Huxtable 1990; Schuller-Levis and Park 2003; Lambert et al. 2015), and as anti-inflammatory, antiapoptotic, and antioxidant (Jong et al. 2012; Marcinkiewicz and Kontny 2014; Oja and Saransaari 2015; Shimada et al. 2015). In most cells, including brain cells, taurine is an osmolyte playing a major role to preserve brain volume facing acute and chronic conditions of hyponatremia and in other pathologies concurrent with cerebral edema (Verbalis and Gullans 1991; Schousboe and Pasantes-Morales 1992; Pasantes-Morales et al. 1998; Estévez et al. 1999). Taurine interacts with some neurotransmitter receptors (Albrecht and Schousboe 2005; Chan et al. 2014) and exerts neuroprotective actions in a number of injuring conditions (Sun and Xu 2008; Sun et al. 2011; Ye et al. 2013; Prentice et al. 2015). Taurine enhances proliferation and promotes neuronal specification of murine and human neural stem/progenitor cells (Hernández-Benitez et al. 2010a, 2013) and increases neurogenesis in the hippocampus (Gebara et al. 2015). Severe taurine deficiency results in heart and skeletal muscle dysfunction, retinal degeneration and blindness, and deficits in brain development (Hayes et al. 1975; Sturman et al. 1991; Sturman 1992; Warskulat et al. 2007). Taurine interaction with mitochondrial tRNAs improves the synthesis of mitochondrial proteins, an effect which relates taurine with mitochondrial neuropathologies (Suzuki et al. 2002). The interplay of biosynthesis, transport from external or internal sources, leak, and excretion preserves brain taurine homeostasis in physiological conditions (Fig. 1). This interplay also permits the proper adjustments upon deviations of normal in the internal/external environment.

2 Taurine Content in the Brain

Taurine concentration in the brain in most vertebrates is high, but large variations are found between species. In the human brain, reported taurine levels are around 1–2 mM, whereas in the mouse brain, this is 5–10 times higher. Carnivores like cat have higher amounts of brain taurine than omnivores as human or monkeys, and herbivores have lower amounts (Table 1) (Worden and Stipanuk 1985). Taurine concentration shows variation between brain areas, but the species difference persists (Frahm et al. 1989; Dawson et al. 1990). These large interspecies variations have not an obvious explanation. Taurine in the brain as in most tissues is part of the pool of organic osmolytes in charge of the cell volume control facing changes in external osmolality (Junankar and Kirk 2000). The relative

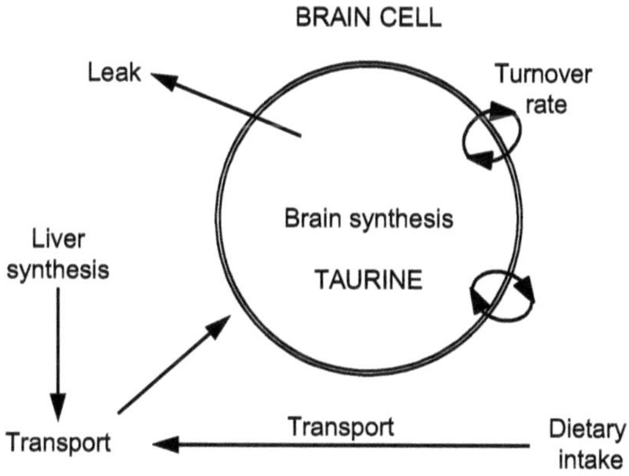

Fig. 1 Mechanisms preserving taurine homeostasis in the brain. Taurine biosynthesis from cysteine via the cysteine sulfinic acid pathway occurs in the brain of some but not all species. Taurine from dietary intake or from that synthesized in the liver and carried by the Na^{+}-dependent taurine transporter (TAUT) is the main supply of taurine pools. Taurine leaves the cell by a leak pathway and is excreted without metabolic degradation in urine and as taurocholic acid in feces. Taurine turnover in the brain is of over 7 days

Table 1 Taurine concentration in mammalian brain and retina

	Brain (μmol/g wet weight)	Retina (μmol/g wet weight)
Human	1.4–3.0	24
Monkey	1.4–2.0	29
Mouse	8.3–10.7	16
Rat	3.7–6.6	33
Cat	2.0–6.1	43
Guinea pig	0.9–1.4	32

Data from Voaden et al. (1977); Mandel and Pasantes-Morales (1978); Worden and Stipanuk (1985)

contribution to the pool of each one of these osmolytes may be different in different tissues or species, according to their particular accessibility and metabolism. N-acetyl aspartate, for instance, is found in high levels in the brain of species where taurine is low (Birken and Oldendorf 1989). Interestingly, the large species differences in brain taurine content are not found in the retina. Taurine retinal levels are very high around 20–43 mM and similar between species, including human (Voaden et al. 1977) (Table 1). Regardless of the content found in the adult brain, in all species the developing brain has significantly higher taurine levels, which decline as development progresses (Sturman and Gaull 1975). This suggested a taurine requirement for an optimal brain development, a notion later supported by studies showing disturbed maturation and migration of neurons and decreased number of astrocytes in some brain areas in taurine-deficient cats and monkeys (Sturman 1986; Sturman et al. 1991).

3 Taurine Biosynthesis, Turnover, and Transport

3.1 *Biosynthesis*

The main pathway for taurine biosynthesis in most species involves the sequence of the following reactions: (*i*) oxidation of cysteine by cysteine dioxygenase (CDO; EC 1.13.11.20) to form cysteinsulfinic acid, (*ii*) decarboxylation of the cysteine sulfinic acid by the cysteinsulfinic acid decarboxylase (CSAD; EC4.1.1.29), and (*iii*) oxidation of the resulting hypotaurine to taurine (Fig. 2). While CDO and CSAD are well characterized, the putative enzyme for hypotaurine oxidation, the hypotaurine dehydrogenase (EC 1.8.1.3), remains uncharacterized. Hypotaurine and taurine can be formed alternatively via cysteamine oxidation by the 2-aminoethanethiol dioxygenase, but this pathway is of low significance in the brain.

CSAD is the rate-limit enzyme for taurine synthesis. It is found as a dimer with a molecular mass of 115 kDa. It is a pyridoxal-phosphate-dependent enzyme and its crystal structure has been recently reported (http://www.rcsb.org/pdb/explore/explore.do?structureId=2JIS). CSAD activity is markedly higher in the rodent brain than in the cat, monkey, or human brain or in herbivores such as guinea pig and sheep (Rassin 1981; Worden and Stipanuk 1985). Interestingly, in herbivores CSAD activity in liver is notably high and then the liver supplies taurine to other organs with low synthesis capacity, like the brain. The cellular localization of brain CSAD has been studied in rodents, in vivo or in cultured cells.

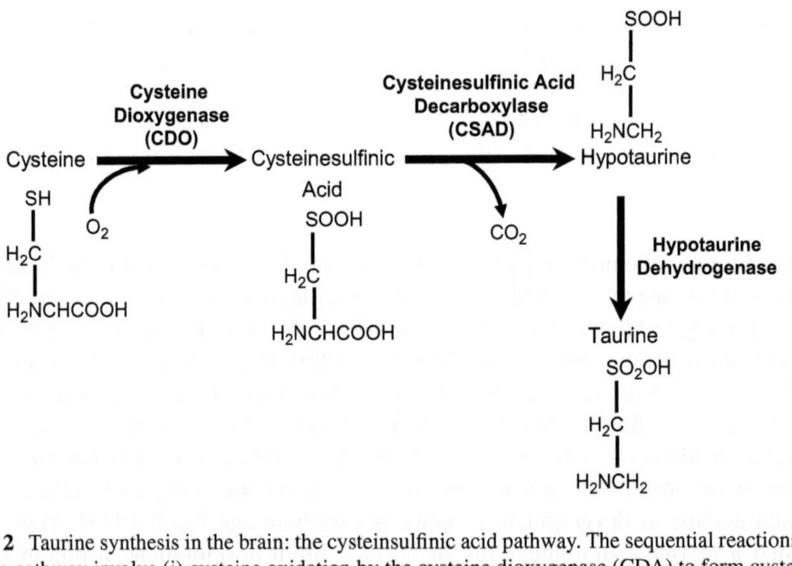

Fig. 2 Taurine synthesis in the brain: the cysteinsulfinic acid pathway. The sequential reactions of these pathway involve (i) cysteine oxidation by the cysteine dioxygenase (CDA) to form cysteine sulfinic acid (CSA), (ii) decarboxylation of CSA by the cysteinsulfinic acid decarboxylase (CSAD) to hypotaurine, and (iii) subsequent oxidation to taurine. A minor synthesis pathway via cysteamine, found in other tissues, is of no relevance in the brain

Results in vivo are controversial, with some reports showing CSAD expression in neurons only, in astrocytes only, or in both cell types (Reymond et al. 1996; Dominy et al. 2004; Tappaz 2004). Differences in antibodies for immunolabeling prepared from different sources sera could explain this discrepancy. More neat results come from studies in neurons and astrocytes in culture. Incorporation of ^{35}S from cysteine into taurine, activity assays, protein, and gene expression all confirm that cultured astrocytes are fully capable of taurine synthesis (Reymond et al. 1996; Dominy et al. 2004; Vitvitsky et al. 2011). In contrast, low CSAD activity, if any, is found in neurons. Taurine formation from ^{35}S from cysteine which is strong in astrocytes is barely found in neurons (Vitvitsky et al. 2011). Also, neurons have only traces of hypotaurine, in contrast to the high levels in astrocytes. Neural progenitor/stem cells obtained from the embryonic mice brain or from the human fetal brain are unable to synthesize taurine and become taurine depleted when cultured in a taurine-free medium, but addition of taurine restores completely the levels in intact brain (Hernández-Benitez et al. 2010a, 2013) (Fig. 3a). Based on these results, a cooperative metabolic interaction between astrocytes and neurons has been suggested (Dominy et al. 2004; Vitvitsky et al. 2011). It should be noticed that these are studies in rodent brain cells, but other species such as cat, rhesus monkey, and human with lower capability to synthesize taurine in the brain rely more on transport than on synthesis, as discussed below.

3.2 Dietary Sources of Taurine

Taurine is present in high concentration in skeletal muscles of vertebrates and invertebrate. Therefore, meat is the main source of taurine for carnivore and omnivore species. Beef, pork, poultry (dark meat), and fish meat are rich in taurine (10–60 µmoles/g). Shrimps, oysters, and other seafood contain even higher concentration of taurine (Pasantes-Morales et al. 1989; Laidlaw et al. 1990). Thus, one portion of 100 g of this food is largely sufficient to meet the daily requirement of taurine which is calculated in 1000–1500 µmoles. Dairy products contain low amounts of taurine. Taurine content in milk from mammalian species, including human, ranges 14–75 µmoles/dl. Exceptions are the very high levels in cat milk (234 µmoles/dl) and the low levels in cow milk (1 µmole/dl) (Rassin et al. 1978). Early infant formulas were prepared following the cow milk components and, consequently, contained very low amounts of taurine. This was later corrected, and now commercial formulas have taurine levels similar to those of human milk. Taurine is largely absent in vegetables and fruits, but it is present though in low amounts, in most bean varieties including soybeans (25–100 nmol/g). Other leguminous seeds containing taurine in similar concentrations are lentils, chick peas, and horse beans (Pasantes-Morales et al. 1989). Taurine is also present in comparable concentrations in oil seeds (Pasantes-Morales et al. 1989). All the so-called energy drinks in the market contain 1 g of taurine per can.

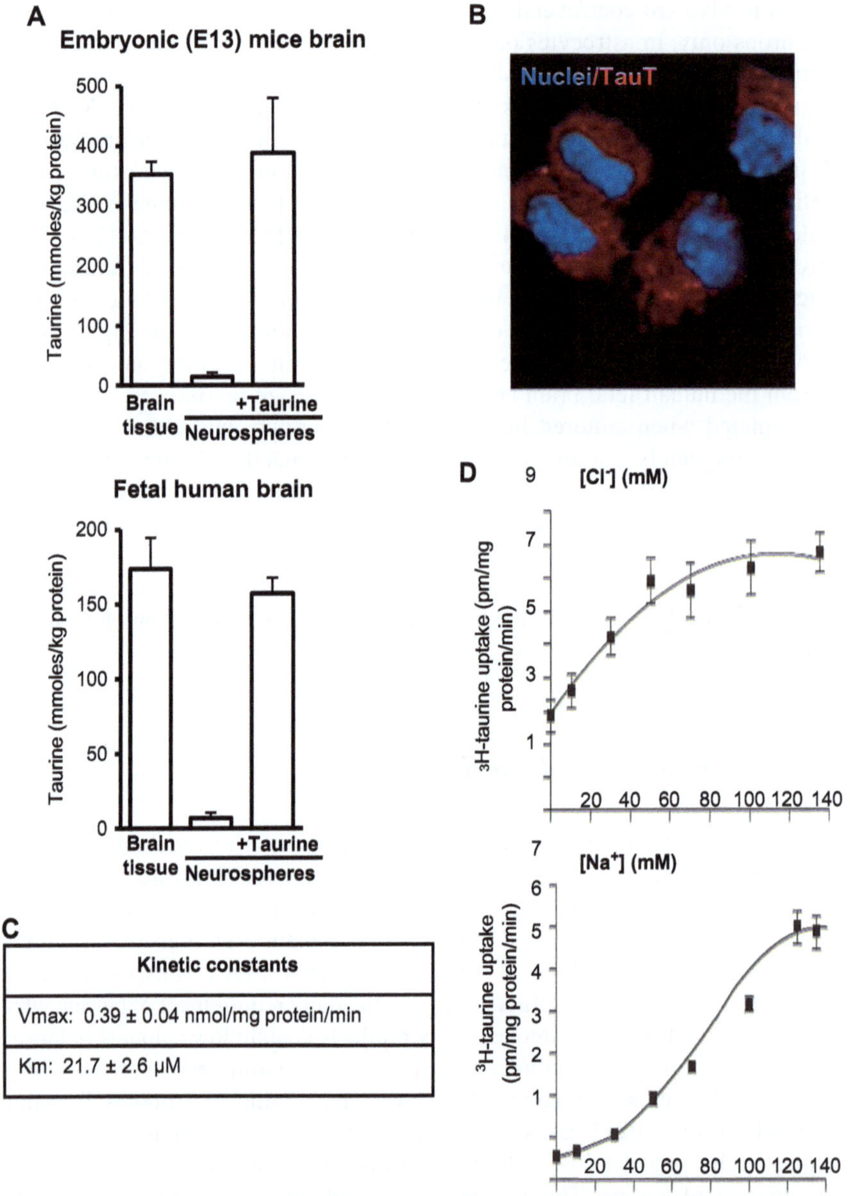

Fig. 3 The taurine pool in neural stem/progenitor cells. (**a**) Cells obtained from the embryonic (E13) mice brain or from the fetal human brain, cultured without fetal serum, in a taurine-free medium necessary for neurosphere formation and stem/progenitor cells proliferation become taurine depleted. Levels comparable to those found in the intact tissue are obtained by addition of taurine to the culture medium. (**b**) Cells accumulate taurine by the taurine transporter TAUT localized at the cell membrane. (**c**) TAUT kinetic values and (**d**) Na^+/Cl^- dependence curves (Details in Hernández-Benitez et al. 2010b)

3.3 *Taurine Turnover Rate and Metabolic Degradation*

Turnover of taurine in the brain as in other excitable tissues is very low, in the range of 3–18 days (Spaeth and Schneider 1974; Hayes and Sturman 1981). Taurine is not catabolized but excreted as such in urine and as bile salt in feces. Taurine turnover is of high relevance facing a decrease in dietary intake, particularly in species like cats and human with low capacity of biosynthesis and depending mostly of the external supply for preservation of the taurine tissue pools. As compensatory mechanisms, the leak taurine pathway is arrested, and taurocholic acid formation switches to only glycocholic acid synthesis.

3.4 *Taurine Transport: The Taurine Transporter TAUT*

Albeit taurine biosynthesis occurs in the brain, a significant amount of the taurine pool in the nervous tissue comes from either the dietary intake or from taurine synthesized in the liver and taken up from plasma by the specific taurine transporter (TAUT). TAUT (SLC6A6) is part of the family of Na+/Cl-dependent transporters formed by about 16 members, including transporters for creatine, GABA, glycine, and neurotransmitter biogenic amines. TAUT carries essentially taurine and other β-amino acids such as β-alanine and hypotaurine. It may also transport GABA, although with very low affinity (GABA is preferentially transported by the specific, high affinity transporters GAT). TAUT has been cloned from the rat and mouse brain (Smith et al. 1992; Liu et al. 1992). A cDNA clone from a mouse brain library encodes TAUT with a nucleotide sequence corresponding to a protein of 590 amino acids and estimated molecular mass of 65.9 KDa. The hydropathy analysis predicts an integral membrane protein with 12 membrane-spanning domains, with the amino and carboxyl termini exposed on the cytoplasmic side of the plasma membrane (Liu et al. 1992). TAUT cloned from rat brain shows a nucleotide sequence corresponding to a protein formed by 621 amino acids and a molecular mass estimated in 70 KDa (Smith et al. 1992). Hydropathy analysis is indicative of a molecule with 12 transmembrane domains. The amino acid sequence of the taurine transporter from rat brain shows 40% identity with the glycine transporter and 52% and 62% identity with the GABA transporters GAT1 and GAT2, respectively (Smith et al. 1992).

TAUT has been characterized in neurons and astrocytes in culture and in neural progenitors (Fig. 3b–d). Kinetic values reported are for Km in the range of 20–39 μM and for V max (in nmoles/mg protein/min) of 0.20–0.67 (Larsson et al. 1986; Sánchez-Olea et al. 1991; Schousboe et al. 1991; Fujita et al. 2006; Hernández-Benitez et al. 2010b; Oenarto et al. 2014). TAUT is dependent on the energy from the Na^+ transmembrane gradient to accumulate taurine against a large concentration gradient. TAUT is also Cl^--dependent. The proposed stoichiometry of Na^+/Cl^- and taurine in most brain cell types is of 1 taurine, 2Na^+, and 1 Cl^-. The Na^+-concentration dependence of TAUT is defined by a sigmoid curve and that for Cl^- is hyperbolic (see as an illustrating example Fig.3c in neural progenitors) (Hernández-Benitez et al. 2010b). The role of Cl^- is not so well defined, but it is proposed to facilitate Na^+ binding to the transporter molecule. The TAUT structure contains sites for glycosylation and for PKC and PKA phosphorylation (Han and Chesney 2006; Lambert et al. 2015). Under physiological conditions, TAUT is in charge of the transport into the brain of dietary taurine or of that synthesized in the liver. TAUT also has a key role in the cell adaptive mechanisms facing changes in cell volume, particularly in hypertonicity, as later discussed.

4 Brain and Taurine Deficiency: Findings from Taurine Dietary Restriction, CSD, and TAUT Genetic Ablation

Brain taurine pools originate from biosynthesis and/or intake from dietary sources. The contribution of these two mechanisms varies in different species. In rats and mice, dietary taurine and biosynthesis contribute with about half each, to the brain taurine pool (Huxtable and Lippincott 1982a, b; Gottschall-Pass et al. 1995). Species with high taurine dietary intake and low biosynthetic capacity as cat and human rely mostly in exogenous supplies. Conversely, herbivores such as rabbit, guinea pig, or sheep have a very low taurine intake and good biosynthesis capacity, though not in the brain but in the liver (Worden and Stipanuk 1985; Huxtable 1992). Knowledge about the relative contribution of synthesis and transport to brain taurine homeostasis has been strengthened by the recent generation of mice with genetic ablation of CSAD or TAUT.

4.1 *Taurine Dietary Restriction*

Effects of a taurine-free diet have been studied in monkeys and cats. In monkeys fed 3–5 weeks with a taurine-deficient diet, the brain taurine content decreases by 20–60% depending on the area, being the retina particularly resistant to taurine depletion. Compensatory changes are observed facing the taurine dietary restriction: CSAD activity is unchanged in the brain but increases markedly in the liver; the synthesis of taurocholic acid is minimal, essentially all urine excretion ceases; and the turnover rate is reduced (Sturman et al. 1991). Taurine restriction results in some reversible changes in the electroretinogram and developmental disturbances in some brain areas (Hayes and Sturman 1981; Neuringer et al. 1990). The effect of taurine deficiency in cats is more drastic, since as carnivores, they depend mostly on the dietary intake for preserving taurine pools in all organs and tissues. Furthermore, cats cannot synthesize glycocholic acid, and facing the dietary restriction, they must mobilize large amounts of taurine from tissue pools to form the bile salt. Taurine deficiency had strong effects in reproductive performance. Feeding a taurine-free diet during 10 weeks causes a dramatic decrease of 80–95% in brain taurine in most areas (Hayes and Sturman 1981). As in monkeys, the retina tenaciously retains its taurine content, mainly by ceasing completely the cell loss, but if the dietary restriction persists, and taurine levels further drop, photoreceptors degenerate and the animal becomes blind. Oxidative stress, shrinkage, and apoptosis and mitochondrial dysfunction are proposed as the mechanisms leading to photoreceptor death. Taurine deficiency also results in profound alterations in brain development (Hayes et al. 1975; Sturman 1986). The consequences of taurine dietary restriction in human have not been examined in detail. This is a topic of interest due to the extended practice of veganism in which the taurine intake is considerably reduced.

4.2 TAUT and CSAD Knockout (KO) Mice

Taurine levels in most tissues decreased markedly in the TAUT KO mice, with reductions of 70–74% in the liver and kidney and of 80–90% in the skeletal and cardiac muscle (Warskulat et al. 2004, 2007; Ito et al. 2010). A more drastic reduction of over 90% occurs in the brain stem, cerebellum, hippocampus, striatum, and cerebral cortex. Remarkably, there is no evidence of overt gross neurological dysfunctions. Being taurine a major brain osmolyte, it would be of interest to investigate how these taurine-deficient brains respond to osmotic challenges. It cannot be excluded that reduction in the brain taurine content as in TAUT null mice is compensated by upregulation of other organic osmolytes, as has been found in the heart (Warskulat et al. 2004). Specific alterations are detected in striatopallidal neurons from TAUT KO mice characterized by disinhibition of the synchronized neuronal activity. This may be due to changes in the neuronal $GABA_A$ receptor in the absence of taurine that weakens the GABAergic inhibition. The normal striatopallidal network activity is recovered by taurine supplementation. Similar studies in other areas are warranted, since taurine effects on GABA and glutamate receptors are well documented (Albrecht and Schousboe 2005; Chan et al. 2014). Similar to taurine-deficient cats, the TAUT KO mice shows progressive retinal taurine depletion, degenerative photoreceptor death, and loss of vision (Heller-Stilb et al. 2002). Disturbances in the developing brain found in cats are not so far described in TAUT KO mice.

A CSAD KO mouse was developed which has very low levels of brain and liver taurine. Taurine depletion in CSAD null mice occurs only at the second homozygous generation. However, these animals have a very poor reproductive performance and most died within 24 h (Park et al. 2014).

The interest of these TAUT/CSAD KO mice is that compared to the cat model, it represents a simpler and faster method to generate taurine-deficient animals. Indeed, the numerous reports on the consequences of TAUT ablation and the subsequent depletion of taurine pools have enlarged our knowledge about the importance of taurine in various tissues and organs. The similitude in the consequences of taurine deficiency in cats and monkeys and those in TAUT null mice suggests that they may be extrapolated to other species, including humans.

5 Taurine Homeostasis Facing Volume Changes in the Brain

Changes in cell volume represent a challenge for every organ or tissue, but this is even more critical for the brain. Encased in the rigid cranium, brain tissue expansion causes compression and rupture of the vasculature that results in ischemia, infarct, excitotoxicity, and neuronal death. Swelling also increases brain excitability and seizure activity. In extreme cases of acute and severe swelling, caudal herniation compresses neural nuclei at the brain stem, with high danger of mortality by cardiac and respiratory arrest. Besides these extreme consequences, minor uncontrolled changes in cell water content are also disruptive of the brain physiology, by affecting the concentration of signaling molecules and the organization and size of intracellular compartments with adverse consequences for synaptic functional contacts.

Volume changes modify the brain taurine homeostasis. Redistribution of osmolytes between intracellular and extracellular compartments is the core of the cell adaptive mechanisms to preserve volume in the brain. As a major osmolyte, taurine content is then reduced or increased in cells as part of the mechanisms to attain an osmotic balance facing deviations in normotonicity. The plasma osmolarity is commonly well controlled and so are the extracellular fluids, but this is affected in pathologies associated with hypo- or hypernatremia or in neuropathologies disturbing the normal ion gradients and water concentration at the extracellular/intracellular compartments. When this occurs, water moves in the direction imposed by its concentration gradient and cells swell or shrink (Hoffmann et al. 2009) (Fig. 4a). These volume changes are sensed by the cell and activate a plethora of intracellular signals some of which ultimately lead to translocation of osmolytes in the direction required to attain the osmotic equilibrium (Fig. 4a, b). This adaptive mechanism is highly preserved during evolution and is present in most animal cells, even in those highly specialized neurons (Pasantes-Morales et al. 1993). Taurine features as its high cell content, free in the cytosol and its relative metabolic inertness, permit the osmotically directed fluxes of taurine to occur in any direction, with minor disturbance on the cell physiology.

The involvement of taurine in volume regulation comprises from systemic level to nerve endings and synaptic vesicles. Taurine is related to systemic volume regulation at the supraoptic nucleus, via an interaction with neuronal glycine receptors (Hussy et al. 2001). Taurine and TAUT are present in isolated nerve endings, where a volume-sensitive efflux of taurine has been characterized (Tuz et al. 2004). Taurine is found in high levels in synaptic vesicles obtained from the bovine brain (Kontro et al. 1980) and is present in the cholinergic vesicles of the Torpedo electric organ (Vyas and Bradford 1987). The coexistence of taurine with different neurotransmitters in synaptic vesicles suggests a regulatory role in the volume control within the limited space of a vesicle.

Volume-sensitive efflux of taurine from brain cells is broadly documented. Swelling-evoked taurine release has been described in the brain in vivo (Estévez et al. 1999; Haskew-Layton et al. 2008), in brain slices (Oja and Saransaari 1992; Law 1994), excised retina (Pasantes-Morales et al. 1999), neurons and astrocytes in culture (Schousboe and Pasantes-Morales 1992; Cardin et al. 2003), brain cell lines (Basavappa et al. 1996; Ordaz et al. 2004), and isolated nerve endings (Tuz et al.

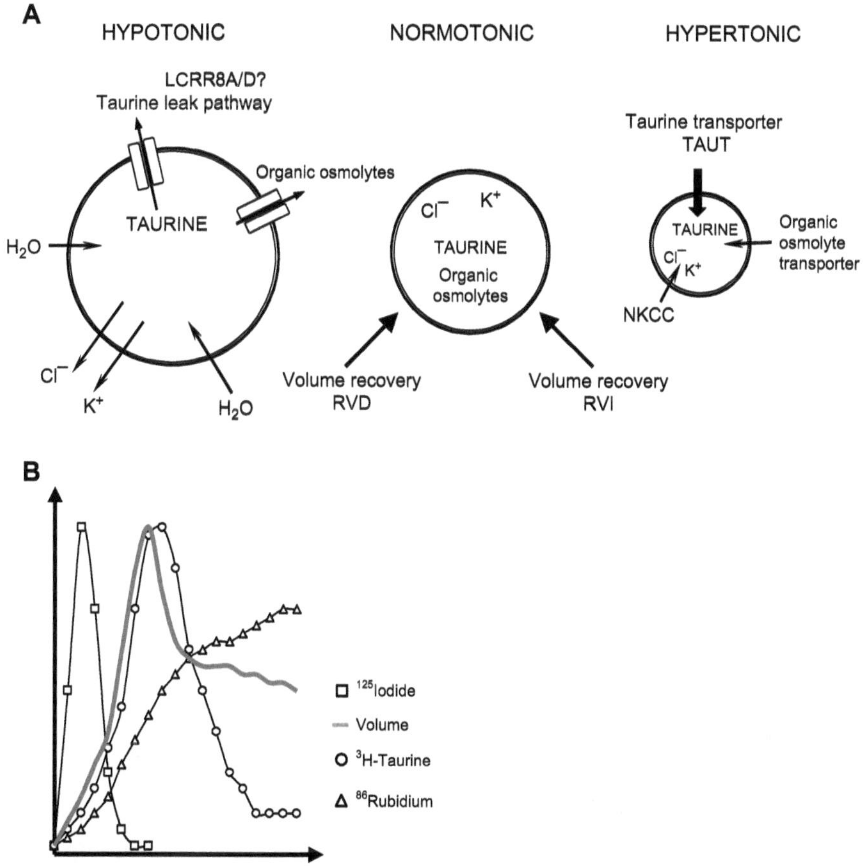

Fig. 4 Schematic representation of the pathways involved in cell volume adjustment after changes in external osmolarity. (**a**) Upon variations in external tonicity, osmotically obligated water fluxes move in the direction required to attain a new osmotic balance, and cells swell or shrink. The change in cell volume activates the volume regulatory processes accomplished by translocation of Na^+, K^+, Cl^-, and organic osmolytes, followed by corrective osmotic water fluxes. Taurine is part of the group of organic osmolytes and plays this role in most animal cells, including brain cells. (**b**) Time course of the change in cell volume, taurine, Cl^-, and K^+ efflux from cultured astrocytes. The time course of taurine efflux closely parallels the change in cell volume (Pasantes-Morales, unpublished)

2004). The time course and features of this release are similar in the different in vitro preparations. Reduction in external osmolarity evokes an immediate taurine efflux, which attains a peak within 1–3 min. Then, the taurine efflux rate decreases, and in most cases, its time course parallels closely to that of the volume regulatory decrease (Cardin et al. 2003). Inactivation is more rapid in cultured cells than in vivo or in brain slices, a difference likely due to the slow diffusion of the released taurine in the organized tissue. The volume-sensitive efflux of taurine is dependent on the magnitude of the osmolarity reduction.

Moreover, taurine contributes to protect neurons facing volume changes. It is known that astrocytes selectively swell by hypotonicity while neurons are spared. Studies from Nagelhus et al. (1993) in rat cerebellum illustrate how taurine acts in preventing neuronal swelling. In hypotonicity, the large taurine pool in Purkinje cells is completely lost and captured by astrocytes, which eventually also release it in response to swelling. This strategy protects neurons and their intricate cytoarchitecture. This contrasts with other osmolytes such as K^+, Cl^-, and glutamate, which disturb neuroexcitability and even cause neuronal death, when their extracellular levels are deregulated.

5.1 *Mechanisms for Taurine Translocation Following Cell Volume Changes*

The fluxes of taurine following cell volume changes occur by different mechanisms depending on whether cell volume is increased or decreased. Swelling-induced taurine efflux occurs via an energy-independent, bidirectional leak pathway (Hoffmann and Lambert 1983; Sánchez-Olea et al. 1991), while taurine accumulation into a dehydrated cell is mediated by the increased TAUT activity (Sánchez-Olea et al. 1992; Beetsch and Olson 1996).

The leak pathway which mediates the swelling-activated taurine efflux is Ca^{2+}-independent and temperature and pH-sensitive. It is regulated by non-hydrolytic ATP binding and inhibited by most blockers of Cl^- channels (Shennan 2008). Its molecular identity is as yet undefined. The sensitivity of taurine efflux to Cl^- channel blockers has raised the question of whether taurine moves across the volume-regulated anion channel (VRAC). The pore size of VRAC is sufficiently large as to permeate osmolytes as taurine and glutamate, and when taurine is switched from neutral to anionic form by changes in cell pH, it indeed permeates through VRAC (Banderali and Roy 1992). However, at physiological pH, taurine as well as most organic osmolytes has no net charge and cannot transit across an anion channel. Until recently, even the molecular identity of VRAC was undefined, but two reports in 2014 (Qiu et al. 2014; Voss et al. 2014) provided strong evidence relating VRAC with the multispan transmembrane protein LCRR8, encoded by a gene of unknown function named *leucine-rich repeat containing 8A* (*LRRC8A*). LCRR8 has five isoforms, LCRR8A-E. The referred studies plus that by Planells-Cases et al. (2015) propose that particular combinations of the LRRC8 protein isoforms could form channels preferentially permeable to Cl^- or to taurine (Voss et al. 2014; Qiu et al. 2014). Genetic ablation showed that the LRRC8A subunit is essential for the transport of Cl and taurine, since both the $I_{Cl(swell)}$ and volume-sensitive taurine fluxes are abolished in *LRRC8A*$^{-/-}$ cells. Regarding other subunits, the volume-activated Cl^- fluxes require LRRC8CE, while in contrast, the channel which permeates taurine has a strict requirement of LCRR8D, whereas LCRR8BCE subunits appear unnecessary (Planells-Cases et al. 2015). It is proposed then that various combinations of LCRR8 subunits may form channels with a different pore structure as to permeate anion or neutral molecules. Other intrinsic properties such as the inactivation rate may differ according to the subunit composition of the channel, explaining the marked difference in this parameter between the volume-sensitive Cl^- and taurine fluxes. These studies were carried out in HEK, HCT116, and HeLa cells, but results were similar to those in Hyzinski-Garcia et al. (2014) in cultured mouse cortical astrocytes, showing abolished hypotonic taurine efflux by LCRR8A siRNA knockdown.

In hypertonicity, the water concentration gradient drives water outflow cells to shrink. This triggers the active process of volume regulation known as regulatory volume increase, achieved by accumulation of intracellular osmolytes, taurine among them, as to create osmotically obligated inward water fluxes and cell volume recovery. The intracellular increase of osmolytes occurs in most cases (sorbitol is an exception), by the operation of electroneutral ion cotransporters and by Na^+-dependent transporters for organic osmolytes. The concentration of ion osmolytes increases rapidly, but even if this might be convenient for a rapid volume control, the rise in ionic strength perturbs the structure and function of macromolecules. Hence, with time, replacement of inorganic osmolyte for compatible osmolytes reduces the risk of macromolecular damage while preserving normal volume. Studies in vivo as well as in cultured astrocytes demonstrate that taurine accumulation in brain cells shrunken by hypertonicity is mediated by TAUT (Trachtman 1992; Sánchez-Olea et al. 1992; Beetsch and Olson 1996). Taurine uptake rate and cell content increase by almost twofold within the first 12 h in hypertonicity. Analysis of TAUT kinetics showed a significant increase in Vmax with no change in the Km value. At the same time, hypertonicity almost abolishes the isotonic leak efflux of taurine (Sánchez-Olea et al. 1992; Beetsch and Olson 1996). Interestingly, after 48 h, the uptake rate returns to initial values, but the taurine content in cells remains unchanged (Beetsch and Olson 1996). This is suggestive of a concerted uptake/biosynthesis interaction, which has not as yet well studied (Vitvitsky et al. 2011). Indeed, increased biosynthesis is shown to contribute in part to the elevation of taurine content following hypertonicity in mice-cultured astrocytes and neurons (Vitvitsky et al. 2011).

6 Taurine Homeostasis in Neuropathologies and Aging

6.1 Cell Swelling in Acute and Chronic Hyponatremia

Hyponatremia is a common electrolyte disorder that in the acute condition concurs with significant brain edema. Astrocytes are the cells that predominantly swell in cerebral edema (cytotoxic swelling). Astrocyte swelling is followed by the adaptive process of regulatory volume decrease. However, if sudden and severe hyponatremia exceeds the capacity of astrocytes to regulate volume, the persistent edema results in often serious neurological problems (Giuliani and Peri 2014). In contrast to the injuring effect of cerebral edema in acute hyponatremia, brain swelling is minimal in chronic hyponatremia. When changes in external osmolarity are gradual and small, the cell volume regulation operates efficiently, and cell volume remains unchanged. This isovolumetric regulation is an active process accomplished by the efflux of K^+, Cl^-, and organic osmolytes. Taurine plays an important role in this process (Ordaz et al. 2004). The efflux of organic osmolytes is sustained as long as hyponatremia persists, becoming an essential adaptive mechanism in chronic hyponatremia (Fig. 5b). However, these long-term compensatory changes of brain cells to the chronic condition modify substantially the distribution of intracellular osmolytes tending to reach an osmotic balance according to the modified plasma tonicity. As a result, the osmolyte taurine pools are significantly reduced. In rat brain, Na^+ and K^+ levels decrease by 7–11% and Cl^- by 22–26%. Organic osmolyte reduction is notably more drastic. The decrease of glutamate, glutamine, and creatine is of around 50% and that of myo-inositol is of 60%. Taurine is the osmolyte showing the most pronounced reduction in brain levels, of over 80% (Verbalis and Gullans 1991) (Fig. 5b). Such reduction in taurine concentration during hyponatremia is observed in essentially all brain regions (Massieu et al. 2004). Therefore, the brain homeostasis in terms of osmolarity has substantially changed. This becomes important during correction of the Na^+ plasma levels as an active therapy. An overly rapid correction that exceeds the brain capacity to restore the pools of lost osmolytes generates an inverse osmotic gradient with consequent cell dehydration (Fig. 5a). The cell shrinkage, of astrocytes primarily, is proposed to contribute to the pathophysiology of osmotic demyelination syndrome, a highly deleterious complication of inappropriate hyponatremia intervention. Although the pathogenesis of osmotic demyelination syndrome is not fully understood, the changes in astrocyte volume are considered crucial in this process (Fig. 5c). These changes in cell volume may disturb the astrocyte-oligodendrocyte gap junction complex and the integrity of the glial syncytium, necessary for myelination processes (Kengne et al. 2011; Giuliani and Peri 2014) (Fig. 5c). Another possibility is that the osmotic shrinkage of cells at the blood-brain barrier permits the entry of complement and other cytotoxic plasma components (Fig. 5c). The osmotic demyelination syndrome occurs mainly in the pons and in some other extrapontine regions but not in all brain regions. There is at present no obvious explanation for this selective occurrence of demyelination, since the volume-related changes in astrocytes are likely to be similar. Understanding the role of volume changes in the astrocyte-oligodendrocyte interaction in the osmotic demyelination syndrome is a topic of high interest for future research. The risks of inadvertent sudden correction of chronic hyponatremia due to changes in the intracellular osmolyte concentrations should be emphasized in the clinical practice.

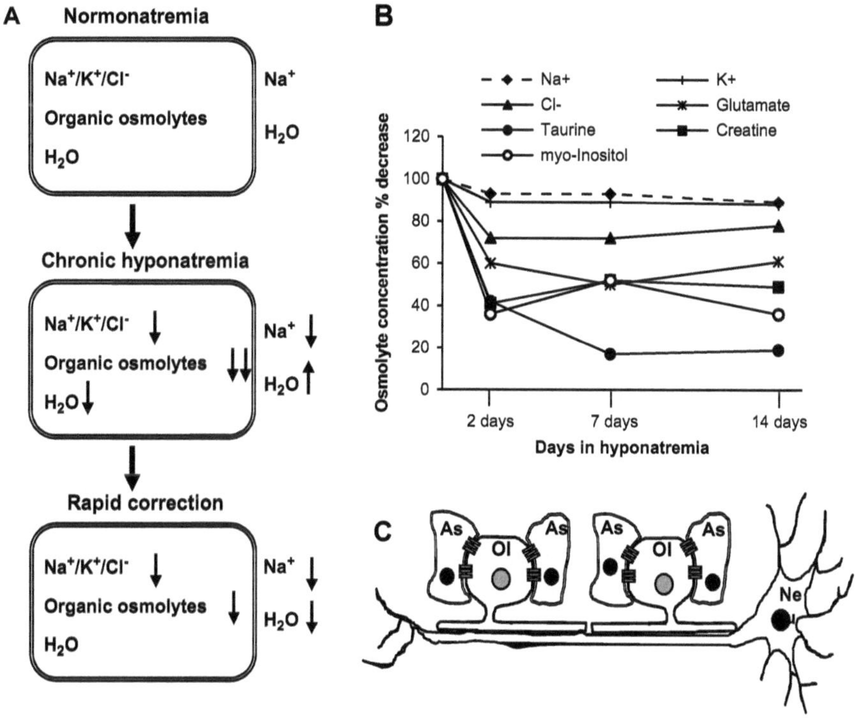

Fig. 5 Changes in brain osmolyte pool in chronic hyponatremia. (**a**) Ion/organic osmolytes and water in normonatremia, chronic hyponatremia, and after rapid correction of chronic hyponatremia. (**b**). Decrease with time of the main osmolyte content in the brain during hyponatremia. (**c**) Proposed cellular correlate of the demyelination syndrome. Astrocyte/oligodendrocyte connexions are altered by rapid correction of hyponatremia affecting myelination

6.2 Cell Swelling and Taurine Homeostasis in Ischemia

Cerebral edema is a serious complication following brain ischemia. Cellular (cytotoxic) swelling and vasogenic edema evolve sequentially, as a result of the energy failure, arrest of the Na^+/K^+ ATPase activity, disturbance of the transmembrane Na^+/K^+ exchanges, and dissipation of ion gradients. Cytotoxic swelling is due to Na^+, K^+, and Cl^- overload, followed by inwardly directed water fluxes. This is the first step in a cascade of ion gradient disturbances between the intravascular and extracellular compartments that culminates in blood-brain barrier injury and vasogenic edema (Annunziato et al. 2013, Khanna et al. 2014; Kahle et al. 2015). Since the early report by Benveniste et al. (1984), the ischemia-induced efflux of taurine and other organic osmolytes from the brain has been well documented (Phillis and O'Regan 2003; Oja and Saransaari 2013). This exit is at least in part a response to cell swelling, but other mechanisms participate as well, such as the reverse operation of the Na^+-dependent transporters for organic osmolytes. With time, and corresponding to this elevation in the interstitial space, the intracellular concentration of glutamate, aspartate, N-acetylaspartate, GABA, glutamine, taurine, and creatine decreases markedly, about 41% in average. This osmolyte redistribution, particularly the increased extracellular levels of excitatory amino acids, may further aggravate the brain injury by overactivation of neuronal glutamate receptors. In contrast, changes in the intracellular/extracellular taurine levels do not modify the brain cell function; on the contrary, besides contributing to alleviate swelling, the increased extracellular taurine levels may be beneficial during ischemia and ischemia/reperfusion. A neuroprotectant action of taurine has been reported in in vivo and in vitro models of brain ischemia and ischemia/reperfusion (Ricci et al. 2009; Sun et al. 2011; Menzie et al. 2013; Gharibani et al. 2013). The mechanism of this protective effect is not well defined. Among the possibilities considered are (*i*) a membrane stabilizer action preventing ion overload and Ca^{2+} influx during ischemia reperfusion, (*ii*) a direct antiapoptotic effect, and (*iii*) a prevention of mitochondrial dysfunction. This later effect may be related to taurine interaction with mtRNA that makes more efficient the production of mitochondrial proteins of the respiratory chain (Suzuki et al. 2002). These protective effects of taurine require of high extracellular concentrations. Although the taurine level in the extracellular space is only around 100 μM, this may increase exponentially upon activation by swelling of the volume-sensitive taurine efflux.

6.3 *Taurine and Aging*

Neurochemical changes occur in the brain with age, including some related to metabolism and to synaptic transmission. A decline in brain taurine was first reported in aging F344 rats (Eppler and Dawson 1998), but this was not confirmed, however, in other rat strains (Paban et al. 2010). Similarly a recent longitudinal analysis in vivo by ^{1}H magnetic resonance spectroscopy in the aging mouse brain, which detected significant decreases in glutamate, aspartate, and GABA in hippocampus, cortex, and striatum, showed no change in taurine levels in any of these regions (Duarte et al. 2014). A significant reduction (28–40%) of taurine is reported in serum in old rats, together with a decrease in urine taurine excretion (Wallace and Dawson 1990). At present, knowledge about taurine homeostasis in the aging brain requires, in the first place, to firmly establish whether taurine content in the brain and in brain regions is changing with age or not. Also, a possible correlation between changes with age in the dietary pattern and the observed differences in taurine content in plasma and urine should be investigated. Understanding the brain taurine homeostasis in aging is of interest considering the high prevalence of chronic hyponatremia in the elderly and the increase with age of oxidative stress and other deleterious conditions that may be protected by taurine (Aydın et al. 2016). Taurine supplementation is claimed to protect from cognitive deficits in old mice, with a parallel increase in glutamate and GABA levels (El Idrissi et al. 2013). Also of interest is that taurine increases hippocampal neurogenesis in aging mice (Gebara et al. 2015). Clearly, the homeostasis of taurine in the aging brain is an interesting field for research in the future.

Acknowledgments Work in the author's laboratory is supported by Dirección General de Asuntos del Personal Académico (DGAPA) at the Universidad Nacional Autónoma de México (UNAM), Grant No. PAPIIT IN205916.

Conflict of Interest The author declares no conflicts of interest.

References

Albrecht J, Schousboe A. Taurine interaction with neurotransmitter receptors in the CNS: an update. Neurochem Res. 2005;30(12):1615–21.

Annunziato L, Boscia F, Pignataro G. Ionic transporter activity in astrocytes, microglia, and oligodendrocytes during brain ischemia. J Cereb Blood Flow Metab. 2013;33(7):969–82.

Aydın AF, Çoban J, Doğan-Ekici I, Betül-Kalaz E, Doğru-Abbasoğlu S, Uysal M. Carnosine and taurine treatments diminished brain oxidative stress and apoptosis in D-galactose aging model. Metab Brain Dis. 2016;31(2):337–45.

Banderali U, Roy G. Anion channels for amino acids in MDCK cells. Am J Phys. 1992;263(6 Pt 1):C1200–7.

Basavappa S, Huang CC, Mangel AW, Lebedev DV, Knauf PA, Ellory JC. Swelling-activated amino acid efflux in the human neuroblastoma cell line CHP-100. J Neurophysiol. 1996;76(2):764–9.

Beetsch JW, Olson JE. Hyperosmotic exposure alters total taurine quantity and cellular transport in rat astrocyte cultures. Biochim Biophys Acta. 1996;1290(2):141–8.

Benveniste H, Drejer J, Schousboe A, Diemer NH. Elevation of the extracellular concentrations of glutamate and aspartate in rat hippocampus during transient cerebral ischemia monitored by intracerebral microdialysis. J Neurochem. 1984;43(5):1369–74.

Birken DL, Oldendorf WH. N-acetyl-L-aspartic acid: a literature review of a compound prominent in 1H-NMR spectroscopic studies of brain. Neurosci Biobehav Rev. 1989;13(1):23–31.

Cardin V, Lezama R, Torres-Márquez ME, Pasantes-Morales H. Potentiation of the osmosensitive taurine release and cell volume regulation by cytosolic Ca2+ rise in cultured cerebellar astrocytes. Glia. 2003;44(2):119–28.

Chan CY, Sun HS, Shah SM, Agovic MS, Friedman E, Banerjee SP. Modes of direct modulation by taurine of the glutamate NMDA receptor in rat cortex. Eur J Pharmacol. 2014;728:167–75.

Dawson R, Wallace DR, King MJ. Monoamine and amino acid content in brain regions of Brattleboro rats. Neurochem Res. 1990;15(7):755–61.

Dominy J, Eller S, Dawson R. Building biosynthetic schools: reviewing compartmentation of CNS taurine synthesis. Neurochem Res. 2004;29(1):97–103.

Duarte JM, Do KQ, Gruetter R. Longitudinal neurochemical modifications in the aging mouse brain measured in vivo by 1H magnetic resonance spectroscopy. Neurobiol Aging. 2014;35(7):1660–8.

El Idrissi A, Shen CH, L'amoreaux WJ. Neuroprotective role of taurine during aging. Amino Acids. 2013;45(4):735–50.

Eppler B, Dawson R Jr. The effects of aging on taurine content and biosynthesis in different strains of rats. Adv Exp Med Biol. 1998;442:55–61.

Estévez AY, O'Regan MH, Song D, Phillis JW. Effects of anion channel blockers on hyposmotically induced amino acid release from the in vivo rat cerebral cortex. Neurochem Res. 1999;24:447–52.

Frahm J, Michaelis T, Merboldt KD, Hanicke W, Gyngell ML, Chien D, et al. Localized NMR spectroscopy in vivo. Progress and problems. NMR Biomed. 1989;2(5–6):188–95.

Fujita T, Shimada A, Wada M, Miyakawa S, Yamamoto A. Functional expression of taurine transporter and its up-regulation in developing neurons from mouse cerebral cortex. Pharm Res. 2006;23(4):689–96.

Gebara E, Udry F, Sultan S, Toni N. Taurine increases hippocampal neurogenesis in aging mice. Stem Cell Res. 2015;14(3):369–79.

Gharibani PM, Modi J, Pan C, Menzie J, Ma Z, Chen PC, Tao R, Prentice H, Wu JY. The mechanism of taurine protection against endoplasmic reticulum stress in an animal stroke model of cerebral artery occlusion and stroke-related conditions in primary neuronal cell culture. Adv Exp Med Biol. 2013;776:241–58.

Giuliani C, Peri A. Effects of hyponatremia on the brain. J Clin Med. 2014;3(4):1163–77.

Gottschall-Pass KT, Gorecki DK, Paterson PG. Effect of taurine deficiency on tissue taurine concentrations and pregnancy outcome in the rat. Can J Physiol Pharmacol. 1995;73(8):1130–5.

Han X, Chesney RW. Mechanisms of regulation of taurine transporter activity: a complex interplay of regulatory systems. Adv Exp Med Biol. 2006;583:79–90.

Haskew-Layton RE, Rudkouskaya A, Jin Y, Feustel PJ, Kimelberg HK, Mongin AA. Two distinct modes of hypoosmotic medium-induced release of excitatory amino acids and taurine in the rat brain in vivo. PLoS One. 2008;3(10):e3543.

Hayes KC, Sturman JA. Taurine in metabolism. Annu Rev Nutr. 1981;1:401–25.

Hayes KC, Carey RE, Schmidt SY. Retinal degeneration associated with taurine deficiency in the cat. Science. 1975;188(4191):949–51.

Heller-Stilb B, van Roeyen C, Rascher K, Hartwig HG, Huth A, Seeliger MW, Warskulat U, Häussinger D. Disruption of the taurine transporter gene (taut) leads to retinal degeneration in mice. FASEB J. 2002;16(2):231–3.

Hernández-Benitez R, Pasantes-Morales H, Saldana IT, Ramos-Mandujano G. Taurine stimulates proliferation of mice embryonic cultured neural progenitor cells. J Neurosci Res. 2010a;88(8):1673–81.

Hernández-Benitez R, Pasantes-Morales H, Pinzon-Estrada E, Ramos-Mandujano G. Functional expression and subcellular localization of the taurine transporter TauT in murine neural precursors. Dev Neurosci. 2010b;32(4):321–8.

Hernández-Benitez R, Vangipuram SD, Ramos-Mandujano G, Lyman WD, Pasantes-Morales H. Taurine enhances the growth of neural precursors derived from fetal human brain and promotes neuronal specification. Dev Neurosci. 2013;35(1):40–9.

Hoffmann EK, Lambert IH. Amino acid transport and cell volume regulation in Ehrlich ascites tumour cells. J Physiol. 1983;338:613–25.

Hoffmann EK, Lambert IH, Pedersen SF. Physiology of cell volume regulation in vertebrates. Physiol Rev. 2009;89(1):193–277.

Hussy N, Brès V, Rochette M, Duvoid A, Alonso G, Dayanithi G, Moos FC. Osmoregulation of vasopressin secretion via activation of neurohypophysial nerve terminals glycine receptors by glial taurine. J Neurosci. 2001;21(18):7110–6.

Huxtable RJ. The interaction between taurine, calcium and phospholipids: further investigations of a trinitarian hypothesis. Prog Clin Biol Res. 1990;351:185–96.

Huxtable RJ. Physiological actions of taurine. Physiol Rev. 1992;72(1):101–63.

Huxtable RJ, Lippincott SE. Diet and biosynthesis as sources of taurine in the mouse. J Nutr. 1982a;112(5):1003–10.

Huxtable RJ, Lippincott SE. Relative contribution of diet and biosynthesis to the taurine content of the adult rat. Drug Nutr Interact. 1982b;1(2):153–68.

Hyzinski-Garcia MC, Rudkouskaya A, Mongin AA. LRRC8A protein is indispensable for swelling-activated and ATP-induced release of excitatory amino acids in rat astrocytes. J Physiol. 2014;592(22):4855–62.

Ito T, Oishi S, Takai M, Kimura Y, Uozumi Y, Fujio Y, Schaffer SW, Azuma J. Cardiac and skeletal muscle abnormality in taurine transporter-knockout mice. J Biomed Sci. 2010;17(Suppl 1):S20.

Jong CJ, Azuma J, Schaffer S. Mechanism underlying the antioxidant activity of taurine: prevention of mitochondrial oxidant production. Amino Acids. 2012;42(6):2223–32.

Junankar PR, Kirk K. Organic osmolyte channels: a comparative view. Cell Physiol Biochem. 2000;10(5–6):355–60.

Kahle KT, Khanna AR, Alper SL, Adragna NC, Lauf PK, Sun D, et al. K-Clcotransporters, cell volume homeostasis, and neurological disease. Trends Mol Med. 2015;21(8):513–23.

Kengne G, Nicaise F, Soupart C, Boom A, Schiettecatte J, Pochet R, Brion JP, Decaux G. Astrocytes are an early target in osmotic demyelination syndrome. J Am Soc Nephrol. 2011;22(10):1834–45.

Khanna A, Kahle KT, Walcott BP, Gerzanich V, Simard JM. Disruption of ion homeostasis in the neurogliovascular unit underlies the pathogenesis of ischemic cerebral edema. Transl Stroke Res. 2014;5(1):3–16.
Kontro P, Marnela KM, Oja SS. Free amino acids in the synaptosome and synaptic vesicle fractions of different bovine brain areas. Brain Res. 1980;184(1):129–41.
Laidlaw SA, Grosvenor M, Kopple JD. The taurine content of common foodstuff. J Parent Nutr. 1990;14(2):183–8.
Lambert IH, Kristensen DM, Holm JB, Mortensen OH. Physiological role of taurine--from organism to organelle. Acta Physiol. 2015;213(1):191–212.
Larsson OM, Griffiths R, Allen IC, Schousboe A. Mutual inhibition kinetic analysis of gamma-aminobutyric acid, taurine, and beta-alanine high-affinity transport into neurons and astrocytes: evidence for similarity between the taurine and beta-alanine carriers in both cell types. J Neurochem. 1986;47(2):426–32.
Law RO. Taurine efflux and the regulation of cell volume in incubated slices of rat cerebral cortex. Biochim Biophys Acta. 1994;1221(1):21–8.
Liu QR, Lopez-Corcuera B, Nelson H, Mandiyan S, Nelson N. Cloning and expression of a cDNA encoding the transporter of taurine and beta-alanine in mouse brain. Proc Natl Acad Sci U S A. 1992;89(24):12145–9.
Mandel P, Pasantes-Morales H. Taurine in the nervous system. In: Ehrenpreis S, Kopin I, editors. Reviews in neuroscience, vol. 3. New York: Raven Press; 1978. p. 158–93.
Marcinkiewicz J, Kontny E. Taurine and inflammatory diseases. Amino Acids. 2014;46(1):7–20.
Massieu L, Montiel T, Robles G, Quesada O. Brain amino acids during hyponatremia in vivo: clinical observations and experimental studies. Neurochem Res. 2004;29(1):73–81.
Menzie J, Prentice H, Wu JY. Neuroprotective mechanisms of taurine against ischemic stroke. Brain Sci. 2013;3(2):877–907.
Nagelhus EA, Lehmann A, Ottersen OP. Neuronal-glial exchange of taurine during hypo-osmotic stress: a combined immunocytochemical and biochemical analysis in rat cerebellar cortex. Neuroscience. 1993;54(3):615–31.
Neuringer M, Palackal T, Kujawa M, Moretz RC, Sturman JA. Visual cortex development in rhesus monkeys deprived of dietary taurine. Prog Clin Biol Res. 1990;351:415–22.
Oenarto J, Görg B, Moos M, Bidmon HJ, Häussinger D. Expression of organic osmolyte transporters in cultured rat astrocytes and rat and human cerebral cortex. Arch Biochem Biophys. 2014;560:59–72.
Oja SS, Saransaari P. Taurine release and swelling of cerebral cortex slices from adult and developing mice in media of different ionic compositions. J Neurosci Res. 1992;32(4):551–61.
Oja SS, Saransaari P. Ischemia induces release of endogenous amino acids from the cerebral cortex and cerebellum of developing and adult mice. J Amino Acids. 2013;2013:839036.
Oja SS, Saransaari P. Open questions concerning taurine with emphasis on the brain. AdvExp Med Biol. 2015;803:409–13.
Ordaz B, Vaca L, Franco R, Pasantes-Morales H. Volume changes and whole cell membrane currents activated during gradual osmolarity decrease in C6 glioma cells: contribution of two types of K+ channels. Am J Phys Cell Phys. 2004;286(6):C1399–409.
Paban V, Fauvelle F, Alescio-Lautier B. Age-related changes in metabolic profiles of rat hippocampus and cortices. Eur J Neurosci. 2010;31(6):1063–73.
Park E, Park SY, Dobkin C, Schuller-Levis G. Development of a novel cysteine sulfinic Acid decarboxylase knockout mouse: dietary taurine reduces neonatal mortality. J Amino Acids. 2014;34:68–09.
Pasantes-Morales H, Quesada O, Alcocer L, Sánchez Olea R. Taurine content in foods. Nutr Report Intern. 1989;40:793–801.
Pasantes-Morales H, Alavez S, Sánchez Olea R, Morán J. Contribution of organic and inorganic osmolytes to volume regulation in rat brain cells in culture. Neurochem Res. 1993;18(4):445–52.
Pasantes-Morales H, Quesada O, Morán J. Taurine: an osmolyte in mammalian tissues. Adv Exp Med Biol. 1998;442:209–17.

Pasantes-Morales H, Ochoa de la Paz LD, Sepúlveda J, Quesada O. Amino acids as osmolytes in the retina. Neurochem Res. 1999;24(11):1339–46.
Phillis JW, O'Regan MH. Characterization of modes of release of amino acids in the ischemic/reperfused rat cerebral cortex. Neurochem Int. 2003;43(4–5):461–7.
Planells-Cases R, Lutter D, Guyader C, Gerhards NM, Ullrich F, Elger DA, et al. Subunit composition of VRAC channels determines substrate specificity and cellular resistance to Pt-based anti-cancer drugs. EMBO J. 2015;34(24):2993–3008.
Prentice H, Modi JP, Wu JY. Mechanisms of neuronal protection against excitotoxicity, endoplasmic reticulum stress, and mitochondrial dysfunction in stroke and neurodegenerative diseases. Oxidative Med Cell Longev. 2015;2015:964518.
Qiu Z, Dubin AE, Mathur J, Tu B, Reddy K, Miraglia LJ, et al. SWELL1, a plasma membrane protein, is an essential component of volume-regulated anion channel. Cell. 2014;157(2):447–58.
Rassin DK. Taurine, cysteinesulfinic acid decarboxylase and glutamic acid in brain. Adv Exp Med Biol. 1981;139:257–71.
Rassin DK, Sturman JA, Gaull GE. Taurine and other free amino acids in milk of man and other mammals. Early Hum Dev. 1978;2(1):1–13.
Reymond I, Almarghini K, Tappaz M. Immunocytochemical localization of cysteine sulfinate decarboxylase in astrocytes in the cerebellum and hippocampus: a quantitative double immunofluorescence study with glial fibrillary acidic protein and S-100 protein. Neuroscience. 1996;75(2):619–33.
Ricci L, Valoti M, Sgaragli G, Frosini M. Protection by taurine of rat brain cortical slices against oxygen glucose deprivation- and reoxygenation-induced damage. Eur J Pharmacol. 2009;621(1–3):26–32.
Sánchez-Olea R, Moran J, Schousboe A, Pasantes-Morales H. Hyposmolarity-activated fluxes of taurine in astrocytes are mediated by diffusion. Neurosci Lett. 1991;130(2):233–6.
Sánchez-Olea R, Moran J, Pasantes-Morales H. Changes in taurine transport evoked by hyperosmolarity in cultured astrocytes. J Neurosci Res. 1992;32(1):86–92.
Schousboe A, Pasantes-Morales H. Role of taurine in neural cell volume regulation. Can J Physiol Pharmacol. 1992;70(Suppl):S356–61.
Schousboe A, Sánchez Olea R, Moran J, Pasantes-Morales H. Hyposmolarity-induced taurine release in cerebellar granule cells is associated with diffusion and not with high-affinity transport. J Neurosci Res. 1991;30(4):661–5.
Schuller-Levis GB, Park E. Taurine: new implications for an old amino acid. FEMS Microbiol Lett. 2003;226(2):195–202.
Shennan DB. Swelling-induced taurine transport: relationship with chloride channels, anion-exchangers and other swelling-activated transport pathways. Cell Physiol Biochem. 2008;21(1-3):15–28.
Shimada K, Jong CJ, Takahashi K, Schaffer SW. Role of ROS production and turnover in the antioxidant activity of taurine. Adv Exp Med Biol. 2015;803:581–96.
Smith KE, Borden LA, Wang CH, Hartig PR, Branchek TA, Weinshank RL. Cloning and expression of a high affinity taurine transporter from rat brain. Mol Pharmacol. 1992;42(4):563–9.
Spaeth DG, Schneider DL. Turnover of taurine in rat tissues. J Nutr. 1974;104(2):179–86.
Sturman J. Review: taurine deficiency and the cat. Adv Exp Med Biol. 1992;315:1–5.
Sturman JA. Nutritional taurine and central nervous system development. Ann N Y Acad Sci. 1986;477:196–213.
Sturman JA, Gaull GE. Taurine in the brain and liver of the developing human and monkey. J Neurochem. 1975;25(6):831–5.
Sturman JA, Messing JM, Rossi SS, Hofmann AF, Neuringer M. Tissue taurine content, activity of taurine synthesis enzymes and conjugated bile acid composition of taurine-deprived and taurine-supplemented rhesus monkey infants at 6 and 12 mo of age. J Nutr. 1991;121(6):854–62.
Sun M, Xu C. Neuroprotective mechanism of taurine due to up-regulating calpastatin and down-regulating calpain and caspase-3 during focal cerebral ischemia. Cell Mol Neurobiol. 2008;28(4):593–611.
Sun M, Gu Y, Zhao Y, Xu C. Protective functions of taurine against experimental stroke through depressing mitochondria-mediated cell death in rats. Amino Acids. 2011;40(5):1419–29.

Suzuki T, Wada T, Saigo K, Watanabe K. Taurine as a constituent of mitochondrial tRNAs: new insights into the functions of taurine and human mitochondrial diseases. EMBO J. 2002;21(23):6581–9.
Tappaz ML. Taurine biosynthetic enzymes and taurine transporter: molecular identification and regulations. Neurochem Res. 2004;29(1):83–96.
Trachtman H. Cell volume regulation: a review of cerebral adaptive mechanisms and implications for clinical treatment of osmolal disturbances: II. Pediatr Nephrol. 1992;6(1):104–12.
Tuz K, Peña-Segura C, Franco R, Pasantes-Morales H. Depolarization, exocytosis and amino acid release evoked by hyposmolarity from cortical synaptosomes. Eur J Neurosci. 2004;19(4):916–24.
Verbalis JG, Gullans SR. Hyponatremia causes large sustained reductions in brain content of multiple organic osmolytes in rats. Brain Res. 1991;567(2):274–82.
Vitvitsky V, Garg SK, Banerjee R. Taurine biosynthesis by neurons and astrocytes. J Biol Chem. 2011;286(37):32002–10.
Voaden MJ, Lake N, Marshall J, Morjaria B. Studies on the distribution of taurine and other neuroactive amino acids in the retina. Exp Eye Res. 1977;25(3):249–57.
Voss FK, Ullrich F, Munch J, Lazarow K, Lutter D, Mah N, et al. Identification of LRRC8 heteromers as an essential component of the volume-regulated anion channel VRAC. Science. 2014;344(6184):634–8.
Vyas S, Bradford HF. Co-release of acetylcholine, glutamate and taurine from synaptosomes of Torpedo electric organ. Neurosci Lett. 1987;82(1):58–64.
Wallace DR, Dawson R Jr. Decreased plasma taurine in aged rats. Gerontology. 1990;36(1):19–27.
Warskulat U, Flögel U, Jacoby C, Hartwig HG, Thewissen M, Merx MW, Molojavyi A, Heller-Stilb B, Schrader J, Häussinger D. Taurine transporter knockout depletes muscle taurine levels and results in severe skeletal muscle impairment but leaves cardiac function uncompromised. FASEB J. 2004;18(3):577–9.
Warskulat U, Heller-Stilb B, Oermann E, Zilles K, Haas H, Lang F, et al. Phenotype of the taurine transporter knockout mouse. Methods Enzymol. 2007;428:439–58.
Worden JA, Stipanuk MH. A comparison by species, age and sex of cysteinesulfinate decarboxylase activity and taurine concentration in liver and brain of animals. Comp Biochem Physiol B. 1985;82(2):233–9.
Ye HB, Shi HB, Yin SK. Mechanisms underlying taurine protection against glutamate-induced neurotoxicity. Can J Neurol Sci. 2013;40(5):628–34.

Glycine Transporters and Its Coupling with NMDA Receptors

Francisco Zafra, Ignacio Ibáñez, David Bartolomé-Martín, Dolores Piniella, Marina Arribas-Blázquez, and Cecilio Giménez

Abstract Glycine plays two roles in neurotransmission. In caudal areas like the spinal cord and the brainstem, it acts as an inhibitory neurotransmitter, but in all regions of the CNS, it also works as a co-agonist with L-glutamate at N-methyl-D-aspartate receptors (NMDARs). The glycine fluxes in the CNS are regulated by two specific transporters for glycine, GlyT1 and GlyT2, perhaps with the cooperation of diverse neutral amino acid transporters like Asc-1 or SNAT5/SN2. While GlyT2 and Asc-1 are neuronal proteins, GlyT1 and SNAT5 are mainly astrocytic, although neuronal forms of GlyT1 also exist. GlyT1 has attracted considerable interest from the medical community and the pharmaceutical industry since compelling evidence indicates a clear association with the functioning of NMDARs, whose activity is decreased in various psychiatric illnesses. By controlling extracellular glycine, transporter inhibitors might potentiate the activity of NMDARs without activating excitotoxic processes. Physiologically, GlyT1 is a central actor in the cross talk between glutamatergic, glycinergic, dopaminergic, and probably other neurotransmitter systems. Many of these relationships begin to be unraveled by studies performed in recent years using genetic and pharmacological models. These studies are also clarifying the interactions between glycine, glycine transporters, and other co-agonists of the glycine site of NMDARs like D-serine. These findings are also relevant to understand the pathophysiology of devastating diseases like schizophrenia, depression, anxiety, epilepsy, stroke, and chronic pain.

Keywords Glycine • Transport • Glutamate • NMDA receptors • Astrocytes • Schizophrenia • GlyT1

F. Zafra (✉) • I. Ibáñez • D. Bartolomé-Martín • D. Piniella
M. Arribas-Blázquez • C. Giménez
Centro de Biología Molecular Severo Ochoa, Facultad de Ciencias,
Consejo Superior de Investigaciones Científicas, Universidad Autónoma de Madrid,
C / Nicolás Cabrera, 1, 28049 Madrid, Spain

Centro de Investigación Biomédica en Red de Enfermedades Raras and IdiPAZ,
Instituto de Salud Carlos III, Madrid, Spain
e-mail: fzafra@cbm.csic.es

A. Ortega, A. Schousboe (eds.), *Glial Amino Acid Transporters*,
Advances in Neurobiology 16, DOI 10.1007/978-3-319-55769-4_4

List of Abbreviation

AMPA	α-amino-3-hydroxy-5-methylisoxazole-4-propionic acid
GlyR	Glycine receptor
GlyT1	Glycine transporter-1
GlyT2	Glycine transporter-2
LTP	Long-term potentiation
NFPS	(±)-N-[3-(4′fluorophenyl)-3-(4′-phenylphenoxy)propyl]sarcosine
NMDA	N-methyl-D-aspartate
NMDAR	NMDA receptor
SCZ	Schizophrenia
SNAT	Sodium-coupled neutral amino acid transporters

1 Introduction

A simple search in PubMed database reveals that one subtype of glutamate receptor, the NMDA receptor (NMDAR), has maintained a remarkable interest for the last 25 years, accounting for about 40,000 references. This is more than double the interest aroused altogether by the two other subtypes of ionotropic glutamate receptors (AMPA and kainate receptors) and quadruplicates the number of references to metabotropic glutamate receptors. The reason for this decided interest may be related to the fact that NMDARs are located at a crossroads where fast excitatory neurotransmission converges with plasticity processes that are on the basis of extraordinary brain properties such as learning, memory, and cognition (Nakazawa et al. 2004; Collingridge et al. 2013). This strategic situation makes NMDARs not only hubs for physiology but also for pathology. Thus, acute overstimulation of NMDARs during ischemic stroke initiates the so-called excitotoxic process that ends up in neuronal death. Chronic excitotoxicity has been hypothesized to play a role in numerous neurodegenerative diseases including Alzheimer's disease, Huntington's disease, or amyotrophic lateral sclerosis. In addition, hypofunction of NMDARs seem to be detrimental, and evidence accumulates that this is associated to schizophrenia (SCZ) and other psychiatric diseases. Accordingly, these receptors have concentrated a lot of effort to develop drugs that might modulate their activity.

2 NMDAR Subunits and the Glycine Binding Site

The atomic structure of NMDAR has been recently resolved at 4 Å resolution (Karakas and Furukawa 2014; Lee et al. 2014). NMDARs are obligatory heterotetramers mainly composed of two GluN1 subunits and two GluN2 and/or GluN3 subunits. There are eight different splice variants of the GluN1 genes, four GluN2 genes (GluN2A-D), and two GluN3 genes (GluN3A-B) (for a review, Paoletti et al. 2013). The subunit composition varies regionally and developmentally. For instance, GluN2A and GluN2B subunits are especially relevant in the adult forebrain and present different function and distribution in neurons. In general GluN2B is expressed at both synaptic and extrasynaptic sites, whereas GluN2A is mainly expressed at the synapse (Lavezzari et al. 2004; Papouin et al. 2012). This subunit composition undergoes developmental variations, and, as the brain develops, synaptic NMDARs containing GluN2A subunits are targeted to synapses, and those with GluN2B subunits, which dominate in the neonatal brain, are displaced to extrasynaptic sites (Williams et al. 1993; Sheng et al. 1994). However, at least in the developing brain, this distribution is dynamic and is subjected to modifications by synaptic activity. For instance, NMDARs containing GluN2B are redistributed away from glutamate synapses through increased lateral diffusion during LTP in immature neurons (Dupuis et al. 2014). Opening of NMDAR channels involves the relief of the Mg^{2+} blockade of the ionic pore by membrane depolarization and results in an influx of calcium ions that activates diverse signal transduction cascades that control strength of neural connectivity or neuroplasticity.

A unique property of NMDARs among ionotropic receptors is that opening the channel requires the presence of a co-agonist. While glutamate binds to a bilobulated cavity located in the extracellular domain of GluN2 subunits, the other ligand, that was identified as glycine by Johnson and Ascher (1987), binds to a similar cavity located in GluN1 or GluN3 subunits (called the glycine-B site, as opposed to the glycine-A site on the strychnine-sensitive glycine receptor GlyR). However, the physiological role of the glycine-B site was largely controversial because the affinity for glycine is very high (in the low micromolar range), and it was thought that the site would be tonically saturated at the concentrations of glycine assumed to be present in the synaptic cleft (Kemp et al. 1988). Nonetheless, subsequent studies revealed that it is not necessarily the case, supporting the existence of subpopulations of NMDARs regulated by glycine in vivo. This is probably dependent on several different factors. Firstly, there are a number of transport systems for glycine that precisely control the glycine concentration in the synaptic and the perisynaptic space (Berger et al. 1998; Bergeron et al. 1998; Li et al. 2009; Wilcox et al. 1996; Chen et al. 2003). Neurons and glial cells express different glycine transporters that are located strategically and regulate the concentration of glycine in the neighborhood of NMDARs in a dynamic manner. Among them, glycine transporter GlyT1 might play a pivotal role, but there are also a number of low-affinity transporters for neutral amino acids, like Asc-1 or SNATs, that might contribute significantly to the process. Secondly, NMDARs have a heterogeneous subunit composition, which determines the affinity for glycine as well as the subcellular localization and developmental stage of the synapses. Heterodimeric NMDARs containing GluN2B subunits display ten times higher affinity for glycine than those containing the GluN2A subunits (EC_{50} ~0.1 μM and 1 μM, respectively). Occupancy of the glycine binding site not only governs the opening of the channel but also increases the affinity of the receptor for glutamate (glutamate and glycine sites are allosterically coupled) and modulates the receptor function by decreasing its desensitization (Mayer et al. 1984; Lester et al. 1993). In addition, glycine primes NMDARs for endocytosis thereby controlling the levels of the receptor in the cell surface (Nong et al. 2003). Further complexity is added by the fact that D-amino acids, especially D-serine, are almost as effective as glycine in activating the receptor by binding to the glycine-B site (Kleckner and Dingledine 1988; Mothet et al. 2000; Yang et al. 2003; Panatier et al. 2006; Henneberger et al. 2010). Immunohistochemical localization of D-serine demonstrated that this amino acid is mainly localized in astrocytes and that its distribution matches quite extensively with the expression of NMDARs (Schell et al. 1995), thus raising the possibility that D-serine could be the physiological ligand of the glycine-B site. In fact, some studies suggested that D-serine would be the preferred ligand (Mothet et al. 2000). However, subsequent reports have come to draw a more complex picture, where it seems to emerge a regional and temporal differentiation in the preference for one or the other ligand. The extracellular levels of D-serine in the CNS are controlled by two members of the SLC1 family, the Na^+-dependent alanine-serine-cysteine transporters 1 and 2 (ASCT1 and ASCT2) (Martineau et al. 2014). These are expressed by both astrocytes and neurons. Additionally, D-serine can be transported by the neuronal Na^+-independent antiporter alanine-serine-cysteine-1 (Asc-1), a member of the SLC7 family, that can work in the reverse manner to release D-serine and glycine from neurons (Fukasawa et al. 2000; Helboe et al. 2003, Rosenberg et al. 2013). D-serine is also released from astrocytes via mechanisms implicating Ca^{2+} and SNARE-dependent exocytosis (Yang et al. 2003; Henneberger et al. 2010; Martineau et al. 2013) and sustained by the activity of a glia-specific vesicular transporter for D-serine (Martineau et al. 2013; although see Agulhon et al. 2010). Thus, an open question is how dynamics are the fluxes of D-serine in the neighborhood of NMDARs and how they compare with those of glycine to cooperate/compete in determining the responses of the receptor. Glycine appears to be the preferred co-agonist in receptors containing GluN2B subunits, while it would be D-serine for those containing GluN2A (Fossat et al. 2012; Papouin et al. 2012; Le Bail et al. 2015). This might explain the developmental change in the ligand preference described in the connection between the Schaffer collaterals and CA1 pyramidal neurons, turning from glycine to D-serine in parallel to the replacement of GluN2B by GluN2A that occurs between weeks 1 and 3 after birth (Le Bail et al. 2015). Nevertheless exceptions might exist to this rule, like synapses of the hypothalamic supraoptic nucleus that robustly express GluN2B but seem to depend only on D-serine – although inhibitors of glycine transporter were not used in these studies (Panatier et al. 2006; Doherty and Sladek 2011). These observations suggest a complex synapse-specific cross talk between both ligands. As an example, two recent articles show that both, glycine and D-serine, are necessary for induction of LTP in CA1 hippocampal area and dentate gyrus (Le Bail et al. 2015), as well as in the lateral nucleus of the amygdala (LA) (Li et al. 2013). In LA, the prevalence of D-serine or glycine at synaptic NMDARs would be determined by synaptic activity. Ambient D-serine may maintain activation of NMDARs in LA neurons in the absence of evoked synaptic events, while activity-dependent release of glycine from astrocytes is implicated in the activation of NMDARs during afferent stimulation (Li et al. 2013). In addition, electrophysiological measures both in the inner retina and in the hypoglossal nucleus also suggested that GlyT1 activity keeps glycine levels near NMDARs at sufficiently low concentrations so as to allow D-serine to play a major role as an NMDAR co-agonist (Berger et al. 1998; Stevens et al. 2010). Blockade of GlyT1 with a specific inhibitor increased the extracellular levels of glycine to saturating levels and canceled the effect of added D-serine (Stevens et al. 2010). So, the question is if there are any mechanisms (physiological or pathological) that regulate in a concerted manner the diverse enzymes and transporters that control the fluxes of glycine and D-serine in the synapse to determine the activity profiles of NMDARs (reviewed by Mothet et al. 2015).

3 Regulation of the Extracellular Glycine Concentration by Glycine Transporters

The fluxes of glycine at inhibitory and excitatory synapses are controlled by two glycine transporters, GlyT1 and GlyT2, which belong to the sodium- and chloride-dependent neurotransmitter transporter family and are encoded by genes SLC6A9 and SLC6A5, respectively (Aragón and López-Corcuera 2005). GlyT1 and GlyT2 have different regional and cellular expression patterns in the CNS. Initially, mRNA for GlyT1 was localized at high concentrations in glutamatergic neurons and also in glial cells (Smith et al. 1992; Borowsky et al. 1993; Zafra et al. 1995b). However, early immunohistochemical studies detected the glial protein but failed in recognizing the neuronal forms of GlyT1, probably due to some kind of epitope occlusion of the neuronal protein (Zafra et al. 1995a). The expression in astrocytes is especially high in glycinergic areas, where GlyT1-immunoreactive glial profiles ensheath glycinergic synapses. Its essential role in these synapses is to lower extracellular glycine concentration as shown in GlyT1-deficient mice, where the decay time constant of glycinergic mIPSCs recorded in hypoglossal motoneurons was longer than that in wild-type mice, indicating an increased synaptic concentration of neurotransmitter (Gomeza et al. 2003a). Antibodies developed later allowed the detection of GlyT1 immunoreactivity not only in astrocytes but also in neuronal elements, mainly in glutamatergic terminals along the forebrain. Lower levels of immunoreactivity were also observed in the postsynaptic membrane of asymmetric synapses, forming immunoprecipitable complexes with NMDAR (Cubelos et al. 2005a). As we will discuss later in more detail, this distribution is compatible with a role of GlyT1 in modulating NMDARs. Additionally, GlyT1 has been found not only in the plasma membrane of glutamatergic terminals but also in synaptic vesicles (Cubelos et al. 2014). It is unknown if this localization is related to recent evidence indicating that glycine is released from glutamatergic terminals in hippocampal neurons in a Ca^{2+}-dependent manner upon depolarization (Muller et al. 2013). This release would require an accumulative mechanism of glycine into synaptic vesicles that is not characterized (VIATT is absent in these terminals). It was suggested that GlyT1 present in vesicular membrane might mediate the accumulation of glycine into the lumen either by a diffusive equilibration process or, even, by some type of uncharacterized active transport (Cubelos et al. 2014). Independently of the Ca^{2+}-dependent release of glycine, presumably vesicular, there is also evidence for a Ca^{2+}-independent release component that might be attributed to the reversal operation of GlyT1 located in the astrocytic or the neuronal membrane (Galli et al. 1993; Luccini et al. 2008). The GlyT1-dependent uptake of glycine is an electrochemical process coupled to the movement of sodium and chloride ions, with a stoichiometry 1 Gly/2 Na^+/1 Cl^-, that under depolarizing conditions might be reverted, allowing the efflux of glycine, either from the glial cells or from the neurons (Roux and Supplisson 2000; Huang et al. 2004; Aubrey et al. 2005). Diverse estimations suggest that GlyT1 activity is not far from equilibrium. Roux and Supplison (2000) calculated, assuming that intracellular glycine is 2 mM in astrocytes (in the hippocampus) and 10 mM in glycinergic neurons, that GlyT1 is close to equilibrium for an extracellular glycine concentration of 100 nM and a resting potential of −70 mV. Non-vesicular glycine release via GlyT1 reversal may occur under high-frequency stimulation (HFS) and probably ischemia, conditions that depolarize the astrocyte. HFS triggers AMPA receptor activation in astrocytes and an increase in intracellular [Na^+], which combined with plasma membrane depolarization may be sufficient to induce the reversion of GlyT1 during HFS (Attwell et al. 1993; Rose and Ransom 1996; Roux and Supplisson 2000; Marcaggi and Attwell 2004; Huang et al. 2004). Similarly, during ischemia, glial cells are depolarized up to −50 mV and intracellular [Na^+] may rise up to 39 mM (Attwell et al. 1993), conditions that might allow the reversion of GlyT1. Indeed, this seems to be the case in hypoxic retina (Hanuska et al. 2016) and in hippocampal brain slices in various cell-damaging conditions (Saransaari and Oja 2001). Although the reversal of GlyT1 under physiological condition has less experimental support, a recent article shows the existence of dopamine-induced release of glycine from cortical astrocytes in primary culture that is blocked by a specific GlyT1 inhibitor (Shibasaki et al. 2016). However, if this also occurs in native tissue is insufficiently documented. As we will discuss later, there is other transporters for glycine that might contribute to glycine release.

The other glycine transporter, GlyT2, is a neuronal protein, associated only to glycinergic neurons in the spinal cord, the cerebellum, and diverse nuclei of the brainstem, like the lateral superior olive, the inferior colliculi, and the dorsal and ventral cochlear nuclei, among others (Zafra et al. 1995a, b; Friauf et al. 1999). Minor populations of glycinergic interneurons immunoreactive for GlyT2 have been also described in the hippocampus (Danglot et al. 2004; Song et al. 2006), supporting the existence of functional glycinergic synapses in this region. Electron microscopy shows that GlyT2 is enriched in presynaptic terminals containing high concentrations of glycine. Within glycinergic boutons, GlyT2 immunostain was associated with the plasma membrane but often appeared as discrete clumps, generally excluded from the region of the active sites of synapses, suggesting that it may be excluded from synaptic clefts (Spike et al. 1997). Importantly, while most of the transporters of this family have a stoichiometric coupling of two sodium ions transported with every glycine molecule (including GlyT1), GlyT2 is coupled to the electrochemical movement of three sodium ions, favoring the maintenance of a high concentration gradient along the presynaptic membrane and supplying enough glycine for presynaptic vesicle refilling, a process that seems necessary to preserve quantal glycine content in synaptic vesicles (Rousseau et al. 2008; Pérez-Siles et al. 2012; Apostolides and Trussell 2013). GlyT2 activity dysfunctions reduce presynaptic glycine release and cause a significant decrease of inhibitory glycinergic neurotransmission that results in spasticity (James et al. 2012). Nevertheless, some evidence suggests that GlyT2 also participates in reuptake of glycine from the synaptic cleft, and the blockade of the transporter function could enhance glycinergic inhibitory neurotransmission in some situations, although to a lesser extent than GlyT1: for instance, the pharmacological blockade of GlyT2 in lamina X neurons of rat spinal cord slices increases glycinergic neurotransmission in the spinal cord (Bradaïa et al. 2004). Similarly, the GlyT2 inhibitor Org 25543 increases the extracellular glycine concentration, as detected by microdialysis perfusion of the lumbar dorsal spinal cord of rats (Whitehead et al. 2004), suggesting that in caudal regions of the CNS, glial GlyT1 and neuronal GlyT2 closely cooperate in the regulation of extracellular glycine at inhibitory synaptic sites.

Together, these studies indicate that neuronal GlyT1 has an optimal distribution to regulate the binding of glycine to NMDARs in forebrain regions, while GlyT2 and the glial GlyT1 are better situated to participate in inhibitory glycinergic neurotransmission in caudal areas. Nevertheless, in these caudal regions (and probably other areas), where there is a coexistence of neurons expressing NMDAR and GlyR, this division of labor cannot be so strict. It is known the existence of a cross talk is between these receptors. Glycine released from glycinergic terminals might reach nearby glutamatergic synapses by spillover under some circumstances, overcoming the barrier imposed by glycine transporters to the diffusion of the neurotransmitter (Ahmadi et al. 2003). Normally, glial processes surrounding glutamatergic terminals also contain high levels of GlyT1. These transporters are properly positioned to allow control of the glycine fluxes and NMDA-mediated neurotransmission. Reverse transport of glycine through the glial GlyT1 might also play a role at inhibitory synapses. It has been hypothesized that at times of low-synaptic activity, glycine concentration in the synaptic cleft would decrease allowing the reversal operation of GlyT1 and the release of glycine which is taken up by GlyT2 on postsynaptic neurons for packaging into vesicles (Aubrey et al. 2005). However, the rigorous testing of these hypotheses and an accurate measurement of glycine fluxes between glia and neurons clearly require new tools that enable real-time determination of the oscillations in the concentration of glycine in synaptic and extrasynaptic sites in response to different physiological or pathological stimuli. Examples of these tools might be fluorescent probes with similar characteristics to those already available to measure glutamate fluxes (Marvin et al. 2013).

4 Pharmacological and Genetic Models to Study GlyT1 Function

4.1 GlyT1 Inhibitors

The above-described localization experiments suggested a number of potential roles for glycine transporters. However, as it has been pointed in the preceding paragraphs, definitive proofs about the real physiological meaning of these proteins were only obtained after developing specific inhibitors and genetically modified mice. Of course the idea of developing compounds that increase the availability of glycine at glutamatergic synapses (and perhaps at glycinergic ones) was appealing since it might provide novel therapeutic avenues to treat cognitive impairments in a number of psychiatric conditions and perhaps be relevant in the treatment of pain, epilepsy, or even be useful in enhancing specific cognitive functions in healthy subjects. In theory, these compounds would have lower excitotoxicity than direct NMDAR agonists, and, moreover, they may offer greater spatial and temporal selectivity, since selective inhibition of GlyT1 may result in potentiation of NMDAR only in those specific brain regions activated by social stimuli and cognitive challenges where synaptically released glutamate and pharmacologically increased glycine meet each other. Nevertheless, a potential drawback of this strategy is the enhancement of the inhibitory glycinergic neurotransmission with affectation of motor and sensorial pathways regulated by GlyRs. Even so, there are situations where the stimulation of GlyRs could be therapeutically favorable as might be the case for treating neuropathic pain.

A detailed and updated review about GlyT1 pharmacology has been published recently (Cioffi and Guzzo 2016). In brief, early work had shown that sarcosine (N-Methyl-glycine) inhibits the high-affinity glial transporter of glycine (Zafra and Giménez 1989), and, later, Liu et al. (1993) showed that this compound inhibits GlyT1 but not GlyT2. Thus, sarcosine was used as a lead compound to develop the first generation of GlyT1 inhibitors. These include compounds like (±)-N-[3-(4′fluorophenyl)-3-(4′-phenylphenoxy)propyl]sarcosine (NFPS) (the R-enantiomer is named ALX5407) that showed a non-competitive binding with high affinity and slow dissociation, leading to prolonged elevation in synaptic glycine concentrations that were able to activate GlyRs and probably were the cause of decreased motor and respiratory activity when administered in vivo to rodents. Several sarcosine derivatives were developed by Organon (Org 24461, Org 24598, and others), Pfizer (N-[3-phenyl-3(4′-(4-toluoyl)phenoxy)-propyl]sarcosine or NPTS), Lundbeck (2-arylsulfanylphenyl-1-oxyalkylamino derivatives), Merck (indandione sarcosine derivatives), and Amgen (benzhydryl piperazine analogue AMG 747), among others. Several of these inhibitors increased glycine levels in the CSF without having the negative motor effects of NFPS but keeping the positive effects in animal models of schizophrenia. Four of them were assayed in clinical trials, being Org 25935 as the most advanced one. In preclinical assays Org 25935 attenuated the scopolamine-induced deficits in the object retrieval/detour task or the ketamine-induced working memory deficits in monkeys, with an inverted U-shaped dose-response curve indicating that the maximum efficacy is achieved at submaximal occupancy levels (Castner et al. 2014). In humans, it was well tolerated and reduced ketamine-induced psychomimetic and perceptual alterations. However, in clinical trials Org 25935 did not differ significantly from placebo in reducing negative symptoms of SCZ or improving cognitive functioning when administered as adjunctive treatment to atypical antipsychotics (Schoemaker et al. 2014).

After the sarcosine derivatives, a second generation of non-sarcosine-based compounds was developed by several companies, and some of them entered into clinical trials for SCZ. Chemically, they can be divided into several categories: methylphenidate-derived (SSR504734, SSR103800, GSK1018921, GSK931145), alkyl and heteroaromatic substituted sulfonamides and sulfones (ACPPB, DCCCyB, and several others), heteroaryl amides (PF-03463275 and others), benzoylpiperazines (several compounds by Hoffmann-La Roche, including bitopertin), benzoylisoindolines (derived from benzoylpiperazines), and several others (Cioffi and Guzzo 2016). Most of the assayed compounds in these categories are competitive inhibitors, although bitopertin is a non-competitive one. Thus, although it was suggested that competitive inhibition of GlyT1 might impart pharmacological advantages, the mechanism of action of bitopertin, which is the most advanced compound in clinical trials, questions this hypothesis (Mezler et al. 2008). In general, all these compounds were designed and assayed for improved selectivity for GlyT1 over GlyT2, better water solubility or shorter residence time on the transporter, aspects that were expected to improve their therapeutic outcome.

4.2 *Pharmacological Models*

The availability of all these compounds has allowed answering the question of whether GlyT1 is capable of controlling the activity of NMDARs. In fact, the results show that this is the case, by controlling both the channel opening and those processes triggered by the gating (e.g., the long-term potentiation, LTP, or the long-term depression, LTD). And this is a general phenomenon throughout many regions of the brain, as evidenced by multiple observations from rat cortical and hippocampal neurons, hypoglossal motoneurons, or spinal cord lamina X neurons. In all cases, blockade of GlyT1 promotes an elevation of glycine levels and this impacts in NMDAR activity (Chen et al. 2003; Kinney et al. 2003; Martina et al. 2004; Lim et al. 2004; Bradaïa et al. 2004; Zhang et al. 2008, 2014). This is finally reflected in an effect on learning and memory in diverse experimental paradigms. Thus, a number of first and second generation of GlyT1 inhibitors reversed many of the cognitive deficits observed in animal models of acute or neonatal NMDAR antagonist treatment (phencyclidine, MK801, ketamine) including impairments on reference memory, object and social recognition memory, and working memory. As we will discuss below, this is especially relevant in the context of psychotic diseases. And also in normal rats, GlyT1 inhibitors improved working memory and social recognition, suggesting that these compounds might be useful as enhancers of cognitive functions in healthy humans (Singer et al. 2009a).

GlyT1 inhibitors have been also instrumental in clarifying basic mechanism of the brain circuitry. In this sense, two seemingly contradictory functions of glycine and GlyT1, both in excitatory and inhibitory neurotransmission, could be more tightly linked than it could be suspected, in areas where NMDARs and GlyRs coexist, through a GlyT1-mediated control of the overall excitability of neural networks. The link is provided by the magnitude of the glycine concentration in the synaptic cleft. For instance, Zhang et al. (2014) altered the concentrations of glycine by controlling the doses and the time of incubation of hippocampal slices with the GlyT1 inhibitor NFPS. While moderate levels of glycine promoted LTP, higher doses induced LTD. These changes in synaptic plasticity were dependent on trafficking of NMDARs to (LTP) and from (LTD) the membrane. In this way, low levels of glycine seem to act mainly on the glycine-B site of NMDARs to induce LTP by promoting the insertion of NMDAR in the synapse (by uncharacterized trafficking mechanisms). However, when levels of glycine get higher and cross the threshold of the NMDAR, endocytosis mechanism (Nong et al. 2003) and glycine may activate GlyRs. As a result, the inhibitory effect may be stronger than the excitatory one, and the net effect mediated by glycine is a depression of the NMDA response (Zhang et al. 2014). Another consequence in this scenario of a stepwise increase in the concentrations of extracellular glycine is the triggering of homeostatic mechanisms in the neuronal network to avoid runaway excitation after LTP induction. In hippocampal slices this compensatory mechanism was dependent on GlyT1 located in glial cells since fluoroacetate, a glia-specific metabolic inhibitor, blocked the effect as also did strychnine, indicating its dependence on hippocampal GlyRs (Zhang et al. 2008). Similarly, during the induction and expression of NMDAR-dependent LTP in pyramidal neurons of the visual cortex, GlyT1 controls the activity of extrasynaptic GlyRs, and this resulted in a shunting inhibition of afferent inputs which thus displayed a depression (a LTD-like effect) at the soma after dendritic integration. In this case, it seems that the NMDAR co-agonist is synaptic D-serine rather than glycine, illustrating again the complex cross talk between D-serine and glycine (Meunier et al. 2016).

The glutamatergic system also interacts at different levels with the dopaminergic system. Recent observations support a role of GlyT1 in this cross talk, evidenced by the GlyT1 inhibitor ACPPB. Using a model of unilateral 6-OHDA-induced lesions that spared the mesoaccumbens projection, ACPPB promoted dopaminergic reinnervation of the dorsal striatum and normalized 6-OHDA-induced lateralization of sensorimotor behavior (Schmitz et al. 2013). Both effects were dependent on the presence of NMDARs in dopamine neurons. An important consequence of these findings is that if functional sprouting could be induced in dopamine axons in areas that are spared from denervation in Parkinson's disease (i.e. the caudate nucleus, medial portions of the putamen, and the nucleus accumbens), GlyT1 inhibitors might point toward future therapeutic treatments for these patients.

However, all these observations rely on the pharmacological manipulation of GlyT1 and other proteins of the system. In view of the important role of this transporter in regulating the excitatory/inhibitory balance of the neuronal networks, the question is whether the activity of GlyT1 can be modulated under physiological or pathological states by endogenous mechanisms. As discussed above, an increase of the glycine level could occur under some pathophysiological states, such as seizure and ischemia. Due to the dissipation of ionic gradients, the activity of GlyTs is downregulated or even reversed in ischemia (Huang et al. 2004; Baker et al. 1991). GlyT1 can also be silenced by certain regulatory factors like Zn^{2+} (Ju et al. 2004) and protons (Aubrey et al. 2000), both of which are stored in transmitter vesicles, and transient changes in extracellular pH or Zn^{2+} occur during synaptic transmission (Qian and Noebels 2005; Krishtal et al. 1987). Also, the rate of glycine reuptake by GlyT1 can be downregulated by intracellular factors and signaling pathways such as arachidonic acid (Zafra et al. 1990; Pearlman et al. 2003), protein kinase C activation (Gomeza et al. 1995; Sato et al. 1995; Fernández-Sánchez et al. 2009), Ca^{2+}/calmodulin-dependent enzymes (Lopez-Colome and Gadea 1999) or GSK3-β (Jimenez et al. 2015). Additionally, the intracellular membrane trafficking of GlyT1 can be regulated by the SNARE protein syntaxin-1A, which decreases the concentration of GlyT1 protein on the plasma membrane (Geerlings et al. 2000), or by interactions with the exocyst (Cubelos et al. 2005b). There are also complex interactions with other neurotransmitters or neuromodulators, like the purinergic system that enhances the activity of GlyT1 (Jimenez et al. 2011). Accordingly, the function of GlyT1 seems to be effectively regulated under physiological and pathological conditions, and, thus, the concentration of extracellular glycine can be tightly controlled. The balance between stimulatory and inhibitory signaling pathways will finally determine the activity of the transporter, and, consequently, the glycine concentration might fluctuate in the synaptic cleft under physiological conditions. Because of this, GlyT1 remains a target of substantial pharmacological interest for intervening in diseases associated with dysfunction of either NMDARs or GlyR.

4.3 Genetic Models

Additional source of information on the physiological and pathological role of glycine transporters are the various genetically modified mice that have been produced for more than one decade, since the leading work of H. Betz and collaborators (Gomeza et al. 2003a, b). These advances have been extensively reviewed by Mohler et al. (2011). Relative to GlyT1, perhaps the most relevant aspect that cannot be addressed by pharmacological models is the dissection of the physiological role of glial and neuronal forms of the transporter. The three initial genetic models were unable to deal with this question since they were obtained by global deletion of the Slc6a9 (GlyT1) gene, and, moreover, they were neonatally lethal (Gomeza et al. 2003a; Coyle and Tsai 2004; Gabernet et al. 2005). However, heterozygous were viable and at least two lines (GlyT1tm1.$1^{+/-}$ and GlyT$1^{+/- \, Tsai}$) displayed altered NMDAR responses that were compatible with the saturation of de glycine-B site when GlyT1 was reduced to about a half, including an enhanced NMDA/AMPA response ratio, a resistance to the disruptive effect of amphetamine on prepulse inhibition and a tendency for improved memory retention (Tsai et al. 2004b; Gabernet et al. 2005; Martina et al. 2005). Heterozygous mice also show morphologic and physiologic alterations including an increased number of synapses and an enhanced neuronal excitability, changes that might be attributed to the chronic high levels of glycine in glutamatergic synapses (Bakkar et al. 2011).

Later, the first conditional GlyT1 mouse was generated and termed GlyT1tm1.2$^{fl/fl}$, which contained two pLox sites flanking exons 4 and 11 of Slc6a9 (Yee et al. 2006). These floxed mice were bred with transgenic mice containing Cre recombinase under the CaMKII promoter and resulted in the ablation of GlyT1 in forebrain neurons. As a consequence, both the concentration and the activity of GlyT1 decreased by approximately 30% in forebrain at postnatal day 21. The diminished GlyT1-selective glycine uptake was accompanied by an important increase in the NMDA/AMPA response ratio. Another conditional strain designed to suppress the expression of GlyT1 simultaneously in neurons and astrocytes of the forebrain was obtained by breeding the GlyT1tm1.2$^{fl/fl}$ with the Emx1Cre/Cre mice (EMX/GlyT1-KO) (Singer et al. 2009b). The simultaneous disruption of GlyT1 in neurons and glia resulted in a near-complete absence of response to the acute phencyclidine challenge. This suggests that NMDAR function in EMX/GlyT1-KO mice is altered (being more resistant to systemic pharmacological blockade of NMDAR by phencyclidine), presumably due to increased levels of synaptic glycine. This observation is in good agreement with the finding that GlyT1 inhibitors are highly effective in attenuating the motor stimulant effect of NMDAR blockers (Harsing et al. 2003; Depoortere et al. 2005; Boulay et al. 2008; Singer et al. 2009b). Surprisingly, the EMX/GlyT1-KO mice did not show alterations in the NMDA-mediated EPSC in the hippocampus. The reason for this difference with the neuronal model is unclear, but several possibilities were suggested. For instance, a more drastic increase in the extracellular glycine in the neuronal/glial-deficient mice might have primed NMDAR for endocytosis (Nong et al. 2003). Also, a tonic stimulation of inhibitory GlyRs in the glial/neuronal depleted model might alter the activity of hippocampal network, indirectly compensating the NMDAR current alterations observed in the neuronal model (Singer et al. 2009b). Additional differences between these mice were found in their capability for associative learning that was potentiated in the neuronal model but not in the neuronal/glial. Differences were also observed in the working memory paradigms that were not affected in the neuronal model in contrast to the promnesic effects observed in mutant mice with GlyT1 deletion extended to cortical glial cells (Dubroqua et al. 2012), suggestive of an important and specific role of glial cells in regulating cognitive functions. Both models, however, displayed a similarly increased recognition memory in tests for object familiarity judgment. Interestingly this was also observed in a pharmacological model using GlyT1 inhibitors (Depoortere et al. 2005; Boulay et al. 2008; Karasawa et al. 2008). In general, the different models display a wide spectrum of procognitive effects that support the idea that GlyT1 is a promising target for the treatment of cognitive symptoms in psychotic diseases. An additional conditional strain was developed again in the lab of H. Betz to suppress the expression of GlyT1 in neurons or in astrocytes (Eulenburg et al. 2010). The targeting vector was designed to enable Cre recombinase-mediated inactivation of the GlyT1 gene through deletion of exons 3 and 4, and these mice were bred with two strains of transgenic mice expressing Cre recombinase. One was under the control of the neuron-specific synapsin 1 promoter, the other under the control of the mouse glial fibrillary acidic protein (GFAP) promoter. The neuronal line did not show motor or respiratory deficits, and the authors conclude that in caudal regions of the CNS, neuronal GlyT1 does not contribute significantly to the regulation of inhibitory glycinergic neurotransmission. However, most of the mice of the glial line developed a strong hypotonic phenotype, which finally resulted in premature death between postnatal day 1 and 10, indicating that glial GlyT1 is the major player in regulating glycinergic neurotransmission. Unfortunately there are no reports related to NMDA-dependent behaviors in these mice. Interestingly, few mice of the glial strain survived and developed to adult age, indicating that Glyt1 is essential in the perinatal period but not in adults where perhaps GlyT1 function could be assumed by other glycine transporters.

5 Other Glycine Transporters

The preceding paragraphs summarize an important set of evidence supporting an unquestionable role of GlyT1 in the regulation of NMDARs and GlyRs, especially when GlyT1 operates in the forward direction. Less support has the idea that under physiological conditions GlyT1 might contribute to a fast non-vesicular release of glycine. If this has to occur, glycine might reach glutamatergic synapses by spillover from neighboring glycinergic synapses (Berger and Isaacson 1999; Turecek and Trussell 2001; Ahmadi et al. 2003). However, this might be an alternative in caudal areas of the brain but it seems unlikely in forebrain areas where glycinergic terminals are sparse or absent. A more likely source of glycine would be the reversal operation of a transporter with some electrochemical characteristics better suited to work in the efflux mode. The brain contains several amino acid transporters that belong to the SLC7 and SLC38 family and that might participate in the control of the neuronal-glial fluxes of glycine and other amino acids. Some of them, especially the already mentioned Asc-1 and SNAT5, might fulfill these requirements, showing potential reversion under physiological conditions.

Asc-1 is a plasma membrane antiporter present in neurons that has high affinity for small neutral amino acids, such as glycine, L-serine, D-serine, alanine, and cysteine (Fukasawa et al. 2000; Helboe et al. 2003). It has a widespread distribution throughout the brain, and it is located exclusively in presynaptic terminals. Data obtained with Asc-1 knock-out (KO) mice indicate that this protein is the main D-serine transporter in the brain (Rutter et al. 2007), able to secrete D-serine in an exchange reaction with other endogenous substrates. In addition, Acs-1 can also extrude glycine affecting NMDAR responses at low-frequency stimulation in hippocampal slices (Rosenberg et al. 2013). However, a recent article reveals that Asc-1 participates mainly in glycinergic transmission in the spinal cord (Safory et al. 2015). The KO mice show a marked decrease in glycine concentrations in the brain and spinal cord along with impairment of glycinergic inhibitory transmission and a hyperekplexia-like phenotype that results in postnatal death, but it is rescued by replenishing brain glycine levels (Xie et al. 2005; Safory et al. 2015). If these KO mice have phenotypes associated to NMDAR, malfunction was not reported, and it is possible that the strong glycine-related phenotype obscures more subtle changes in NMDAR-dependent processes mediated by D-serine or glycine via the Asc-1 transporter.

Another potential mediator of the glycine efflux is SNAT5 (also known as SN2), a member of the gene family SLC38 (Mackenzie and Erickson 2004) that includes diverse transporters for neutral amino acids and notably for glutamine. Some of the members of this family might take part of the glutamate-glutamine cycle between neurons and glial cells. SNAT5 recognizes not only glutamine but also glycine and some other neutral amino acids, including alanine, serine, histidine, or asparagine. Indeed, in vitro studies have shown that glycine is one of the preferred substrates (Nakanishi et al. 2001). Studies on SNAT5 are handicapped by the absence of both, specific inhibitors and animal models, but immunohistochemical and electrophysiological studies indicate that this is a protein with appropriate characteristics to regulate NMDARs. Light and electron microscopy shows that SNAT5 is a glial protein enriched in glutamatergic areas, where immunoreactive processes ensheath glutamatergic terminals (Cubelos et al. 2005c). Moreover, this expression pattern emerges during the postnatal development in parallel to the expression of essential proteins of the glutamatergic system like the vesicular glutamate transporter vGLUT1 and the glial glutamate transporter GLT-1,and to the functional maturation of these synapses (Rodriguez et al. 2014). The electrophysiological characterization indicates that transport mechanism involves Na^+ co-transport and the simultaneous exchange of H^+, resulting in an electroneutral movement of glycine across the astrocyte membrane. Furthermore, SNAT5 mediates the glycine-gated uncoupled flow of H^+ that, together with the coupled one, seems to favor the release of glycine rather than its uptake during neural activity (Hamdani et al. 2012). Neuronal activation results in accumulation of Na^+ and glutamate in perisynaptic glial cells (Chaudhry et al. 1995). In addition, the extracellular concentration of K^+ increases, which activates Na^+/HCO_3^- cotransporters at astrocytic membranes. This further increases the intracellular sodium concentration and the pH (Brookes 2000). The outwardly directed glycine gradient and the inwardly directed gradient for H^+ can now override the gradient of Na^+ so that SNAT5 can readily release glycine. Indeed, the Km of SNAT5 for glycine (about 7 mM) better concurs with the cytoplasmic concentrations of glycine (from 4 up to 11 mM in cultured primary astrocytes) than with the extracellular one, implicating a preference for release mode of SNAT5 (Verleysdonk et al. 1999; Hamdani et al. 2012).

6 GlyT1 as a Target for Psychiatric and Neurologic Diseases

6.1 Schizophrenia

SCZ is a severe chronic and disabling brain disorder affecting approximately 1% of the world population. Clinical symptoms fall into three broad categories: positive, negative, and cognitive symptoms. Current antipsychotic medication primarily improves positive symptoms basically by acting on D_2 dopamine receptors. However, these treatments have a limited value for the other categories of symptoms. An imbalance in the complex, interrelated chemical signaling that allows brain cells to communicate with each other is assumed to underlie SCZ, a disease with a slow gestation along the neurodevelopmental process that usually manifests itself in adolescence and early youth. NMDARs play a key role for shaping neuronal connections during brain development, and evidence has accumulated indicating that hypofunction of NMDARs underlies a number of alterations observed in schizophrenia (review by Moghaddam and Javitt 2012; Coyle 2006, 2012). Initially, this hypothesis is derived from the observation that various antagonists of the NMDARs, like ketamine, phencyclidine, or MK801, mimic numerous symptoms of SCZ in healthy adults and potentiate the positive, negative, and cognitive symptoms in patients. Based on these observations, administration of these NMDAR antagonists has been widely used to induce NMDAR hypofunction in animals as a pharmacological model of SCZ, and several laboratories have demonstrated schizophrenia-like changes on a number of behavioral measures relevant to positive, negative, and cognitive symptoms (Wiescholleck and Manahan-Vaughan 2013). Additionally, different KO mice also model to some extent the disease. Mice that express mutated GluN1 subunits with lowered glycine affinity display cognitive and learning defects including non-habituating hyperactivity, increased stereotyped behavior, disruptions of nest-building activity, and poor performance in the Morris water maze (Ballard et al. 2002). Also transgenic mice expressing reduced levels of the GluN1 subunit display behavioral abnormalities similar to those observed in pharmacologically induced models of SCZ (Mohn et al. 1999), and mice lacking the GluN2A subunit exhibit an increased spontaneous locomotor activity in novel environments and an impairment of latent learning in a water-finding task besides deficit in hippocampal LTP and spatial learning (Miyamoto et al. 2001). These behavioral phenotypes resemble some of the positive and negative symptoms displayed by SCZ patients thereby supporting the hypothesis (Coyle 2012; Ramsey 2009). Further support to the hypothesis is provided by recent large-scale, genome-wide investigations that have recognized that SCZ is a heterogeneous disease entity involving a large number of genes and noncoding risk loci. These studies identified several genes encoding synaptic proteins including NMDAR-associated downstream and upstream signaling proteins that play a central role in the pathogenesis of SCZ (Fromer et al. 2014; Schizophrenia Working Group 2014; Peykov et al. 2015; Balu and Coyle 2015).

Consequently, the stimulation of the glycine-B site on the NMDARs has been considered to be an effective way of indirectly enhancing NMDAR function avoiding excitotoxicity. Indeed, GlyT1 inhibitors exhibit antipsychotic activity in several animal models, as we have exposed in previous paragraphs (Alberati et al. 2012; Boulay et al. 2008; Depoortère et al. 2005; Harada et al. 2012; Chaki et al. 2015). Moreover, sarcosine has been proven to alleviate both negative symptoms and cognitive dysfunction, in addition to positive symptoms, when administered as an adjunctive therapy in small-scale clinical trials (Tsai et al. 2004a). Also, the addition of the second-generation GlyT1 inhibitor bitopertin (RG1678) to standard antipsychotics resulted in a significant reduction of negative symptoms in a randomized, double-blind study in patients with predominant negative symptoms (Umbricht et al. 2014). However, other trials could not replicate these results, together with the negative results mentioned above, for Org 25935 produced a substantial reduction of expectations (Schoemaker et al. 2014). Indeed, a phase III clinical trial carried out by Hofmann-LaRoche for bitopertin failed to reach its endpoints to improve negative symptoms, and the assay was stopped. Consequently, the efficacy of GlyT1 inhibitors against negative symptoms and several cognitive domains needs further investigations. Excellent reviews on the different GlyT1 inhibitors assayed in preclinical and clinical studies have been published (Harvey and Yee 2013; Singer et al. 2015). The neural mechanisms underlying the improvement of cognitive and social deficits caused by GlyT1 inhibitors in animal models are not fully understood but probably involve complex interaction between the glutamatergic and the dopaminergic systems, with additional intervention of the glycinergic and GABAergic systems in diverse areas of the forebrain. For instance, the GlyT1 inhibitor SSR504734 potentiates dopaminergic signaling since it increases the release of dopamine in the prefrontal cortex and in the nucleus accumbens (Depoortère et al. 2005; Leoneti et al. 2006). At least in the nucleus accumbens, these dopamine-enhancing effects could result from increased glycinergic inhibition, indicating that in the effect of GlyT1 inhibitors there are more underlying factors than just the potentiation of NMDARs (Lidö et al. 2011).

6.2 Drug Addiction

One of the many derivatives of the pathological manifestations of SCZ is the frequency with which the schizophrenic patient is addicted to different drugs of abuse including cocaine or ethanol, with a prevalence between three to five times higher than in healthy controls (Coyle 2006). NMDARs also have been implicated in the aberrant regulation of synaptic plasticity that is critical for substance abuse and addiction since glutamatergic inputs from cortical and subcortical regions modulate the mesolimbic dopamine system thereby regulating aspects of drug-seeking behaviors (see Carlezon and Thomas 2009, for a review). Indeed, these behaviors have been modified in animal models by treatments with the partial glycine-B site agonist D-cycloserine or D-serine (Paolone et al. 2009; Kelamangalath and Wagner 2010). Consistently, GlyT1+/-heterozygote mice, with overactive NMDARs, show some phenotypes consistent with that idea (Puhl et al. 2015). Indeed, inhibitors of GlyT1 (Org 25935 and Org 24598) are effective in reducing relapse-like compulsive drinking and alcohol preference in rodents (Molander, et al. 2007; Vengeliene et al. 2010; Lidö et al. 2011). However, the results in humans have been disappointing, and phase II clinical trials were stopped before conclusion (Bejczy et al. 2014).

Therefore, all these assays in humans either for the treatment of SCZ or for drug addictions illustrate the difficulties of the task, probably due to difficulties in setting the adequate doses for a system affecting simultaneously to excitatory and inhibitory neurotransmission and the involvement of multiple circuits in the regulation of these psychiatric dysfunctions. Also it is difficult to predict the response of these systems to long-term treatments that might result in a complex remodeling of the NMDAR system. Further complexity is added by the fact that frequently the clinical assays have been performed with GlyT1 inhibitors as a coadjuvant therapy with dopaminergic antipsychotics. Nevertheless, still there are evidence for a beneficial influence of sarcosine in humans (Lane et al. 2008; Singh and Singh 2011; Strzelecki et al. 2015), thereby maintaining the hope that one day both the right type of inhibitor and form of administration will be found to achieve a significant therapeutic effect.

6.3 Depression

A consequence derived from recent genetic studies is that genetic risk does not map neatly on psychiatric clinical diagnoses, which is perhaps not surprising given the degree of genetic complexity and the continuous nature of many psychiatric traits. Therefore, there is evidence for shared genetic risk between SCZ, bipolar disorder, autism spectrum disorders, intellectual disability, and attention-deficit hyperactivity disorder. Consequently all these diseases might have a glutamatergic substrate that might be rescued by GlyT1 inhibitors. Indeed, several studies performed in animal models of depression support that GlyT1 inhibitors display an antidepressant effect (Boulay et al. 2008; Depoortère et al. 2005), and in a recent double-blinded trial performed in 40 patients with major depression, sarcosine was found to result in greater improvements in several scores than citalopram (Huang et al. 2013). Although these observations are apparently contradictory to the findings that ketamine, a NMDAR antagonist, is a potent antidepressant (Krystal et al. 2013), a recent study proves that the antidepressant effect of ketamine is NMDAR independent (Zanos et al. 2016).

6.4 Anxiety

Treatment of anxiety with GlyT1 inhibitors has been also considered as a therapeutic possibility. In principle, NMDAR activation induces anxiety-like behavior in mice (Miguel and Nunes-de-Souza 2008), and this seems contradictory to the mode of action of these compounds. However, SSR504734, a GlyT1 inhibitor, has anxiolytic actions since it attenuates both the acquisition and the expression of contextual conditioned fear in rats (Nishikawa et al. 2010) and decreases maternal separation-induced ultrasonic vocalization (USV) in rat pups (Depoortère et al. 2005). However, these effects were reversed by administration of strychnine indicating that they are mediated by GlyRs instead of NMDARs. Indeed, GlyT1-induced decreases in USV were not reversed by administration of the glycine-B antagonist L-687,414 (Komatsu et al. 2015). An anxiolytic effect of GlyT1 inhibitors was also observed by infusion in the amygdala of NFPS, potentiating the fear extinction in a paradigm of conditioned fear in rats. Experimental data in this case suggest that the molecular mechanism acts via an enhancement of NMDA-mediated AMPA receptor endocytosis in the amygdala (Mao et al. 2009).

6.5 Ischemia and Exotoxicity

The control of NMDAR activation is crucial for neuronal function and viability. Overstimulation of NMDAR triggers excitotoxic cell death processes. However, this is critically dependent on the NMDAR subunit composition. Activation of GluN2B-containing NMDAR have been more associated with a death signal than GluN2A-containing NMDAR, inducing Ca^{2+} accumulation, mitochondrial swelling, and neuronal degeneration (Martel et al. 2012). Interestingly, sublethal doses of NMDA activate a neuroprotective mechanism named brain preconditioning. Thus, the question is whether a moderate increase in the NMDAR activity by GlyT1 inhibitors might emulate the effect of low doses of NMDA. Recent evidence indicates that this might be the case since both sarcosine and NFPS induced preconditioning in vivo. The preconditioning protocol by GlyT1 inhibitors reduced the expression of GluN2B subunits, whereas did not change the expression of GluN1 or GluN2A (Pinto et al. 2014, 2015).

6.6 Neuropathic Pain

Neuropathic pain is another pathological condition in which glycine transporters have attracted considerable interest as targets for drug intervention. In this case it comes into play the balance between excitatory and inhibitory neurotransmission in the spinal cord. The pain information is transported from the periphery to the thalamus through spinal centers using, among others, glutamatergic mechanisms. But GABAergic and glycinergic interneurons of the spinal cord filter and modulate the flow of information. An imbalance between spinal inhibitory and excitatory neurotransmission leads to increased responses to noxious stimuli (Costigan et al. 2009). A reduction of inhibitory neurotransmission as well as an exaggeration of excitatory processes in the spinal cord contributes to the development of increased pain sensitivity. In the spinal cord, it is evident that GlyT1 plays an essential role in controlling both the receptor activity of strychnine-sensitive glycine (mainly GlyR3a) and, secondly, the diffusion or spillover of glycine to the glycine-B site of NMDARs. GlyT1 inhibitors, Org 25935, sarcosine, and NFPS108, contribute to pain relief in different model of neuropathic pain, probably by potentiating the activity of GlyRs in the spinal cord (Tanabe et al. 2008; Morita et al. 2008; Barthel et al. 2014). Recent observations also involved GlyT1 in the mode of action of lidocaine, an anesthetic that has been used to treat neuropathic pain. The lidocaine metabolite, N-ethylglycine, was shown to be a specific inhibitor of GlyT1 that in rodent models of inflammatory and neuropathic pain resulted in an efficient amelioration of hyperalgesia and allodynia without affecting acute pain. N-ethylglycine reduced the increase in neuronal firing of wide-dynamic-range neurons caused by inflammatory pain induction. This effect probably was due to an enhancement of the spinal inhibition, secondary to the increase of glycine concentration at glycinergic inhibitory synapses (Werdehausen et al. 2015). A similar GlyR-mediated mechanism is involved in pain relief in a mouse model of bone cancer (Motoyama et al. 2014). These studies reinforce the idea that GlyT1 substrates may be useful therapeutic agents in chronic pain states involving spinal disinhibition. It is unclear whether spillover of glycine to NMDARs after treatments with glycine transporter inhibitors has positive or negative effects on suppression of pain. Thus, Morita et al. (2008) reported a lag time of 1–2 h after administration of GlyT1 inhibitors before developing the anti-allodynia effect, a lag that was suppressed by administration of glycine-B site antagonists and, therefore, attributable to NMDAR operation. But perhaps a permanent exposure to GlyT1 inhibitor could decrease the amount of NMDARs by priming their endocytosis and decreasing the glutamatergic signaling that might reinforce the pain signal (Nong et al. 2003; Barthel et al. 2014).

6.7 Epilepsy

Epileptic seizures are a major neurological disorder with a particular high incidence in children. In addition to GABA, the glycinergic system is crucially involved in the regulation of neuronal excitability. Inhibition of glycine receptors can evoke epileptiform discharges in the adult and in the immature brain (Straub et al. 1997; Chen et al. 2014). Relative to the involvement of glycine transporters in epileptogenesis, Socala et al. (2010) found that sarcosine exhibits anticonvulsive activity. However, due to the high doses of sarcosine (800–1000 mg/kg) required for effective activity, the possibility of taking sarcosine as a potential antiepileptic drug may remain elusive. Moreover, it is worth noting that in addition to antagonizing the glycine transporter, sarcosine also directly potentiates NMDAR function as a co-agonist. However, a couple of studies support the idea that pharmacological manipulation of GlyT1 might constitute a valuable treatment for epilepsy. Shen et al. (2015), using two different rodent models of temporal lobe epilepsy (TLE), demonstrated robust overexpression of GlyT1 in the hippocampal formation, suggesting dysfunctional glycine signaling in epilepsy. In support of a role of dysfunctional glycine signaling in the pathophysiology of epilepsy, both the genetic deletion of GlyT1 in hippocampus and the GlyT1 inhibitor LY2365109 increased seizure thresholds in mice. Importantly, chronic seizures in the mouse model of TLE were robustly suppressed by systemic administration of the GlyT1 inhibitor LY2365109. A second study (Zhao et al. 2016) showed that an effective inhibitor of GlyT1, termed M22, elevated the tonic seizure threshold in the mouse model of maximal electroshock seizure threshold and did not impair motor function. Given that current epilepsy treatment is limited by poor responses to available antiepileptic drugs and limited tolerance due to major cognitive side effects, both studies conclude that the GlyT1 inhibitors have potential as new anticonvulsive drugs or as the lead compounds for antiepileptic drugs development.

7 Conclusions nd Prospects

For nearly 25 years, experimental evidence has accumulated which clearly contradicts the initial proposals indicating that the glycine-B site on NMDARs was chronically saturated. Data obtained from pharmacological models and from genetically modified mice indicate that this is not the case as the glycine transporter GlyT1 occupies a strategic position, especially in glial cells, where its expression is higher but also in glutamatergic terminals. This gives the ability to modulate the glycine concentration in the vicinity of NMDARs. Studies have shown that inhibitors of GlyT1 are able to affect various psychical and neurological functions and could be useful in treating deleterious conditions such as schizophrenia, depression, or anxiety, but also pain or epilepsy, despite preliminary clinical studies in humans have been rather disappointing. Perhaps the evolution of the chemical properties of these inhibitors or the discovery of new signaling pathways that may modulate the activity or gene expression (e.g., through specific microRNAs or epigenetic mechanisms) could provide new pharmacological tools to modify the activity of GlyT1 and the associated functions.

Acknowledgments We would like to thank Dr. FJ Díez-Guerra for critical comments on the manuscript. This work was supported by grants from the Spanish MINECO (SAF2014-55686-R) and from "Fundación Ramón Areces." The latter also provides support to the CBMSO through an institutional grant.

Conflict of Interest The author declares no conflicts of interest.

References

Agulhon C, Fiacco TA, McCarthy KD. Hippocampal short- and long-term plasticity are not modulated by astrocyte Ca^{2+} signaling. Science. 2010;327(5970):1250–4.

Ahmadi S, Muth-Selbach U, Lauterbach A, Lipfert P, Neuhuber WL, Zeilhofer HU. Facilitation of spinal NMDA receptor currents by spillover of synaptically released glycine. Science. 2003;300(5628):2094–7.

Alberati D, Moreau JL, Lengyel J, Hauser N, Mory R, Borroni E, et al. Glycine reuptake inhibitor RG1678: a pharmacologic characterization of an investigational agent for the treatment of schizophrenia. Neuropharmacology. 2012;62(2):1152–61.

Apostolides PF, Trussell LO. Rapid, activity-independent turnover of vesicular transmitter content at a mixed glycine/GABA synapse. J Neurosci. 2013;33(11):4768–81.

Aragón C, López-Corcuera B. Glycine transporters: crucial roles of pharmacological interest revealed by gene deletion. Trends Pharmacol Sci. 2005;26(6):283–6.

Attwell D, Barbour B, Szatkowski M. Nonvesicular release of neurotransmitter. Neuron. 1993 Sep;11(3):401–7.

Aubrey KR, Mitrovic AD, Vandenberg RJ. Molecular basis for proton regulation of glycine transport by glycine transporter subtype 1b. Mol Pharmacol. 2000;58(1):129–35.

Aubrey KR, Vandenberg RJ, Clements JD. Dynamics of forward and reverse transport by the glial glycine transporter, glyt1b. Biophys J. 2005;89(3):1657–68.

Baker AJ, Zornow MH, Scheller MS, Yaksh TL, Skilling SR, Smullin DH, et al. Changes in extracellular concentrations of glutamate, aspartate, glycine, dopamine, serotonin, and dopamine metabolites after transient global ischemia in the rabbit brain. J Neurochem. 1991;57(4):1370–9.

Bakkar W, Ma CL, Pabba M, Khacho P, Zhang YL, Muller E, et al. Chronically saturating levels of endogenous glycine disrupt glutamatergic neurotransmission and enhance synaptogenesis in the CA1 region of mouse hippocampus. Synapse. 2011;65(11):1181–95.

Ballard TM, Pauly-Evers M, Higgins GA, Ouagazzal AM, Mutel V, Borroni E, et al. Severe impairment of NMDA receptor function in mice carrying targeted point mutations in the glycine binding site results in drug-resistant nonhabituating hyperactivity. J Neurosci. 2002;22(15):6713–23.

Balu DT, Coyle JT. The NMDA receptor 'glycine modulatory site' in schizophrenia: D-serine, glycine, and beyond. Curr Opin Pharmacol. 2015;20:109–15.

Barthel F, Urban A, Schlösser L, Eulenburg V, Werdehausen R, Brandenburger T, et al. Long-term application of glycine transporter inhibitors acts antineuropathic and modulates spinal N-methyl-D-aspartate receptor subunit NR-1 expression in rats. Anesthesiology. 2014;121(1):160–9.

Bejczy A, Nations KR, Szegedi A, Schoemaker J, Ruwe F, Söderpalm B. Efficacy and safety of the glycine transporter-1 inhibitor org 25935 for the prevention of relapse in alcohol-dependent patients: a randomized, double-blind, placebo-controlled trial. Alcohol Clin Exp Res. 2014;38(9):2427–35.

Berger AJ, Isaacson JS. Modulation of motoneuron N-methyl-D-aspartate receptors by the inhibitory neurotransmitter glycine. J Physiol Paris. 1999;93(1-2):23–7.

Berger AJ, Dieudonné S, Ascher P. Glycine uptake governs glycine site occupancy at NMDA receptors of excitatory synapses. J Neurophysiol. 1998;80(6):3336–40.

Bergeron R, Meyer TM, Coyle JT, Greene RW. Modulation of N-methyl-D-aspartate receptor function by glycine transport. Proc Natl Acad Sci U S A. 1998;95(26):15730–4.

Borowsky B, Mezey E, Hoffman BJ. Two glycine transporter variants with distinct localization in the CNS and peripheral tissues are encoded by a common gene. Neuron. 1993;10(5):851–63.

Boulay D, Pichat P, Dargazanli G, Estenne-Bouhtou G, Terranova JP, Rogacki N, et al. Characterization of SSR103800, a selective inhibitor of the glycine transporter-1 in models predictive of therapeutic activity in schizophrenia. Pharmacol Biochem Behav. 2008;91(1):47–58.

Bradaïa A, Schlichter R, Trouslard J. Role of glial and neuronal glycine transporters in the control of glycinergic and glutamatergic synaptic transmission in lamina X of the rat spinal cord. J Physiol. 2004;559(Pt 1):169–86.

Brookes N. Functional integration of the transport of ammonium, glutamate and glutamine in astrocytes. Neurochem Int. 2000;37(2-3):121–9.
Carlezon WA Jr, Thomas MJ. Biological substrates of reward and aversion: a nucleus accumbens activity hypothesis. Neuropharmacology. 2009;56(Suppl 1):122–32.
Castner SA, Murthy NV, Ridler K, Herdon H, Roberts BM, Weinzimmer DP, et al. Relationship between glycine transporter 1 inhibition as measured with positron emission tomography and changes in cognitive performances in nonhuman primates. Neuropsychopharmacology. 2014;39(12):2742–9.
Chaki S, Shimazaki T, Karasawa J, Aoki T, Kaku A, Iijima M, et al. Efficacy of a glycine transporter 1 inhibitor TASP0315003 in animal models of cognitive dysfunction and negative symptoms of schizophrenia. Psychopharmacology. 2015;232(15):2849–61.
Chaudhry FA, Lehre KP, van Lookeren Campagne M, Ottersen OP, Danbolt NC, Storm-Mathisen J. Glutamate transporters in glial plasma membranes: highly differentiated localizations revealed by quantitative ultrastructural immunocytochemistry. Neuron. 1995;15(3):711–20.
Chen L, Muhlhauser M, Yang CR. Glycine tranporter-1 blockade potentiates NMDA-mediated responses in rat prefrontal cortical neurons in vitro and in vivo. J Neurophysiol. 2003;89(2):691–703.
Chen R, Okabe A, Sun H, Sharopov S, Hanganu-Opatz IL, Kolbaev SN, et al. Activation of glycine receptors modulates spontaneous epileptiform activity in the immature rat hippocampus. J Physiol. 2014;592(10):2153–68.
Cioffi CL, Guzzo PR. Inhibitors of glycine transporter-1: potential therapeutics for the treatment of CNS disorders. Curr Top Med Chem. 2016;16(29):3404–37.
Collingridge GL, Volianskis A, Bannister N, France G, Hanna L, Mercier M, et al. The NMDA receptor as a target for cognitive enhancement. Neuropharmacology. 2013;64:13–26.
Costigan M, Scholz J, Woolf CJ. Neuropathic pain: a maladaptive response of the nervous system to damage. Annu Rev Neurosci. 2009;32:1–32.
Coyle JT. Substance use disorders and Schizophrenia: a question of shared glutamatergic mechanisms. Neurotox Res. 2006;10(3-4):221–33.
Coyle JT. NMDA receptor and schizophrenia: a brief history. Schizophr Bull. 2012;38(5):920–6.
Coyle JT, Tsai G. The NMDA receptor glycine modulatory site: a therapeutic target for improving cognition and reducing negative symptoms in schizophrenia. Psychopharmacology. 2004;174(1):32–8.
Cubelos B, Giménez C, Zafra F. Localization of the GLYT1 glycine transporter at glutamatergic synapses in the rat brain. Cereb Cortex. 2005a;15(4):448–59.
Cubelos B, Giménez C, Zafra F. The glycine transporter GLYT1 interacts with Sec3, a component of the exocyst complex. Neuropharmacology. 2005b;49(6):935–44.
Cubelos B, González-González IM, Giménez C, Zafra F. Amino acid transporter SNAT5 localizes to glial cells in the rat brain. Glia. 2005c;49(2):230–44.
Cubelos B, Leite C, Giménez C, Zafra F. Localization of the glycine transporter GLYT1 in glutamatergic synaptic vesicles. Neurochem Int. 2014;73:204–10.
Danglot L, Rostaing P, Triller A, Bessis A. Morphologically identified glycinergic synapses in the hippocampus. Mol Cell Neurosci. 2004;27(4):394–403.
Depoortère R, Dargazanli G, Estenne-Bouhtou G, Coste A, Lanneau C, Desvignes C, et al. Neurochemical, electrophysiological and pharmacological profiles of the selective inhibitor of the glycine transporter-1 SSR504734, a potential new type of antipsychotic. Neuropsychopharmacology. 2005;30(11):1963–85.
Doherty FC, Sladek CD. NMDA receptor subunit expression in the supraoptic nucleus of adult rats: dominance of NR2B and NR2D. Brain Res. 2011;1388:89–99.
Dubroqua S, Serrano L, Boison D, Feldon J, Gargiulo PA, Yee BK. Intact working memory in the absence of forebrain neuronal glycine transporter 1. Behav Brain Res. 2012;230(1):208–14.
Dupuis JP, Ladépêche L, Seth H, Bard L, Varela J, Mikasova L, et al. Surface dynamics of GluN2B-NMDA receptors controls plasticity of maturing glutamate synapses. EMBO J. 2014;33(8):842–61.

Eulenburg V, Retiounskaia M, Papadopoulos T, Gomeza J, Betz H. Glial glycine transporter 1 function is essential for early postnatal survival but dispensable in adult mice. Glia. 2010;58(9):1066–73.

Fernández-Sánchez E, Martínez-Villarreal J, Giménez C, Zafra F. Constitutive and regulated endocytosis of the glycine transporter GLYT1b is controlled by ubiquitination. J Biol Chem. 2009;284(29):19482–92.

Fossat P, Turpin FR, Sacchi S, Dulong J, Shi T, Rivet JM, et al. Glial D-serine gates NMDA receptors at excitatory synapses in prefrontal cortex. Cereb Cortex. 2012;22(3):595–606.

Friauf E, Aragón C, Löhrke S, Westenfelder B, Zafra F. Developmental expression of the glycine transporter GLYT2 in the auditory system of rats suggests involvement in synapse maturation. J Comp Neurol. 1999;412(1):17–37.

Fromer M, Pocklington AJ, Kavanagh DH, Williams HJ, Dwyer S, Gormley P, et al. De novo mutations in schizophrenia implicate synaptic networks. Nature. 2014;506(7487):179–84.

Fukasawa Y, Segawa H, Kim JY, Chairoungdua A, Kim DK, Matsuo H, et al. Identification and characterization of a Na(+)-independent neutral amino acid transporter that associates with the 4F2 heavy chain and exhibits substrate selectivity for small neutral D- and L-amino acids. J Biol Chem. 2000;275(13):9690–8.

Gabernet L, Pauly-Evers M, Schwerdel C, Lentz M, Bluethmann H, Vogt K, et al. Enhancement of the NMDA receptor function by reduction of glycine transporter-1 expression. Neurosci Lett. 2005;373(1):79–84.

Galli A, Mori F, Bargellini M, Coppini L. Sodium-dependent release of exogenous glycine from preloaded rat hippocampal synaptosomes. J Neural Transm Gen Sect. 1993;93(3):167–79.

Geerlings A, López-Corcuera B, Aragón C. Characterization of the interactions between the glycine transporters GLYT1 and GLYT2 and the SNARE protein syntaxin 1A. FEBS Lett. 2000;470(1):51–4.

Gomeza J, Zafra F, Olivares L, Giménez C, Aragón C. Regulation by phorbol esters of the glycine transporter (GLYT1) in glioblastoma cells. Biochim Biophys Acta. 1995;1233(1):41–6.

Gomeza J, Hülsmann S, Ohno K, Eulenburg V, Szöke K, Richter D, Betz H. Inactivation of the glycine transporter 1 gene discloses vital role of glial glycine uptake in glycinergic inhibition. Neuron. 2003a;40(4):785–96.

Gomeza J, Ohno K, Hülsmann S, Armsen W, Eulenburg V, Richter DW, et al. Deletion of the mouse glycine transporter 2 results in a hyperekplexia phenotype and postnatal lethality. Neuron. 2003b;40(4):797–806.

Hamdani el H, Gudbrandsen M, Bjørkmo M, Chaudhry FA. The system N transporter SN2 doubles as a transmitter precursor furnisher and a potential regulator of NMDA receptors. Glia. 2012;60(11):1671–83.

Hanuska A, Szénási G, Albert M, Koles L, Varga A, Szabo A, et al. Some operational characteristics of glycine release in rat retina: the role of reverse mode operation of glycine transporter type-1 (GlyT-1) in ischemic conditions. Neurochem Res. 2016;41(1-2):73–85.

Harada K, Nakato K, Yarimizu J, Yamazaki M, Morita M, Takahashi S, et al. A novel glycine transporter-1 (GlyT1) inhibitor, ASP2535 (4-[3-isopropyl-5-(6-phenyl-3-pyridyl)-4H-1,2,4-triazol-4-yl]-2,1,3-benzoxadiazol e), improves cognition in animal models of cognitive impairment in schizophrenia and Alzheimer's disease. Eur J Pharmacol. 2012;685(1-3):59–69.

Harsing LG Jr, Gacsalyi I, Szabo G, Schmidt E, Sziray N, Sebban C, et al. The glycine transporter-1 inhibitors NFPS and Org 24461: a pharmacological study. Pharmacol Biochem Behav. 2003;74(4):811–25.

Harvey RJ, Yee BK. Glycine transporters as novel therapeutic targets in schizophrenia, alcohol dependence and pain. Nat Rev Drug Discov. 2013;12(11):866–85.

Helboe L, Egebjerg J, Møller M, Thomsen C. Distribution and pharmacology of alanine-serine-cysteine transporter 1 (asc-1) in rodent brain. Eur J Neurosci. 2003;18(8):2227–38.

Henneberger C, Papouin T, Oliet SH, Rusakov DA. Long-term potentiation depends on release of D-serine from astrocytes. Nature. 2010;463(7278):232–6.

Huang H, Barakat L, Wang D, Bordey A. Bergmann glial GlyT1 mediates glycine uptake and release in mouse cerebellar slices. J Physiol. 2004;560(Pt 3):721–36.

Huang CC, Wei IH, Huang CL, Chen KT, Tsai MH, Tsai P, et al. Inhibition of glycine transporter-I as a novel mechanism for the treatment of depression. Biol Psychiatry. 2013;74(10):734–41.

James VM, Gill JL, Topf M, Harvey RJ. Molecular mechanisms of glycine transporter GlyT2 mutations in startle disease. Biol Chem. 2012;393(4):283–9.

Jiménez E, Zafra F, Pérez-Sen R, Delicado EG, Miras-Portugal MT, Aragón C, et al. P2Y purinergic regulation of the glycine neurotransmitter transporters. J Biol Chem. 2011;286(12):10712–24.

Jiménez E, Núñez E, Ibáñez I, Zafra F, Aragón C, Giménez C. Glycine transporters GlyT1 and GlyT2 are differentially modulated by glycogen synthase kinase 3β. Neuropharmacology. 2015;89:245–54.

Johnson JW, Ascher P. Glycine potentiates the NMDA response in cultured mouse brain neurons. Nature. 1987;325(6104):529–31.

Ju P, Aubrey KR, Vandenberg RJ. Zn^{2+} inhibits glycine transport by glycine transporter subtype 1b. J Biol Chem. 2004;279(22):22983–91.

Karakas E, Furukawa H. Crystal structure of a heterotetrameric NMDA receptor ion channel. Science. 2014;344(6187):992–7.

Karasawa J, Hashimoto K, Chaki S. D-Serine and a glycine transporter inhibitor improve MK-801-induced cognitive deficits in a novel object recognition test in rats. Behav Brain Res. 2008;186(1):78–83.

Kelamangalath L, Wagner JJ. D-serine treatment reduces cocaine-primed reinstatement in rats following extended access to cocaine self-administration. Neuroscience. 2010;169(3):1127–35.

Kemp JA, Foster AC, Leeson PD, Priestley T, Tridgett R, Iversen LL, et al. 7-Chlorokynurenic acid is a selective antagonist at the glycine modulatory site of the N-methyl-D-aspartate receptor complex. Proc Natl Acad Sci U S A. 1988;85(17):6547–50.

Kinney GG, Sur C, Burno M, Mallorga PJ, Williams JB, Figueroa DJ, et al. The glycine transporter type 1 inhibitor N-[3-(4′-fluorophenyl)-3-(4′-phenylphenoxy)propyl]sarcosine potentiates NMDA receptor-mediated responses in vivo and produces an antipsychotic profile in rodent behavior. J Neurosci. 2003;23(20):7586–91.

Kleckner NW, Dingledine R. Requirement for glycine in activation of NMDA-receptors expressed in Xenopus oocytes. Science. 1988;241(4867):835–7.

Komatsu H, Furuya Y, Sawada K, Asada T. Involvement of the strychnine-sensitive glycine receptor in the anxiolytic effects of GlyT1 inhibitors on maternal separation-induced ultrasonic vocalization in rat pups. Eur J Pharmacol. 2015;746:252–7.

Krishtal OA, Osipchuk YV, Shelest TN, Smirnoff SV. Rapid extracellular pH transients related to synaptic transmission in rat hippocampal slices. Brain Res. 1987;436(2):352–6.

Krystal JH, Sanacora G, Duman RS. Rapid-acting glutamatergic antidepressants: the path to ketamine and beyond. Biol Psychiatry. 2013;73(12):1133–41.

Lane HY, Liu YC, Huang CL, Chang YC, Liau CH, Perng CH, et al. Sarcosine (N-methylglycine) treatment for acute schizophrenia: a randomized, double-blind study. Biol Psychiatry. 2008;63(1):9–12.

Lavezzari G, McCallum J, Dewey CM, Roche KW. Subunit-specific regulation of NMDA receptor endocytosis. J Neurosci. 2004;24(28):6383–91.

Le Bail M, Martineau M, Sacchi S, Yatsenko N, Radzishevsky I, Conrod S, et al. Identity of the NMDA receptor coagonist is synapse specific and developmentally regulated in the hippocampus. Proc Natl Acad Sci U S A. 2015;112(2):E204–13.

Lee CH, Lü W, Michel JC, Goehring A, Du J, Song X, et al. NMDA receptor structures reveal subunit arrangement and pore architecture. Nature. 2014;511(7508):191–7.

Leonetti M, Desvignes C, Bougault I, Souilhac J, Oury-Donat F, Steinberg R. 2-Chloro-N-[(S)-phenyl [(2S)-piperidin-2-yl] methyl]-3-trifluoromethyl benzamide, monohydrochloride, an inhibitor of the glycine transporter type 1, increases evoked-dopamine release in the rat nucleus accumbens in vivo via an enhanced glutamatergic neurotransmission. Neuroscience. 2006;137(2):555–64.

Lester RA, Tong G, Jahr CE. Interactions between the glycine and glutamate binding sites of the NMDA receptor. J Neurosci. 1993;13(3):1088–96.

Li Y, Krupa B, Kang JS, Bolshakov VY, Liu G. Glycine site of NMDA receptor serves as a spatiotemporal detector of synaptic activity patterns. J Neurophysiol. 2009;102(1):578–89.

Li Y, Sacchi S, Pollegioni L, Basu AC, Coyle JT, Bolshakov VY. Identity of endogenous NMDAR glycine site agonist in amygdala is determined by synaptic activity level. Nat Commun. 2013;4:1760.

Lidö HH, Ericson M, Marston H, Söderpalm B. A role for accumbal glycine receptors in modulation of dopamine release by the glycine transporter-1 inhibitor org25935. Front Psychiatry. 2011;2:8.

Lim R, Hoang P, Berger AJ. Blockade of glycine transporter-1 (GLYT-1) potentiates NMDA receptor-mediated synaptic transmission in hypoglossal motorneurons. J Neurophysiol. 2004;92(4):2530–7.

Liu QR, López-Corcuera B, Mandiyan S, Nelson H, Nelson N. Cloning and expression of a spinal cord- and brain-specific glycine transporter with novel structural features. J Biol Chem. 1993;268(30):22802–8.

López-Colomé AM, Gadea A. Regulation of glycine transport in cultured Müller cells by Ca^{2+}/calmodulin-dependent enzymes. Ann N Y Acad Sci. 1999;868:685–8.

Luccini E, Romei C, Raiteri L. Glycinergic nerve endings in hippocampus and spinal cord release glycine by different mechanisms in response to identical depolarizing stimuli. J Neurochem. 2008;105(6):2179–89.

Mackenzie B, Erickson JD. Sodium-coupled neutral amino acid (System N/A) transporters of the SLC38 gene family. Pflugers Arch. 2004;447(5):784–95.

Mao SC, Lin HC, Gean PW. Augmentation of fear extinction by infusion of glycine transporter blockers into the amygdala. Mol Pharmacol. 2009;76(2):369–78.

Marcaggi P, Attwell D. Role of glial amino acid transporters in synaptic transmission and brain energetics. Glia. 2004;47(3):217–25.

Martel MA, Ryan TJ, Bell KF, Fowler JH, McMahon A, Al-Mubarak B, et al. The subtype of GluN2 C-terminal domain determines the response to excitotoxic insults. Neuron. 2012;74(3):543–56.

Martina M, Gorfinkel Y, Halman S, Lowe JA, Periyalwar P, Schmidt CJ, et al. Glycine transporter type 1 blockade changes NMDA receptor-mediated responses and LTP in hippocampal CA1 pyramidal cells by altering extracellular glycine levels. J Physiol. 2004;557(Pt 2):489–500.

Martina M, B-Turcotte ME, Halman S, Tsai G, Tiberi M, Coyle JT, Bergeron R. Reduced glycine transporter type 1 expression leads to major changes in glutamatergic neurotransmission of CA1 hippocampal neurones in mice. J Physiol. 2005;563(3):777–93.

Martineau M, Shi T, Puyal J, Knolhoff AM, Dulong J, Gasnier B, et al. Storage and uptake of D-serine into astrocytic synaptic-like vesicles specify gliotransmission. J Neurosci. 2013;33(8):3413–23.

Martineau M, Parpura V, Mothet JP. Cell-type specific mechanisms of D-serine uptake and release in the brain. Front Synaptic Neurosci. 2014;6:12.

Marvin JS, Borghuis BG, Tian L, Cichon J, Harnett MT, Akerboom J, et al. An optimized fluorescent probe for visualizing glutamate neurotransmission. Nat Methods. 2013 Feb;10(2):162–70.

Mayer ML, Westbrook GL, Guthrie PB. Voltage-dependent block by Mg^{2+} of NMDA responses in spinal cord neurones. Nature. 1984;309:261.

Meunier CN, Dallérac G, Le Roux N, Sacchi S, Levasseur G, Amar M, et al. D-serine and glycine differentially control neurotransmission during visual cortex critical period. PLoS One. 2016;11(3):e0151233.

Mezler M, Hornberger W, Mueller R, Schmidt M, Amberg W, Braje W, et al. Inhibitors of GlyT1 affect glycine transport via discrete binding sites. Mol Pharmacol. 2008;74(6):1705–15.

Miguel TT, Nunes-de-Souza RL. Anxiogenic-like effects induced by NMDA receptor activation are prevented by inhibition of neuronal nitric oxide synthase in the periaqueductal gray in mice. Brain Res. 2008;1240:39–46.

Miyamoto Y, Yamada K, Noda Y, Mori H, Mishina M, Nabeshima T. Hyperfunction of dopaminergic and serotonergic neuronal systems in mice lacking the NMDA receptor epsilon1 subunit. J Neurosci. 2001;21(2):750–7.

Moghaddam B, Javitt D. From revolution to evolution: the glutamate hypothesis of schizophrenia and its implication for treatment. Neuropsychopharmacology. 2012;37(1):4–15.

Möhler H, Boison D, Singer P, Feldon J, Pauly-Evers M, Yee BK. Glycine transporter 1 as a potential therapeutic target for schizophrenia-related symptoms: evidence from genetically modified mouse models and pharmacological inhibition. Biochem Pharmacol. 2011;81(9):1065–77.

Mohn AR, Gainetdinov RR, Caron MG, Koller BH. Mice with reduced NMDA receptor expression display behaviors related to schizophrenia. Cell. 1999;98(4):427–36.

Molander A, Lidö HH, Löf E, Ericson M, Söderpalm B. The glycine reuptake inhibitor Org 25935 decreases ethanol intake and preference in male wistar rats. Alcohol Alcohol. 2007;42(1):11–8.

Morita K, Motoyama N, Kitayama T, Morioka N, Kifune K, Dohi T. Spinal antiallodynia action of glycine transporter inhibitors in neuropathic pain models in mice. J Pharmacol Exp Ther. 2008;326(2):633–45.

Mothet JP, Parent AT, Wolosker H, Brady RO Jr, Linden DJ, Ferris CD, et al. D-serine is an endogenous ligand for the glycine site of the N-methyl-D-aspartate receptor. Proc Natl Acad Sci U S A. 2000;97(9):4926–31.

Mothet JP, Le Bail M, Billard JM. Time and space profiling of NMDA receptor co-agonist functions. J Neurochem. 2015;135(2):210–25.

Motoyama N, Morita K, Shiraishi S, Kitayama T, Kanematsu T, Uezono Y, et al. Relief of cancer pain by glycine transporter inhibitors. Anesth Analg. 2014;119(4):988–95.

Muller E, Bakkar W, Martina M, Sokolovski A, Wong AY, Legendre P, et al. Vesicular storage of glycine in glutamatergic terminals in mouse hippocampus. Neuroscience. 2013;242:110–27.

Nakanishi T, Kekuda R, Fei YJ, Hatanaka T, Sugawara M, Martindale RG, et al. Cloning and functional characterization of a new subtype of the amino acid transport system N. Am J Phys Cell Physiol. 2001;281(6):C1757–68.

Nakazawa K, McHugh TJ, Wilson MA, Tonegawa S. NMDA receptors, place cells and hippocampal spatial memory. Nat Rev Neurosci. 2004;5(5):361–72.

Nishikawa H, Inoue T, Izumi T, Nakagawa S, Koyama T. SSR504734, a glycine transporter-1 inhibitor, attenuates acquisition and expression of contextual conditioned fear in rats. Behav Pharmacol. 2010;21(5-6):576–9.

Nong Y, Huang YQ, Ju W, Kalia LV, Ahmadian G, Wang YT, et al. Glycine binding primes NMDA receptor internalization. Nature. 2003;422(6929):302–7.

Panatier A, Theodosis DT, Mothet JP, Touquet B, Pollegioni L, Poulain DA, et al. Glia-derived D-serine controls NMDA receptor activity and synaptic memory. Cell. 2006;125(4):775–84.

Paoletti P, Bellone C, Zhou Q. NMDA receptor subunit diversity: impact on receptor properties, synaptic plasticity and disease. Nat Rev Neurosci. 2013;14(6):383–400.

Paolone G, Botreau F, Stewart J. The facilitative effects of D-cycloserine on extinction of a cocaine-induced conditioned place preference can be long lasting and resistant to reinstatement. Psychopharmacology. 2009;202(1-3):403–9.

Papouin T, Ladépêche L, Ruel J, Sacchi S, Labasque M, Hanini M, et al. Synaptic and extrasynaptic NMDA receptors are gated by different endogenous coagonists. Cell. 2012;150(3):633–46.

Pearlman RJ, Aubrey KR, Vandenberg RJ. Arachidonic acid and anandamide have opposite modulatory actions at the glycine transporter, GLYT1a. J Neurochem. 2003;84(3):592–601.

Pérez-Siles G, Núñez E, Morreale A, Jiménez E, Leo-Macías A, Pita G, et al. An aspartate residue in the external vestibule of GLYT2 (glycine transporter 2) controls cation access and transport coupling. Biochem J. 2012;442(2):323–34.

Peykov S, Berkel S, Degenhardt F, Rietschel M, Nöthen MM, Rappold GA. Rare SHANK2 variants in schizophrenia. Mol Psychiatry. 2015;20(12):1487–8.

Pinto MC, Simão F, da Costa FL, Rosa DV, de Paiva MJ, Resende RR, et al. Sarcosine preconditioning induces ischemic tolerance against global cerebral ischemia. Neuroscience. 2014;271:160–9.

Pinto MC, Lima IV, da Costa FL, Rosa DV, Mendes-Goulart VA, Resende RR, et al. Glycine transporters type 1 inhibitor promotes brain preconditioning against NMDA-induced excitotoxicity. Neuropharmacology. 2015;89:274–81.

Puhl MD, Berg AR, Bechtholt AJ, Coyle JT. Availability of N-methyl-d-aspartate receptor coagonists affects cocaine-induced conditioned place preference and locomotor sensitization: implications for comorbid schizophrenia and substance abuse. J Pharmacol Exp Ther. 2015;353(3):465–70.

Qian J, Noebels JL. Visualization of transmitter release with zinc fluorescence detection at the mouse hippocampal mossy fibre synapse. J Physiol. 2005;566(Pt 3):747–58.

Ramsey AJ. NR1 knockdown mice as a representative model of the glutamate hypothesis of schizophrenia. Prog Brain Res. 2009;179:51–8. doi:10.1016/S0079-6123(09)17906-2.

Rodríguez A, Ortega A, Berumen LC, García-Alcocer MG, Giménez C, Zafra F. Expression of the System N transporter (SNAT5/SN2) during development indicates its plausible role in glutamatergic neurotransmission. Neurochem Int. 2014;73:166–71.

Rose CR, Ransom BR. Mechanisms of H+ and Na+ changes induced by glutamate, kainate, and D-aspartate in rat hippocampal astrocytes. J Neurosci. 1996;16(17):5393–404.

Rosenberg D, Artoul S, Segal AC, Kolodney G, Radzishevsky I, Dikopoltsev E, et al. Neuronal D-serine and glycine release via the Asc-1 transporter regulates NMDA receptor-dependent synaptic activity. J Neurosci. 2013;33(8):3533–44.

Rousseau F, Aubrey KR, Supplisson S. The glycine transporter GlyT2 controls the dynamics of synaptic vesicle refilling in inhibitory spinal cord neurons. J Neurosci. 2008;28(39):9755–68.

Roux MJ, Supplisson S. Neuronal and glial glycine transporters have different stoichiometries. Neuron. 2000;25(2):373–83.

Rutter AR, Fradley RL, Garrett EM, Chapman KL, Lawrence JM, Rosahl TW, et al. Evidence from gene knockout studies implicates Asc-1 as the primary transporter mediating d-serine reuptake in the mouse CNS. Eur J Neurosci. 2007;25(6):1757–66.

Safory H, Neame S, Shulman Y, Zubedat S, Radzishevsky I, Rosenberg D, et al. The alanine-serine-cysteine-1 (Asc-1) transporter controls glycine levels in the brain and is required for glycinergic inhibitory transmission. EMBO Rep. 2015;16(5):590–8.

Saransaari P, Oja SS. Characteristics of hippocampal glycine release in cell-damaging conditions in the adult and developing mouse. Neurochem Res. 2001;26(7):845–52.

Sato K, Adams R, Betz H, Schloss P. Modulation of a recombinant glycine transporter (GLYT1b) by activation of protein kinase C. J Neurochem. 1995;65(5):1967–73.

Schell MJ, Molliver ME, Snyder SH. D-serine, an endogenous synaptic modulator: localization to astrocytes and glutamate-stimulated release. Proc Natl Acad Sci U S A. 1995;92(9):3948–52.

Schizophrenia Working Group of the Psychiatric Genomics Consortium. Biological insights from 108 schizophrenia-associated genetic loci. Nature. 2014;511(7510):421–7.

Schmitz Y, Castagna C, Mrejeru A, Lizardi-Ortiz JE, Klein Z, Lindsley CW, et al. Glycine transporter-1 inhibition promotes striatal axon sprouting via NMDA receptors in dopamine neurons. J Neurosci. 2013;33(42):16778–89.

Schoemaker JH, Jansen WT, Schipper J, Szegedi A. The selective glycine uptake inhibitor org 25935 as an adjunctive treatment to atypical antipsychotics in predominant persistent negative symptoms of schizophrenia: results from the GIANT trial. J Clin Psychopharmacol. 2014;34(2):190–8.

Shen HY, van Vliet EA, Bright KA, Hanthorn M, Lytle NK, Gorter J, et al. Glycine transporter 1 is a target for the treatment of epilepsy. Neuropharmacology. 2015;99:554–65.

Sheng M, Cummings J, Roldan LA, Jan YN, Jan LY. Changing subunit composition of heteromeric NMDA receptors during development of rat cortex. Nature. 1994;368(6467):144–7.

Shibasaki K, Hosoi N, Kaneko R, Tominaga M, Yamada K. Glycine release from astrocytes via functional reversal of GlyT1. J Neurochem. 2016; doi:10.1111/jnc.13741. [Epub ahead of print]

Singer P, Feldon J, Yee BK. The glycine transporter 1 inhibitor SSR504734 enhances working memory performance in a continuous delayed alternation task in C57BL/6 mice. Psychopharmacology. 2009a;202(1-3):371–84.

Singer P, Yee BK, Feldon J, Iwasato T, Itohara S, Grampp T, et al. Altered mnemonic functions and resistance to N-METHYL-d-aspartate receptor antagonism by forebrain conditional knockout of glycine transporter 1. Neuroscience. 2009b;161(2):635–54.

Singer P, Dubroqua S, Yee BK. Inhibition of glycine transporter 1: the yellow brick road to new schizophrenia therapy? Curr Pharm Des. 2015;21(26):3771–87.

Singh SP, Singh V. Meta-analysis of the efficacy of adjunctive NMDA receptor modulators in chronic schizophrenia. CNS Drugs. 2011;25(10):859–85.

Smith KE, Borden LA, Hartig PR, Branchek T, Weinshank RL. Cloning and expression of a glycine transporter reveal colocalization with NMDA receptors. Neuron. 1992;8(5):927–35.

Socała K, Nieoczym D, Rundfeldt C, Wlaź P. Effects of sarcosine, a glycine transporter type 1 inhibitor, in two mouse seizure models. Pharmacol Rep. 2010;62(2):392–7.

Song W, Chattipakorn SC, McMahon LL. Glycine-gated chloride channels depress synaptic transmission in rat hippocampus. J Neurophysiol. 2006;95(4):2366–79.

Spike RC, Watt C, Zafra F, Todd AJ. An ultrastructural study of the glycine transporter GLYT2 and its association with glycine in the superficial laminae of the rat spinal dorsal horn. Neuroscience. 1997;77(2):543–51.

Stevens ER, Gustafson EC, Miller RF. Glycine transport accounts for the differential role of glycine vs. D-serine at NMDA receptor coagonist sites in the salamander retina. Eur J Neurosci. 2010;31(5):808–16.

Straub H, Köhling R, Speckmann EJ. Strychnine-induced epileptiform activity in hippocampal and neocortical slice preparations: suppression by the organic calcium antagonists verapamil and flunarizine. Brain Res. 1997;773(1-2):173–80.

Strzelecki D, Podgórski M, Kałużyńska O, Gawlik-Kotelnicka O, Stefańczyk L, Kotlicka-Antczak M, et al. Supplementation of antipsychotic treatment with sarcosine – GlyT1 inhibitor – causes changes of glutamatergic (1)NMR spectroscopy parameters in the left hippocampus in patients with stable schizophrenia. Neurosci Lett. 2015;606:7–12.

Tanabe M, Takasu K, Yamaguchi S, Kodama D, Ono H. Glycine transporter inhibitors as a potential therapeutic strategy for chronic pain with memory impairment. Anesthesiology. 2008;108(5):929–37.

Tsai G, Lane HY, Yang P, Chong MY, Lange N. Glycine transporter I inhibitor, N-methylglycine (sarcosine), added to antipsychotics for the treatment of schizophrenia. Biol Psychiatry. 2004a;55(5):452–6.

Tsai G, Ralph-Williams RJ, Martina M, Bergeron R, Berger-Sweeney J, Dunham KS, Jiang Z, Caine SB, Coyle JT. Gene knockout of glycine transporter 1: characterization of the behavioral phenotype. Proc Natl Acad Sci U S A. 2004b;101(22):8485–90.

Turecek R, Trussell LO. Presynaptic glycine receptors enhance transmitter release at a mammalian central synapse. Nature. 2001;411(6837):587–90.

Umbricht D, Alberati D, Martin-Facklam M, Borroni E, Youssef EA, Ostland M, et al. Effect of bitopertin, a glycine reuptake inhibitor, on negative symptoms of schizophrenia: a randomized, double-blind, proof-of-concept study. JAMA Psychiatry. 2014;71(6):637–46.

Vengeliene V, Leonardi-Essmann F, Sommer WH, Marston HM, Spanagel R. Glycine transporter-1 blockade leads to persistently reduced relapse-like alcohol drinking in rats. Biol Psychiatry. 2010;68(8):704–11.

Verleysdonk S, Martin H, Willker W, Leibfritz D, Hamprecht B. Rapid uptake and degradation of glycine by astroglial cells in culture: synthesis and release of serine and lactate. Glia. 1999;27(3):239–48.

Werdehausen R, Mittnacht S, Bee LA, Minett MS, Armbruster A, Bauer I, et al. The lidocaine metabolite N-ethylglycine has antinociceptive effects in experimental inflammatory and neuropathic pain. Pain. 2015;156(9):1647–59.

Whitehead KJ, Pearce SM, Walker G, Sundaram H, Hill D, Bowery NG. Positive N-methyl-D-aspartate receptor modulation by selective glycine transporter-1 inhibition in the rat dorsal spinal cord in vivo. Neuroscience. 2004;126(2):381–90.
Wiescholleck V, Manahan-Vaughan D. Long-lasting changes in hippocampal synaptic plasticity and cognition in an animal model of NMDA receptor dysfunction in psychosis. Neuropharmacology. 2013;74:48–58.
Wilcox KS, Fitzsimonds RM, Johnson B, Dichter MA. Glycine regulation of synaptic NMDA receptors in hippocampal neurons. J Neurophysiol. 1996;76(5):3415–24.
Williams K, Russell SL, Shen YM, Molinoff PB. Developmental switch in the expression of NMDA receptors occurs in vivo and in vitro. Neuron. 1993;10(2):267–78.
Xie X, Dumas T, Tang L, Brennan T, Reeder T, Thomas W, et al. Lack of the alanine-serine-cysteine transporter 1 causes tremors, seizures, and early postnatal death in mice. Brain Res. 2005;1052(2):212–21.
Yang Y, Ge W, Chen Y, Zhang Z, Shen W, Wu C, et al. Contribution of astrocytes to hippocampal long-term potentiation through release of D-serine. Proc Natl Acad Sci U S A. 2003;100(25):15194–9.
Yee BK, Balic E, Singer P, Schwerdel C, Grampp T, Gabernet L, et al. Disruption of glycine transporter 1 restricted to forebrain neurons is associated with a procognitive and antipsychotic phenotypic profile. J Neurosci. 2006;26(12):3169–81.
Zafra F, Giménez C. Characteristics and adaptive regulation of glycine transport in cultured glial cells. Biochem J. 1989;258(2):403–8.
Zafra F, Alcantara R, Gomeza J, Aragon C, Gimenez C. Arachidonic acid inhibits glycine transport in cultured glial cells. Biochem J. 1990;271(1):237–42.
Zafra F, Aragón C, Olivares L, Danbolt NC, Giménez C, Storm-Mathisen J. Glycine transporters are differentially expressed among CNS cells. J Neurosci. 1995a;15(5 Pt 2):3952–69.
Zafra F, Gomeza J, Olivares L, Aragón C, Giménez C. Regional distribution and developmental variation of the glycine transporters GLYT1 and GLYT2 in the rat CNS. Eur J Neurosci. 1995b;7(6):1342–52.
Zanos P, Moaddel R, Morris PJ, Georgiou P, Fischell J, Elmer GI, et al. NMDAR inhibition-independent antidepressant actions of ketamine metabolites. Nature. 2016;533(7604):481–6.
Zhang LH, Gong N, Fei D, Xu L, Xu TL. Glycine uptake regulates hippocampal network activity via glycine receptor-mediated tonic inhibition. Neuropsychopharmacology. 2008;33(3):701–11.
Zhang XY, Ji F, Wang N, Chen LL, Tian T, Lu W. Glycine induces bidirectional modifications in N-methyl-D-aspartate receptor-mediated synaptic responses in hippocampal CA1 neurons. J Biol Chem. 2014;289(45):31200–11.
Zhao J, Tao H, Xian W, Cai Y, Cheng W, Yin M, et al. A highly selective inhibitor of glycine transporter-1 elevates the threshold for maximal electroshock-induced tonic seizure in mice. Biol Pharm Bull. 2016;39(2):174–80.

Revised Ion/Substrate Coupling Stoichiometry of GABA Transporters

Sepehr Eskandari, Samantha L. Willford, and Cynthia M. Anderson

Abstract The purpose of this review is to highlight recent evidence in support of a 3 Na^+: 1 Cl^-: 1 GABA coupling stoichiometry for plasma membrane GABA transporters (SLC6A1, SLC6A11, SLC6A12, SLC6A13) and how the revised stoichiometry impacts our understanding of the contribution of GABA transporters to GABA homeostasis in synaptic and extrasynaptic regions in the brain under physiological and pathophysiological states. Recently, our laboratory probed the GABA transporter stoichiometry by analyzing the results of six independent measurements, which included the shifts in the thermodynamic transporter reversal potential caused by changes in the extracellular Na^+, Cl^-, and GABA concentrations, as well as the ratio of charge flux to substrate flux for Na^+, Cl^-, and GABA under voltage-clamp conditions. The shifts in the transporter reversal potential for a tenfold change in the external concentration of Na^+, Cl^-, and GABA were 84 ± 4, 30 ± 1, and 29 ± 1 mV, respectively. Charge flux to substrate flux ratios were 0.7 ± 0.1 charges/Na^+, 2.0 ± 0.2 charges/Cl^-, and 2.1 ± 0.1 charges/GABA. We then compared these experimental results with the predictions of 150 different transporter stoichiometry models, which included 1–5 Na^+, 0–5 Cl^-, and 1–5 GABA per transport cycle. Only the 3 Na^+: 1 Cl^-: 1 GABA stoichiometry model correctly predicts the results of all six experimental measurements. Using the revised 3 Na^+: 1 Cl^-: 1 GABA stoichiometry, we propose that the GABA transporters mediate GABA uptake under most physiological conditions. Transporter-mediated GABA release likely takes place under pathophysiological or extreme physiological conditions.

Keywords GABA • Transport • Transporter • Stoichiometry • Forward transport • Reverse transport • Uptake • Release • SLC6 • SLC6A1 • SLC6A11 • SLC6A12 • SLC6A13

S. Eskandari (✉) • S.L. Willford • C.M. Anderson
Biological Sciences Department, California State Polytechnic University, Pomona, CA 91768, USA
e-mail: seskandari@cpp.edu

A. Ortega, A. Schousboe (eds.), *Glial Amino Acid Transporters*, Advances in Neurobiology 16, DOI 10.1007/978-3-319-55769-4_5

1 Introduction

Soon after the discovery of naturally occurring γ-aminobutyric acid (GABA) in the mammalian brain (Awapara et al. 1950; Roberts and Frankel 1950; Udenfriend 1950), several studies showed an important role for GABA as an inhibitory neurotransmitter in invertebrate and vertebrate nervous systems (Florey 1954; Bazemore et al. 1956; Boistel and Fatt 1958; Curtis and Phillis 1958; Elliott and van Gelder 1958; Kuffler and Edwards 1958; Florey and McLennan 1959; Krnjevic and Phillis 1963; Otsuka et al. 1966; Krnjevic and Schwartz (1967); Obata et al. 1967). At around the same time, there was a growing body of evidence for the ability of the brain tissue to accumulate GABA against a concentration gradient (Elliott and Van Gelder 1958; Varon et al. 1965a, b, 1967; Weinstein et al. 1965, 1967; Strasberg and Elliott 1967; Iversen and Neal 1968), and this GABA uptake system came to be accepted as the main mechanism by which GABAergic signal transmission is terminated in the nervous system (Borden 1996; Nelson 1998; Kanner and Zomot 2008; Schousboe et al. 2014; Scimemi 2014a, b). Following these early studies and continuing to the present, intense effort was devoted to understand the mechanism by which neurons and glia transport GABA from the surrounding medium, ultimately leading to our current understanding that four GABA transporter (GAT) isoforms are responsible for modulating GABAergic inhibitory neurotransmission by regulating the synaptic and extrasynaptic concentrations of GABA in the brain (Borden 1996; Nelson 1998; Gadea and López-Colomé 2001; Dalby 2003; Richerson and Wu 2003; Conti et al. 2004, 2011; Scimemi 2014a). In this brief review, we focus on the functional properties of plasma membrane GABA transporters (belonging to the solute carrier 6, SLC6, family) and how these properties may lead to significant consequences for synaptic neurotransmission. We place particular focus on the transporter ion/substrate stoichiometry, as our laboratory has recently resolved a decades-old debate over the transport coupling ratio of ions and GABA (Willford et al. 2015).

2 GABA Transporters and Na^+- and Cl^--Coupled GABA Transport

Plasma membrane GABA transporters are electrogenic, secondary active transport proteins that couple the cotranslocation of Na^+, Cl^-, and GABA across the plasma membrane of neurons and glia (Fig. 1) (Borden 1996; Nelson 1998; Gadea and López-Colomé 2001; Dalby 2003; Richerson and Wu 2003; Conti et al. 2004, 2011; Kristensen et al. 2011). GATs regulate the extracellular concentration of GABA during resting and active conditions in synaptic as well as extrasynaptic regions in the brain, and, thus, they modulate tonic and phasic inhibitory GABAergic signaling in the nervous system (Borden 1996; Nelson 1998; Gadea and López-Colomé 2001; Dalby 2003; Richerson and Wu 2003; Conti et al. 2004, 2011; Kristensen et al. 2011; Scimemi 2014a). Following synaptic release of GABA, GABA transporter activity

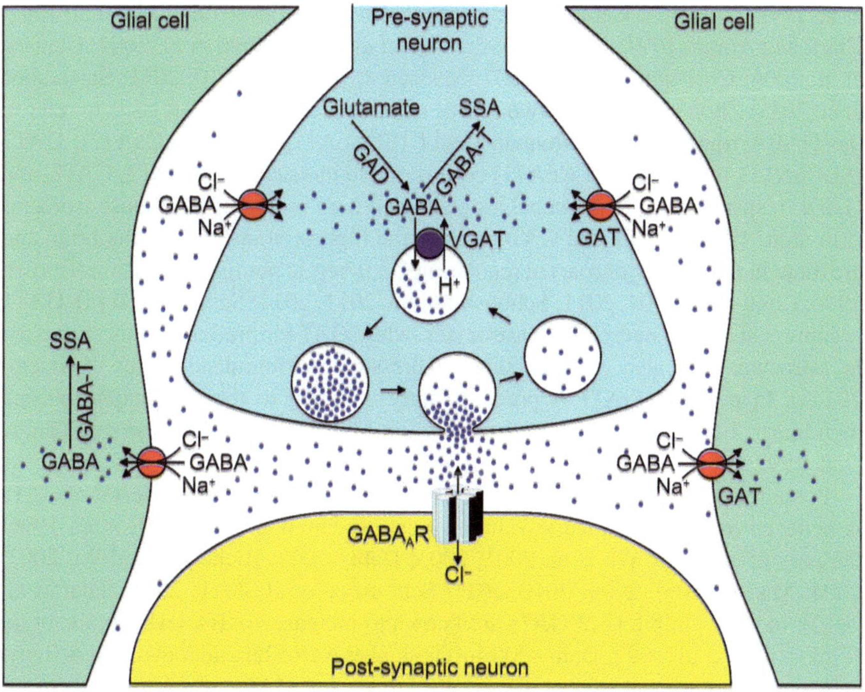

Fig. 1 *GABAergic synapse and GABA transporters*. Most localization studies have demonstrated the presence of GABA transporters in the presynaptic plasma membrane as well as in the plasma membrane of surrounding glial cells (astrocytes). The GABA transporters regulate the extracellular concentration of GABA under resting and active conditions in synaptic and extrasynaptic regions within the brain. We suggest that under most physiological conditions, the GABA transporters work in the forward mode to couple the transport of three Na^+ ions, one Cl^- ion, and one GABA molecule into the cell. Reverse transport, leading to GABA release, occurs under pathophysiological or extreme physiological conditions (see Figs. 5, 6, and 7). *GABA* γ-aminobutyric acid; *$GABA_AR$* $GABA_A$ receptor; *GABA-T* GABA transaminase; *GAD* glutamic acid decarboxylase; *GAT* GABA transporter; *VGAT* vesicular GABA transporter; *SSA* succinic semialdehyde

shapes the postsynaptic response by regulating the concentration and lifetime of GABA in the synaptic cleft (Scimemi 2014a, b). The GABA transporters also serve to minimize spillover of synaptically released GABA to surrounding synapses (Alle and Geiger 2007; Conti et al. 2011).

The GABA transporters belong to the neurotransmitter/Na^+ symporter family (NSS; 2.A.22 according to the transporter classification system; SLC6 according to the Human Genome Organization classification) (Nelson 1998; Busch and Saier 2002; Chen et al. 2004; Beuming et al. 2006; Saier et al. 2006, 2009; Livesay et al. 2007; Bröer and Gether 2012; Scimemi 2014a). The SLC6 family includes four GABA transporter isoforms: GAT1, GAT2, GAT3, and GAT4 corresponding to SLC6A1, SLC6A12, SLC6A13, and SLC6A11, respectively. All four isoforms are present in the mammalian brain and exhibit significant differences in function, pharmacology, and localization (Guastella et al. 1990; Nelson et al. 1990; Borden et al. 1992, 1994; Clark et al. 1992; Liu et al. 1992a, b, 1993; Yamauchi et al. 1992; Clark and Amara 1994; Borden 1996; Chen et al. 2004; Conti et al. 2004; Clausen et al. 2006; Christiansen et al. 2007; Madsen et al. 2009, 2010, 2011; Schousboe et al. 2011; Zhou et al. 2012). Two sets of nomenclature systems are used to refer to the GABA transporters: rat/human GAT1 (SLC6A1), BGT1 (SLC6A12), GAT2 (SLC6A13), and GAT3 (SLC6A11) correspond to mouse GAT1, GAT2, GAT3, and GAT4, respectively. The nomenclature adopted here is that of the mouse isoforms (Liu et al. 1993). GAT1 and GAT4 exhibit the highest abundance in the brain and are thought to play major roles in regulating GABAergic neurotransmission (Borden 1996; Conti et al. 2004, 2011; Schousboe et al. 2011, 2014; Scimemi 2014a). GAT1 is expressed in both neurons and astrocytes, while GAT4 is predominantly expressed in astrocytes and also exhibits some expression in oligodendrocytes (Scimemi 2014a). In neurons, GAT1 is predominantly confined to the presynaptic plasma membrane; however, postsynaptic expression has also been demonstrated (Snow et al. 1992).

Potentiation of GABAergic neurotransmission via inhibition or reversal of GATs is believed to have therapeutic value in treating epileptic seizures (Borden 1996; Gaspary et al. 1998; Wu et al. 2001, 2003; Dalby 2003; Richerson and Wu 2003, 2004; Madsen et al. 2009, 2010, 2011; Schousboe et al. 2011, 2014; Salat et al. 2012). Indeed, inhibitors of GATs are known to increase GABA levels in the brain (Fink-Jensen et al. 1992; Dalby 2000). These agents exhibit anticonvulsant activity, and one (tiagabine), which preferentially targets the GAT1 isoform, has been in clinical use since 1997 (Yunger et al. 1984; Nielsen et al. 1991; Swinyard et al. 1991; Suzdak et al. 1992; Suzdak anb Jansen 1995; Dalby et al. 1997; Morimoto et al. 1997; Dalby 2000, 2003; Madsen et al. 2009, 2010, 2011; Salat et al. 2012). The physiological importance of GATs is further highlighted in GAT1 knockout mice, which exhibits a number of abnormalities including tremor, ataxia, and nervousness (Chiu et al. 2002, 2005; Jensen et al. 2003; Xu et al. 2007, 2008).

The discovery of a GABA uptake system in the brain was followed by a number of studies to elucidate the energetic requirements of GABA transport, ionic dependence of GABA transport, the nature of ions cotransported with GABA, stoichiometry of transport, and electrogenicity of transport (Kanner and Schuldiner 1987, Borden 1996). GABA transport was determined not to rely directly on metabolism but instead was driven by ion gradients (i.e., secondary active transport) (Kanner 1978). It was determined that Na^+ and Cl^- play an important role in GABA transport and, moreover, the involvement of other ions was ruled out (Weinstein et al. 1965; Strasberg and Elliott 1967; Gottesfeld and Elliott 1971; Martin and Smith 1972; Kanner 1978; Kanner and Kifer 1981; Kanner and Schuldiner 1987; Kavanaugh et al. 1992; Mager et al. 1993, 1996). GABA transport was shown to be electrogenic, demonstrating higher uptake rates at hyperpolarized membrane potentials (Kanner 1978; Pastuszko et al. 1982; Kanner et al. 1983; Kavanaugh et al. 1992; Mager et al. 1993).

GABA transport mediated by GATs is coupled to the cotranslocation of Na^+ and Cl^- across the plasma membrane (Radian and Kanner 1983; Keynan and Kanner 1988; Matskevitch et al. 1999; Loo et al. 2000; Willford et al. 2015). GABA transport is absolutely Na^+ dependent (Borden et al. 1992; Mager et al. 1993, 1996; Lu and Hilgemann 1999a; Matskevitch et al. 1999; Loo et al. 2000; Sacher et al. 2002; Whitlow et al. 2003; Karakossian et al. 2005; Gonzales et al. 2007; Omoto et al. 2012); however, dependence on Cl^- is less strict (Borden et al. 1992; Clark et al. 1992; Keynan et al. 1992; Mager et al. 1993; Matskevitch et al. 1999; Loo et al. 2000; Sacher et al. 2002; Whitlow et al. 2003; Karakossian et al. 2005; Christiansen et al. 2007; Gonzales et al. 2007; Omoto et al. 2012). In the absence of Cl^-, transport rates are 30–50% of the rates in the presence of Cl^- (at a membrane potential of −50 mV). The degree of dependence on Cl^- varies with the GAT isoform, and there is evidence that membrane hyperpolarization can partially compensate for the absence of Cl^- (Mager et al. 1993; Loo et al. 2000; Giovannardi et al. 2003). Therefore, the GABA transporters can be referred to as Na^+-dependent and Cl^--facilitated transporters (Karakossian et al. 2005). The Na^+/Cl^-/GABA transport coupling ratio (i.e., stoichiometry) will be discussed in the next Sect. (3).

The GABA transporters are electrogenic, and their activity has been studied in native cells as well as in cells of expression systems by using electrophysiological methods (Kavanaugh et al. 1992; Mager et al. 1993, 1996, 1998; Cammack et al. 1994; Lu and Hilgemann, 1999a, b; Biedermann et al. 2002; Bicho and Grewer 2005; Wu et al. 2007). In voltage-clamp studies, five main types of electrophysiological signals are obtained from GATs: (1) cotransport current, also referred to as transport-associated current (Kavanaugh et al. 1992; Mager et al. 1993, 1998; Loo et al. 2000); (2) uncoupled leak or channel currents (Mager et al. 1996; Bismuth et al. 1997; MacAulay et al. 2002; Grossman and Nelson 2003; Kanner 2003; Karakossian et al. 2005); (3) voltage-induced presteady-state charge movements (Mager et al. 1993, 1996; Lu and Hilgemann 1999b; Li et al. 2000; Loo et al. 2000; Fesce et al. 2002; Sacher et al. 2002; Soragna et al. 2005; Cherubino et al. 2012; Meinild and Forster 2012); (4) transient currents evoked by GABA binding (Cammack et al. 1994; Bicho and Grewer 2005; Karakossian et al. 2005); and (5) charge movements induced by Na^+, Cl^-, and inhibitor binding to and dissociation from the transporter (Mager et al. 1996; Lu and Hilgemann 1999b). Each of the abovementioned electrophysiological signals has contributed immensely to our understanding of transporter function (Mager et al. 1998; Lu and Hilgemann 1999a, b; Bicho and Grewer 2005; Karakossian et al. 2005; Meinild et al. 2009; Meinild and Forster 2012; Willford et al. 2015). In this review, we focus only on steady-state inward and outward cotransport currents that result from Na^+/Cl^-/GABA cotransport into and out of the cell and how they can be used to determine the transporter stoichiometry (see Sect. 4).

Based on the direction of the imposed transmembrane electrochemical driving force, the GABA transporters can work in the forward (GABA uptake) or reverse (GABA release) mode (Martin 1973; Blaustein and King 1976; Kuhar and Zarbin 1978; Moscowitz and Cutler 1980; Yazulla and Kleinschmidt 1983; Schwartz 1987; Bernath and Zigmond 1988; Pin and Bockaert 1989; Gallo et al. 1991; Taylor and Gordon-Weeks 1991; O'Malley et al. 1992; Belhage et al. 1993; Gaspary et al. 1998; Lu and Hilgemann 1999a; Wu et al. 2001, 2003, 2006, 2007; Barakat and Bordey 2002; Wang et al. 2003; Allen et al. 2004a, b; Bertram et al. 2011; Cherubino et al. 2012; Héja et al. 2012; Milanese et al. 2014; Romei et al. 2014; Willford et al. 2015). The forward mode mediates the cotransport of Na^+/Cl^-/GABA into the cell (GABA uptake) and, because of the electrogenic nature of the transport cycle, simultaneously results in the movement of net positive charge into the cell (Pastuszko et al. 1982; Kavanaugh et al. 1992; Mager et al. 1993, 1996; Matskevitch et al. 1999; Loo et al. 2000; Sacher et al. 2002; Whitlow et al. 2003; Bicho and Grewer 2005; Karakossian et al. 2005; Krause and Schwarz 2005; Gonzales et al. 2007; Matthews et al. 2009; Christiansen et al. 2007; Meinild and Forster 2012; Omoto et al. 2012). In electrophysiological measurements, the forward mode is recorded as an inward (i.e., negative) current (Figs. 2a, c, e and 3a), which is directly proportional to influxes of Na^+, Cl^-, and GABA and, therefore, is an excellent assay of GABA transporter function (Loo et al. 2000; Sacher et al. 2002; Whitlow et al. 2003; Karakossian et al. 2005; Gonzales et al. 2007; Matthews et al. 2009; Omoto et al. 2012; Willford et al. 2015). Indeed, the cotransport current has been used extensively to study various features of the GABA transporters including the kinetics of transport, stoichiometry, turnover rate, substrate selectivity, and pharmacology. The magnitude of the inward current is proportional to the total number of functional transporters in the plasma membrane (Gonzales et al. 2007). In the absence of Cl^-, a smaller Na^+/GABA current is observed (Mager et al. 1993; Matskevitch et al. 1999; Loo et al. 2000; Sacher et al. 2002; Whitlow et al. 2003; Karakossian et al. 2005; Christiansen et al. 2007; Gonzales et al. 2007; Omoto et al. 2012). At least in the *Xenopus laevis* oocyte expression system, the cotransport current is tightly coupled to Na^+/Cl^-/GABA cotransport (Loo et al. 2000; Sacher et al. 2002; Karakossian et al. 2005; Gonzales et al. 2007; Matthews et al. 2009; Omoto et al. 2012; Willford et al. 2015).

The reverse mode of transporter function leads to Na^+/Cl^-/GABA cotransport out of the cell, simultaneously resulting in the movement of net positive charge out of the cell (Lu and Hilgemann 1999a; Wang et al. 2003; Bertram et al. 2011; Cherubino et al. 2012; Willford et al. 2015). In electrophysiological measurements, the reverse mode is recorded as an outward (i.e., positive) current (Fig. 2a, c, and e). In response to a series of voltage pulses, steady-state inward and outward currents can be measured in the same cell allowing for the determination of the reversal potential of the GABA-evoked transporter-mediated current (see Fig. 2a, c, and e). The reversal potential is the membrane potential at which the transporter-mediated current is zero, indicating that there is no net inward or outward Na^+/Cl^-/GABA cotransport across the plasma membrane. Importantly, the reversal potential identifies the thermodynamic equilibrium condition for the transporter, where the sum of chemical and electrical potential gradients for all co-substrates is zero (see Eq. 1). The reversal potential can be used to assess the ion/substrate stoichiometry of transport (see Sect. 4).

The Na^+/Cl^-/GABA coupling stoichiometry is a fundamental property of the transporter, and an accurate knowledge of the stoichiometry is essential to understanding the role the GABA transporters play in GABAergic signaling. Beyond a simple elucidation of the number of co-substrates involved in the transport cycle, the stoichiometry sets a thermodynamic limit for GABA accumulation into cells, and, in addition, it sets a limit for the lowest GABA concentration that can be achieved in the

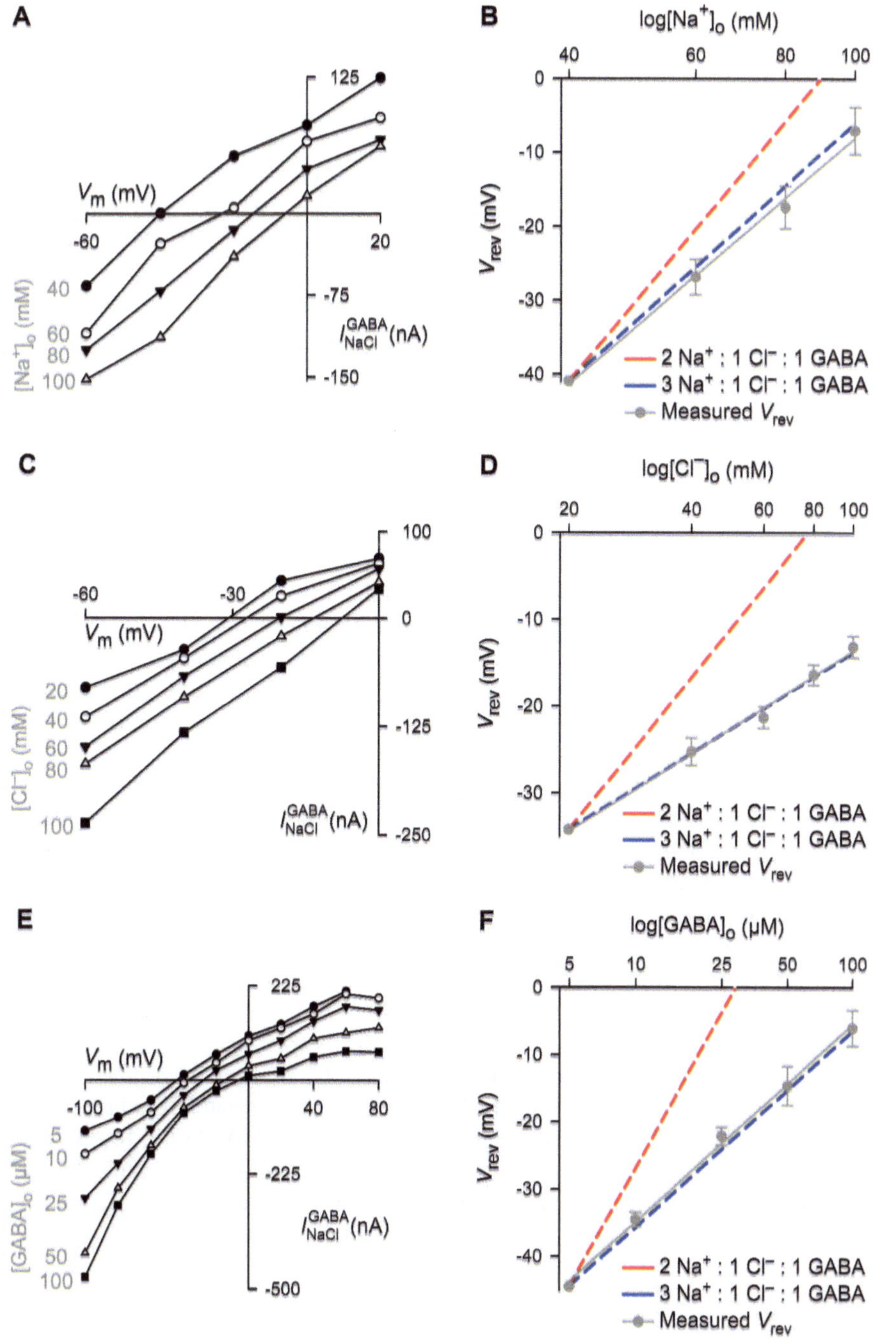

Fig. 2 *GABA transporters thermodynamically couple cotranslocation of* Na^+*,* Cl^-*, and GABA across the plasma membrane*. Reversal potential (V_{rev}) of GAT1-mediated current is a function of the extracellular concentrations of Na^+ (**a**), Cl^- (**c**), and GABA (**e**). There was a predictable shift in V_{rev} for a tenfold change in the extracellular concentrations of Na^+ (84 ± 4 mV; **b**), Cl^- (30 ± 1 mV; **d**), and GABA (29 ± 1 mV; **f**). In all cases, the data are consistent with the predictions of the 3 Na^+: 1 Cl^-: 1 GABA stoichiometry model (*blue line*) (Reproduced by permission from Willford et al. (2015). See Willford et al. (2015) for experimental details)

extracellular fluid (Richerson and Wu 2003). Therefore, the stoichiometry sets a thermodynamic limit for the transporter concentrative capacity (ratio of intracellular to extracellular GABA concentration). The stoichiometry also determines whether the transmembrane concentration and electrochemical gradients favor the forward or reverse mode of the transporter (Schwartz 1987; Lu and Hilgemann 1999a; Richerson and Wu 2003; Allen et al. 2004a; Wu et al. 2007; Bertram et al. 2011; Cherubino et al. 2012).

The rate at which the GABA transporters transport GABA across the plasma membrane (i.e., turnover rate) contributes to shaping postsynaptic events at fast and slow GABAergic synapses. Therefore, an accurate knowledge of the transporter turnover rate is essential to understanding the role of GABA transporters in synaptic physiology. The transporter turnover rate is defined as the number of GABA molecules translocated across the plasma membrane per unit time and is generally expressed in cycles per second. Most estimates of GABA transporter turnover rates have been obtained at nonphysiological temperatures (room temperature) and at subsaturating Na^+ and Cl^- concentrations (Mager et al. 1993; Sacher et al. 2002; Bicho and Grewer 2005; Karakossian et al. 2005). For the forward mode of transport, turnover rate estimates have ranged from 5 to 13 s^{-1} at membrane potentials ranging from −40 to −80 mV. The low turnover rate estimated by these studies raises questions about the effectiveness of the GABA transporters in shaping the GABA concentration profile at fast GABAergic synapses, where the inhibitory postsynaptic currents decay with time constants of ≤10 ms. In a series of experiments, we determined the unitary turnover rate of GAT1 by correlating functional assays of transporter operation with electron microscopic determination of transporter densities in the plasma membrane. When the turnover rate was adjusted for all relevant physiological parameters, such as Na^+ and Cl^- concentrations, temperature, and membrane potential, the GAT1 unitary turnover rate was estimated to be 79–93 s^{-1} at 37 °C and membrane potentials ranging from −50 mV to −90 mV (Gonzales et al. 2007). Indeed, steady-state transport for all GABA transporter isoforms is very temperature dependent, as assessed by the temperature coefficient (Q_{10}) of the transporter-mediated, GABA-evoked steady-state current. Q_{10} is defined as the factor by which the rate of transport increases when the temperature is increased by 10 °C. We have determined the Q_{10} values for GAT1 (2.8; Gonzales et al. 2007), GAT2 (3.4; Lee and Eskandari, unpublished), GAT3 (3.9; Lee and Eskandari, unpublished), and GAT4 (4.3; Karakossian et al. 2005). The high-temperature dependence suggests that all isoforms have higher physiological turnover rates than previously thought (Mager et al. 1993; Kanner and Zomot 2008; Kristensen et al. 2011; Scimemi 2014a).

3 A Brief History of GABA Transporter Stoichiometry

Early studies of GABA transport narrowed the stoichiometry of the transport cycle to an estimate of 2 or 3 Na^+: 1 Cl^-: 1 GABA, which is inclusive of what we now believe the stoichiometry to be (3 Na^+: 1 Cl^-: 1 GABA). Steady-state kinetic studies demonstrated a Hill coefficient of ~1.5 to ~3 for Na^+ activation of transport and a Hill coefficient of ~1 for Cl^- and GABA activation of transport, suggesting the existence of two or more Na^+ binding sites, one Cl^- binding site, and one GABA binding site (e.g., Martin and Smith 1972; Martin 1973; Kuhar and Zarbin 1978; Blaustein and King 1976; Kavanaugh et al. 1992; Rasola et al. 1995; Matskevitch et al. 1999). Based on these results, the stoichiometry postulated was 2 or 3 Na^+: 1 Cl^-: 1 GABA. However, ion activation of transport is not equivalent to ion translocation, and, moreover, the Hill coefficient can only serve as a reliable indicator of transport stoichiometry if a very high degree of cooperativity exists between the binding sites (Weiss 1997; Rudnick 1998). Therefore, further work was needed to demonstrate correlative fluxes of Na^+ and Cl^- with GABA.

Using $^{22}Na^+$, $^{36}Cl^-$, and radiolabeled GABA ([^{3}H]-GABA), cotransport of GABA with Na^+ and Cl^- was demonstrated directly in several studies (Radian and Kanner 1983; Keynan and Kanner 1988; Matskevitch et al. 1999; Loo et al. 2000; Willford et al. 2015). The pioneering studies of Kanner and colleagues with membrane vesicles containing the GABA transporter, or with purified GAT1 reconstituted in proteoliposomes, demonstrated direct flux coupling of Na^+ and Cl^- with GABA (Radian and Kanner 1983; Keynan and Kanner 1988). The flux ratios obtained led the authors to propose a stoichiometry of 2 or 3 Na^+: 1 Cl^-: 1 GABA.

Following the cloning of the GABA transporters (Guastella et al. 1990; Nelson et al. 1990), it became possible to carry out detailed studies of transporter isoforms in cells of expression systems. In particular, it was now possible to perform uptake under voltage clamp for the GABA transporter isoforms expressed in *Xenopus laevis* oocytes in order to correlate transporter-mediated charge fluxes and co-substrate (Na^+, Cl^-, and GABA) fluxes (Mager et al. 1996; Matskevitch et al. 1999; Loo et al. 2000; Sacher et al. 2002; Whitlow et al. 2003; Krause and Schwarz 2005; Karakossian et al. 2005; Gonzales et al. 2007; Matthews et al. 2009; Meinild and Forster 2012; Omoto et al. 2012; Willford et al. 2015).

Matskevitch et al. (1999) expressed the betaine/GABA transporter (BGT-1; SLC6A12) in *Xenopus* oocytes and used tracer fluxes and GABA-evoked current measurements to propose a 3 Na^+: 1 or 2 Cl^-: 1 GABA stoichiometry for BGT1. The uncertainty in the Cl^-/GABA ratio may be due to the fact that tracer and charge flux measurements were obtained from separate groups of oocytes from the same batch and not from individual oocytes under voltage clamp. Indeed, the variability in transporter expression levels even in oocytes from the same batch necessitates that tracer uptake and transporter currents be measured in individual voltage-clamp oocytes.

A common observation of uptake under voltage-clamp studies has been that 2 net positive charges enter the cell for every GABA molecule transported into the cell, and, interestingly, this observation had also been reported in cerebral cortex synaptosomes (Pastuszko et al. 1982). For Na^+ transport mediated by the GABA transporters, the ratio of charge flux to Na^+ flux was less than unity: 0.7 charge/Na^+ (Willford et al. 2015) or 0.9 charge/Na^+ (Loo et al. 2000). For Cl^-, the flux ratio was 2 charges/Cl^- (Loo et al. 2000; Willford et al. 2015).

It is helpful to compare the experimental results noted above to the predictions of the 2 Na^+: 1 Cl^-: 1 GABA and 3 Na^+: 1 Cl^-: 1 GABA stoichiometry models (Table 1). The charge flux to co-substrate flux ratios predicted by the 2 Na^+: 1 Cl^-: 1 GABA stoichiometry model (0.5 charge/Na^+, 1 charge/Cl^-, 1 charge/GABA) are entirely

Table 1 2 Na^+: 1 Cl^-: 1 GABA and 3 Na^+: 1 Cl^-: 1 GABA stoichiometry models

n_{Na}[a]	n_{Cl}[a]	n_{GABA}[a]	V_{rev} Equation[b]	e/Na^+[c]	e/Cl^-[c]	e/GABA[c]	ΔV_{rev} per 10× Δ in $[Na^+]_o$ (mV)[d]	ΔV_{rev} per 10× Δ in $[Cl^-]_o$ (mV)[d]	ΔV_{rev} per 10× Δ in $[GABA]_o$ (mV)[d]
2	1	1	$V_{rev} = \frac{RT}{F}\ln\left(\frac{Na_o^2 Cl_o G_o}{Na_i^2 Cl_i G_i}\right)$	0.50	1.0	1.0	116.7	58.3	58.3
3	1	1	$V_{rev} = \frac{RT}{2F}\ln\left(\frac{Na_o^3 Cl_o G_o}{Na_i^3 Cl_i G_i}\right)$	0.67	2.0	2.0	87.5	29.2	29.2

Reproduced by permission from Willford et al. (2015).

[a]Number of co-substrate ions/molecules transported across the plasma membrane per transport cycle (n_{Na}, n_{Cl}, and n_{GABA})

[b]V_{rev}, reversal potential; *R*, gas constant; *T*, absolute temperature; *F*, Faraday's constant; *G*, GABA; *o* and *i* refer to extracellular and intracellular concentrations, respectively. V_{rev} equations were derived from Eq. 1

[c]Predicted net charge translocated across the plasma membrane per co-substrate per transport cycle (e/Na^+, e/Cl^-, and *e*/GABA)

[d]ΔV_{rev} calculations were performed using unique equations derived for each of the stoichiometry models examined. See Willford et al. (2015) for the equations used for the two stoichiometry models shown in this table. $T = 294$ K

inconsistent with the observed experimental data, whereas the values predicted by the 3 Na^+: 1 Cl^-: 1 GABA stoichiometry model (0.67 charge/Na^+, 2 charges/Cl^-, 2 charges/GABA) are entirely consistent with the observed experimental data. Nevertheless, prior to the study of Willford et al. (2015), investigators assumed a 2 Na^+: 1 Cl^-: 1 GABA stoichiometry model.

To account for the discrepancy between the experimentally measured 2 charges/GABA ratio and the prediction of the 2 Na^+: 1 Cl^-: 1 GABA stoichiometry model (1 charge/GABA), Loo et al. (2000) proposed a Cl^-/Cl^- exchange model, which hypothesizes that the Cl^- ion cotransported into the cell with Na^+ and GABA is stoichiometrically exchanged for an internal Cl^-. This model predicts that there is no net GAT-mediated Cl^- flux across the plasma membrane and that chloride does not provide a thermodynamic contribution to the transport process (Loo et al. 2000). However, Willford et al. (2015) showed that there is a predictable and tight thermodynamic coupling between the transmembrane Cl^- electrochemical gradient and the reversal potential of GAT1 cotransport current (see next section and Fig. 3), thus, making the Cl^-/Cl^- exchange mechanism unlikely.

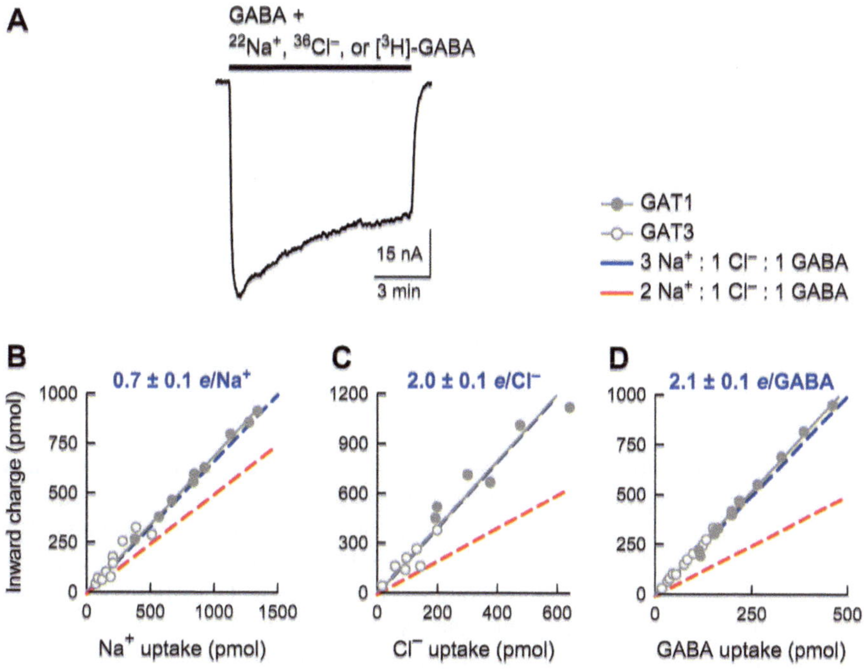

Fig. 3 *Net charge transported across the plasma membrane per Na^+, Cl^-, and GABA*. Uptake under voltage clamp was used to determine the net charge translocated across the plasma membrane per Na^+, Cl^-, and GABA. (**a**) The GABA-evoked current was recorded in the presence of $^{22}Na^+$, $^{36}Cl^-$, or [^{3}H]-GABA in *Xenopus laevis* oocytes expressing GAT1 or GAT3. The net inward charge translocated into the cell was obtained from the time integral of the GABA-evoked inward current and correlated with tracer influx in the same cell. For both GAT1 and GAT3, the charge flux to substrate flux ratios were 0.7 ± 0.1 elementary charge (*e*) per Na^+ (**b**), 2.1 ± 0.1 *e*/Cl^- (**c**), and 2.1 ± 0.1 *e*/GABA (**d**). In all cases, the data are consistent with the predictions of the 3 Na^+: 1 Cl^-: 1 GABA stoichiometry model (*blue line*) (Reproduced by permission from Willford et al. (2015). See Willford et al. (2015) for experimental details)

It has also been suggested that GABA-gated channel (i.e., uncoupled) modes of transporter function can explain charge translocation in excess of that predicted by the 2 Na^+: 1 Cl^-: 1 GABA stoichiometry model. Using mammalian expression systems, three studies proposed the existence of GABA-gated channel modes of conduction in GAT1 (Cammack et al. 1994; Cammack and Schwartz 1996; Risso et al. 1996). However, when expressed in *Xenopus* oocytes, Lu and Hilgemann (1999a) could not find any evidence for GAT1 channel activity, and this observation is consistent with ours (unpublished observations). Studying GAT1 expressed in the *Xenopus* oocyte expression system, Krause and Schwarz (2005) suggested that approximately half of the total GABA-evoked transporter-mediated current is carried by Na^+ ions via a channel mode of GAT1 and that the channel mode can be differentially abolished by very low concentrations of the specific transporter blocker SKF-89976A. However, the results of other studies of GATs expressed in *Xenopus* oocytes suggested that a GABA-evoked channel mode of conductance is unlikely because the net charge transported per GABA molecule is two regardless of the membrane potential, temperature, GABA concentration, or partial transporter inhibition by the competitive inhibitors SKF-89976A and NO-711 (Loo et al. 2000; Gonzales et al. 2007; Matthews et al. 2009). Thus, there is tight coupling of charge flux and GABA flux mediated by the GABA transporters.

We have also considered other hypotheses that could potentially explain the discrepancy between the measured charge flux and the value predicted by the 2 Na^+: 1 Cl^-: 1 GABA stoichiometry. While the involvement of other ions in transport has been ruled out in several studies (Weinstein et al. 1965; Strasberg and Elliott 1967; Gottesfeld and Elliott 1971; Martin and Smith 1972; Kanner 1978; Kanner and Schuldiner 1987; Kavanaugh et al. 1992; Mager et al. 1993, 1996), the observation that protons can interact with the GABA transporters to alter voltage-induced presteady-state charge movements (Grossman and Nelson 2003) prompted us to further examine the possible involvement of protons. GABA is a zwitterion at physiological pH of the extracellular and intracellular fluid compartments, and docking studies based on high-resolution structure templates suggest that GABA binds to the binding pocket as a zwitterion (Skovstrup et al. 2010, 2012). However, it is conceivable that GABA is not transported as a zwitterion. For example, it is possible that GABA may be transported in a form that carries a net negative or positive charge. If this is the case, upon entry into the cytoplasm (pH $\approx$ 7.2), it is expected that GABA would again assume a zwitterion form by accepting or releasing a proton. If this scenario is at play, a change in cytoplasmic pH is expected under high-capacity transport conditions. We tested this hypothesis by measuring the GABA-evoked current concurrently with cytoplasmic pH for several minutes in *Xenopus* oocytes expressing GAT1 at a very high level (cotransport current of ~1 μA). The results showed that the cytoplasmic pH was constant during high-capacity transport conditions, suggesting the absence of GAT-mediated proton transport, and that GABA is most likely transported across the plasma membrane as a zwitterion (unpublished observations).

Other studies used the GAT1 cotransport current reversal potential to determine the transporter stoichiometry. In an attempt to show the validity of the 2 Na^+: 1 Cl^-: 1 GABA stoichiometry model, Lu and Hilgemann (1999a) measured the reversal potential of GAT1 cotransport current in giant excised *Xenopus* oocyte plasma membrane patches. However, the reported reversal potential measurements exhibited considerable variability (even under identical experimental conditions), making it difficult to use that data set to draw a reliable conclusion regarding the stoichiometry (Lu and Hilgemann 1999a).

Wu et al. (2007) used an indirect approach to estimate the GAT1 reversal potential by using a sniffer cell expressing the $GABA_A$ receptor positioned close to another cell expressing GAT1. A depolarizing voltage ramp protocol was applied to the cell expressing GAT1, and GAT1-mediated GABA release was detected by the nearby sniffer cell via GABA activation of the $GABA_A$ receptor. We believe this method is unlikely to provide a reliable estimate of the reversal potential because (1) the reversal potential of the GAT1 transporter current is not measured directly and (2) the depolarizing ramp protocol used leads to a reduction in GAT1 turnover rate during the course of the experiment (Gonzales et al. 2007). Thus, this method does not account for the baseline contribution of GAT1 to establishing the extracellular GABA concentration, which may confound the results obtained.

In summary, prior to the study of Willford et al. (2015), the 2 Na^+: 1 Cl^-: 1 GABA stoichiometry model was broadly adopted even though it lacked solid experimental support and, moreover, was entirely inadequate in accounting for key experimental observations obtained for the GABA transporters. As noted above, experimentally measured ratios of charge flux and co-substrate flux are entirely inconsistent with the predictions of the 2 Na^+: 1 Cl^-: 1 GABA stoichiometry model, but rather are consistent with the predictions of the 3 Na^+: 1 Cl^-: 1 GABA stoichiometry model (Table 1). Moreover, as we have recently shown, the shifts in the thermodynamic reversal potential of the GABA-evoked current are consistent with the 3 Na^+: 1 Cl^-: 1 GABA stoichiometry model (Willford et al. 2015). In the next section, we review recently published evidence in support of the revised 3 Na^+: 1 Cl^-: 1 GABA stoichiometry model for all four GABA transporter isoforms.

4 Recent Evidence in Support of a Revised Ion/Substrate Coupling Stoichiometry for the GABA Transporters

Given that the accepted 2 Na^+: 1 Cl^-: 1 GABA stoichiometry model was inconsistent with several experimental results, our laboratory reexamined the stoichiometry of the GABA transporters by performing thermodynamic reversal potential measurements for GAT1 (Willford et al. 2015), as well as charge flux to substrate flux ratio measurements for GAT1, GAT3, and GAT4 under a variety of experimental conditions (Sacher et al. 2002; Whitlow et al. 2003; Karakossian et al. 2005; Gonzales et al. 2007; Matthews et al. 2009; Omoto et al. 2012; Willford et al. 2015). In all, the data provided six independent measurements that collectively suggest a revised 3 Na^+: 1 Cl^-: 1 GABA. Below, we briefly review the evidence supporting the revised stoichiometry model (for details, see Willford et al. 2015).

The thermodynamic approach involved measuring the shift in the GAT-mediated GABA-evoked steady-state current reversal potential as the extracellular concentration of Na^+, Cl^-, or GABA was changed (Fig. 2), and the results were compared against the predictions of 150 different GABA transporter stoichiometry models, which included 1–5 Na^+, 0–5 Cl^-, and 1–5 GABA per transport cycle (see Table 1 and supplementary Table S1 in Willford et al. 2015). The reversal potential is the membrane potential (V_m) at which transport is at thermodynamic equilibrium (i.e., no net inward or outward transport) and can be described by the following equation:

$$V_{\mathrm{rev}} = \frac{RT}{F\left(n_{\mathrm{Na}} z_{\mathrm{Na}} + n_{\mathrm{Cl}} z_{\mathrm{Cl}}\right)} \times ln \left\{ \left(\frac{\left[\mathrm{Na}^+\right]_o}{\left[\mathrm{Na}^+\right]_i} \right)^{n_{\mathrm{Na}}} \left(\frac{\left[\mathrm{Cl}^-\right]_o}{\left[\mathrm{Cl}^-\right]_i} \right)^{n_{\mathrm{Cl}}} \left(\frac{\left[\mathrm{GABA}\right]_o}{\left[\mathrm{GABA}\right]_i} \right)^{n_{\mathrm{GABA}}} \right\} \quad (1)$$

where V_{rev} is the reversal potential of the GAT-mediated current, n is the number of co-substrate ions/molecules transported per cycle, z is the ion valence, F is Faraday's constant, R is the gas constant, T is the absolute temperature, and the subscripts o and i refer to the extracellular and intracellular concentrations of co-substrates (Na^+, Cl^-, and GABA).

The predictions for two stoichiometry models are shown in Table 1 and demonstrate that the experimental data closely match the predictions of the 3 Na^+: 1 Cl^-: 1 GABA stoichiometry model (Willford et al. 2015).

Uptake under voltage-clamp experiments has been used to determine charge flux and substrate flux ratios for many electrogenic cotransporters expressed in *Xenopus* oocytes (Mager et al. 1996; Eskandari et al. 1997; Forster et al. 1999; Mackenzie et al. 1998; Loo et al. 2000; Sacher et al. 2002; Whitlow et al. 2003; Karakossian et al. 2005; Gonzales et al. 2007; Matthews et al. 2009; Omoto et al. 2012; Willford et al. 2015). They provide a direct measure of the electrogenic behavior of the transporter and, moreover, suggest the coupling ratio between the transported co-substrates. For a direct comparison, as well as to minimize experimental errors, it is essential that charge flux and substrate flux measurements are done in the same cells (Fig. 3).

For the GABA transporters GAT1, GAT3, and GAT4, several studies have shown the number of charges transported per GABA molecule to be 2 charges/GABA, and, importantly, this ratio appears to be fixed under a variety of experimental conditions including at different membrane voltages; Na^+, Cl^-, and GABA concentrations; temperatures; and also in the presence of subsaturating concentrations of transporter inhibitors (Mager et al. 1996; Loo et al. 2000; Sacher et al. 2002; Whitlow et al. 2003; Karakossian et al. 2005; Gonzales et al. 2007; Matthews et al. 2009; Omoto et al. 2012; Willford et al. 2015). This ratio is consistent with the prediction of the 3 Na^+: 1 Cl^-: 1 GABA stoichiometry model and not that of the 2 Na^+: 1 Cl^-: 1 GABA model (Fig. 3d and Table 1).

Applying a similar strategy to GAT1 and GAT3, the ratio of charge flux to Na^+ flux was shown to be 0.7 charge/Na^+, and the ratio of charge flux to Cl^- flux was shown to be 2 charges/Cl^- (Willford et al. 2015), and the results were consistent with the previous measurements (Loo et al. 2000). Taken together, the charge/Na^+ and charge/Cl^- ratios are also consistent with the predictions of the 3 Na^+: 1 Cl^-: 1 GABA stoichiometry model and not those of the 2 Na^+: 1 Cl^-: 1 GABA model (Fig. 3b, c and Table 1). As mentioned above, Matskevitch et al. (1999) also used tracer fluxes and GABA-evoked current measurements to propose a 3 Na^+: 1 GABA transport coupling ratio for BGT1. Matskevitch et al. (1999) narrowed the Cl^-/GABA ratio to 1 or 2.

It is also important to emphasize that the charge/co-substrate (Na^+, Cl^-, and GABA) ratios observed are independent of the membrane potential, co-substrate concentration, and the GABA transporter isoform examined. This provides strong evidence in favor of tight coupling of charge flux and Na^+, Cl^-, and GABA fluxes and suggests that, at least in the *Xenopus* oocyte system used to obtain these measurements, uncoupled ion fluxes are not likely to provide a significant contribution to the measured macroscopic currents mediated by the GABA transporters.

Collectively, the charge/flux ratios for GAT1 (SLC6A1), BGT1 (SLC6A12), GAT3 (SLC6A13), and GAT4 (SLC6A11) suggest that all GABA transporter isoforms operate with a 3 Na^+: 1 Cl^-: 1 GABA stoichiometry (Fig. 4) (Mager et al. 1996; Matskevitch et al. 1999; Loo et al. 2000; Sacher et al. 2002; Whitlow et al. 2003; Karakossian et al. 2005; Gonzales et al. 2007; Matthews et al. 2009; Omoto et al. 2012; Willford et al. 2015). We note that the supporting data for this revised stoichiometry are provided by high-resolution measurements of the GABA transporters expressed in *Xenopus laevis* oocytes. At least in this system, uncoupled GABA-evoked currents are nonexistent or are insignificant in magnitude to contribute to the measured macroscopic signals from a large population of transporters.

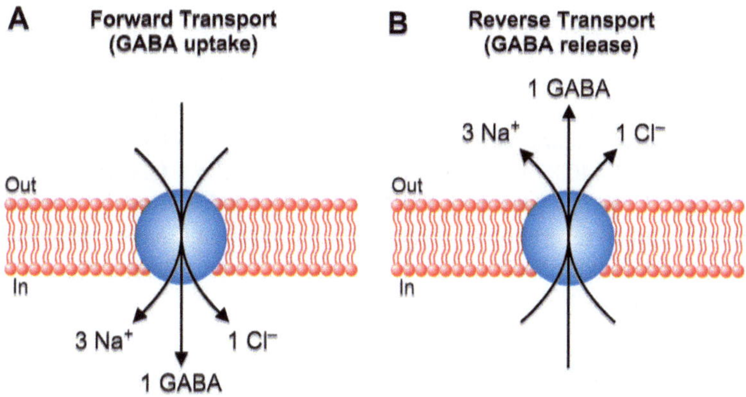

Fig. 4 *Revised ion/substrate coupling stoichiometry of the GABA transporters*. Recent data (see Figs. 2 and 3) provide evidence in support of a 3 Na^+: 1 Cl^-: 1 GABA stoichiometry model for the GABA transporters. This coupling ratio applies to both the forward (**a**) and reverse (**b**) modes of transporter operation. (**a**) The forward mode of transporter operation leads to GABA uptake by neurons and glia. (**b**) The reverse mode of transporter operation leads to GABA release from cells (Reproduced by permission from Willford et al. (2015))

5 Implications of the Revised GABA Transporter Stoichiometry for Synaptic and Extrasynaptic GABA Concentrations

By removing GABA from the extracellular space (forward mode of transport), the GABA transporters are responsible for maintaining sub-micromolar GABA concentrations in the brain extracellular fluid at rest, and, in addition, they shape the post-synaptic response following GABA release from axon terminals (Dingledine and Korn 1985; Solís and Nicoll 1992; Thompson and Gähwiler 1992; Isaacson et al. 1993; Borden 1996; Overstreet et al. 2000; Kinney and Spain 2002; Dalby 2003; Overstreet and Westbrook 2003; Richerson and Wu 2003; Allen et al. 2004a; Keros and Hablitz 2005; Kinney 2005; Bragina et al. 2008; Kirmse et al. 2009; Kristensen et al. 2011; Ransom et al. 2013). In contrast, there is also good evidence that membrane depolarization and conditions that lead to elevated cytoplasmic GABA or reduced transmembrane Na^+ gradients promote transporter-mediated GABA release from neurons and glia (reverse mode of transport) (Moscowitz and Cutler 1980; Tapia and Arias 1982; Yazulla and Kleinschmidt 1983; Schwartz 1987; Bernath and Zigmond 1988; Pin and Bockaert 1989; Gallo et al. 1991; Taylor and Gordon-Weeks 1991; O'Malley et al. 1992; Belhage et al. 1993; Borden 1996; Gaspary et al. 1998; Wu et al. 2001, 2003, 2006, 2007; Barakat and Bordey 2002; Richerson and Wu 2003; Wang et al. 2003; Allen et al. 2004a, b; Angulo et al. 2008; Koch and Magnusson 2009; Héja et al. 2012; Milanese et al. 2014; Romei et al. 2014).

In light of the revised stoichiometry, the equations shown in Table 2 can be used to compare the predicted concentrative capacities of the GABA transporters at equilibrium conditions. Concentrative capacity can be defined as the ratio of extracellular to intracellular GABA concentration and is a measure of the effectiveness of the GABA transporters in accumulating GABA in the cytoplasm of neurons and glia against a concentration gradient. It can be seen that the 3 Na^+: 1 Cl^-: 1 GABA stoichiometry model predicts a much steeper transmembrane GABA concentration gradient. For example, at −60 mV and assuming physiological concentrations and temperature, the concentrative capacity value for the 3 Na^+: 1 Cl^-: 1 GABA stoichiometry model is 1,440,874 and that for the 2 Na^+: 1 Cl^-: 1 GABA stoichiometry model is 15,771. We have developed an online tool for calculating the concentrative capacity (as well as several other parameters) for different GABA transporter stoichiometry models: http://www.cpp.edu/~seskandari/gaba_transporter_stoichiometry_and_reversal_potential.html.

Table 2 Concentrative capacity equations

2 Na^+: 1 Cl^-: 1 GABA	3 Na^+: 1 Cl^-: 1 GABA
$\frac{[G]_i}{[G]_o} = \frac{[Na^+]_o^2[Cl^-]_o}{[Na^+]_i^2[Cl^-]_i} \times e^{-\left(\frac{V_{rev}F(2z_{Na}+z_{Cl})}{RT}\right)}$	$\frac{[G]_i}{[G]_o} = \frac{[Na^+]_o^3[Cl^-]_o}{[Na^+]_i^3[Cl^-]_i} \times e^{-\left(\frac{V_{rev}F(3z_{Na}+z_{Cl})}{RT}\right)}$

V_{rev} reversal potential; R gas constant; T absolute temperature; F Faraday's constant; G GABA; o and i refer to extracellular and intracellular concentrations, respectively

While it is essential to assume thermodynamic equilibrium conditions when deriving the reversal potential equations or when calculating the concentrative capacity, we note that it is unlikely that under most physiological or pathophysiological conditions, GAT-mediated Na^+/Cl^-/GABA cotransport across the plasma membrane is at equilibrium. Given the large fluctuations that take place in key parameters such as the membrane potential and the extracellular GABA concentration during synaptic release, it is more likely that the Na^+/Cl^-/GABA cotransport mediated by the GABA transporters is far from equilibrium under most physiological conditions. Therefore, we believe it is more instructive to examine the driving force that acts on Na^+/Cl^-/GABA cotransport across the plasma membrane in order to determine the direction of transport mediated by the GABA transporters (i.e., uptake or release). The driving force acting on Na^+/Cl^-/GABA cotransport is defined by Eq. 2:

$$V_{\mathrm{DF}} = V_{\mathrm{m}} - V_{rev} \tag{2}$$

where V_{DF} is the driving force, V_{m} is the membrane potential, and V_{rev} is the reversal potential of the GABA-evoked steady-state current mediated by the GABA transporters. It can be readily seen that no driving force will act on transport (i.e., $V_{\mathrm{DF}} = 0$) if the membrane potential is at the reversal potential (i.e., $V_{\mathrm{m}} = V_{\mathrm{rev}}$), indicating no net Na^+/Cl^-/GABA cotransport into or out of the cell. On the other hand, at any membrane potential other than the reversal potential, a driving force acts to drive Na^+/Cl^-/GABA cotransport into or out of the cell. If the value of the driving force is less than zero (i.e., negative), the GABA transporters work in the forward mode to transport Na^+/Cl^-/GABA into the cell (i.e., GABA uptake). If the value of the driving force is greater than zero (i.e., positive), the GABA transporters work in the reverse mode to transport Na^+/Cl^-/GABA out of the cell (i.e., GABA release).

A plot of driving force as a function of membrane potential is useful because it immediately reveals the reversal potential and, in addition, highlights the membrane potentials that favor GABA uptake or release (Fig. 5). Under physiological Na^+, Cl^-, and GABA concentrations, the 2 Na^+: 1 Cl^-: 1 GABA stoichiometry model predicts a V_{rev} of −66 mV, whereas the 3 Na^+: 1 Cl^-: 1 GABA stoichiometry model predicts a V_{rev} of −3 mV (Fig. 5). Therefore, the revised 3 Na^+: 1 Cl^-: 1 GABA stoichiometry implies that the GABA transporters are more resistant to reversing the direction of transport than is currently believed. In the same manner, we have plotted V_{rev} and V_{DF} for parameters that are likely to experience significant fluctuations under physiological or pathophysiological conditions ($[\mathrm{GABA}]_o$, $[\mathrm{GABA}]_i$, $[Na^+]_i$, and $[Cl^-]_i$) (Figs. 6 and 7). Close examination of these plots also reveals that the driving force favors Na^+/Cl^-/GABA cotransport into the cell under most physiological conditions. Release occurs under extreme physiological and/or pathophysiological conditions (Héja et al. 2009; Luccini et al. 2010). Indeed, there is debate over the ease with which the GABA transporters reverse under physiological conditions (Héja et al. 2012; Egawa et al. 2013; Wojtowicz et al. 2013; Savtchenko et al. 2015), suggesting that additional studies are needed to probe this issue further.

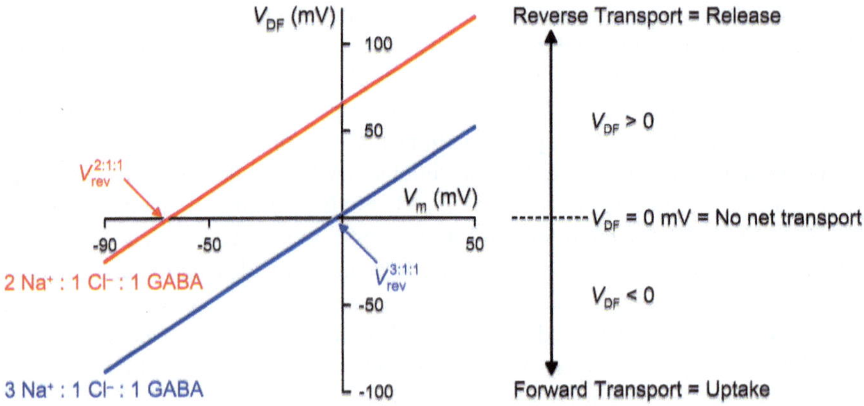

Fig. 5 *Reversal potential and driving force predictions for two stoichiometry models*. Using Eq. 1 and physiological extracellular and intracellular concentrations of Na^+, Cl^-, and GABA, the reversal potential (V_{rev}) was determined for the 2 Na^+: 1 Cl^-: 1 GABA ($V_{rev}^{2:1:1}$ = −66.3 mV) and 3 Na^+: 1 Cl^-: 1 GABA ($V_{rev}^{3:1:1}$ = −2.9 mV) stoichiometry models. The following values were used: T = 310 K, R = 8.314 J.K^{-1}.mol^{-1}, F = 96,485 C.mol^{-1}, n_{Na} = 2 or 3, n_{Cl} = 1, n_{GABA} = 1, z_{Na} = +1, z_{Cl} = −1, $[Na^+]_o$ = 145 mM, $[Na^+]_i$ = 15 mM, $[Cl^-]_o$ = 125 mM, $[Cl^-]_i$ = 7 mM, $[GABA]_o$ = 0.1 μM, and $[GABA]_i$ = 2 mM. Using Eq. 2 and the calculated V_{rev}, the driving force (V_{DF}) acting on Na^+/Cl^-/GABA cotransport across the plasma membrane was calculated at membrane potentials ranging from −90 mV to +50 mV for the 2 Na^+: 1 Cl^-: 1 GABA (*red line*) and 3 Na^+: 1 Cl^-: 1 GABA (*blue line*) stoichiometry models. Note that a negative driving force results in Na^+/Cl^-/GABA cotransport into the cell (i.e., GABA uptake) and a positive driving force causes Na^+/Cl^-/GABA cotransport out of the cell (i.e., GABA release). For the 2 Na^+: 1 Cl^-: 1 GABA stoichiometry model (*red line*), reverse transport (i.e., GABA release) occurs at membrane potentials more positive than $V_{rev}^{3:1:1}$ = −66.3 mV. For the revised 3 Na^+: 1 Cl^-: 1 GABA stoichiometry model (*blue line*), GABA release would be caused by membrane depolarization to values more positive than $V_{rev}^{3:1:1}$ = −2.9 mV. We have developed an online tool for calculating V_{rev} for different stoichiometry models: http://www.cpp.edu/~seskandari/gaba_transporter_stoichiometry_and_reversal_potential.html

6 Implications of the Revised GABA Transporter Stoichiometry for the Mechanism of Na^+/Cl^-/GABA Cotransport

The 3 Na^+: 1 Cl^-: 1 GABA stoichiometry necessitates a reexamination of the kinetic model for GAT-mediated Na^+/Cl^-/GABA cotransport across the plasma membrane (Hilgemann and Lu 1999; Li et al. 2000; Fesce et al. 2002; Sacher

Fig. 6 (continued) cotransport across the plasma membrane. A negative driving force results in GABA uptake, whereas a positive driving force causes GABA release. With the exception of the changing concentration values shown on the x-axes, the following values were used: T = 310 K, R = 8.314 J.K^{-1}.mol^{-1}, F = 96,485 C.mol^{-1}, n_{Na} = 3, n_{Cl} = 1, n_{GABA} = 1, z_{Na} = +1, z_{Cl} = −1, $[Na^+]_o$ = 145 mM, $[Na^+]_i$ = 15 mM, $[Cl^-]_o$ = 125 mM, $[Cl^-]_i$ = 7 mM, $[GABA]_o$ = 0.1 μM, and $[GABA]_i$ = 2 mM. The revised 3 Na^+: 1 Cl^-: 1 GABA stoichiometry model predicts that Na^+/Cl^-/GABA cotransport is in the forward direction (i.e., GABA uptake) under most physiological conditions

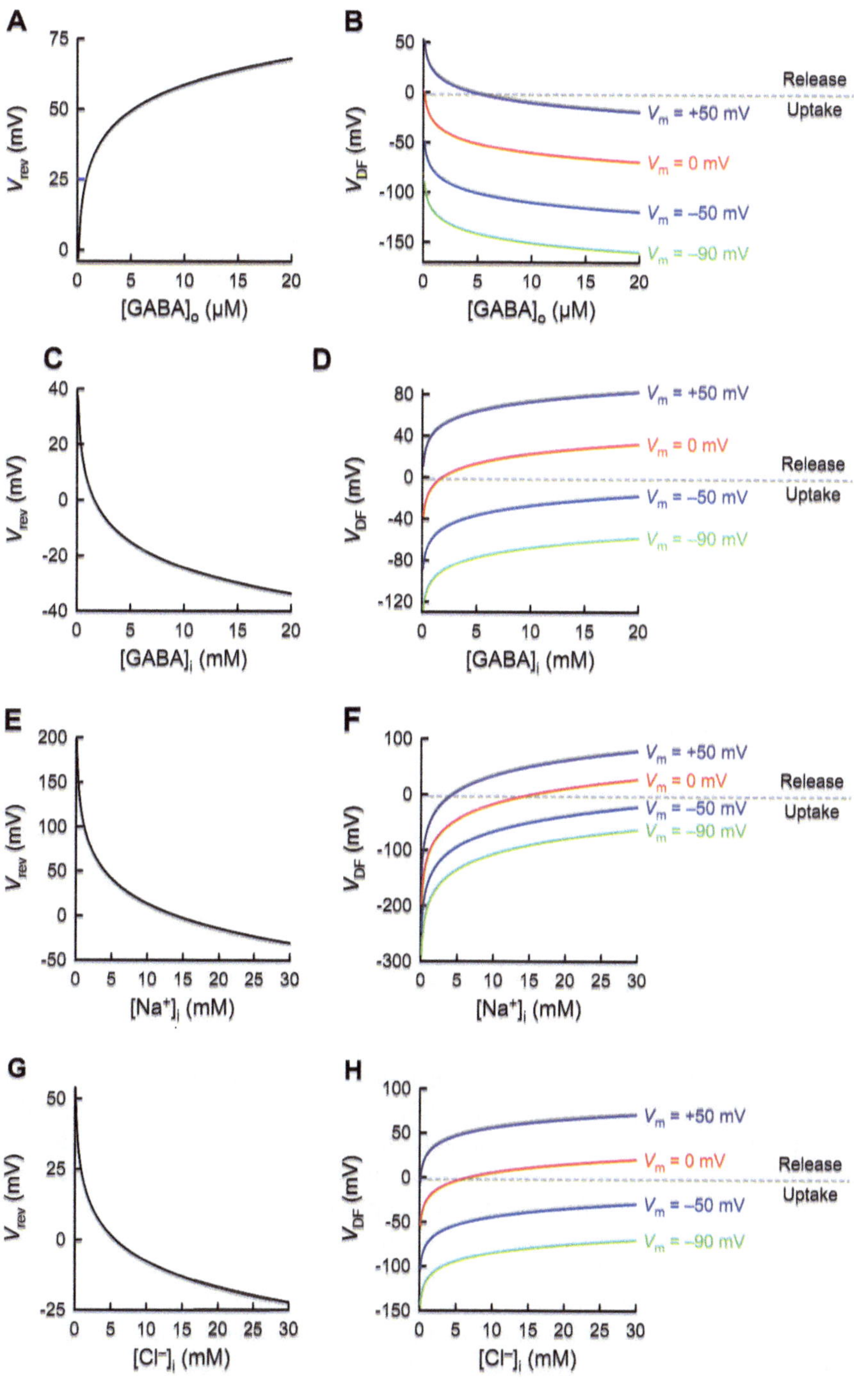

Fig. 6 *Model predictions*. Using the 3 Na^+: 1 Cl^-: 1 GABA stoichiometry and physiological extracellular and intracellular concentrations of Na^+, Cl^-, and GABA, Eq. 1 was used to calculate the transporter reversal potential (V_{rev}) as a function of changing $[GABA]_o$ (**a**), $[GABA]_i$ (**c**), $[Na^+]_i$ (**e**), and $[Cl^-]_i$ (**g**). In panels **b**, **d**, **f**, and **h**, the calculated V_{rev} was used to determine the driving force (V_{DF}), at the indicated membrane potentials, acting on transporter-mediated Na^+/Cl^-/GABA Fig. 6

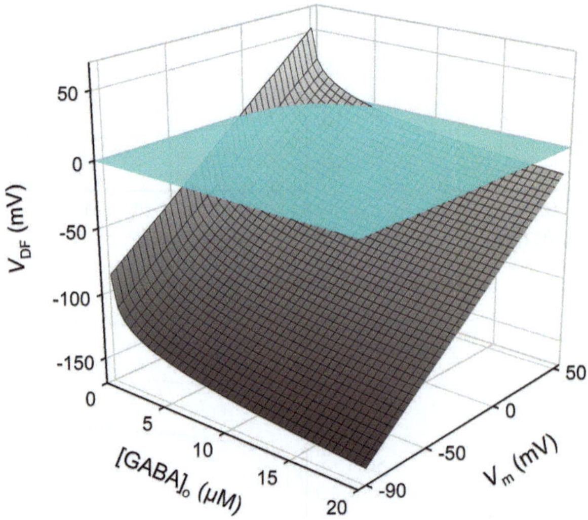

Fig. 7 *Driving force acting on $Na^+/Cl^-/GABA$ cotransport as a function of membrane potential and extracellular GABA concentration.* Similar to the model predictions shown in Fig. 6b, driving force (V_{DF}) values were calculated as a function of the extracellular GABA concentration ($[GABA]_o$) as well as the membrane potential (V_m). $[GABA]_o$ and V_m are the most likely parameters to fluctuate significantly under physiological conditions. A negative driving force results in GABA uptake, whereas a positive driving force causes GABA release. The plane represents the point where the driving force is zero ($V_{DF} = 0$ mV; i.e., no net transport). GABA uptake occurs at driving force values below the zero plane. GABA release occurs at driving force values above the zero plane. Note that even at highly depolarized membrane potentials, small elevations in $[GABA]_o$ lead to conditions that favor forward $Na^+/Cl^-/GABA$ cotransport into the cell (i.e., GABA uptake). $T = 310$ K, $R = 8.314$ $J.K^{-1}.mol^{-1}$, $F = 96{,}485$ $C.mol^{-1}$, $n_{Na} = 3$, $n_{Cl} = 1$, $n_{GABA} = 1$, $z_{Na} = +1$, $z_{Cl} = -1$, $[Na^+]_o = 145$ mM, $[Na^+]_i = 15$ mM, $[Cl^-]_o = 125$ mM, $[Cl^-]_i = 7$ mM, and $[GABA]_i = 2$ mM

et al. 2002; Richerson and Wu 2003; Whitlow et al. 2003; Peres et al. 2004; Bicho and Grewer 2005; Karakossian et al. 2005; Soragna et al. 2005; Kanner and Zomot 2008; Cherubino et al. 2012; Meinild and Forster 2012). To date, all proposed kinetic models of GABA transporters have assumed a 2 Na^+: 1 Cl^-: 1 GABA stoichiometry and, moreover, rarely have these models accounted for the fact that the transporter is not absolutely dependent on Cl^-. While the dependence of the transporter on Na^+ is absolute, it is well documented that Na^+/GABA cotransport across the plasma membrane can proceed in the absence of Cl^-, albeit at significantly lower turnover rates and with a reduced voltage-dependence profile (Borden et al. 1992; Clark et al. 1992; Keynan et al. 1992; Mager et al. 1993; Matskevitch et al. 1999; Loo et al. 2000; Sacher et al. 2002; Whitlow et al. 2003; Karakossian et al. 2005; Christiansen et al. 2007; Gonzales et al. 2007; Omoto et al. 2012). Here, we use the data available for the GABA transporters to propose a kinetic scheme that addresses all of the known functional features of these proteins. Our laboratory is currently validating the predictions of this proposed model against the available experimental data by performing simulations under a

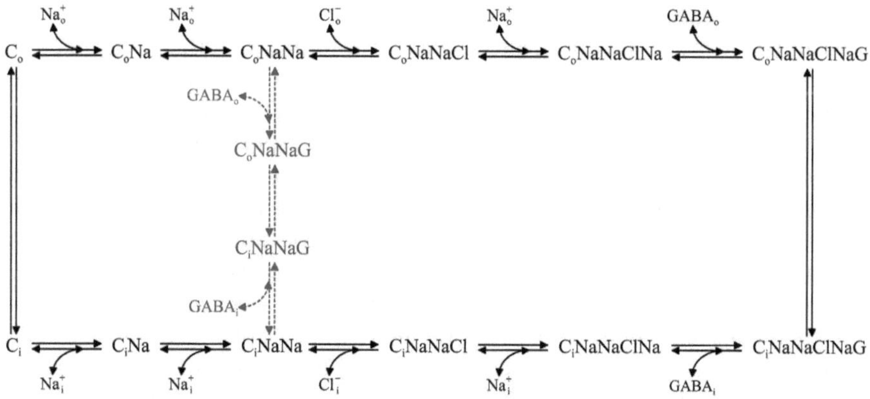

Fig. 8 *Proposed transport kinetic scheme for the GABA transporters.* The kinetic scheme shown incorporates the revised 3 Na^+: 1 Cl^-: 1 GABA stoichiometry, as well as other known functional features of the GABA transporters. *C* denotes carrier; *Na*, Na^+; *Cl*, Cl^-; and G, GABA. The subscripts "o" and "i" refer to the outward- and inward-facing carrier binding sites. Clockwise transitions of the transport cycle result in forward Na^+/Cl^-/GABA cotransport into the cell (i.e., GABA uptake). Counterclockwise transitions result in reverse Na^+/Cl^-/GABA cotransport out of the cell (i.e., GABA release). The scheme allows for tightly coupled 3 Na^+: 1 Cl^-: 1 GABA cotransport across the plasma membrane, leading to the translocation of 0.67 charge/Na^+, 2 charges/Cl^-, and 2 charges/GABA per transport cycle (see Fig. 3 and Willford et al. 2015). The model also accounts for Cl^--independent Na^+/GABA cotransport (gray states and transitions), which leads to the translocation of 1 charge/Na^+ and 2 charges/GABA per transport cycle (Loo et al. 2000)

variety of conditions and in response to perturbations in membrane voltage and ion and substrate concentrations (Eskandari and Anderson, unpublished).

The transport cycle can be represented by a series of partial reactions involving Na^+, Cl^-, and GABA binding and dissociation at the external or internal membrane surfaces, as well as reorientation of the loaded and empty carrier binding sites to face the extracellular or intracellular side of the plasma membrane (Fig. 8). Based on steady-state and presteady-state kinetic studies, it is thought that Na^+ binding to the transporter constitutes the first and rate-limiting step in the transport cycle (Mager et al. 1993, 1996; Cammack et al. 1994; Lu and Hilgemann 1999a, b; Hilgemann and Lu 1999; Li et al. 2000; Loo et al. 2000; Fesce et al. 2002; Karakossian et al. 2005; Soragna et al. 2005). Importantly, results from voltage-induced presteady-state charge movements have suggested that, even in the absence of GABA, two Na^+ ions bind to the GABA transporter from the extracellular space (Mager et al. 1993, 1996, 1998; Loo et al. 2000; Sacher et al. 2002; Karakossian et al. 2005; Meinild and Forster 2012). Moreover, steady-state kinetic data and concentration jump experiments suggest that GABA is likely the last co-substrate to bind from the extracellular fluid before the fully loaded carrier undergoes a conformational change to expose the binding sites to the cytoplasmic domain (Mager et al. 1993, 1996; Cammack et al. 1994; Giovannardi et al. 2003; Bicho and Grewer 2005; Karakossian et al. 2005).

In the presence of all three co-substrates (Na^+, Cl^-, and GABA), clockwise transitions of the proposed kinetic scheme lead to the cotransport of three Na^+ ions, one Cl^- ion, and one GABA molecule into the cell for every transport cycle. If the direction of transport is reversed (counterclockwise transitions), three Na^+ ions, one Cl^- ion, and one GABA molecule are transported out of the cell for every transport cycle. Both clockwise and counterclockwise transitions of the transport cycle highlight the electrogenic nature of the transport process which, consistent with experimental data, leads to the translocation of two net positive charges across the plasma membrane for every transport cycle.

As noted above and reflected in the proposed kinetic scheme, the GABA transporters can mediate Na^+/GABA cotranslocation across the plasma membrane in the absence of Cl^-. In the proposed kinetic scheme, in the absence of Cl^-, the relevant Na^+- and GABA-bound transporter states and transitions are shown in gray (Fig. 8). The strict thermodynamic dependence of Na^+/Cl^-/GABA cotransport on Cl^- suggests that Na^+/GABA cotransport is nonexistent or insignificant under physiological intracellular and extracellular concentrations of Cl^- (see Fig. 2c and d). Nevertheless, this mode of transporter operation provides an opportunity to refine the kinetic scheme further. We suggest that in the absence of Cl^-, GABA can bind to the transporter after two Na^+ binding steps are completed. This decision was made to account for the finding of Loo et al. (2000) who demonstrated that the net charge translocated per GABA (1.9 charges/GABA) was the same in the presence and absence of extracellular Cl^-.

Therefore, the proposed kinetic scheme not only accounts for the known features of steady-state Na^+/Cl^-/GABA cotransport across the plasma membrane, but it also provides a mechanism for the existence as well as the electrogenic behavior of Cl^--independent Na^+/GABA cotransport.

7 Conclusions and Future Directions

In order to achieve an integrative understanding of the physiological functions of the GABA transporters in nervous and non-nervous tissues, it is essential that we know the (1) Na^+/Cl^-/GABA cotransport stoichiometry, (2) rate at which GABA is transported across the plasma membrane per unit time (i.e., transport cycle turnover rate), (3) precise transporter localization in cells (e.g., presynaptic terminals and surrounding glial cells), (4) transporter density in the neuronal and glial plasma membranes (i.e., how many transporters per square micron of axon terminal or glial cell plasma membrane), and (5) dynamic regulation of the abovementioned parameters under physiological, pharmacological, and pathophysiological conditions.

In recent years, our laboratory has devoted considerable effort to elucidate the ion/substrate coupling stoichiometry of the GABA transporters. The results of substrate flux measurements under voltage-clamp conditions, as well as thermodynamic reversal potential measurements, provide strong evidence in support of a 3 Na^+: 1 Cl^-: 1 GABA stoichiometry for the GABA transporters (Willford et al. 2015).

The free energy for Na^+/Cl^-/GABA cotransport mediated by the GABA transporters is provided by the combined electrochemical gradients of Na^+ and Cl^-. The results of experiments performed under a variety of conditions suggest that the stoichiometry is fixed. We note that the data for this conclusion are provided by studies of GATs expressed in *Xenopus laevis* oocytes, in which GABA-evoked uncoupled currents mediated by the GABA transporters are nonexistent or insignificant in magnitude.

At physiological membrane potentials and extracellular and intracellular concentrations of Na^+, Cl^-, and GABA, the 3 Na^+: 1 Cl^-: 1 GABA stoichiometry places significant constraints on the direction of transport under non-equilibrium conditions, which perhaps represent the vast majority of physiological conditions under which the transporter operates. Under most physiological conditions, the calculated driving forces suggest that the GABA transporters work in the forward mode, transporting Na^+/Cl^-/GABA into neurons and glia (see Fig. 6). During neuronal spike activity, the axon terminal experiences sustained depolarization which could potentially cause reversal of presynaptic GABA transporters. However, even under these conditions, the rise in extracellular GABA caused by synaptic release may bring about conditions that favor uptake rather than release (see Figs. 6 and 7). Reverse transport is even less likely for GABA transporters located in astrocytes which, compared to neurons, generally have more negative and more stable membrane potentials. These predictions are consistent with the ability of GABA transporters to remove GABA from the extracellular space even during synaptic activity when the extracellular GABA concentration is elevated and the membrane potential is depolarized (e.g., Savtchenko et al. 2015). Altogether, the findings suggest that significant transporter-mediated GABA release likely occurs under pathophysiological or extreme physiological conditions where the membrane potential is depolarized to extreme positive voltages and/or the Na^+ electrochemical gradient is dissipated (see Fig. 6).

It is important to note that the proposed 3 Na^+: 1 Cl^-: 1 GABA stoichiometry for the GABA transporters is not without precedent in the SLC6 family of transporters (Kristensen et al. 2011; Bröer and Gether 2012; Scimemi 2014a). The stoichiometry of the glycine transporter, GlyT2a, has been shown to be 3 Na^+: 1 Cl^-: 1 glycine, and the stoichiometry is believed to be fixed resulting in a tightly coupled transport cycle (Roux and Supplisson 2000; Supplisson and Roux 2002; Rousseau et al. 2008). We also note that the evidence for GlyT2a stoichiometry was obtained in *Xenopus* oocytes.

The crystal structure of a bacterial member of NSS family, the leucine transporter from *Aquifex aeolicus* ($LeuT_{Aa}$), and other recently solved structures of related transporters have dramatically advanced our understanding of the structure and function of SLC6 transporters (Yamashita et al. 2005; Penmatsa et al. 2013; Wang et al. 2013, 2015; Penmatsa and Gouaux 2014; Rudnick et al. 2014; Singh and Pal 2015; Coleman et al. 2016). In spite of the wealth of structure-function data available, it is difficult to speculate about the location of the third Na^+ binding site in the three-dimensional structure of the GABA transporters. All current structural models of the GABA transporters are based on the high-resolution crystal structure of LeuT, which has two Na^+ binding sites (e.g., Zomot et al. 2007; Skovstrup et al. 2010, 2012). It is hoped that as additional SLC6 transporter structures are solved, a better picture of Na^+ binding sites will emerge for transporters that operate with a 3 Na^+: 1 Cl^-: 1 substrate stoichiometry.

Our laboratory has also been concerned with the question of transporter turnover rate, which is defined as the number of GABA molecules translocated across the plasma membrane per unit time. When adjusted for all relevant physiological parameters, we estimated the GAT1 unitary turnover to be 79–93 s^{-1} at 37 °C and membrane potentials ranging from −50 mV to −90 mV (Gonzales et al. 2007). This turnover rate is nearly an order of magnitude higher than previous estimates.

In summary, to better understand the contributions of the GABA transporters to synaptic physiology, we suggest that the revised 3 Na^+: 1 Cl^-: 1 GABA ion/substrate transport stoichiometry as well as higher physiological transporter turnover rates should be incorporated in future functional models of GABAergic synapses.

Acknowledgments The authors wish to acknowledge support from the National Institute of General Medical Sciences (NIGMS) of the US National Institutes of Health (NIH), the California State University Program for Education and Research in Biotechnology (CSUPERB), and the California State University Agricultural Research Institute (CSU ARI).

Conflict of Interest The author declares no conflicts of interest.

References

Alle H, Geiger JR. GABAergic spill-over transmission onto hippocampal mossy fiber boutons. J Neurosci. 2007;27:942–50.

Allen NJ, Káradóttir R, Attwell D. Reversal or reduction of glutamate and GABA transport in CNS pathology and therapy. Pflugers Arch. 2004a;449:132–42.

Allen NJ, Rossi DJ, Attwell D. Sequential release of GABA by exocytosis and reversed uptake leads to neuronal swelling in simulated ischemia of hippocampal slices. J Neurosci. 2004b;24:3837–49.

Angulo MC, Le Meur K, Kozlov AS, Charpak S, Audinat E. GABA, a forgotten gliotransmitter. Prog Neurobiol. 2008;86:297–303.

Awapara J, Landua AJ, Fuerst R, Seale B. Free γ-aminobutyric acid in brain. J Biol Chem. 1950;187:35–9.

Barakat L, Bordey A. GAT-1 and reversible GABA transport in Bergmann glia in slices. J Neurophysiol. 2002;88:1407–19.

Bazemore A, Elliott KA, Florey E. Factor I and gamma-aminobutyric acid. Nature. 1956;178:1052–3.

Belhage B, Hansen GH, Schousboe A. Depolarization by K^+ and glutamate activates different neurotransmitter release mechanisms in GABAergic neurons: vesicular versus non-vesicular release of GABA. Neuroscience. 1993;54:1019–34.

Bernath S, Zigmond MJ. Characterization of [^{3}H]GABA release from striatal slices: evidence for a calcium-independent process via the GABA uptake system. Neuroscience. 1988;27:563–70.

Bertram S, Cherubino F, Bossi E, Castagna M, Peres A. GABA reverse transport by the neuronal cotransporter GAT1: influence of internal chloride depletion. Am J Physiol Cell Physiol. 2011;301:C1064–73.

Beuming T, Shi L, Javitch JA, Weinstein H. A comprehensive structure-based alignment of prokaryotic and eukaryotic neurotransmitter/Na^+ symporters (NSS) aids in the use of the LeuT structure to probe NSS structure and function. Mol Pharmacol. 2006;70:1630–42.
Bicho A, Grewer C. Rapid substrate-induced charge movements of the GABA transporter GAT1. Biophys J. 2005;89:211–31.
Biedermann B, Bringmann A, Reichenbach A. High-affinity GABA uptake in retinal glial (Müller) cells of the guinea pig: electrophysiological characterization, immunohistochemical localization, and modeling of efficiency. Glia. 2002;39:217–28.
Bismuth Y, Kavanaugh MP, Kanner BI. Tyrosine 140 of the γ-aminobutyric acid transporter GAT-1 plays a critical role in neurotransmitter recognition. J Biol Chem. 1997;272:16096–102.
Blaustein MP, King AC. Influence of membrane potential on the sodium-dependent uptake of γ-aminobutyric acid by presynaptic nerve terminals: experimental observations and theoretical considerations. J Membr Biol. 1976;30:153–73.
Boistel J, Fatt P. Membrane permeability change during inhibitory transmitter action in crustacean muscle. J Physiol Lond. 1958;144:176–91.
Borden LA. GABA transporter heterogeneity: pharmacology and cellular localization. Neurochem Int. 1996;29:335–56.
Borden LA, Smith KE, Hartig PR, Branchek TA, Weinshank RL. Molecular heterogeneity of the γ-aminobutyric acid (GABA) transport system. Cloning of two novel high affinity GABA transporters from rat brain. J Biol Chem. 1992;267:21098–104.
Borden LA, Dhar TG, Smith KE, Branchek TA, Gluchowski C, Weinshank RL. Cloning of the human homologue of the GABA transporter GAT-3 and identification of a novel inhibitor with selectivity for this site. Recept Channels. 1994;2:207–13.
Bragina L, Marchionni I, Omrani A, Cozzi A, Pellegrini-Giampietro DE, Cherubini E, Conti F. GAT-1 regulates both tonic and phasic $GABA_A$ receptor-mediated inhibition in the cerebral cortex. J Neurochem. 2008;105:1781–93.
Bröer S, Gether U. The solute carrier 6 family of transporters. Br J Pharmacol. 2012;167:256–78.
Busch W, Saier MH Jr. The transporter classification (TC) system, 2002. Crit Rev Biochem Mol Biol. 2002;37:287–337.
Cammack JN, Schwartz EA. Channel behavior in a γ-aminobutyrate transporter. Proc Natl Acad Sci U S A. 1996;93:723–7.
Cammack JN, Rakhilin SV, Schwartz EA. A GABA transporter operates asymmetrically and with variable stoichiometry. Neuron. 1994;13:949–60.
Chen NH, Reith ME, Quick MW. Synaptic uptake and beyond: the sodium- and chloride-dependent neurotransmitter transporter family SLC6. Pflugers Arch. 2004;447:519–31.
Cherubino F, Bertram S, Bossi E, Peres A. Pre-steady-state and reverse transport currents in the GABA transporter GAT1. Am J Physiol Cell Physiol. 2012;302:C1096–108.
Chiu CS, Jensen K, Sokolova I, Wang D, Li M, Deshpande P, Davidson N, Mody I, Quick MW, Quake SR, Lester HA. Number, density, and surface/cytoplasmic distribution of GABA transporters at presynaptic structures of knock-in mice carrying GABA transporter subtype 1-green fluorescent protein fusions. J Neurosci. 2002;22:10251–66.
Chiu CS, Brickley S, Jensen K, Southwell A, Mckinney S, Cull-Candy S, Mody I, Lester HA. GABA transporter deficiency causes tremor, ataxia, nervousness, and increased GABA-induced tonic conductance in cerebellum. J Neurosci. 2005;25:3234–45.
Christiansen B, Meinild AK, Jensen AA, Brauner-Osborne H. Cloning and characterization of a functional human γ-aminobutyric acid (GABA) transporter, human GAT-2. J Biol Chem. 2007;282:19331–41.
Clark JA, Amara SG. Stable expression of a neuronal γ-aminobutyric acid transporter, GAT-3, in mammalian cells demonstrates unique pharmacological properties and ion dependence. Mol Pharmacol. 1994;46:550–7.
Clark JA, Deutch AY, Gallipoli PZ, Amara SG. Functional expression and CNS distribution of a β-alanine-sensitive neuronal GABA transporter. Neuron. 1992;9:337–48.

Clausen RP, Madsen K, Larsson OM, Frølund B, Krogsgaard-Larsen P, Schousboe A. Structure-activity relationship and pharmacology of γ-aminobutyric acid (GABA) transport inhibitors. Adv Pharmacol. 2006;54:265–84.

Coleman JA, Green EM, Gouaux E. X-ray structures and mechanism of the human serotonin transporter. Nature. 2016;532:334–9.

Conti F, Minelli A, Melone M. GABA transporters in the mammalian cerebral cortex: localization, development and pathological implications. Brain Res Brain Res Rev. 2004;45:196–212.

Conti F, Melone M, Fattorini G, Bragina L, Ciappelloni S. A role for GAT-1 in presynaptic GABA homeostasis? Front Cell Neurosci. 2011;5:2.

Curtis DR, Phillis JW. gamma-Amino-n-butyric acid and spinal synaptic transmission. Nature. 1958;182:323.

Dalby NO. GABA-level increasing and anticonvulsant effects of three different GABA uptake inhibitors. Neuropharmacology. 2000;39:2399–407.

Dalby NO. Inhibition of γ-aminobutyric acid uptake: anatomy, physiology and effects against epileptic seizures. Eur J Pharmacol. 2003;479:127–37.

Dalby NO, Thomsen C, Fink-Jensen A, Lundbeck J, Søkilde B, Man CM, Sørensen PO, Meldrum B. Anticonvulsant properties of two GABA uptake inhibitors NNC 05-2045 and NNC 05-2090, not acting preferentially on GAT-1. Epilepsy Res. 1997;28:51–61.

Dingledine R, Korn SJ. γ-Aminobutyric acid uptake and the termination of inhibitory synaptic potentials in the rat hippocampal slice. J Physiol Lond. 1985;366:387–409.

Egawa K, Yamada J, Furukawa T, Yanagawa Y, Fukuda A. Cl^- homeodynamics in gap junction-coupled astrocytic networks on activation of GABAergic synapses. J Physiol Lond. 2013;591:3901–17.

Elliott KA, van Gelder NM. Occlusion and metabolism of γ-aminobutyric acid by brain tissue. J Neurochem. 1958;3:28–40.

Eskandari S, Loo DD, Dai G, Levy O, Wright EM, Carrasco N. Thyroid Na^+/I^- symporter. Mechanism, stoichiometry, and specificity. J Biol Chem. 1997;272:27230–8.

Fesce R, Giovannardi S, Binda F, Bossi E, Peres A. The relation between charge movement and transport-associated currents in the rat GABA cotransporter rGAT1. J Physiol Lond. 2002;545:739–50.

Fink-Jensen A, Suzdak PD, Swedberg MD, Judge ME, Hansen L, Nielsen PG. The γ-aminobutyric acid (GABA) uptake inhibitor, tiagabine, increases extracellular brain levels of GABA in awake rats. Eur J Pharmacol. 1992;220:197–201.

Florey E. An inhibitory and an excitatory factor of mammalian central nervous system, and their action of a single sensory neuron. Arch Int Physiol. 1954;62:33–53.

Florey E, McLennan H. The effects of factor I and of gamma-aminobutyric acid on smooth muscle preparations. J Physiol Lond. 1959;145:66–76.

Forster IC, Loo DD, Eskandari S. Stoichiometry and Na^+ binding cooperativity of rat and flounder renal type II Na^+-P_i cotransporters. Am J Phys. 1999;276:F644–9.

Gadea A, López-Colomé AM. Glial transporters for glutamate, glycine, and GABA: II. GABA transporters. J Neurosci Res. 2001;63:461–8.

Gallo V, Patrizio M, Levi G. GABA release triggered by the activation of neuron-like non-NMDA receptors in cultured type 2 astrocytes is carrier-mediated. Glia. 1991;4:245–55.

Gaspary HL, Wang W, Richerson GB. Carrier-mediated GABA release activates GABA receptors on hippocampal neurons. J Neurophysiol. 1998;80:270–81.

Giovannardi S, Fesce R, Bossi E, Binda F, Peres A. Cl^- affects the function of the GABA cotransporter rGAT1 but preserves the mutal relationship between transient and transport currents. Cell Mol Life Sci. 2003;60:550–6.

Gonzales AL, Lee W, Spencer SR, Oropeza RA, Chapman JV, Ku JY, Eskandari S. Turnover rate of the γ-aminobutyric acid transporter GAT1. J Membr Biol. 2007;220:33–51.

Gottesfeld Z, Elliott KA. Factors that affect the binding and uptake of GABA by brain tissue. J Neurochem. 1971;18:683–90.

Grossman TR, Nelson N. Effect of sodium lithium and proton concentrations on the electrophysiological properties of the four mouse GABA transporters expressed in *Xenopus* oocytes. Neurochem Int. 2003;43:431–43.

Guastella J, Nelson N, Nelson H, Czyzyk L, Keynan S, Miedel MC, Davidson N, Lester HA, Kanner BI. Cloning and expression of a rat brain GABA transporter. Science. 1990;249:1303–6.

Héja L, Barabás P, Nyitrai G, Kékesi KA, Lasztóczi B, Töke O, Tárkányi G, Madsen K, Schousboe A, Dobolyi A, Palkovits M, Kardos J. Glutamate uptake triggers transporter-mediated GABA release from astrocytes. PLoS One. 2009;4:e7153.

Héja L, Nyitrai G, Kékesi O, Dobolyi A, Szabó P, Fiáth R, Ulbert I, Pál-Szenthe B, Palkovits M, Kardos J. Astrocytes convert network excitation to tonic inhibition of neurons. BMC Biol. 2012;10:26.

Hilgemann DW, Lu CC. GAT1 ($GABA:Na^+:Cl^-$) cotransport function. Database reconstruction with an alternating access model. J Gen Physiol. 1999;114:459–75.

Isaacson JS, Solís JM, Nicoll RA. Local and diffuse synaptic actions of GABA in the hippocampus. Neuron. 1993;10:165–75.

Iversen LL, Neal MJ. The uptake of [^{3}H]GABA by slices of rat cerebral cortex. J Neurochem. 1968;15:1141–9.

Jensen K, Chiu CS, Sokolova I, Lester HA, Mody I. GABA transporter-1 (GAT1)-deficient mice: differential tonic activation of $GABA_A$ versus $GABA_B$ receptors in the hippocampus. J Neurophysiol. 2003;90:2690–701.

Kanner BI. Active transport of γ-aminobutyric acid by membrane vesicles isolated from rat brain. Biochemistry. 1978;17:1207–11.

Kanner BI. Transmembrane domain I of the γ-aminobutyric acid transporter GAT-1 plays a crucial role in the transition between cation leak and transport modes. J Biol Chem. 2003;278:3705–12.

Kanner BI, Kifer L. Efflux of γ-aminobutyric acid by synaptic plasma membrane vesicles isolated from rat brain. Biochemistry. 1981;20:3354–8.

Kanner BI, Schuldiner S. Mechanism of transport and storage of neurotransmitters. CRC Crit Rev Biochem. 1987;22:1–38.

Kanner BI, Zomot E. Sodium-coupled neurotransmitter transporters. Chem Rev. 2008;108:1654–68.

Kanner BI, Bendahan A, Radian R. Efflux and exchange of γ-aminobutyric acid and nipecotic acid catalysed by synaptic plasma membrane vesicles isolated from immature rat brain. Biochim Biophys Acta. 1983;731:54–62.

Karakossian MH, Spencer SR, Gomez AQ, Padilla OR, Sacher A, Loo DD, Nelson N, Eskandari S. Novel properties of a mouse γ-aminobutyric acid transporter (GAT4). J Membr Biol. 2005;203:65–82.

Kavanaugh MP, Arriza JL, North RA, Amara SG. Electrogenic uptake of γ-aminobutyric acid by a cloned transporter expressed in *Xenopus* oocytes. J Biol Chem. 1992;267:22007–9.

Keros S, Hablitz JJ. Subtype-specific GABA transporter antagonists synergistically modulate phasic and tonic $GABA_A$ conductances in rat neocortex. J Neurophysiol. 2005;94:2073–85.

Keynan S, Kanner BI. γ-Aminobutyric acid transport in reconstituted preparations from rat brain: coupled sodium and chloride fluxes. Biochemistry. 1988;27:12–7.

Keynan S, Suh YJ, Kanner BI, Rudnick G. Expression of a cloned γ-aminobutyric acid transporter in mammalian cells. Biochemistry. 1992;31:1974–9.

Kinney GA. GAT-3 transporters regulate inhibition in the neocortex. J Neurophysiol. 2005;94:4533–7.

Kinney GA, Spain WJ. Synaptically evoked GABA transporter currents in neocortical glia. J Neurophysiol. 2002;88:2899–908.

Kirmse K, Kirischuk S, Grantyn R. Role of GABA transporter 3 in GABAergic synaptic transmission at striatal output neurons. Synapse. 2009;63:921–9.

Koch U, Magnusson AK. Unconventional GABA release: mechanisms and function. Curr Opin Neurobiol. 2009;19:305–10.

Krause S, Schwarz W. Identification and selective inhibition of the channel mode of the neuronal GABA transporter 1. Mol Pharmacol. 2005;68:1728–35.

Kristensen AS, Andersen J, Jørgensen TN, Sørensen L, Eriksen J, Loland CJ, Strømgaard K, Gether U. SLC6 neurotransmitter transporters: structure, function, and regulation. Pharmacol Rev. 2011;63:585–640.

Krnjevic K, Phillis JW. Iontophoretic studies of neurones in the mammalian cerebral cortex. J Physiol Lond. 1963;165:274–304.
Krnjevic K, Schwartz S. The action of γ-aminobutyric acid on cortical neurones. Exp Brain Res. 1967;3:320–36.
Kuffler SW, Edwards C. Mechanism of gamma aminobutyric acid (GABA) action and its relation to synaptic inhibition. J Neurophysiol. 1958;21:589–610.
Kuhar MJ, Zarbin MA. Synaptosomal transport: a chloride dependence for choline, GABA, glycine and several other compounds. J Neurochem. 1978;31:251–6.
Li M, Farley RA, Lester HA. An intermediate state of the γ-aminobutyric acid transporter GAT1 revealed by simultaneous voltage clamp and fluorescence. J Gen Physiol. 2000;115:491–508.
Liu QR, Mandiyan S, Nelson H, Nelson N. A family of genes encoding neurotransmitter transporters. Proc Natl Acad Sci U S A. 1992a;89:6639–43.
Liu QR, Nelson H, Mandiyan S, López-Corcuera B, Nelson N. Cloning and expression of a glycine transporter from mouse brain. FEBS Lett. 1992b;305:110–4.
Liu QR, López-Corcuera B, Mandiyan S, Nelson H, Nelson N. Molecular characterization of four pharmacologically distinct γ-aminobutyric acid transporters in mouse brain. J Biol Chem. 1993;268:2106–12.
Livesay DR, Kidd PD, Eskandari S, Roshan U. Assessing the ability of sequence-based methods to provide functional insight within membrane integral proteins: a case study analyzing the neurotransmitter/Na^+ symporter family. BMC Bioinf. 2007;8:397.
Loo DD, Eskandari S, Boorer KJ, Sarkar HK, Wright EM. Role of Cl^- in electrogenic Na^+-coupled cotransporters GAT1 and SGLT1. J Biol Chem. 2000;275:37414–22.
Lu CC, Hilgemann DW. GAT1 (GABA:Na^+:Cl^-) cotransport function. Steady state studies in giant Xenopus oocyte membrane patches. J Gen Physiol. 1999a;114:429–44.
Lu CC, Hilgemann DW. GAT1 (GABA:Na^+:Cl^-) cotransport function. Kinetic studies in giant Xenopus oocyte membrane patches. J Gen Physiol. 1999b;114:445–57.
Luccini E, Romei C, Di Prisco S, Raiteri M, Raiteri L. Ionic dysregulations typical of ischemia provoke release of glycine and GABA by multiple mechanisms. J Neurochem. 2010;114:1074–84.
MacAulay N, Zeuthen T, Gether U. Conformational basis for the Li^+-induced leak current in the rat γ-aminobutyric acid (GABA) transporter-1. J Physiol Lond. 2002;544:447–58.
Mackenzie B, Loo DD, Wright EM. Relationships between Na^+/glucose cotransporter (SGLT1) currents and fluxes. J Membr Biol. 1998;162:101–6.
Madsen KK, Clausen RP, Larsson OM, Krogsgaard-Larsen P, Schousboe A, White HS. Synaptic and extrasynaptic GABA transporters as targets for anti-epileptic drugs. J Neurochem. 2009;109(Suppl 1):139–44.
Madsen KK, White HS, Schousboe A. Neuronal and non-neuronal GABA transporters as targets for antiepileptic drugs. Pharmacol Ther. 2010;125:394–401.
Madsen KK, Ebert B, Clausen RP, Krogsgaard-Larsen P, Schousboe A, White HS. Selective GABA transporter inhibitors tiagabine and EF1502 exhibit mechanistic differences in their ability to modulate the ataxia and anticonvulsant action of the extrasynaptic $GABA_A$ receptor agonist gaboxadol. J Pharmacol Exp Ther. 2011;338:214–9.
Mager S, Naeve J, Quick M, Labarca C, Davidson N, Lester HA. Steady states, charge movements, and rates for a cloned GABA transporter expressed in Xenopus oocytes. Neuron. 1993;10:177–88.
Mager S, Kleinberger-Doron N, Keshet GI, Davidson N, Kanner BI, Lester HA. Ion binding and permeation at the GABA transporter GAT1. J Neurosci. 1996;16:5405–14.
Mager S, Cao Y, Lester HA. Measurement of transient currents from neurotransmitter transporters expressed in *Xenopus* oocytes. Methods Enzymol. 1998;296:551–66.
Martin DL. Kinetics of the sodium-dependent transport of gamma-aminobutyric acid by synaptosomes. J Neurochem. 1973;21:345–56.
Martin DL, Smith AA 3rd. Ions and the transport of gamma-aminobutyric acid by synaptosomes. J Neurochem. 1972;19:841–55.
Matskevitch I, Wagner CA, Stegen C, Bröer S, Noll B, Risler T, Kwon HM, Handler JS, Waldegger S, Busch AE, Lang F. Functional characterization of the betaine/γ-aminobutyric acid transporter BGT-1 expressed in *Xenopus* oocytes. J Biol Chem. 1999;274:16709–16.

Matthews E Jr, Rahnama-Vaghef A, Eskandari S. Inhibitors of the γ-aminobutyric acid transporter 1 (GAT1) do not reveal a channel mode of conduction. Neurochem Int. 2009;55:732–40.
Meinild AK, Forster IC. Using lithium to probe sequential cation interactions with GAT1. Am J Physiol Cell Physiol. 2012;302:C1661–75.
Meinild AK, Loo DD, Skovstrup S, Gether U, Macaulay N. Elucidating conformational changes in the γ-aminobutyric acid transporter-1. J Biol Chem. 2009;284:16226–35.
Milanese M, Romei C, Usai C, Oliveri M, Raiteri L. A new function for glycine GlyT2 transporters: stimulation of γ-aminobutyric acid release from cerebellar nerve terminals through GAT1 transporter reversal and Ca^{2+}-dependent anion channels. J Neurosci Res. 2014;92:398–408.
Morimoto K, Sato H, Yamamoto Y, Watanabe T, Suwaki H. Antiepileptic effects of tiagabine, a selective GABA uptake inhibitor, in the rat kindling model of temporal lobe epilepsy. Epilepsia. 1997;38:966–74.
Moscowitz JA, Cutler RW. Bidirectional movement of γ-aminobutyric acid in rat spinal cord slices. J Neurochem. 1980;35:1394–9.
Nelson N. The family of Na^+/Cl^- neurotransmitter transporters. J Neurochem. 1998;71:1785–803.
Nelson H, Mandiyan S, Nelson N. Cloning of the human brain GABA transporter. FEBS Lett. 1990;269:181–4.
Nielsen EB, Suzdak PD, Andersen KE, Knutsen LJ, Sonnewald U, Braestrup C. Characterization of tiagabine (NO-328), a new potent and selective GABA uptake inhibitor. Eur J Pharmacol. 1991;196:257–66.
O'Malley DM, Sandell JH, Masland RH. Co-release of acetylcholine and GABA by the starburst amacrine cells. J Neurosci. 1992;12:1394–408.
Obata K, Ito M, Ochi R, Sato N. Pharmacological properties of the postsynaptic inhibition by Purkinje cell axons and the action of γ-aminobutyric acid on deiters neurones. Exp Brain Res. 1967;4:43–57.
Omoto JJ, Maestas MJ, Rahnama-Vaghef A, Choi YE, Salto G Jr, Sanchez RV, Anderson CM, Eskandari S. Functional çonsequences of sulfhydryl modification of the γ-aminobutyric acid transporter 1 at a single solvent-exposed cysteine residue. J Membr Biol. 2012;245:841–57.
Otsuka M, Iversen LL, Hall ZW, Kravitz EA. Release of gamma-aminobutyric acid from inhibitory nerves of lobster. Proc Natl Acad Sci U S A. 1966;56:1110–5.
Overstreet LS, Westbrook GL. Synapse density regulates independence at unitary inhibitory synapses. J Neurosci. 2003;23:2618–26.
Overstreet LS, Jones MV, Westbrook GL. Slow desensitization regulates the availability of synaptic $GABA_A$ receptors. J Neurosci. 2000;20:7914–21.
Pastuszko A, Wilson DF, Erecinska M. Energetics of γ-aminobutyrate transport in rat brain synaptosomes. J Biol Chem. 1982;257:7514–9.
Penmatsa A, Gouaux E. How LeuT shapes our understanding of the mechanisms of sodium-coupled neurotransmitter transporters. J Physiol Lond. 2014;592:863–9.
Penmatsa A, Wang KH, Gouaux E. X-ray structure of dopamine transporter elucidates antidepressant mechanism. Nature. 2013;503:85–90.
Peres A, Giovannardi S, Bossi E, Fesce R. Electrophysiological insights into the mechanism of ion-coupled cotransporters. News Physiol Sci. 2004;19:80–4.
Pin JP, Bockaert J. Two distinct mechanisms, differentially affected by excitatory amino acids, trigger GABA release from fetal mouse striatal neurons in primary culture. J Neurosci. 1989;9:648–56.
Radian R, Kanner BI. Stoichiometry of sodium- and chloride-coupled γ-aminobutyric acid transport by synaptic plasma membrane vesicles isolated from rat brain. Biochemistry. 1983;22:1236–41.
Ransom CB, Tao W, Wu Y, Spain WJ, Richerson GB. Rapid regulation of tonic GABA currents in cultured rat hippocampal neurons. J Neurophysiol. 2013;109:803–12.
Rasola A, Galietta LJ, Barone V, Romeo G, Bagnasco S. Molecular cloning and functional characterization of a GABA/betaine transporter from human kidney. FEBS Lett. 1995;373:229–33.
Richerson GB, Wu Y. Dynamic equilibrium of neurotransmitter transporters: not just for reuptake anymore. J Neurophysiol. 2003;90:1363–74.

Richerson GB, Wu Y. Role of the GABA transporter in epilepsy. Adv Exp Med Biol. 2004;548:76–91.
Risso S, DeFelice LJ, Blakely RD. Sodium-dependent GABA-induced currents in GAT1-transfected HeLa cells. J Physiol Lond. 1996;490(Pt 3):691–702.
Roberts E, Frankel S. γ-Aminobutyric acid in brain: its formation from glutamic acid. J Biol Chem. 1950;187:55–63.
Romei C, Sabolla C, Raiteri L. GABA release provoked by disturbed Na^+, K^+ and Ca^{2+} homeostasis in cerebellar nerve endings: roles of Ca^{2+} channels, Na^+/Ca^{2+} exchangers and GAT1 transporter reversal. Neurochem Int. 2014;72:1–9.
Rousseau F, Aubrey KR, Supplisson S. The glycine transporter GlyT2 controls the dynamics of synaptic vesicle refilling in inhibitory spinal cord neurons. J Neurosci. 2008;28:9755–68.
Roux MJ, Supplisson S. Neuronal and glial glycine transporters have different stoichiometries. Neuron. 2000;25:373–83.
Rudnick G. Ion-coupled neurotransmitter transport: thermodynamic vs. kinetic determinations of stoichiometry. Methods Enzymol. 1998;296:233–47.
Rudnick G, Krämer R, Blakely RD, Murphy DL, Verrey F. The SLC6 transporters: perspectives on structure, functions, regulation, and models for transporter dysfunction. Pflugers Arch. 2014;466:25–42.
Sacher A, Nelson N, Ogi JT, Wright EM, Loo DD, Eskandari S. Presteady-state and steady-state kinetics and turnover rate of the mouse γ-aminobutyric acid transporter (mGAT3). J Membr Biol. 2002;190:57–73.
Saier MH Jr, Tran CV, Barabote RD. TCDB: the transporter classification database for membrane transport protein analyses and information. Nucleic Acids Res. 2006;34:D181–6.
Saier MH Jr, Yen MR, Noto K, Tamang DG, Elkan C. The transporter classification database: recent advances. Nucleic Acids Res. 2009;37:D274–8.
Salat K, Wieckowska A, Wieckowski K, Höfner GC, Kaminski J, Wanner KT, Malawska B, Filipek B, Kulig K. Synthesis and pharmacological properties of new GABA uptake inhibitors. Pharmacol Rep. 2012;64:817–33.
Savtchenko L, Megalogeni M, Rusakov DA, Walker MC, Pavlov I. Synaptic GABA release prevents GABA transporter type-1 reversal during excessive network activity. Nat Commun. 2015;6:6597.
Schousboe A, Madsen KK, White HS. GABA transport inhibitors and seizure protection: the past and future. Future Med Chem. 2011;3:183–7.
Schousboe A, Madsen KK, Barker-Haliski ML, White HS. The GABA synapse as a target for antiepileptic drugs: a historical overview focused on GABA transporters. Neurochem Res. 2014;39:1980–7.
Schwartz EA. Depolarization without calcium can release γ-aminobutyric acid from a retinal neuron. Science. 1987;238:350–5.
Scimemi A. Structure, function, and plasticity of GABA transporters. Front Cell Neurosci. 2014a;8:161.
Scimemi A. Plasticity of GABA transporters: an unconventional route to shape inhibitory synaptic transmission. Front Cell Neurosci. 2014b;8:128.
Singh SK, Pal A. Biophysical approaches to the study of LeuT, a prokaryotic homolog of neurotransmitter sodium symporters. Methods Enzymol. 2015;557:167–98.
Skovstrup S, Taboureau O, Bräuner-Osborne H, Jørgensen FS. Homology modelling of the GABA transporter and analysis of tiagabine binding. ChemMedChem. 2010;5:986–1000.
Skovstrup S, David L, Taboureau O, Jørgensen FS. A steered molecular dynamics study of binding and translocation processes in the GABA transporter. PLoS One. 2012;7:e39360.
Snow H, Lowrie MB, Bennett JP. A postsynaptic GABA transporter in rat spinal motor neurones. Neurosci Lett. 1992;143:119–22.
Solís JM, Nicoll RA. Postsynaptic action of endogenous GABA released by nipecotic acid in the hippocampus. Neurosci Lett. 1992;147:16–20.
Soragna A, Bossi E, Giovannardi S, Pisani R, Peres A. Relations between substrate affinities and charge equilibration rates in the rat GABA cotransporter GAT1. J Physiol Lond. 2005;562:333–45.

Strasberg P, Elliott KA. Further studies on the binding of γ-aminobutyric acid by brain. Can J Biochem. 1967;45:1795–807.

Supplisson S, Roux MJ. Why glycine transporters have different stoichiometries. FEBS Lett. 2002;529:93–101.

Suzdak PD, Jansen JA. A review of the preclinical pharmacology of tiagabine: a potent and selective anticonvulsant GABA uptake inhibitor. Epilepsia. 1995;36:612–26.

Suzdak PD, Frederiksen K, Andersen KE, Sørensen PO, Knutsen LJ, Nielsen EB. NNC-711, a novel potent and selective gamma-aminobutyric acid uptake inhibitor: pharmacological characterization. Eur J Pharmacol. 1992;224:189–98.

Swinyard EA, White HS, Wolf HH, Bondinell WE. Anticonvulsant profiles of the potent and orally active GABA uptake inhibitors SK&F 89976-A and SK&F 100330-A and four prototype antiepileptic drugs in mice and rats. Epilepsia. 1991;32:569–77.

Tapia R, Arias C. Selective stimulation of neurotransmitter release from chick retina by kainic and glutamic acids. J Neurochem. 1982;39:1169–78.

Taylor J, Gordon-Weeks PR. Calcium-independent γ-aminobutyric acid release from growth cones: role of gamma-aminobutyric acid transport. J Neurochem. 1991;56:273–80.

Thompson SM, Gähwiler BH. Effects of the GABA uptake inhibitor tiagabine on inhibitory synaptic potentials in rat hippocampal slice cultures. J Neurophysiol. 1992;67:1698–701.

Udenfriend S. Identification of γ-aminobutyric acid in brain by the isotope derivative method. J Biol Chem. 1950;187:65–9.

Varon S, Weinstein H, Baxter CF, Roberts E. Uptake and metabolism of exogenous γ-aminobutyric acid by subcellular particles in a sodium-containing medium. Biochem Pharmacol. 1965a;14:1755–64.

Varon S, Weinstein H, Kakefuda T, Roberts E. Sodium-dependent binding of γ-aminobutyric acid by morphologically characterized subcellular brain particles. Biochem Pharmacol. 1965b;14:1213–24.

Varon S, Weinstein H, Roberts E. Sodium-dependent metabolism and transport of γ-aminobutyric acid in subcellular particles from brain. Protoplasma. 1967;63:318–21.

Wang D, Deken SL, Whitworth TL, Quick MW. Syntaxin 1A inhibits GABA flux, efflux, and exchange mediated by the rat brain GABA transporter GAT1. Mol Pharmacol. 2003;64:905–13.

Wang H, Goehring A, Wang KH, Penmatsa A, Ressler R, Gouaux E. Structural basis for action by diverse antidepressants on biogenic amine transporters. Nature. 2013;503:141–5.

Wang KH, Penmatsa A, Gouaux E. Neurotransmitter and psychostimulant recognition by the dopamine transporter. Nature. 2015;521:322–7.

Weinstein H, Varon S, Muhleman DR, Roberts E. A carrier-mediated transfer model for the accumulation of ^{14}C-γ-aminobutyric acid by subcellular brain particles. Biochem Pharmacol. 1965;14:273–88.

Weinstein H, Varon S, Roberts E. Evidence for a carrier-mediated transfer model for the accumulation of ^{14}C-γ-aminobutyric acid by brain particles at zero degrees centigrade. Protoplasma. 1967;63:315–7.

Weiss JN. The Hill equation revisited: uses and misuses. FASEB J. 1997;11:835–41.

Whitlow RD, Sacher A, Loo DD, Nelson N, Eskandari S. The anticonvulsant valproate increases the turnover rate of γ-aminobutyric acid transporters. J Biol Chem. 2003;278:17716–26.

Willford SL, Anderson CM, Spencer SR, Eskandari S. Evidence for a revised ion/substrate coupling stoichiometry of GABA transporters. J Membr Biol. 2015;248:795–810.

Wójtowicz AM, Dvorzhak A, Semtner M, Grantyn R. Reduced tonic inhibition in striatal output neurons from Huntington mice due to loss of astrocytic GABA release through GAT-3. Front Neural Circuits. 2013;7:188.

Wu Y, Wang W, Richerson GB. GABA transaminase inhibition induces spontaneous and enhances depolarization-evoked GABA efflux via reversal of the GABA transporter. J Neurosci. 2001;21:2630–9.

Wu Y, Wang W, Richerson GB. Vigabatrin induces tonic inhibition via GABA transporter reversal without increasing vesicular GABA release. J Neurophysiol. 2003;89:2021–34.

Wu Y, Wang W, Richerson GB. The transmembrane sodium gradient influences ambient GABA concentration by altering the equilibrium of GABA transporters. J Neurophysiol. 2006;96:2425–36.

Wu Y, Wang W, Díez-Sampedro A, Richerson GB. Nonvesicular inhibitory neurotransmission via reversal of the GABA transporter GAT-1. Neuron. 2007;56:851–65.

Xu Y, Cai Y, Gong N, Chen C, Wu Y, Zhang-Nunes S, Wang Z, Xu T, Fei J. Homeostatic plasticity of GABAergic synaptic transmission in mice lacking GAT1. Biochem Biophys Res Commun. 2007;361:499–504.

Xu YF, Cai YQ, Cai GQ, Jiang J, Sheng ZJ, Wang ZG, Fei J. Hypoalgesia in mice lacking GABA transporter subtype 1. J Neurosci Res. 2008;86:465–70.

Yamashita A, Singh SK, Kawate T, Jin Y, Gouaux E. Crystal structure of a bacterial homologue of Na^+/Cl^--dependent neurotransmitter transporters. Nature. 2005;437:215–23.

Yamauchi A, Uchida S, Kwon HM, Preston AS, Robey RB, Garcia-Perez A, Burg MB, Handler JS. Cloning of a Na^+- and Cl^--dependent betaine transporter that is regulated by hypertonicity. J Biol Chem. 1992;267:649–52.

Yazulla S, Kleinschmidt J. Carrier-mediated release of GABA from retinal horizontal cells. Brain Res. 1983;263:63–75.

Yunger LM, Fowler PJ, Zarevics P, Setler PE. Novel inhibitors of γ-aminobutyric acid (GABA) uptake: anticonvulsant actions in rats and mice. J Pharmacol Exp Ther. 1984;228:109–15.

Zhou Y, Holmseth S, Hua R, Lehre AC, Olofsson AM, Poblete-Naredo I, Kempson SA, Danbolt NC. The betaine-GABA transporter (BGT1, slc6a12) is predominantly expressed in the liver and at lower levels in the kidneys and at the brain surface. Am J Physiol Renal Physiol. 2012;302:F316–28.

Zomot E, Bendahan A, Quick M, Zhao Y, Javitch JA, Kanner BI. Mechanism of chloride interaction with neurotransmitter:sodium symporters. Nature. 2007;449:726–30.

EAAT2 and the Molecular Signature of Amyotrophic Lateral Sclerosis

Lauren Taylor Rosenblum and Davide Trotti

Abstract Amyotrophic lateral sclerosis (ALS) is a rapid and fatal neurodegenerative disease, primarily affecting upper and lower motor neurons. It is an extremely heterogeneous disease in both cause and symptom development, and its mechanisms of pathogenesis remain largely unknown. Excitotoxicity, a process caused by excessive glutamate signaling, is believed to play a substantial role, however. Excessive glutamate release, changes in postsynaptic glutamate receptors, and reduction of functional astrocytic glutamate transporters contribute to excitotoxicity in ALS. Here, we explore the roles of each, with a particular emphasis on glutamate transporters and attempts to increase them as therapy for ALS. Screening strategies have been employed to find compounds that increase the functional excitatory amino acid transporter EAAT2 (GLT1), which is responsible for the vast majority of glutamate clearance. One such compound, ceftriaxone, was recently tested in clinical trials but unfortunately did not modify disease course, though its effect on EAAT2 expression in patients was not measured.

Keywords EAAT2 • ALS • Excitotoxicity • Astrocyte • Motor neuron • Glutamate • GLT1 • GluR

Amyotrophic lateral sclerosis (ALS) is a rapidly progressive, fatal neurodegenerative disease that primarily affects motor pathways. Upper and lower motor neurons of the corticospinal tract selectively degenerate leading to muscle spasticity, weakness, muscle wasting, and death. With an incidence of around 1.75 per 100,000 people per year, ALS typically leads to death from respiratory failure in only 3–5 years after symptom onset, though about 10% of patients live substantially longer (Marin et al. 2016; Turner et al. 2013). Diagnosis is made at an average age

L.T. Rosenblum (✉) • D. Trotti
Jefferson Weinberg ALS Center, Vickie and Jack Farber Institute for Neuroscience, Department of Neuroscience, Thomas Jefferson University, 900 Walnut St, Philadelphia, PA 19107, USA
e-mail: lauren.rosenblum@jefferson.edu; davide.trotti@jefferson.edu

A. Ortega, A. Schousboe (eds.), *Glial Amino Acid Transporters*, Advances in Neurobiology 16, DOI 10.1007/978-3-319-55769-4_6

of 55 and is somewhat more common in men. ALS is a highly variable disease: Patients can present with limb (70%), bulbar (25%), or trunk/respiratory (5%) onset, with a combination of upper and motor neuron symptoms such as spasticity, muscle fasciculations, cramps, weakness, and muscle atrophy (Ross and Tabrizi 2011; Turner et al. 2013). About 10–20% of cases are familial, caused by mutations in over a dozen genes (Robberecht and Philips 2013). Around 20% of familial cases are caused by mutations in the *SOD1* gene, one of the first discovered causes. Mutations in the RNA-binding proteins TDP-43 and FUS account for 1–5% of familial cases each. A six-hexanucleotide (GGGGCC) repeat sequence in *C9orf72* was found in 2011 and causes about 40% of familial ALS, as well as a portion of sporadic cases. The majority of inciting factors for sporadic ALS remain unknown.

SOD1 mutations have been used to generate many transgenic animal models of the disease, including the first human SOD1-G93A transgenic mouse model, and those have been used extensively over the last two decades (Gurney et al. 1994; Rosen et al. 1993). The SOD1-G93A mouse model of ALS recapitulates many of the symptoms of ALS, as mice first demonstrate a hind limb tremor, followed by loss of the hind limb splaying reflex, and eventually paralysis and death. Generation of chimeric mice comprised of mixtures of normal and SOD1 mutant-expressing cells demonstrated that glia cells contribute to the death of motor neurons in a process known as non-cell autonomous toxicity (Robberecht and Philips 2013). Astrocytes expressing mutant SOD1 and astrocytes derived from sporadic ALS patients release factors toxic to motor neurons (Haidet-Phillips et al. 2011; Nagai et al. 2007). Unfortunately, while the SOD1 models have dramatically increased our understanding of ALS, they are based off a small fraction of patients with an extremely heterogeneous disease.

Perhaps because of this heterogeneity, the exact pathways of motoneuron toxicity have yet to be understood. Excitotoxicity, however, is believed to play a role. Critical to central nervous system function, the neurotransmitter glutamate communicates the majority of excitatory signals across neuronal synapses. Postsynaptic neurons process its message through a variety of ionotropic and metabotropic glutamate receptors, while glutamate transporters located primarily on astrocytes surrounding the synapse remove it, ending the signal. When this process breaks down, glutamate overstimulates the postsynaptic neurons. This in turn leads to excessive Ca^{2+} influx through NMDA receptors, some AMPA and kainate receptors, and voltage-gated Ca^{2+} channels (Van Den Bosch et al. 2006). The Ca^{2+} burden can cause mitochondrial damage, enzyme activation, reactive oxygen species generation, and other damaging processes. The neuronal damage caused by excessive stimulation, known as excitotoxicity, has been implicated in a variety of acute and chronic conditions, including ALS, epilepsy, cerebral ischemia, schizophrenia, mood disorders, and anxiety (Lauriat and McInnes 2007).

Impaired glutamate transport leads to increased synaptic glutamate, which can lead to excitotoxicity. Toxicity is primarily caused by excessive stimulation of receptors, leading to Ca^{2+} influx and dysregulation. This leads to mitochondrial uptake and dysfunction as well as other organelle dysregulation. Additionally, excessive glutamate signaling leads to TF and IEG activation, protease and calpain activation, cytoskeletal alterations, and ROS production. For a detailed review of its mechanisms, the readers are referred to an earlier review by Wang and Qin (Wang and Qin 2010).

Ample evidence suggests that excitotoxicity, especially of motor neurons, can be induced by impairment of the glutamate transport system. Loss of EAAT1 (GLAST) or EAAT2 (GLT1) by antisense oligonucleotide knockdown in organotypic spinal cord cultures causes a progressive decline in motor neuron viability as measured by ChAT activity (Rothstein 1996). Knockdown of the same astroglial transporters in vivo led to a rapid, progressive motor neuron syndrome (Rothstein 1996). Mice lacking EAAT2 were more susceptible to edema following cold-induced brain injury, with 68% greater edema (Tanaka et al. 1997). Riluzole, the only FDA-approved medication for the treatment of ALS, primarily functions by inhibiting persistent Na^+ currents, repetitive neuronal firing, and neurotransmitter release (Bellingham 2011). Those functions would presumably act to attenuate excitotoxicity by decreasing the excitation of the postsynaptic neurons.

While glutamate is normally cleared from the synapse by excitatory amino acid transporters, the highly predominant transporter, EAAT2 is substantially decreased in ALS (Bruijn et al. 1997). Motor neurons also appear to be particularly vulnerable to excitotoxicity (Kuner et al. 2005). Furthermore, riluzole, the only FDA-approved drug for treating ALS, has been shown to inhibit excitotoxic pathways (Bellingham 2011). EAAT2 may play a role in other neurotoxic pathways in ALS, as the formation and accumulation of a sumoylated fragment has been shown to induce the secretion of neurotoxic factors by astrocytes (Foran et al. 2011). In the last decade, substantial research has been directed toward developing pharmaceuticals that block excitotoxicity and/or increase EAAT2. Specifically, a variety of screening platforms have been employed to detect compounds that are capable of increasing EAAT2 with the goal of improving glutamate clearance and preventing excessive signaling. A number of compounds, including the blood-brain barrier (BBB)-permeable ceftriaxone, have shown promise in cell-based assays and in animal models of ALS (Colton et al. 2010; Li et al. 2011; Rothstein et al. 2005). The recent failure of ceftriaxone in clinical trials was a significant setback, but as EAAT2 changes were not actually measured, EAAT2 remains a viable clinical target for the treatment of ALS (Cudkowicz et al. 2014).

1 Glutamate Dysregulation in ALS

Dysregulation of glutamate levels in ALS has been well explored, though studies have offered conflicting reports of plasma and CNS glutamate levels, likely due to differences in methods and subgroup composition. It has also been reported that glutamate measurements are dependent on sampling variables, particularly storage temperature, perhaps accounting for discrepancies (Wuolikainen et al. 2011). Plasma glutamate levels were substantially elevated, and oral glutamate loading resulted in greater levels in early-stage ALS patients when compared to healthy and diseased controls (Plaitakis and Caroscio 1987). Even more compelling, glutamate was elevated in the cerebrospinal fluid of ALS patients by 100–200% (Rothstein et al. 1990). Conversely, glutamate levels were reduced in the CSF of patients as analyzed with GC/TOFMS (Wuolikainen et al. 2011). Glutamate levels were also found to be decreased in the frontal cortex, cerebellar cortex, lumbar spinal cord, and cervical spinal cord of patients who died of ALS, though the decrease may have been a result of neuronal loss or it could demonstrate the dysregulation of glutamate (Plaitakis et al. 1988). The distinction could also reflect a difference in patient composition: Camu et al. found glutamate to be elevated in the plasma and CSF of ALS patients with spinal onset but decreased in the plasma of ALS patients with bulbar onset (Camu et al. 1993). Similarly, plasma glutamate levels were elevated in ALS patients but only in spinal onset patients (Andreadou et al. 2008). In a large cohort, 41% of ALS patients demonstrated high glutamate levels in the spinal cord, which correlated with spinal onset, more impaired limb function, and faster muscle deterioration (Spreux-Varoquaux et al. 2002).

Glutamate release from spinal cord synaptosomes was significantly higher in SOD1-G93A mice, both at basal levels and with KCl- or ionomycin-evoked release (Bonifacino et al. 2016; Milanese et al. 2011). While the evoked response was more pronounced at later disease stages (over 120 days of age), it was observable even at the early presymptomatic stage (at 30–40 days of age). Basal efflux was even more pronounced at the earlier ages, probably due to a larger readily releasable pool of vesicles. Activation of mGlu1/5 receptors (group I receptors) elicited abnormal glutamate release in synaptosomes of SOD1-G93A mice via calcium release through IP3-sensitive channels (Giribaldi et al. 2013).

2 Glutamate Receptor Changes in ALS and Motor Neuron Vulnerability

As the primary excitatory neurotransmitter of the central nervous system, glutamate signals to postsynaptic neurons through both receptor-gated ionotropic and G-protein-coupled metabotropic receptors (mGluRs) (Nicoll et al. 1990). The ionotroptic receptors, including NMDA, AMPA, and kainate receptors, are all permeable to Na^+ and K^+, but only NMDARs are typically Ca^{2+} permeable as well (Dingledine et al. 1999). AMPA receptors are composed of GluA subunits which confer different properties. Usually included, the GluA2 subunit is typically modified at the mRNA level, leading to a glutamine to arginine mutation and switching the final receptor from calcium permeable to impermeable (Kawahara et al. 2004). AMPA and kainate receptors without the GluR2 subunit or unedited in the Q/R site of the M2 region are Ca^{2+} permeable as well, though much less prevalent. NMDARs are also susceptible to external Mg^{2+} blockage relieved by depolarization, which can be caused by activation of AMPA and kainite receptors.

In ALS, motor neurons appear to be selectively vulnerable to excitotoxicity, particularly via AMPA and kainate receptors. While cerebrospinal fluid (CSF) of ALS patients is toxic to motor neuron cultures, that toxicity can be prevented with AMPA and kainate receptor blockers such as CNQX but not by NMDA antagonists (Couratier et al. 1993; Sen et al. 2005). AMPA and kainate antagonists such as NBQX have also been shown to decrease toxicity in cell models and improve disease progression in the SOD1-G93A mouse model of ALS (Anneser et al. 2006; Tortarolo et al. 2006; Van Damme et al. 2003; Yin et al. 2007). In support of the AMPA/kainate receptor-mediated pathway, exogenous AMPA and kainate have been shown to cause motor neuron toxicity (Carriedo et al. 1996; Corona and Tapia 2004). Infusion of kainic acid to the lumbar spinal cord leads to a dose- and time-dependent loss of motor neurons and to alterations and loss of distal neuromuscular junctions (Blizzard et al. 2015). Furthermore, iPS motor neurons derived from C9orf72 patients were 100-fold more sensitive to glutamate excitotoxicity (Donnelly et al. 2013). Toxicity was blocked by GluA and calcium channel inhibitors. This might be due to loss of ADARB2 via sequestration in RNA foci. A compartmentalized model of excitotoxic exposure was used to show that somatodendritic but not axonal exposure to kainic acid was highly toxic to mouse lower motor neurons in vitro (Blizzard et al. 2015).

The selective vulnerability of motor neurons is likely due to changes in AMPA receptor subunit composition seen in ALS. While in normal cells, NMDA receptors are the primary calcium-permeable receptors, AMPA receptors can also be so under certain circumstances. AMPA receptors are composed of a tetramer of subunits which usually includes GluA2. The GluA2 mRNA is posttranscriptionally edited by adenosine deaminase acting on RNA 2 (ADAR2) in the second transmembrane region, causing an R to Q mutation in the coded protein at codon 607, a mutation which changes the subunit from being calcium permeable to impermeable (Kuner et al. 2005). Normally, the editing efficiency is near complete, so that virtually all AMPA receptors containing GluA2 are calcium impermeable. Alternatively, if the receptor includes GluA3 instead of GluA2 or is unedited, it can be calcium permeable. It has been found that in the motor neurons of ALS patients, GluA2 is incompletely edited at the Q/R site (in almost half of patients, ranging from 0% to 100% editing) in motor neurons of ALS patients, leading to increased calcium permeability (Damme et al. 2002; Kawahara et al. 2004; Takuma et al. 1999). The GluA2-editing enzyme ADAR2 was absent in more than half of sporadic patients and none of controls, coinciding with phosphorylated TDP-43 inclusions (frequently observed in ALS) in neuronal nuclei (Aizawa et al. 2010). A later study confirmed ADAR2 was downregulated in all sporadic patients studied, and all had motor neurons expressing unedited GluA2 versus no motor neurons with substantially unedited GluA2 in controls ($66.0 \pm 22.7\%$ versus $99.4 \pm 0.7\%$ of GluA2 edited) (Hideyama et al. 2012). In an unfortunate cycle, excitotoxic glutamate or NMDA exposure caused neurons to cleave ADAR2 in a calcium- and calpain-dependent pathway, preventing GluA2 editing (Mahajan et al. 2011). It is possible that AMPA could have a similar effect once GluA2 becomes unedited, as it would also become calcium permeable. A decrease in GluA2 expression relative to GluA3 has also been seen in SOD1-G93A mice (Tortarolo et al. 2006).

Other glutamate receptors also play a role in ALS. Metabotropic glutamate receptors are categorized into three groups: Group I includes mGlu1 and mGlu5, which are G_q receptors whose activation leads to IP3 formation; group II includes mGlu2 and mGlu3 which are $G_{i/o}$ receptors and largely function by inhibition of adenylyl cyclase; and group III includes mGlu4, mGlu5, mGlu7, and mGlu8, which are also $G_{i/o}$ receptors. It was found that modulation of mGlu group I receptors was protective of chicken embryonic spinal cord cultures from ALS-CSF-induced toxicity (Anneser et al. 2006). Supporting the role of mGlu in excitotoxicity, SOD1-G93A mice with only one copy of the gene coding for mGlu1 exhibited delayed onset and slower progression, associated with a decrease in abnormal glutamate release (Milanese et al. 2011). Stimulation of mGlu3 but not mGlu2 caused GDNF secretion which protected spinal motor neurons in mixed cultures from excitotoxic death (Battaglia et al. 2015). A group II agonist also increased EAAT2 expression in the spinal cord and rescued spinal cord motor neurons in the SOD1-G93A mouse model but had no effect on the life span of the mouse.

3 Glutamate Transporters and Dysregulation in ALS

Astrocytes play a crucial role in neuronal signaling. In one prime example of the integrated neuron-glia relationship, while a presynaptic neuron signals to a postsynaptic neuron with the excitatory amino acid, glutamate, astrocytes are primarily responsible for clearing glutamate from the synapse to terminate the signal. Synapses depend on the neurotransmitter signal, its activation of receptors (both pre- and postsynaptically), and the termination of that signal. Astrocytes clear glutamate from the synapse with excitatory amino acid transporters (EAATs)/glutamate transporters to end the synaptic transmission signal. Five excitatory amino acid transporters can be found in the human CNS, each with distinct regional specificities and physiologic properties (Arriza et al. 1994; Fairman et al. 1995). Isolated and cloned first, rodent homologues demonstrated sodium-dependent and chloride-independent uptake of glutamate and aspartate. EAAT1, or GLAST in rodents, predominates in astroglia of the cerebellum but can be found elsewhere in the cerebrum as well as other tissue including the heart, lung, and muscle (Arriza et al. 1994; Rothstein et al. 1992; Storck et al. 1992). In contrast, EAAT2, or GLT1 in rodents, is found exclusively in astroglia of the CNS and minimally in the cerebellum and other tissues (Arriza et al. 1994; Danbolt et al. 1990; Pines et al. 1992). EAAT3, or EAAC1 in rodents, is distributed throughout neurons of the cortex and non-CNS tissues including the lung and kidney (Arriza et al. 1997; Kanai and Hediger 1992; Rothstein et al. 1994). EAAT4, cloned in 1995, is found predominantly in cerebellar Purkinje cells, while EAAT5 is found almost exclusively in the retina (Arriza et al. 1997; Fairman et al. 1995). In antisense oligonucleotide knockdown experiments, EAAT1 (GLAST), EAAT2 (GLT1), and EAAT3 (EAAC1) were found to be responsible for approximately 30%, 50%, and 20% of glutamate transport in the striatum, respectively (Rothstein et al. 1996). Concordantly, impairment of glutamate transport led to 13 and 32 times the normal glutamate levels with knockdown of only the astroglial transporters, EAAT1 (GLAST) and EAAT2 (GLT1).

EAAT2 (GLT1 in rodents) is responsible for approximately 90% of glutamate clearance from the synapse (Lauriat and McInnes 2007). Because human EAAT2 and rodent GLT1 have 96% homologous sequences, EAAT2 is often used to denote both human and rodent protein, as is the case for this chapter. A similarly high degree of homology is shared by the rodent glutamate transporters, GLAST and EAAC1, with their human counterparts, EAAT1 and EAAT3 (Kirschner et al. 1994). Synaptic glutamate rapidly binds to EAAT2 (with a rate constant of 10^{-7}/M/s) and then is more slowly transported into the astrocytic space (at about 30 molecules/s) (Takahashi et al. 2015). Three Na^+ ions and one H^+ ion are cotransported, and one K^+ ion is countertransported with each glutamate molecule, with transport coupled to the sodium gradient. The EAATs also function as selective anion channels, with chloride permeation gated by the lateral movement of the glutamate transport domain and pore hydration (Machtens et al. 2015). EAAT2 has eight transmembrane domains and a long 3'-UTR for regulation. Multiple isoforms of EAAT2 have been discovered, primarily classified by their various C-termini as the originally isolated astroglial EAAT2a, the less prevalent astroglial EAAT2b, and the retinally expressed EAAT2c (Chen et al. 2002; Holmseth et al. 2009; Lauriat and McInnes 2007). EAAT2b and EAAT2c have truncated C-termini with instead alternative 11 amino acid sequences. All three variants have some differences in the 3'-UTR as well, suggesting differential regulation (Lauriat and McInnes 2007). All increase toward adulthood, though EAAT2a more than others (Holmseth et al. 2009). Complete knockout of EAAT2 in a mouse model caused premature death (50% survival after 6 weeks) following spontaneous epileptic seizures (Tanaka et al. 1997). Cortical synaptosomes had only 5.8% of the glutamate uptake capacity as controls, suggesting that EAAT2 is responsible for over 90% of glutamate clearance in the cortex. Peak concentrations of synaptically released glutamate in tissue slices were increased and remained elevated in the knockout mice, indicating impaired glutamate clearance following neurotransmission.

Essential to the functioning nervous system, the excitatory amino acid glutamate and its transporter EAAT2 are associated with multiple disorders, including amyotrophic lateral sclerosis (ALS), stroke, Parkinson's disease, and epilepsy (Takahashi et al. 2015). Understanding the regulation of EAAT2 expression could yield clues about its dysregulation in disease states and provide targets for altering it in therapy. Similar regulation of EAAT2 expression is seen in rodents and human astrocytes: Modulators include EGF, cAMP, PACAP, TGF-β, TNF-α, ceftriaxone, and estrogen compounds, often functioning through NF-κB (Takahashi et al. 2015). The transcription factor Pax6 also interacts with a distal enhancer element to increase EAAT2 transcription in astrocytes (Ghosh et al. 2016). Neuronal regulation also plays a role via presynaptic terminals (Yang et al. 2009). Neuronal soluble factors have also been found to induce EAAT2 expression in astrocytes via RTK signaling (Gegelashvili et al. 2000). Though the initiator is unclear, another pathway has been shown to regulate EAAT2 expression: PI3K phosphorylates Akt, which in turn phosphorylates mTOR, leading to an increase in EAAT2 protein (Wu et al. 2010). A fraction of

EAAT2 was found to be constitutively sumoylated in mouse CNS (Foran et al. 2014). The sumoylated transporter is localized to intracellular compartments, and promotion of desumoylation led to increased glutamate uptake, indicating a potential target for increasing EAAT2 function. Furthermore, posttranscriptional regulation and external stressors such as oxidative stress can affect EAAT2 expression (Tian et al. 2007; Zagami et al. 2005, 2009).

Glutamate transporters are decreased and dysregulated in cell models, animal models, and in ALS patients. A 50% decrease in EAAT2 protein was seen in spinal cord homogenates of SOD1-G85R mice (Bruijn et al. 1997). Similar decreases are seen in spinal cord homogenate and the ventral horn of the lumbar spinal cord of SOD1-G93A mice (Bendotti et al. 2001). Focal loss of EAAT2 in the ventral horn of the spinal cord was also seen in the SOD1-G93A rat model of ALS, actually preceding neuronal degeneration (Howland et al. 2002). Both gain and loss of TBPH (the TDP-43 homologue in *Drosophila*) altered mRNA expression of EAAT2, suggesting a link between pathological TDP-43 aggregates and mutations and EAAT2 dysregulation (Diaper et al. 2013). TDP-43 also binds to the 3'-UTR of EAAT2 mRNA, but substantially less so in FTLD-TDP brains (Tollervey et al. 2011). A decrease in EAAT2 expression was also suggested in immunohistochemical studies of rats expressing mutant TDP-43 (M337V) in astrocytes, though the data was not quantified and later control time points were not shown (Tong et al. 2013). Interestingly, FUS has also been found to bind to the 3'-UTR of EAAT2, suggesting regulatory functions (Lagier-Tourenne et al. 2012). While the loss of EAAT2 is believed to contribute to excitotoxicity in ALS, glutamate cannot be efficiently cleared from the synapse. Upregulation of EAAT2 in the spinal cord of SOD1-G93A mice via intraspinal delivery of an AAV8-Gfa2 vector to the ventral horn at disease onset did not protect phrenic motor neurons, their innervations, and nerve function nor extend life span (Li et al. 2015a). This could suggest that the decrease of EAAT2 has its most substantial impact before clinical onset. Similarly, a group II mGlu agonist, LY379268, enhanced EAAT2 expression in the spinal cord and rescued motor neurons but did not extend the life span of SOD1-G93A mice (Battaglia et al. 2015).

Similar, yet more nuanced, changes have been observed in human patients as well. In a small patient study, the density of D-aspartate binding, indicative of total glutamate transporter presence and unable to distinguish transporter subtypes, was decreased in the substantia gelatinosa and intermediate gray matter of motor neuron disease patients with a greater decrease in the ALS subgroup (Shaw et al. 1994). EAAT2 specifically was decreased in the motor cortex (71%) and spinal cord (~60%) of ALS patients, with a dramatic decrease (90%) in the motor cortex of a quarter of the patients (Rothstein et al. 1995). Another EAAT2-specific antibody showed that EAAT2 was decreased in the gray matter of the lumbar spinal cord but actually slightly increased in the middle laminae of the motor cortex (Fray et al. 1998). The discrepancy could potentially be explained by differences in the disease progression of patients. EAAT2 was found to decrease in the anterior horns of ALS and LMND patients, correlating with neuronal loss. Sasaki et al. observed an increase in EAAT2-positive granules in the ventral horn of patients with mild neuronal loss but a decrease in EAAT2 expression in patients with severe neuronal loss (Sasaki et al. 2000).

Recently, purer cultures of iPS-derived astrocytes were generated following insertion of GFP driven by the GFAP promoter using zinc-finger nuclease technology (Zhang et al. 2016). Lines from patients with the ALS-causing mutations SOD1-A4V and C9orf72 demonstrated similar or greater basal expression levels of EAAT2 and glutamate uptake when compared to a control patient line. They also exhibited a more dramatic increase in uptake when co-cultured with neurons. Though only one line of each mutation was studied, the results are suggestive that EAAT2 expression in astrocytes of human patients might not be decreased. EAAT2 expression was also used as a confirmation of astrocyte maturity, so lines expression little or no EAAT2 might not have been included.

In a new cell model for C9orf72-linked ALS, poly-dipeptides of PR, GR, and GA were transfected into NSC-34 cells, to mimic the RAN translation from the poly-hexanucleotide repeat (Kanekura et al. 2016). PR_{20} induced cell death and inhibited protein translation, as did GR_{20} to a lesser extent. PR_{20} interacts with the mRNA of proteins including EAAT2, blocking access of translation factors such as eIF4E and eIF4G, impeding their translation.

It is unlikely that the changes seen in EAAT2 at the protein level are due to DNA mutations or novel RNA variants. The discrepancies could be due to the existence of multiple isoforms of EAAT2, which behave differently in disease yet have not historically been distinguished (Lauriat and McInnes 2007). No gross loss of EAAT2 mRNA was evident in the motor cortex of ALS patients, even those with a dramatic loss of EAAT2 protein (Rothstein et al. 1996). ALS was also not associated with genetic linkage to the EAAT2 gene nor with point mutations within it (Aoki et al. 1998; Jackson et al. 1999).

Dumont et al. found that the mRNA levels of EAAT2a and EAAT2b isoforms were differentially regulated during disease course (Dumont et al. 2013). While EAAT2a mRNA was threefold higher than EAAT2b in the cortex of young SOD1-G93A rats, EAAT2a mRNA decreased by 50% and EAAT2b increased by 50%, until they were at almost equivalent levels in adult and end-stage rats, while no changes were seen in wild-type rats. That change coincided with a decrease in excitatory amino acid uptake by EAAT2 in cortical synaptosomes. In the lumbar spinal cord, EAAT2a increased over time in adult wild-type rats and was reduced in the ventral horns of late-stage mice. EAAT2b was elevated in young SOD1-G93A rats as compared to wild-type controls, but gradually decreased in both the ventral and dorsal horns with disease progression. In impressive agreement, overall EAAT2 levels decreased dramatically in the motor cortex of ALS patients accompanied by a loss in function, yet EAAT2b increased more than twofold (Maragakis et al. 2004).

Aberrant mRNA transcripts may play a role, though data have been mixed. Transcripts with partial intron-7 retention and exon 9 deletion were initially found in the motor cortex and spinal cord only of ALS patients and were correlated with EAAT2 protein loss (Lin et al. 1998). Later studies, however, demonstrated the presence of the same aberrant transcripts in the motor cortex of non-neurologically diseased control patients (Meyer et al. 1999). Differential expression of splice variants has been suggested in ALS brain tissue (Honig et al. 2000; Münch et al. 2002). Partial intron 7 retention and exon 9-skipping transcripts led to the production of truncated forms of EAAT2 which were shown in vitro to be rapidly degraded and have a dominant negative effect on normal EAAT2. Similarly, EAAT2 mRNA transcripts were altered over disease course in the SOD1-G93A mouse model (Münch et al. 2002). Later studies, however, found the same variants in normal controls, Alzheimer disease patients and Lewy body dementia patients, albeit at possibly lower (but not statistically significantly so) levels (Honig et al. 2000; Meyer et al. 1999). The same aberrant transcripts were recently produced in a cellular model of the C9orf72 mutation (Kwon et al. 2014). Human astrocytes expressing 15 or 20 repeats of a synthetic version of the C9orf72-associated RAN peptide PR expressed the aberrant EAAT2 mRNA forms after only 36 h, in a repeat-length-dependent manner. Other toxins did not produce the same aberrant splicing. RNA editing of A_{1591} to I in intron 7 occurs significantly more frequently in the spinal cord, motor cortex, and prefrontal cortex but not cerebellum of sporadic ALS patients than in controls (Flomen and Makoff 2011). Editing activates two cryptic polyadenylation sites, which in turn cause intron-7 retention with termination of transcription transcripts.

Perhaps most indicative, glutamate transport was decreased in synaptosomes derived from the spinal cord (59%), motor cortex (70%), and somatosensory cortex (39%) of patients with ALS as compared to healthy and diseased controls (Rothstein et al. 1992). The same decrease was not seen in the unaffected tissues of visual cortex, striatum, and hippocampus, further suggesting a disease-specific decrease. The decrease in transport was likely due to a decrease in overall protein as no change was seen in the affinity of the transporter for glutamate.

Posttranslational pathways might also contribute to the loss of EAAT2 protein or function in ALS. Oxidative reactions catalyzed by SOD1-A4V and SOD1-I113T but not SOD1-WT were found to inactivate EAAT2 and inhibit glutamate uptake (Trotti et al. 1999). Caspase-3 activation has been observed in cell and rodent models of ALS, while both primary isoforms of EAAT2 contain a caspase-3 consensus sequence (Boston-Howes et al. 2006). Cleavage to a truncated EAAT2 (trEAAT2) and a C-terminal fragment/end (CTE) occurs in a dose- and time-dependent manner and leads to a decrease in glutamate uptake but is blocked by a D to N mutation in the consensus site. The sumoylated product accumulates in the nucleus of astrocytes with disease progression, though it should be noted that the sumoylation site exists only on the more prevalent EAAT2a isoform. Transfection of an artificially fused CTE-SUMO1 construct induced astrocytes to become toxic to motor neurons via secreted factors (Foran et al. 2011). Expression of SOD1-G93A in MDCK cells led to the internalization of EAAT2 (but not EAAT3) and degradation in acidic compartments and inhibited synthesis (Vanoni et al. 2004). It is also interesting to note that riluzole enhances the activity of glutamate transporters including EAAT2 (Fumagalli et al. 2008). While riluzole has multiple effects on the CNS, increased reuptake of glutamate by GluTs could inhibit excitotoxic pathways and contribute to the protective effects seen.

4 Therapeutic Targeting of EAAT2

Due to the role of glutamate dysregulation in ALS and its potential consequence of causing excitotoxicity, already believed to play a substantial role in the progression of ALS, substantial research has been directed toward finding and demonstrating the efficacy of therapeutics that upregulate astrocytic EAAT2 expression. Screening processes have enabled the detection of such compounds, which have shown promise in vitro and in animal models of ALS (Kim et al. 2011). Using a library of FDA-approved compounds, researchers have searched for treatments with an already substantial research literature and a faster pathway to clinical trials. Screening strategies have employed immunoblotting of spinal cord slice cultures, ELISA of astrocyte-like PA-EAAT2 cells, and firefly luciferase expression on human fetal-derived immortalized astroglia using the E2 promoter (Colton et al. 2010; Li et al. 2011, 2015b).

In one of the first such screens for an EAAT2 upregulator, organotypic rat spinal cord slices were treated with a library of 1040 FDA-approved compounds (Rothstein et al. 2005). Homogenates of the treated tissues were blotted for EAAT2 protein expression, yielding numerous hits. A disproportionate number of the top 2% of hits fell in the class of β-lactam antibiotics, which were found to increase protein via enhanced promoter activity. Already known to penetrate the blood-brain barrier and to not cause substantial CNS toxicity, ceftriaxone (a β-lactam antibiotic) was found to increase EAAT2 protein expression in treated mice, both substantially and persistently. SOD1-G93A mice treated before onset had a decreased motor neuron loss after 2 weeks. When ceftriaxone was administered daily to SOD1-G93A mice beginning approximately at the time of onset, a delay in the loss of muscle strength and body weight was seen for 4–6 weeks, and a 10-day extension of survival was found. Further suggesting promise as a clinical therapeutic, ceftriaxone increases NF-κB binding to the EAAT2 promoter in primary human fetal astrocytes, increasing transcription (Lee et al. 2008).

Despite significant promise from the aforementioned preclinical studies and early-phase clinical trials, a phase III clinical trial showed that ceftriaxone was not effective for treating ALS. In the phase I branch of the study, 66 patients were randomized to placebo (pediatric multivitamin), 2 g/day, or 4 g/day via central venous catheter (Berry et al. 2013). Ceftriaxone was shown to remain above the target 1 uM in the CSF with both dosages, indicating sufficient penetration. The same patients were followed for 20 weeks, demonstrating sufficient safety and tolerability, other than hepatobiliary adverse events treated with ursodeoxycholic acid (which was given to all ceftriaxone patients in phase III). Promisingly, functional decline was slower in patients in the high-dose group versus the placebo ($p = 0.0416$) (Cudkowicz et al. 2014). Unfortunately, the decrease in functional decline was not replicated in the stage 3 interim or final analysis, nor was a difference seen in survival. Furthermore, adverse event rates were significantly and substantially higher in the treatment group for gastrointestinal, hepatobiliary, and blood- or bone marrow-related events. Participants receiving ceftriaxone also had more hepatobiliary serious adverse events but fewer infection-related serious adverse events.

Though ceftriaxone was ineffective in treating ALS patients, it remains unclear whether EAAT2 upregulation is no longer a viable clinical target: The actual efficacy of ceftriaxone in increasing EAAT2 in the CNS was not determined as no markers existed at the time of the study, though a PET ligand for EAAT2 is being developed (Gerdes et al. 2015). Animal models also demonstrated efficacy of ceftriaxone only when administered before symptom onset. Perhaps ceftriaxone would be more valuable for pre-onset familial ALS carriers. Stratification by mutation status, onset type, or cognition status might also yield interesting results (van den Berg 2014).

In separate studies, beginning with a cell-based ELISA screen of 140,000 small molecules compounds, Colton et al. identified 293 compounds that increased the expression of EAAT2 in PA-EAAT2 cells by at least 70% (3 standard deviations), 61 of which demonstrated a dose-dependent effect. Of those, three were selected for further optimization due to their potency, lack of toxicity, and following confirmatory biochemical and functional studies (Colton et al. 2010). A representative compound of a pyridazine-based series derived from those experiments LDN/OSU-0212320 was further characterized in a mouse model (Kong et al. 2014). The compound protected motor neurons co-cultured with astrocytes from glutamate-induced excitotoxicity. Furthermore, with treatment beginning near onset, the compound slowed motor decline and weight loss and substantially extended survival, suggesting that it is a promising agent for future trials.

Glutamate uptake by a clonal neural hybrid cell line (MN-1) known to express GluTs was assessed after treatment by a library of 1040 FDA-approved compounds (Boston-Howes et al. 2008). NDGA was identified by the screen and found to enhance glutamate transport in a dose-dependent matter and increase EAAT2 transport by threefold. While it also increased glutamate uptake in vivo as measured in synaptosomes of treated mice, it failed to increase glutamate uptake in symptomatic SOD1-G93A mice or to extend their life span.

Using an immortalized astrocytic cell line expressing firefly luciferase on the EAAT2 promoter, Li et al. screened a library of 1040 FDA-approved compounds for EAAT2 upregulation (Li et al. 2011). Harmine, a naturally occurring beta-carboline alkaloid and one of the top hits, was subsequently found to induce EAAT2 and EAAT1 mRNA and protein expression (as well as expression of their counterparts in mouse-derived cells). Furthermore, treatment of early-onset SOD1-G93A mice led to an increase of EAAT2 in the cortex.

Other recent preclinical trials have yielded promising results in the treatment of ALS by EAAT2 modification. Benkler et al. treated SOD1-G93A mice with a cocktail of lentiviruses encoding EAAT2, GDH2, and NRF2, with the objectives of increasing EAAT2 expression, reducing glutamate availability, and minimizing oxidative stress, respectively. While all three genes led to some neuroprotection in vitro, the combination led to the most neuroprotection, and it was found to expand life span by 19–22 days after administration at 65 days of age (Benkler et al. 2015). This finding further suggests that EAAT2-modifying therapies might be more effective in combination with other treatments.

An alternative explanation for the failure of ceftriaxone in clinical trials and of EAAT2-upregulating drugs in symptomatic animal models is the existence of a secondary EAAT2-mediated neurotoxicity pathway: As previously mentioned, EAAT2 is cleaved by caspase-3 in ALS, leading to the generation of CTE, which was found to be sumoylated (Gibb et al. 2007). Expression of an artificially fused CTE-SUMO1 causes astrocytes to secrete factors toxic to motor neurons (Foran et al. 2011). Not only does that suggest an additional pathway for EAAT2 toxicity in ALS, but that pathway would theoretically be upregulated with many of the EAAT2-targeted treatments, potentially preventing their success by coupling a decrease in excitotoxicity with an increase in secreted toxic factors.

5 Conclusions

Despite decades of research and advances in our understanding of ALS, riluzole remains the only FDA-approved disease-altering medication and prolongs life by only a few months. While it has become apparent that many converging pathways are at work, excitotoxicity remains a probable contributor to motor neuron death. The dysregulation of the glutamate transporter EAAT2 presumably leads to increased synaptic glutamate, causing excessive glutamate signaling and death of postsynaptic neurons. EAAT2 is also cleaved by caspase-3 in ALS, leading to the nuclear accumulation of a sumoylated fragment and astrocytic secretion of toxic factors.

Substantial research efforts have been directed toward transcriptional, translational, and functional activators of EAAT2 as therapeutics for ALS. While they have shown great promise in vitro and in preclinical animal models, they have yet to demonstrate efficacy in human patients. New tools which should aid in the testing of EAAT2-modifying agents are in development. While ceftriaxone failed in clinical trials, it remains unknown whether or not it achieved its mechanistic goal of upregulating EAAT2. A novel PET tracer for EAAT2 is being developed which would enable detection of EAAT2 changes in human patients throughout treatment (Gerdes et al. 2015). Such an agent would allow researchers to determine whether ceftriaxone, or other therapeutic agents actually upregulate EAAT2 before asking if EAAT2 upregulation extends life span in human patients.

Mutations in SOD1 account for only 20% of familial ALS cases and 5% of sporadic cases, while most therapeutics are tested in SOD1 mutant models (Robberecht and Philips 2013). The creation of novel models based on the substantially more prevalent C9orf72 repeat expansion and on iPS models derived from sporadic patients may enable more efficient and predictive screening of ALS therapeutics, including EAAT2-upregulating drugs (van Blitterswijk et al. 2012; Dimos et al. 2008). Motor neurons, astrocytes, and fibroblasts derived from iPSCs of ALS patients are becoming an increasingly reliable model of disease and degeneration.

Libraries of cells derived from familial and sporadic patients are being developed (Li et al. 2015b). These should enable more humanized models of disease, such as examination of EAAT2 expression and function and excitotoxicity across the spectrum of sporadic ALS. Unfortunately, EAAT2 is frequently used as a marker for maturation of astrocytes from iPSCs, which could confound studies of its expression and functional levels (Li et al. 2015b; Zhang et al. 2016). Despite the setback of ceftriaxone's failure as a therapeutic, EAAT2 remains a viable and promising target for the treatment of ALS.

Conflict of Interest The author declares no conflicts of interest.

Bibliography

Aizawa H, Sawada J, Hideyama T, Yamashita T, Katayama T, Hasebe N, et al. TDP-43 pathology in sporadic ALS occurs in motor neurons lacking the RNA editing enzyme ADAR2. Acta Neuropathol (Berl). 2010 Apr 7;120(1):75–84.

Andreadou E, Kapaki E, Kokotis P, Paraskevas GP, Katsaros N, Libitaki G, et al. Plasma glutamate and glycine levels in patients with amyotrophic lateral sclerosis: the effect of riluzole treatment. Clin Neurol Neurosurg. 2008 Mar;110(3):222–6.

Anneser JMH, Chahli C, Borasio GD. Protective effect of metabotropic glutamate receptor inhibition on amyotrophic lateral sclerosis–cerebrospinal fluid toxicity in vitro. Neuroscience. 2006;141(4):1879–86.

Aoki M, Lin C-LG, Rothstein JD, Geller BA, Hosler BA, Munsat TL, et al. Mutations in the glutamate transporter EAAT2 gene do not cause abnormal EAAT2 transcripts in amyotrophic lateral sclerosis. Ann Neurol. 1998 May 1;43(5):645–53.

Arriza JL, Fairman WA, Wadiche JI, Murdoch GH, Kavanaugh MP, Amara SG. Functional comparisons of three glutamate transporter subtypes cloned from human motor cortex. J Neurosci. 1994 Sep 1;14(9):5559–69.

Arriza JL, Eliasof S, Kavanaugh MP, Amara SG. Excitatory amino acid transporter 5, a retinal glutamate transporter coupled to a chloride conductance. Proc Natl Acad Sci. 1997 Apr 15;94(8):4155–60.

Battaglia G, Riozzi B, Bucci D, Di Menna L, Molinaro G, Pallottino S, et al. Activation of mGlu3 metabotropic glutamate receptors enhances GDNF and GLT-1 formation in the spinal cord and rescues motor neurons in the SOD-1 mouse model of amyotrophic lateral sclerosis. Neurobiol Dis. 2015 Feb;74:126–36.

Bellingham MC. A review of the neural mechanisms of action and clinical efficiency of riluzole in treating amyotrophic lateral sclerosis: what have we learned in the last decade? CNS Neurosci Ther. 2011 Feb 1;17(1):4–31.

Bendotti C, Tortarolo M, Suchak SK, Calvaresi N, Carvelli L, Bastone A, et al. Transgenic SOD1 G93A mice develop reduced GLT-1 in spinal cord without alterations in cerebrospinal fluid glutamate levels. J Neurochem. 2001 Nov 15;79(4):737–46.

Benkler C, Barhum Y, Ben-Zur T, Offen D. Multifactorial gene therapy enhancing the glutamate uptake system and reducing oxidative stress delays symptom onset and prolongs survival in the SOD1-G93A ALS mouse model. J Mol Neurosci. 2015 Dec 21;58(1):46–58.

van den Berg LH. Therapy of amyotrophic lateral sclerosis remains a challenge. Lancet Neurol. 2014 Nov;13(11):1062–3.

Berry JD, Shefner JM, Conwit R, Schoenfeld D, Keroack M, Felsenstein D, et al. Design and initial results of a multi-phase randomized trial of ceftriaxone in amyotrophic lateral sclerosis. PLoS One. 2013 Apr 17;8(4):e61177.

van Blitterswijk M, DeJesus-Hernandez M, Rademakers R. How do C9ORF72 repeat expansions cause amyotrophic lateral sclerosis and frontotemporal dementia: can we learn from other non-coding repeat expansion disorders? Curr Opin Neurol. 2012 Dec;25(6):689–700.

Blizzard CA, Southam KA, Dawkins E, Lewis KE, King AE, Clark JA, et al. Identifying the primary site of pathogenesis in amyotrophic lateral sclerosis – vulnerability of lower motor neurons to proximal excitotoxicity. Dis Model Mech. 2015 Mar 1;8(3):215–24.

Bonifacino T, Musazzi L, Milanese M, Seguini M, Marte A, Gallia E, et al. Altered mechanisms underlying the abnormal glutamate release in amyotrophic lateral sclerosis at a pre-symptomatic stage of the disease. Neurobiol Dis. 2016 Nov;95:122–33.

Boston-Howes W, Gibb SL, Williams EO, Pasinelli P, Brown RH Jr, Trotti D. Caspase-3 cleaves and inactivates the glutamate transporter EAAT2. J Biol Chem. 2006 May 19;281(20):14076–84.

Boston-Howes W, Williams EO, Bogush A, Scolere M, Pasinelli P, Trotti D. Nordihydroguaiaretic acid increases glutamate uptake in vitro and in vivo: therapeutic implications for amyotrophic lateral sclerosis. Exp Neurol. 2008 Sep;213(1):229–37.

Bruijn LI, Becher MW, Lee MK, Anderson KL, Jenkins NA, Copeland NG, et al. ALS-linked SOD1 mutant G85R mediates damage to astrocytes and promotes rapidly progressive disease with SOD1-containing inclusions. Neuron. 1997 Feb;18(2):327–38.

Camu W, Billiard M, Baldy-Moulinier M. Fasting plasma and CSF amino acid levels in amyotrophic lateral sclerosis: a subtype analysis. Acta Neurol Scand. 1993 Jul 1;88(1):51–5.

Carriedo SG, Yin HZ, Weiss JH. Motor neurons are selectively vulnerable to AMPA/kainate receptor-mediated injury in vitro. J Neurosci. 1996 Jul 1;16(13):4069–79.

Chen W, Aoki C, Mahadomrongkul V, Gruber CE, Wang GJ, Blitzblau R, et al. Expression of a variant form of the glutamate transporter GLT1 in neuronal cultures and in neurons and astrocytes in the rat brain. J Neurosci. 2002 Mar 15;22(6):2142–52.

Colton CK, Kong Q, Lai L, Zhu MX, Seyb KI, Cuny GD, et al. Identification of translational activators of glial glutamate transporter EAAT2 through cell-based high-throughput screening: an approach to prevent excitotoxicity. J Biomol Screen. 2010 Jul 1;15(6):653–62.

Corona JC, Tapia R. AMPA receptor activation, but not the accumulation of endogenous extracellular glutamate, induces paralysis and motor neuron death in rat spinal cord in vivo. J Neurochem. 2004 May 1;89(4):988–97.

Couratier P, Sindou P, Hugon J, Couratier P, Hugon J, Vallat J-M, et al. Originally published as Volume 1, Issue 8840 Cell culture evidence for neuronal degeneration in amyotrophic lateral sclerosis being linked to glutamate AMPA/kainate receptors. Lancet. 1993 Jan 30;341(8840):265–8.

Cudkowicz ME, Titus S, Kearney M, Yu H, Sherman A, Schoenfeld D, et al. Safety and efficacy of ceftriaxone for amyotrophic lateral sclerosis: a multi-stage, randomised, double-blind, placebo-controlled trial. Lancet Neurol. 2014 Nov;13(11):1083–91.

Damme PV, Van Den Bosch L, Houtte EV, Callewaert G, Robberecht W. GluR2-dependent properties of AMPA receptors determine the selective vulnerability of motor neurons to excitotoxicity. J Neurophysiol. 2002 Sep 1;88(3):1279–87.

Danbolt NC, Pines G, Kanner BI. Purification and reconstitution of the sodium- and potassium-coupled glutamate transport glycoprotein from rat brain. Biochemistry (Mosc). 1990 Jul 1;29(28):6734–40.

Diaper DC, Adachi Y, Lazarou L, Greenstein M, Simoes FA, Domenico AD, et al. Drosophila TDP-43 dysfunction in glia and muscle cells cause cytological and behavioural phenotypes that characterize ALS and FTLD. Hum Mol Genet. 2013 May 31;ddt243.

Dimos JT, Rodolfa KT, Niakan KK, Weisenthal LM, Mitsumoto H, Chung W, et al. Induced pluripotent stem cells generated from patients with ALS can be differentiated into motor neurons. Science. 2008 Aug 29;321(5893):1218–21.

Dingledine R, Borges K, Bowie D, Traynelis SF. The glutamate receptor ion channels. Pharmacol Rev. 1999 Mar 1;51(1):7–62.

Donnelly CJ, Zhang P-W, Pham JT, Haeusler AR, Mistry NA, Vidensky S, et al. RNA toxicity from the ALS/FTD C9ORF72 expansion is mitigated by antisense intervention. Neuron. 2013 Oct 16;80(2):415–28.

Dumont AO, Hermans E, Goursaud S. Differential regulation of the glutamate transporter variants GLT-1a and GLT-1b in the cortex and spinal cord of transgenic rats expressing hSOD1G93A. Neurochem Int. 2013 Aug;63(2):61–8.

Fairman WA, Vandenberg RJ, Arriza JL, Kavanaught MP, Amara SG. An excitatory amino-acid transporter with properties of a ligand-gated chloride channel. Nature. 1995 Jun 15;375(6532):599–603.

Flomen R, Makoff A. Increased RNA editing in EAAT2 pre-mRNA from amyotrophic lateral sclerosis patients: involvement of a cryptic polyadenylation site. Neurosci Lett. 2011 Jun 22;497(2):139–43.

Foran E, Bogush A, Goffredo M, Roncaglia P, Gustincich S, Pasinelli P, et al. Motor neuron impairment mediated by a sumoylated fragment of the glial glutamate transporter EAAT2. Glia. 2011 Nov;59(11):1719–31.

Foran E, Rosenblum L, Bogush A, Pasinelli P, Trotti D. Sumoylation of the astroglial glutamate transporter EAAT2 governs its intracellular compartmentalization. Glia. 2014 Apr 1;n/a–n/a.

Fray AE, Ince PG, Banner SJ, Milton ID, Usher PA, Cookson MR, et al. The expression of the glial glutamate transporter protein EAAT2 in motor neuron disease: an immunohistochemical study. Eur J Neurosci. 1998 Aug 1;10(8):2481–9.

Fumagalli E, Funicello M, Rauen T, Gobbi M, Mennini T. Riluzole enhances the activity of glutamate transporters GLAST, GLT1 and EAAC1. Eur J Pharmacol. 2008 Jan 14;578(2–3):171–6.

Gegelashvili G, Dehnes Y, Danbolt NC, Schousboe A. The high-affinity glutamate transporters GLT1, GLAST, and EAAT4 are regulated via different signalling mechanisms. Neurochem Int. 2000 Aug 1;37(2–3):163–70.

Gerdes J, Ahmed S, Braden M, Dannoon S, Blecha J, VanBrocklin H. Radiosynthesis, rodent and non-human primate studies of a novel PET tracer for the excitatory amino acid transporter 2 (EAAT2) in the CNS. J Nucl Med. 2015 May 1;56(suppl 3):1100.

Ghosh M, Lane M, Krizman E, Sattler R, Rothstein JD, Robinson MB. The transcription factor Pax6 contributes to the induction of GLT-1 expression in astrocytes through an interaction with a distal enhancer element. J Neurochem. 2016 Jan 1;136(2):262–75.

Gibb SL, Boston-Howes W, Lavina ZS, Gustincich S, Brown RH Jr, Pasinelli P, et al. A caspase-3-cleaved fragment of the glial glutamate transporter EAAT2 is sumoylated and targeted to promyelocytic leukemia nuclear bodies in mutant SOD1-linked amyotrophic lateral sclerosis. J Biol Chem. 2007 Nov 2;282(44):32480–90.

Giribaldi F, Milanese M, Bonifacino T, Anna Rossi PI, Di Prisco S, Pittaluga A, et al. Group I metabotropic glutamate autoreceptors induce abnormal glutamate exocytosis in a mouse model of amyotrophic lateral sclerosis. Neuropharmacology. 2013 Mar;66:253–63.

Gurney ME, Pu H, Chiu AY, Dal Canto MC, Polchow CY, Alexander DD, et al. Motor neuron degeneration in mice that express a human Cu,Zn superoxide dismutase mutation. Science. 1994 Jun 17;264(5166):1772–5.

Haidet-Phillips AM, Hester ME, Miranda CJ, Meyer K, Braun L, Frakes A, et al. Astrocytes from familial and sporadic ALS patients are toxic to motor neurons. Nat Biotechnol. 2011 Sep;29(9):824–8.

Hideyama T, Yamashita T, Aizawa H, Tsuji S, Kakita A, Takahashi H, et al. Profound downregulation of the RNA editing enzyme ADAR2 in ALS spinal motor neurons. Neurobiol Dis. 2012 Mar;45(3):1121–8.

Holmseth S, Scott HA, Real K, Lehre KP, Leergaard TB, Bjaalie JG, et al. The concentrations and distributions of three C-terminal variants of the GLT1 (EAAT2; slc1a2) glutamate transporter protein in rat brain tissue suggest differential regulation. Neuroscience. 2009 Sep 15;162(4):1055–71.

Honig LS, Chambliss DD, Bigio EH, Carroll SL, Elliott JL. Glutamate transporter EAAT2 splice variants occur not only in ALS, but also in AD and controls. Neurology. 2000 Oct 24;55(8):1082–8.

Howland DS, Liu J, She Y, Goad B, Maragakis NJ, Kim B, et al. Focal loss of the glutamate transporter EAAT2 in a transgenic rat model of SOD1 mutant-mediated amyotrophic lateral sclerosis (ALS). Proc Natl Acad Sci. 2002 Feb 5;99(3):1604–9.

Jackson M, Steers G, Leigh PN, Morrison KE. Polymorphisms in the glutamate transporter gene EAAT2 in European ALS patients. J Neurol. 1999;246(12):1140–4.

Kanai Y, Hediger MA. Primary structure and functional characterization of a high-affinity glutamate transporter. Nature. 1992 Dec 3;360(6403):467–71.

Kanekura K, Yagi T, Cammack AJ, Mahadevan J, Kuroda M, Harms MB, et al. Poly-dipeptides encoded by the C9ORF72 repeats block global protein translation. Hum Mol Genet. 2016 Feb 29;ddw052.

Kawahara Y, Ito K, Sun H, Aizawa H, Kanazawa I, Kwak S. Glutamate receptors: RNA editing and death of motor neurons. Nature. 2004 Feb 26;427(6977):801.

Kim K, Lee S-G, Kegelman TP, Su Z-Z, Das SK, Dash R, et al. Role of excitatory amino acid transporter-2 (EAAT2) and glutamate in neurodegeneration: opportunities for developing novel therapeutics. J Cell Physiol. 2011 Oct 1;226(10):2484–93.

Kirschner MA, Copeland NG, Gilbert DJ, Jenkins NA, Amara SG. Mouse excitatory amino acid transporter EAAT2: isolation, characterization, and proximity to neuroexcitability loci on mouse chromosome 2. Genomics. 1994 Nov 15;24(2):218–24.

Kong Q, Chang L-C, Takahashi K, Liu Q, Schulte DA, Lai L, et al. Small-molecule activator of glutamate transporter EAAT2 translation provides neuroprotection. J Clin Invest. 2014 Mar 3;124(3):1255–67.

Kuner R, Groom AJ, Bresink I, Kornau H-C, Stefovska V, Müller G, et al. Late-onset motoneuron disease caused by a functionally modified AMPA receptor subunit. Proc Natl Acad Sci U S A. 2005 Apr 19;102(16):5826–31.

Kwon I, Xiang S, Kato M, Wu L, Theodoropoulos P, Wang T, et al. Poly-dipeptides encoded by the C9orf72 repeats bind nucleoli, impede RNA biogenesis, and kill cells. Science. 2014 Sep 5;345(6201):1139–45.

Lagier-Tourenne C, Polymenidou M, Hutt KR, Vu AQ, Baughn M, Huelga SC, et al. Divergent roles of ALS-linked proteins FUS/TLS and TDP-43 intersect in processing long pre-mRNAs. Nat Neurosci. 2012 Sep 30;15(11):1488–97.

Lauriat TL, McInnes LA. EAAT2 regulation and splicing: relevance to psychiatric and neurological disorders. Mol Psychiatry. 2007 Aug 7;12(12):1065–78.

Lee S-G, Su Z-Z, Emdad L, Gupta P, Sarkar D, Borjabad A, et al. Mechanism of ceftriaxone induction of excitatory amino acid transporter-2 expression and glutamate uptake in primary human astrocytes. J Biol Chem. 2008 May 9;283(19):13116–23.

Li Y, Sattler R, Yang EJ, Nunes A, Ayukawa Y, Akhtar S, et al. Harmine, a natural beta-carboline alkaloid, upregulates astroglial glutamate transporter expression. Neuropharmacology. 2011 Jun;60(7–8):1168–75.

Li K, Hala TJ, Seetharam S, Poulsen DJ, Wright MC, Lepore AC. GLT1 overexpression in SOD1G93A mouse cervical spinal cord does not preserve diaphragm function or extend disease. Neurobiol Dis. 2015a Jun;78:12–23.

Li Y, Balasubramanian U, Cohen D, Zhang P-W, Mosmiller E, Sattler R, et al. A comprehensive library of familial human amyotrophic lateral sclerosis induced pluripotent stem cells. PLoS One. 2015b Mar 11;10(3):e0118266.

Lin C-LG, Bristol LA, Jin L, Dykes-Hoberg M, Crawford T, Clawson L, et al. Aberrant RNA processing in a neurodegenerative disease: the cause for absent EAAT2, a glutamate transporter, in amyotrophic lateral sclerosis. Neuron. 1998 Mar;20(3):589–602.

Machtens J-P, Kortzak D, Lansche C, Leinenweber A, Kilian P, Begemann B, et al. Mechanisms of anion conduction by coupled glutamate transporters. Cell. 2015 Jan 29;160(3):542–53.

Mahajan SS, Thai KH, Chen K, Ziff E. Exposure of neurons to excitotoxic levels of glutamate induces cleavage of the RNA editing enzyme, adenosine deaminase acting on RNA 2, and loss of GLUR2 editing. Neuroscience. 2011 Aug 25;189:305–15.

Maragakis NJ, Dykes-Hoberg M, Rothstein JD. Altered expression of the glutamate transporter EAAT2b in neurological disease. Ann Neurol. 2004 Apr 1;55(4):469–77.

Marin B, Boumédiene F, Logroscino G, Couratier P, Babron M-C, Leutenegger AL, et al. Variation in worldwide incidence of amyotrophic lateral sclerosis: a meta-analysis. Int J Epidemiol. 2016 May 16;dyw061.

Meyer T, Fromm A, Münch C, Schwalenstöcker B, Fray AE, Ince PG, et al. The RNA of the glutamate transporter EAAT2 is variably spliced in amyotrophic lateral sclerosis and normal individuals. J Neurol Sci. 1999 Nov 15;170(1):45–50.

Milanese M, Zappettini S, Onofri F, Musazzi L, Tardito D, Bonifacino T, et al. Abnormal exocytotic release of glutamate in a mouse model of amyotrophic lateral sclerosis. J Neurochem. 2011 Mar 1;116(6):1028–42.

Münch C, Ebstein M, Seefried U, Zhu B, Stamm S, Landwehrmeyer GB, et al. Alternative splicing of the 5′-sequences of the mouse EAAT2 glutamate transporter and expression in a transgenic model for amyotrophic lateral sclerosis. J Neurochem. 2002 Aug 1;82(3):594–603.

Nagai M, Re DB, Nagata T, Chalazonitis A, Jessell TM, Wichterle H, et al. Astrocytes expressing ALS-linked mutated SOD1 release factors selectively toxic to motor neurons. Nat Neurosci. 2007 May;10(5):615–22.

Nicoll RA, Malenka RC, Kauer JA. Functional comparison of neurotransmitter receptor subtypes in mammalian central nervous system. Physiol Rev. 1990 Apr 1;70(2):513–65.

Pines G, Danbolt NC, Bjørås M, Zhang Y, Bendahan A, Eide L, et al. Cloning and expression of a rat brain L-glutamate transporter. Nature. 1992 Dec 3;360(6403):464–7.

Plaitakis A, Caroscio JT. Abnormal glutamate metabolism in amyotrophic lateral sclerosis. Ann Neurol. 1987 Nov 1;22(5):575–9.

Plaitakis A, Constantakakis E, Smith J. The neuroexcitotoxic amino acids glutamate and aspartate are altered in the spinal cord and brain in amyotrophic lateral sclerosis. Ann Neurol. 1988 Sep 1;24(3):446–9.

Robberecht W, Philips T. The changing scene of amyotrophic lateral sclerosis. Nat Rev Neurosci. 2013 Apr;14(4):248–64.

Rosen DR, Siddique T, Patterson D, Figlewicz DA, Sapp P, Hentati A, et al. Mutations in Cu/Zn superoxide dismutase gene are associated with familial amyotrophic lateral sclerosis. Nature. 1993 Mar 4;362(6415):59–62.

Ross CA, Tabrizi SJ. Huntington's disease: from molecular pathogenesis to clinical treatment. Lancet Neurol. 2011 Jan;10(1):83–98.

Rothstein JD. Excitotoxicity hypothesis. Neurology. 1996 Oct 1;47(4 Suppl 2):19S–26S.

Rothstein JD, Tsai G, Kuncl RW, Clawson L, Cornblath DR, Drachman DB, et al. Abnormal excitatory amino acid metabolism in amyotrophic lateral sclerosis. Ann Neurol. 1990 Jul 1;28(1):18–25.

Rothstein JD, Martin LJ, Kuncl RW. Decreased glutamate transport by the brain and spinal cord in amyotrophic lateral sclerosis. N Engl J Med. 1992 May 28;326(22):1464–8.

Rothstein JD, Martin L, Levey AI, Dykes-Hoberg M, Jin L, Wu D, et al. Localization of neuronal and glial glutamate transporters. Neuron. 1994 Sep;13(3):713–25.

Rothstein JD, Van Kammen M, Levey AI, Martin LJ, Kuncl RW. Selective loss of glial glutamate transporter GLT-1 in amyotrophic lateral sclerosis. Ann Neurol. 1995 Jul 1;38(1):73–84.

Rothstein JD, Dykes-Hoberg M, Pardo CA, Bristol LA, Jin L, Kuncl RW, et al. Knockout of glutamate transporters reveals a major role for astroglial transport in excitotoxicity and clearance of glutamate. Neuron. 1996 Mar;16(3):675–86.

Rothstein JD, Patel S, Regan MR, Haenggeli C, Huang YH, Bergles DE, et al. β-Lactam antibiotics offer neuroprotection by increasing glutamate transporter expression. Nature. 2005 Jan 6;433(7021):73–7.

Sasaki S, Komori T, Iwata M. Excitatory amino acid transporter 1 and 2 immunoreactivity in the spinal cord in amyotrophic lateral sclerosis. Acta Neuropathol (Berl). 2000;100(2):138–44.

Sen I, Nalini A, Joshi NB, Joshi PG. Cerebrospinal fluid from amyotrophic lateral sclerosis patients preferentially elevates intracellular calcium and toxicity in motor neurons via AMPA/kainate receptor. J Neurol Sci. 2005 Aug 15;235(1–2):45–54.

Shaw PJ, Chinnery RM, Ince PG. [3H]d-aspartate binding sites in the normal human spinal cord and changes in motor neuron disease: a quantitative autoradiographic study. Brain Research. 1994;655(1–2):195–201.

Spreux-Varoquaux O, Bensimon G, Lacomblez L, Salachas F, Pradat PF, Le Forestier N, et al. Glutamate levels in cerebrospinal fluid in amyotrophic lateral sclerosis: a reappraisal using a new HPLC method with coulometric detection in a large cohort of patients. J Neurol Sci. 2002 Jan 15;193(2):73–8.

Storck T, Schulte S, Hofmann K, Stoffel W. Structure, expression, and functional analysis of a Na(+)-dependent glutamate/aspartate transporter from rat brain. Proc Natl Acad Sci. 1992 Nov 15;89(22):10955–9.

Takahashi K, Foster JB, Lin C-LG. Glutamate transporter EAAT2: regulation, function, and potential as a therapeutic target for neurological and psychiatric disease. Cell Mol Life Sci. 2015 Jun 2;72(18):3489–506.

Takuma H, Kwak S, Yoshizawa T, Kanazawa I. Reduction of GluR2 RNA editing, a molecular change that increases calcium influx through AMPA receptors, selective in the spinal ventral gray of patients with amyotrophic lateral sclerosis. Ann Neurol. 1999 Dec 1;46(6):806–15.

Tanaka K, Watase K, Manabe T, Yamada K, Watanabe M, Takahashi K, et al. Epilepsy and exacerbation of brain injury in mice lacking the glutamate transporter GLT-1. Science. 1997 Jun 13;276(5319):1699–702.

Tian G, Lai L, Guo H, Lin Y, Butchbach MER, Chang Y, et al. Translational control of glial glutamate transporter EAAT2 expression. J Biol Chem. 2007 Jan 19;282(3):1727–37.

Tollervey JR, Curk T, Rogelj B, Briese M, Cereda M, Kayikci M, et al. Characterizing the RNA targets and position-dependent splicing regulation by TDP-43. Nat Neurosci. 2011 Apr;14(4):452–8.

Tong J, Huang C, Bi F, Wu Q, Huang B, Liu X, et al. Expression of ALS-linked TDP-43 mutant in astrocytes causes non-cell-autonomous motor neuron death in rats. EMBO J. 2013 May 28;32(13):1917–26.

Tortarolo M, Grignaschi G, Calvaresi N, Zennaro E, Spaltro G, Colovic M, et al. Glutamate AMPA receptors change in motor neurons of SOD1G93A transgenic mice and their inhibition by a noncompetitive antagonist ameliorates the progression of amytrophic lateral sclerosis-like disease. J Neurosci Res. 2006 Jan 1;83(1):134–46.

Trotti D, Rolfs A, Danbolt NC, Brown RH, Hediger MA. SOD1 mutants linked to amyotrophic lateral sclerosis selectively inactivate a glial glutamate transporter. Nat Neurosci. 1999 May;2(5):427–33.

Turner MR, Hardiman O, Benatar M, Brooks BR, Chio A, de Carvalho M, et al. Controversies and priorities in amyotrophic lateral sclerosis. Lancet Neurol. 2013 Mar;12(3):310–22.

Van Damme P, Leyssen M, Callewaert G, Robberecht W, Van Den Bosch L. The AMPA receptor antagonist NBQX prolongs survival in a transgenic mouse model of amyotrophic lateral sclerosis. Neurosci Lett. 2003 Jun 5;343(2):81–4.

Van Den Bosch L, Van Damme P, Bogaert E, Robberecht W. The role of excitotoxicity in the pathogenesis of amyotrophic lateral sclerosis. Biochim Biophys Acta (BBA) – Mol Basis Dis. 2006 Nov;1762(11–12):1068–82.

Vanoni C, Massari S, Losa M, Carrega P, Perego C, Conforti L, et al. Increased internalisation and degradation of GLT-1 glial glutamate transporter in a cell model for familial amyotrophic lateral sclerosis (ALS). J Cell Sci. 2004 Oct 15;117(22):5417–26.

Wang Y, Qin Z. Molecular and cellular mechanisms of excitotoxic neuronal death. Apoptosis. 2010 Nov 1;15(11):1382–402.

Wu X, Kihara T, Akaike A, Niidome T, Sugimoto H. PI3K/Akt/mTOR signaling regulates glutamate transporter 1 in astrocytes. Biochem Biophys Res Commun. 2010 Mar 12;393(3):514–8.

Wuolikainen A, Moritz T, Marklund SL, Antti H, Andersen PM. Disease-related changes in the cerebrospinal fluid metabolome in amyotrophic lateral sclerosis detected by GC/TOFMS. PLoS One. 2011 Apr 4;6(4):e17947.

Yang Y, Gozen O, Watkins A, Lorenzini I, Lepore A, Gao Y, et al. Presynaptic regulation of astroglial excitatory neurotransmitter transporter GLT1. Neuron. 2009 Mar 26;61(6):880–94.

Yin HZ, Tang DT, Weiss JH. Intrathecal infusion of a Ca2+-permeable AMPA channel blocker slows loss of both motor neurons and of the astrocyte glutamate transporter, GLT-1 in a mutant SOD1 rat model of ALS. Exp Neurol. 2007 Oct;207(2):177–85.
Zagami CJ, O'shea RD, Lau CL, Cheema SS, Beart PM. Regulation of glutamate transporters in astrocytes: evidence for a relationship between transporter expression and astrocytic phenotype. Neurotox Res. 2005;7(1–2):143–9.
Zagami CJ, Beart PM, Wallis N, Nagley P, O'shea RD. Oxidative and excitotoxic insults exert differential effects on spinal motoneurons and astrocytic glutamate transporters: Implications for the role of astrogliosis in amyotrophic lateral sclerosis. Glia. 2009 Jan 15;57(2):119–35.
Zhang P-W, Haidet-Phillips AM, Pham JT, Lee Y, Huo Y, Tienari PJ, et al. Generation of GFAP::GFP astrocyte reporter lines from human adult fibroblast-derived iPS cells using zinc-finger nuclease technology. Glia. 2016 Jan 1;64(1):63–75.

Glial GABA Transporters as Modulators of Inhibitory Signalling in Epilepsy and Stroke

Maria E.K. Lie, Anas Al-Khawaja, Maria Damgaard, Anne S. Haugaard, Arne Schousboe, Andrew N. Clarkson, and Petrine Wellendorph

Abstract Imbalances in GABA-mediated tonic inhibition are involved in several pathophysiological conditions. A classical way of controlling tonic inhibition is through pharmacological intervention with extrasynaptic $GABA_A$ receptors that sense ambient GABA and mediate a persistent GABAergic conductance. An increase in tonic inhibition may, however, also be obtained indirectly by inhibiting glial GABA transporters (GATs). These are sodium-coupled membrane transport proteins that normally act to terminate GABA neurotransmitter action by taking up GABA into surrounding astrocytes. The aim of the review is to provide an overview of glial GATs in regulating tonic inhibition, especially in epilepsy and stroke. This entails a comprehensive summary of changes known to occur in GAT expression levels and signalling following epileptic and ischemic insults. Further, we discuss the accumulating pharmacological evidence for targeting GATs in these diseases.

Keywords GAT • GAT3 • BGT1 • GAT1 • Ischemic stroke • Epilepsy • Tonic inhibition • $GABA_A$ receptors

1 Introduction

The neurotransmitter γ-aminobutyric acid (GABA) plays an essential role in controlling inhibitory synaptic transmission in the mammalian central nervous system (CNS). This is mediated largely through ligand-gated $GABA_A$ receptors that, depending on subunit composition, are located either synaptically, mediating phasic

Maria E.K. Lie and Anas Al-Khawaja contributed equally to this work.

M.E.K. Lie • A. Al-Khawaja • M. Damgaard • A.S. Haugaard
A. Schousboe • P. Wellendorph (✉)
Department of Drug Design and Pharmacology, Faculty of Health and Medical Sciences, University of Copenhagen, 2100 Copenhagen, Denmark
e-mail: arne.schousboe@sund.ku.dk; pw@sund.ku.dk

A.N. Clarkson
Department of Anatomy, Brain Health Research Centre and Brain Research New Zealand, University of Otago, 9054 Dunedin, New Zealand

Faculty of Pharmacy, The University of Sydney, Sydney, NSW 2006, Australia

A. Ortega, A. Schousboe (eds.), *Glial Amino Acid Transporters*, Advances in Neurobiology 16, DOI 10.1007/978-3-319-55769-4_7

inhibition, or extrasynaptically, mediating a persistent (tonic) inhibition (for reviews, see Farrant and Nusser 2005; Belelli et al. 2009). Within the last two decades, GABA-mediated tonic inhibition has attracted considerable interest in drug discovery as it has been found to be altered in several brain disorders (for reviews, see Orser 2006; Brickley and Mody 2012; Lee and Maguire 2014). The pharmacological modulation of tonic GABA inhibition has so far mostly been achieved via stimulation of extrasynaptic $GABA_A$ receptors, although regulation may also be attained by more indirect means, for example, by promoting changes in ambient GABA levels in the extracellular space (Brickley and Mody 2012). In this regard, GABA transporters (GATs) expressed on the plasma membrane of glial cells have been found to play an essential role (Fig. 1a) (Madsen et al. 2010). This is due to their ability to remove GABA from the extracellular space (Schousboe 2003), but also in their ability to release GABA by nonvesicular means via reverse transport in response to certain conditions (Richerson and Wu 2003). As illustrated in Fig. 1b, the increased ambient GABA levels from either vesicular or nonvesicular sources may, in part, diffuse to the peri- or extrasynaptic areas where extrasynaptic $GABA_A$ receptors (subtypes $\alpha_{4/6}\beta_{1-3}\delta$ or $\alpha_5\beta_3\gamma_2$) mediate increases in tonic inhibition (Belelli et al. 2009). The degree of GABAergic inhibitory signalling depends on the activity of extrasynaptically expressed GATs that can take up GABA into astrocytes, hereby regulating ambient GABA levels.

Based on the isolated role for glial GATs in the metabolic degradation of GABA (Schousboe et al. 1983), it was proposed that pharmacological inhibition of glial GATs would be an attractive way to increase ambient GABA levels and hence affect GABAergic inhibitory neurotransmission (Schousboe 2003). Modulation of tonic inhibition through GATs is now well supported in diseases such as epilepsy and ischemic stroke, which also parallels with reported changes in expression levels of several of the GATs under these conditions (Conti et al. 2004 and see below). In addition, efforts in medicinal chemistry have provided pharmacologists with useful tool compounds to begin dissecting GAT subtype involvement (for reviews, see Høg et al. 2006; Wellendorph et al. 2016 and refer to Damgaard et al. In Press; Schousboe et al. In Press).

1.1 Classification of GATs

To date, four GATs have been cloned. Following the nomenclature given by IUPHAR (Southan et al. 2016), these are named GAT1, BGT1, GAT2 and GAT3, corresponding to the human names. Being members of the solute carriers (subfamily *SLC6A*), the GATs are characterized by a 12 transmembrane-spanning domain and an overall similar Na^+/Cl^--dependent "alternating-access" transport mechanism (Jardetzky 1966 and reviewed by Kristensen et al. 2011). All four subtypes transport GABA with low micromolar affinity, although BGT1 does this with approximately three times lower affinity and additionally has the ability to transport betaine (hence the name betaine/GABA transporter; BGT1). From a phylogenetic point of view,

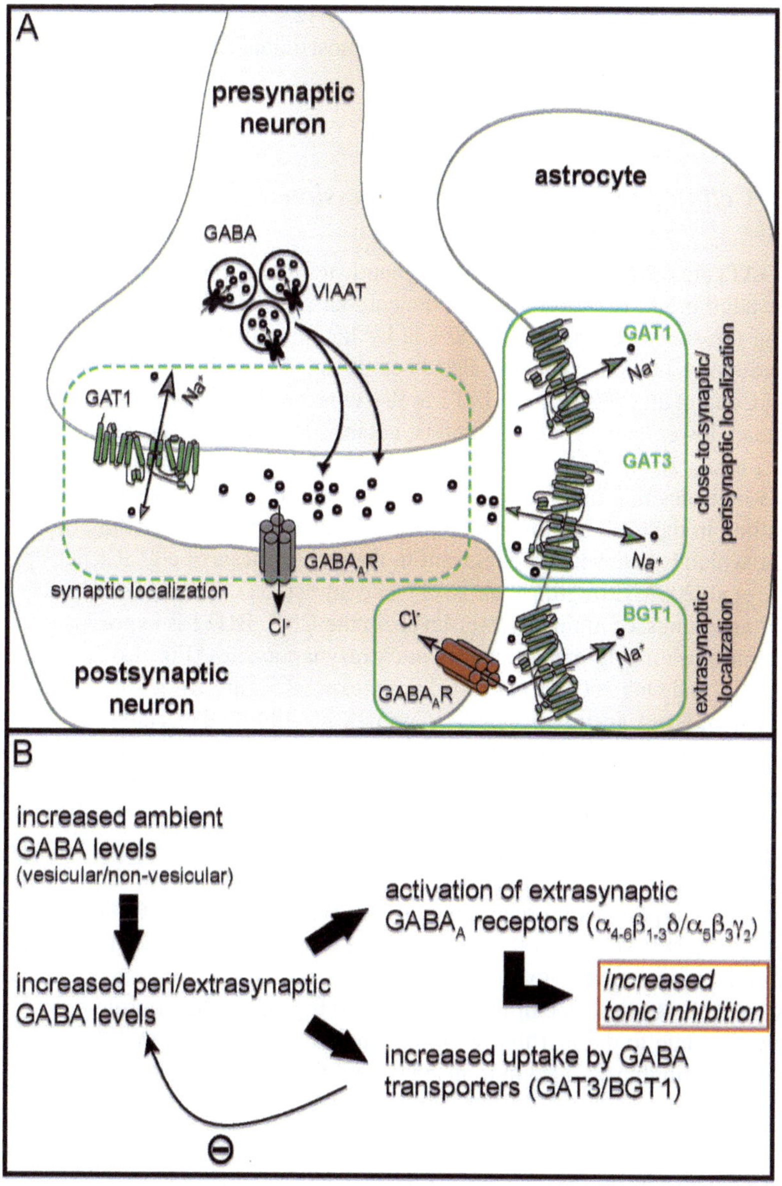

Fig. 1 (**a**) Overview of a GABAergic synapse illustrating the subcellular localization of GATs. The primary expression of GATs is illustrated, and the targets involved in regulating tonic inhibition are framed. (**b**) Illustrative scheme of GABAergic signalling and how it can lead to tonic inhibition. *VIAAT* vesicular inhibitory amino acid transporter (Modified from Wellendorph et al. 2016)

GAT1 diverges the most from the other GATs (e.g. 52% sequence identity with GAT3), whereas BGT1 and GAT2 are the most related (68% sequence identity) (Wellendorph et al. 2016).

1.2 Cellular and Subcellular Expression

The GATs are primarily located on presynaptic terminals and surrounding glial cells in the adult brain. However, the specific cellular and subcellular distribution varies among the subtypes (Fig. 1a). GAT1 is by far the most abundant subtype and is preferentially neuronal but has also been found in astrocytes (Fig. 1a) (Borden 1996; Conti et al. 1998, 2004). GAT3 is also expressed at high levels in the CNS but at a much lower level than GAT1 and is, notably, localized exclusively to astrocytes (Conti et al. 2004; Melone et al. 2005). The abundant expression of GAT1 and GAT3 indicates that these transporters play a significant role in regulating GABA signalling in the CNS (Dalby 2003), including extrasynaptic signalling (Song et al. 2013), which is believed to be relevant in the pathogenesis of e.g. absence epilepsy (Cope et al. 2009; Errington et al. 2011). In contrast to GAT1 and GAT3, BGT1 and GAT2 are expressed at much lower levels in the CNS. BGT1 is expressed on astrocytes and presumably predominantly at extrasynaptic sites (Fig. 1a) (Borden et al. 1995; Zhu and Ong 2004b). GAT2 is faintly expressed throughout the brain, mainly in arachnoid and ependymal cells (Conti et al. 1999). Especially for BGT1, it deserves mention that some controversy exists regarding the exact expression level in the brain under basal conditions. Whereas mRNA levels in some instances suggest widespread and moderate expression (Borden et al. 1995; Borden 1996), more recent studies on genetically engineered BGT1 knockout mice have demonstrated a very low brain expression level (Lehre et al. 2011), which is in fact much more pronounced in peripheral organs such as the liver (Zhou et al. 2012). By contrast, several studies have shown that BGT1 expression can be induced under epileptic conditions or in response to hypertonicity (Zhu and Ong 2004a; Olsen et al. 2005; Rowley et al. 2011). Some of the controversies regarding BGT1 protein expression may well relate to the fact that very few antibodies exist, and that these have not been convincingly validated in terms of specificity and sensitivity.

2 Glial GATs in Epilepsy: Pathophysiological Role and Therapeutic Potential

Epilepsy is one of the most common neurological disorders affecting about 50 million individuals of all ages worldwide (WHO 2016a). It is a chronic and complex condition characterized by an enduring predisposition to unprovoked recurrent seizures (Scharfman 2007). Epilepsy can arise from a genetic cause (Hirose et al. 2005;

Berkovic et al. 2006) or can be acquired after injury to the brain after e.g. stroke, tumour or infections (Lamar et al. 2014). While over 20 antiepileptic drugs (AED) exist, almost one-third of patients remain resistant to sufficient symptomatic treatment leading to a dramatic impairment of life quality as well as severe morbidities and increased mortality (Das et al. 2012; Shaju and Abraham 2013; Pitkänen et al. 2016). More knowledge about the nature of this disorder is therefore needed, and novel therapies are in high demand.

2.1 GABA Signalling Disturbances in Epilepsy

The transition from normal brain function to epilepsy is associated with an imbalance between glutamatergic excitatory and GABAergic inhibitory neurotransmission (Engelborghs et al. 2000; Treiman 2001; Staley 2015). Although epileptogenesis, i.e. the process of creating a persistent increase in the probability of spontaneous seizures (Staley 2015), includes multiple mechanisms such as changes in intrinsic neuronal properties and formation of excitatory recurrent connections, there is substantial evidence that an impaired GABAergic signalling is implicated (Olsen and Avoli 1997; Treiman 2001; Pavlov and Walker 2013). This hypothesis was already supported by several early in vivo studies demonstrating that suppression of GABAergic signalling by $GABA_A$ receptor antagonists, including picrotoxin and bicuculline, induces seizures (Usunoff et al. 1969; Curtis et al. 1970; Meldrum and Horton 1971; Hill et al. 1973; Campbell and Holmes 1984), whereas drugs known to increase GABA-mediated inhibition, such as diazepam and phenobarbital, have anticonvulsant effects (Kaim and Rosenstein 1960; Gastaut et al. 1965; Naquet et al. 1965; Macdonald and Barker 1978; Yasiry and Shorvon 2012).

Today, there is a substantial amount of evidence of alternations at different levels of the GABAergic system that are associated with epilepsy. An altered function and expression of $GABA_A$ receptors (Pavlov and Walker 2013; Lee and Maguire 2014; Schipper et al. 2016), GABA-synthesizing enzymes (Esclapez and Houser 1999; Freichel et al. 2006; Wang et al. 2016) and $GABA_B$ receptors (Princivalle et al. 2003a, b; Bettler et al. 2004) are some examples.

2.2 Regulation of GATs in Epilepsy

Although GATs have received relatively less investigation than other targets of the GABAergic system, there is substantial evidence for a compromised GAT function (During et al. 1995; Patrylo et al. 2001; Pirttimaki et al. 2013) and altered expression patterns in both experimental and human epilepsies. In Table 1 we have summarized existing evidence on the regulation of GAT1, GAT3 and BGT1 expression associated with epilepsy. Dependent on the brain region examined and the time

Table 1 Changes in the expression of GAT1, GAT3 and BGT1 after epilepsy insults in human and animal models

GAT subtype	Model	Technique	Region	Change	Reference
GAT1	Genetically epilepsy-prone (GEP) rats	ISH	Hippocampus	↓ in CA1, CA3–4 and DG (6 days after audiogenic stimulus)	Akbar et al. (1998)
			Cerebral cortex	↓ (6 days after audiogenic stimulus)	
			Inferior colliculus		
			Striatum		
	Lithium-pilocarpine in rats	IHC	Hippocampus	↓ in CA1 and CA3 in the interictal stage (6–12 days after SE) and in the chronic phase (8–18 days after SE) ↑ in the DG inner and outer molecular layer in the acute phase (1 day after SE) ↑ in the outer molecular layer in the interictal and chronic phases ↓ in the hilus in the interictal and chronic phases	André et al. (2001)
	Human TLE	ICC	Hippocampus	↓ in DG and the hippocampal formation No change in subiculum	Arellano et al. (2004)
		IHC	Hippocampus	↓ in CA1, CA3 and hilus ↑ in DG and stratum oriens	Lee et al. (2006)
		WB		No overall change	
		ICC	Hippocampus	In non-HS patients: ↑ in outer molecular layer In HS patients: ↓ in CA3 and stratum granulosum ↑ in outer molecular layer	Mathern et al. (1999)
		IHC	Hippocampus	↓ in severe HS compared to mild HS/control	Schijns et al. (2015)
		WB		No change between mild and severe HS	
		ICC	Hippocampus	↓ in Taylor-type cortical dysplasia	Spreafico et al. (2000)

	Human ganglioglioma-related TLE	IHC	Hippocampus	↓ in peritumoral cortex	Aronica et al. (2007)
	Intrahippocampal KA in mice	IHC	Hippocampus	↓ in CA1 and DG (24 h and 15–30 days after SE)	Bouilleret et al. (2000)
	In utero methylazoxymethanol (MAM) in rats	IHC	Hippocampus	↓ in CA1, CA3 and DG	Calcagnotto et al. (2002)
	Human focal cortical dysplasia (FCD)-related epilepsy	IHC	Neocortex	↓	Calcagnotto et al. (2005)
	Genetic absence epilepsy rats from Strasbourg (GAERS)	ICC and WB	Thalamus	No change	Cope et al. (2009)
			Neocortex		
		Autoradiography	Hippocampus	No change	Sutch et al. (1999)
			Thalamus		
			Cortex		
	Intra-amygdalar $FeCl_3$ in rats	WB	Hippocampus	No change (5 days after injection)	Doi et al. (2005)
		WB	Hippocampus	↑ (5 and 15 days after injection) but returned to control levels 30 days after injection	Ueda and Willmore (2000)
	Pentylenetetrazol (PTZ) kindling in rats	WB	Hippocampus	No change in fully kindled rats (1 and 30 days after last PTZ injection) ↓ in easily kindled rats (30 days after last PTZ injection)	Doi et al. (2009)
	Middle cerebral artery occlusion (MCAO)-evoked non-convulsive seizures in rats	WB	Hippocampus	↑ (1 day after MCAO surgery)	Fu et al. (2015)
			Parietal cortex		
			Amygdala		

(continued)

Table 1 (continued)

GAT subtype	Model	Technique	Region	Change	Reference
	Amygdala-kindled rats	ISH	Hippocampus	↑ in CA3 (1 h after last seizure) ↑ in CA1–4 (4 h after last seizure) but back to control levels 8–24 h after last seizure ↑ in DG (1 and 4 h after last seizure) but back to control levels 8–24 h after last seizure	Hirao et al. (1998)
			Amygdala	No change (1–24 h after last seizure)	
			Cerebral cortex		
			Pyriform cortex		
	Seizure-sensitive gerbils	IHC and WB	Hippocampus	↑ in DG, subiculum and hippocampus proper (pre-seizure and 12 h after seizure onset) No change 30 min after seizure onset	Kang et al. (2001)
	Chronic cyclothiazide-induced epilepsy in rats	IHC	Hippocampus	↓ in CA1, CA3 and DG (6 months after seizure)	Kong et al. (2014)
	Tremor rat	RT-PCR, WB and IHC	Hippocampus	↑ in CA1, CA3 and DG	Mao et al. (2010)
	4-Aminopyridine (4-AP)-induced epilepsy in rats	Immunofluorescence	Hippocampus	No change in CA1 and DG (30 min after 4-AP injection and after 180 min in CA1) ↑ in CA1 (60 min after 4-AP injection) ↑ in DG (60 and 180 min after 4-AP injection)	Medina-Ceja et al. (2012)
	Glutamate transporter knockout (GLAST KO)-kindled mice	WB	Hippocampus	↓ in crude hippocampal membrane fractions	Nagatomo et al. (2007)
	Corticotropin-releasing hormone (CRH)-induced seizures in rats	ICC	Hippocampus	4 h after seizure induction: No change in CA3, DG, hilus and subiculum ↑ in CA1 24 h after seizure induction: ↑ in CA1, DG, hilus and subiculum	Orozco-Suarez et al. (2000)

			Neocortex	No change (4 h after seizure induction) ↑ (1 day after seizure induction)	
			Amygdala	↑ (1 day after seizure induction)	
			Piriform cortex		
	Hyperthermia-induced febrile seizures in rats	ICC	Hippocampus	No change in CA1, CA3, DG and hilus	Orozco-Suarez et al. (2000)
	Pilocarpine-induced epilepsy in mice	qPCR	Hippocampus	↓ (28 days after SE)	Rowley et al. (2011)
	KA-induced epilepsy in rats	ISH and ICC	Hippocampus	↓ (3–4 h, 6 h; 1, 2, 8 and 3 days after KA injection) ↑ in DG (9 h after KA injection)	Sperk et al. (2003)
		WB	Hippocampus	↑ (3 days after KA injection)	Su et al. (2015)
			Parietal cortex		
			Amygdala		
		IPS	Hippocampus	↓ in CA fields (3, 7 and 21 days after KA injection) No change in DG (3, 7 and 21 days after KA injection)	Zhu and Ong (2004a)
GAT3	In utero methylazoxymethanol (MAM) in rats	IHC	Hippocampus	↓ in CA1, CA3 and DG	Calcagnotto et al. (2002)
	Human focal cortical dysplasia (FCD)-related epilepsy	IHC	Neocortex	↓	Calcagnotto et al. (2005)
	Intra-amygdalar $FeCl_3$ in rats	WB	Hippocampus	No change (5 days after injection)	Doi et al. (2005)
		WB	Hippocampus	↑ (5, 15 and 30 days after injection)	Ueda and Willmore (2000)
	Pentylenetetrazol (PTZ) kindling in rats	WB	Hippocampus	No change (1 and 30 days after last PTZ injection)	Doi et al. (2009)

(continued)

Table 1 (continued)

GAT subtype	Model	Technique	Region	Change	Reference
	Middle cerebral artery occlusion (MCAO)-evoked non-convulsive seizures in rats	WB	Hippocampus	↑ (1 day after MCAO surgery)	Fu et al. (2015)
			Parietal cortex		
			Amygdala		
	Amygdala-kindled rats	ISH	Hippocampus	↑ in CA2 (1 and 24 h after last seizure) No change in CA1, contralateral CA2, CA3–4 and DG (1–24 h after last seizure)	Hirao et al. (1998)
			Amygdala	↑ (1 h after last seizure) but back to control levels 2–24 h after last seizure	
			Cerebral cortex		
			Pyriform cortex		
	Seizure-sensitive gerbils	IHC	Hippocampus	No change (pre-seizure, 30 min and 12 h after seizure onset)	Kang et al. (2001)
	Human TLE	IHC	Hippocampus	↑ in DG and hilar regions	
		WB		No change	Lee et al. (2006)
		ICC	Hippocampus	In non-HS patients: ↑ in CA1, CA3 and hilus In HS patients: ↓ in CA2–3 and hilus	Mathern et al. (1999)
		IHC	Hippocampus	↓ in severe HS compared to mild HS/control	Schijns et al. (2015)
		WB		↓ in severe HS compared to mild HS	
	Tremor rat	RT-PCR, WB and IHC	Hippocampus	↑ in CA1 No change in CA3 and DG	Mao et al. (2010)
	Glutamate transporter knockout (GLAST KO)-kindled mice	WB	Hippocampus	No change in crude hippocampal membrane fractions	Nagatomo et al. (2007)
	Pilocarpine-induced epilepsy in mice	qPCR	Hippocampus	↓ (28 days after SE)	Rowley et al. (2011)

	KA-induced epilepsy in rats	ISH and ICC	Hippocampus	↓ (3–4 h, 6 h, 9 h, 12 h; 1, 2, 8 and 30 days after KA injection)	Sperk et al. (2003)
		WB	Hippocampus	↑ (3 days after KA injection)	Su et al. (2015)
			Parietal cortex		
			Amygdala		
		IPS	Hippocampus	↓ in CA fields (3, 7 and 21 days after KA injection) No change in DG (3, 7 and 21 days after KA injection)	Zhu and Ong (2004a)
BGT1	Pilocarpine-induced epilepsy in mice	qPCR	Hippocampus	No change (8 h and 7 days after SE) ↑ (1 day after SE) ↓ (3 days and 28 days after SE)	Rowley et al. (2011)
	KA-induced epilepsy in rats	IPS	Hippocampus	↑ in CA fields (3 and 7 days after KA injection) ↓ in CA fields (21 days after KA injection) compared to 7 days post-KA No change in DG (3, 7 and 21 days after KA injection)	Zhu and Ong (2004a)

DG dentate gyrus; *HS* hippocampal sclerosis; *ICC* immunocytochemistry; *IHC* immunohistochemistry; *IPS* immunoperoxidase staining; *ISH* in situ hybridization; *KA* kainic acid; *qPCR* quantitative polymerase chain reaction; *RT-PCR* real-time polymerase chain reaction; *SE* status epilepticus; *TLE* temporal lobe epilepsy; *WB* Western blot point after the induction of an epileptic insult, the expression of GATs has thus been reported to be either unchanged, up- or downregulated. In addition to regional and time-related differences, this discrepancy could reflect the employment of dissimilar epileptic models that could involve distinct pathophysiological mechanisms giving rise to contradicting results. Therefore, interpreting these studies and explaining the exact role of GATs in epilepsy are difficult. Nevertheless, it stands clear that GATs are involved in the pathophysiology of epilepsy and could therefore be important pharmacological targets, as already demonstrated with the GAT1 inhibitor tiagabine (see below).

2.3 *Pharmacological Inhibition of Glial GATs in Epilepsy*

Enhancing GABA-mediated inhibition in the brain by increasing the extracellular pool of GABA has proven effective in alleviating epileptic symptoms. One successful approach is the pharmacological inhibition of GATs, which has been shown, using in vivo microdialysis, to increase the extracellular GABA concentrations and thereby enhancing inhibitory signalling (Lerma et al. 1984; Fink-Jensen et al. 1992; Juhász et al. 1997; Dalby 2000). Not surprisingly therefore, many GAT inhibitors have been reported to mediate anticonvulsant activity in preclinical epileptic animal models (Yunger et al. 1984; Nielsen et al. 1991; White et al. 1993, 2002, 2005; Bolvig et al. 1999; Dalby 2000), thereby establishing proof of concept that pharmacological inhibition of GATs is a successful anticonvulsive approach. Yet, we have today only one GAT inhibitor, tiagabine, which is clinically approved for the treatment of epileptic disorders. Tiagabine is a GAT1-selective inhibitor (Table 2) that is used in clinic as adjunctive treatment of partial epileptic seizures, albeit poor efficacy in epilepsy management, and substantial side effects such as agitation, sedation and psychotic-like episodes in patients predisposed to psychiatric disorders have been reported (Kalviainen 2001; Schousboe et al. 2011). Interestingly, similar behavioural patterns to the clinical side effects of tiagabine were seen in GAT1 knockout mice (Chiu et al. 2005), suggesting a direct link to the specific GAT1 inhibition. Accordingly, since GAT1 is expressed predominantly in neurons, it has been postulated that inhibiting glial GATs could be a better pharmacological approach. This is supported by the notion that reuptake of GABA into the presynaptic neuron, which primarily occurs via GAT1, allows for recycling of GABA, whereas glial uptake leads to degradation and thereby loss from the neurotransmitter pool (Schousboe et al. 1983). Along this line, early studies reported that neuronal selective GAT inhibitors have pro-convulsive effects that were proposed to derive from a depletion of the synaptic GABA pool (Schousboe et al. 1983; Gonsalves et al. 1989). On the other hand, glial selective GAT inhibitors were demonstrated to increase both the synaptic and extrasynaptic concentrations of GABA (Wood et al. 1983; Juhász et al. 1997). A later study also supported this hypothesis by reporting a better experimental correlation between anticonvulsant activity and glial, rather than neuronal, GAT inhibition (White et al. 2002). Taken together, this could explain the growing interest in the glial GATs as an alternative and perhaps more promising target for seizure management.

2.3.1 GAT3 Inhibition

Being the most predominant astrocytic GAT subtype, GAT3 has gained much attention as a promising glial GAT target in epilepsy. Already 20 years ago, a study by Dalby et al. reported anticonvulsant effects in four rodent models after intraperitoneal (i.p.) injection of two GAT3 inhibitors, NNC 05-2045 and NNC 05-2090 (Dalby et al. 1997), which are derivatives of nipecotic acid (Thomsen et al. 1997). Albeit having affinity for targets outside the GABAergic system, e.g. the α1-adrenoceptors, the authors suggested that the observed seizure protection was primarily mediated by inhibition of GAT3 with a plausible minor contribution of BGT1 (Dalby et al. 1997). Nevertheless, because of their off-target effects, it is difficult to draw a conclusive link between the GAT3/BGT1 inhibition and the observed anticonvulsive effects. This has, however, been suggested with the compound (*S*)-SNAP-5114 (Table 2) (Dhar et al. 1994), which has proven to be the best pharmacological tool, we have to date, for probing the potential of inhibiting astrocytic GAT3 as an anticonvulsive strategy. When examined for anticonvulsive effects after systemic administration in different epileptic models, (*S*)-SNAP-5114 was found to be virtually devoid of any effect except for a weak effect against audiogenic seizures (~100-fold less potent than tiagabine) (Dalby 2000). This was proposed to be a result of a poor pharmacokinetic profile or bioavailability rather than a lack of role in seizure management (Dalby 2000, 2003). This is supported by the fact that (*S*)-SNAP-5114 was able to block electroshock-induced convulsions when administered in the substantia nigra (Dalby 2000). Although new analogues of (*S*)-SNAP-5114, with improved chemical stability and subtype selectivity, have been reported (Pabel et al. 2012), the in vivo implications have not yet been examined.

Table 2 Selectivity overview of ligands at GATs

Compound	mGAT1	mBGT1	mGAT3
	IC_{50} (μM)		
(*R*,*S*)-EF-1502	x (7)	x (26)	–
RPC-425	–	x (45)	–
(*S*)-SNAP-5114	–	–	x (5)
Tiagabine	x (0.8)	–	–

Data from Madsen et al. (2010) and Vogensen et al. (2013)
Refer to (Damgaard et al. In Press; Schousboe et al. In Press) for chemical structures and names
"x" denotes primary site(s) of action at mouse GAT
"–" denotes potency lower than 100 μM

2.3.2 BGT1 Inhibition

BGT1 has not been considered an obvious drug target for managing epileptic seizures, owing to its low expression in the brain and poor affinity for GABA compared to the other GAT subtypes. This view has, however, been challenged by a series of studies employing various combinations of GAT inhibitors in animal models that suggested a functional role of BGT1 in epilepsy (Fig. 2). The selectivity profile of inhibitors has been instrumental in deducing the role of especially BGT1 (Table 2), as shall be detailed below. The first demonstration was with the BGT1-/GAT1-selective inhibitor (*R,S*)-EF-1502 (EF-1502) (Clausen et al. 2005), which was shown to possess a broad spectrum of anticonvulsive effects in animal models of generalized and partial epilepsy after i.p. administration (White et al. 2005). Most importantly, EF-1502 gave a synergistic, rather than additive, anticonvulsive effect when co-applied with either tiagabine (Fig. 2) or LU-32-176, another GAT1-selective inhibitor. On the other hand, when tiagabine was administered in combination with LU-32-176, only an additive effect was seen (White et al. 2005). The synergistic activity of EF-1502 in combination with tiagabine was also reported in an in vitro study with spontaneous electrographic bursting (SB) in brain slices from an epilepsy rat model (Smith et al. 2008). Along this line, a synergistic anticonvulsant effect was reported for the GAT3-selective inhibitor (*S*)-SNAP-5114 in combination with EF-1502, but not with tiagabine (Fig. 2) (Madsen

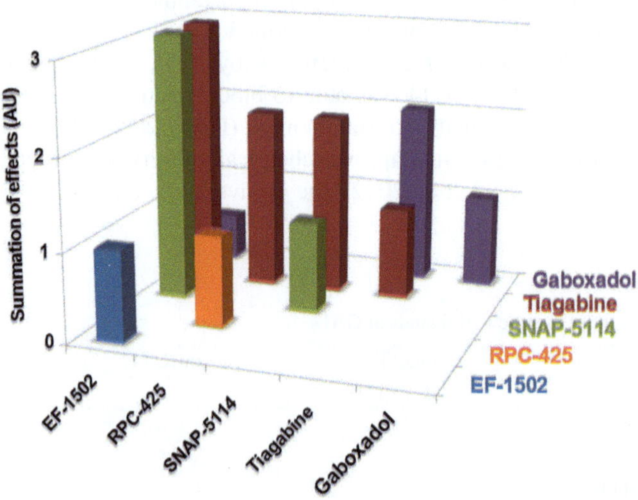

Fig. 2 Summation of anticonvulsive effects from combinations of various pharmacological interventions in animal models of epilepsy (White et al. 2005; Madsen et al. 2009, 2011; Vogensen et al. 2013). *AU* arbitrary units; *0* no effect; *0.5* antagonistic effect; *1* effect; *2* additive effect; *3* synergistic effect. SNAP-5114 means (*S*)-SNAP-5114

et al. 2009). Taken together, these experiments suggested that the observed synergistic anticonvulsant activity required the inhibition of BGT1 along with either GAT1 or GAT3, thereby proposing a functional contribution of BGT1 and a potential utility of this GAT subtype in managing epileptic seizures. Taken the subcellular distribution of GATs into account (Fig. 1a), it has been hypothesized that the synergistic effect arises from an increase in the extrasynaptic GABA levels as a consequence of inhibiting the extrasynaptically expressed BGT1, which results in the activation of extrasynaptic $GABA_A$ receptors (Madsen et al. 2010). This idea was supported by a study with EF-1502, tiagabine and gaboxadol, which is a superagonist of extrasynaptic α4-containing $GABA_A$ receptors (Madsen et al. 2011). When administered in animal models of epilepsy, gaboxadol had anticonvulsive effects alone and additive effects in combination with tiagabine (Fig. 2). Interestingly, gaboxadol was devoid of anticonvulsive effects when co-administered with EF-1502 (Madsen et al. 2011). This suggested that gaboxadol was displaced from its extrasynaptic sites by GABA as a result of an increase in ambient concentrations of the latter upon BGT1 inhibition by EF-1502 (Madsen et al. 2011). Altogether, these studies supported the idea of BGT1 inhibition being a successful anticonvulsive strategy, which is possibly mediated by an increase in the levels of GABA followed by activation of extrasynaptic $GABA_A$ receptor populations.

As opposed to these findings, however, the BGT1-selective inhibitor RPC-425 has been reported to only give an additive anticonvulsive effect when co-administered with tiagabine in the audiogenic seizure-susceptible Frings mouse (Fig. 2) (Vogensen et al. 2013). Although this apparently refutes a functional role of BGT1 based on the reported synergistic interaction between EF-1502 and tiagabine (White et al. 2005), this discrepancy could also be explained by different pharmacokinetic properties, a distinct mechanism of action or pharmacological profile of RPC-425 compared to EF-1502 (Vogensen et al. 2013).

In another study conducted by Lehre and co-workers (Lehre et al. 2011), BGT1 was proposed to have no or only little impact on acute seizure susceptibility in a BGT1 knockout mouse, thereby suggesting no functional role of this GAT subtype in epilepsy. Nevertheless, as GATs do undergo regulation in terms of function and expression (Table 1), the lack of a seizure-related phenotype in BGT1 knockout mice could also likely be the result of compensatory mechanisms involving other GAT subtypes. Hence, taken the plasticity of the GAT system into consideration, it can be difficult to translate the outcome from this knockout study into a non-existing function of BGT1. Likewise, the above-mentioned pharmacological studies involving combinations of different GAT inhibitors (Fig. 2 and Table 2) only present indirect and nonconclusive evidence for a role of BGT1 in seizure management. Therefore, there is a need for the development of more potent BGT1-selective inhibitors that can be used to directly assess the functional and potential therapeutic role of BGT1 in epilepsy.

3 Glial GATs in Ischemic Stroke: Pathophysiological Role and Therapeutic Potential

Ischemic stroke is the second largest killer worldwide (WHO 2016b) and the leading cause of long-term disability with 15 million new stroke incidents every year worldwide (World Health Federation 2016). Ischemic stroke occurs when a brain region is deprived of normal blood flow. The only available treatment today is thrombolytic therapy, which has to be given within 4.5 h after the incidence (Emberson et al. 2014), leaving rehabilitation as the only other option (Dobkin 2008; Dobkin and Dorsch 2013) and new therapies an extremely pressing matter. With the aim of exploring novel therapeutic opportunities for ischemic stroke, the GABA system has recently come into focus, presenting novel drug targets for treatment (Clarkson et al. 2010; Brickley and Mody 2012; Carmichael 2012; Clarkson 2012). This renewed excitement in $GABA_A$ receptor signalling after stroke is devoted towards understanding subunit-specific changes and modulation thereof (Clarkson et al. 2010; Lake et al. 2015), including changes in GATs (Clarkson et al. 2010; Jaenisch et al. 2016), rather than targeting $GABA_A$ receptor subtypes non-selectively, which has failed to translate into positive clinical outcomes in the past (Ginsberg 2008).

3.1 *Ischemic Stroke*

The lack of blood supply to a given brain region ceases energy-dependent processes leading to disruption of the membrane potential and glutamate excitotoxicity. This triggers a plethora of downstream effects referred to as the ischemic cascade, which ultimately leads to extensive cell death and the formation of the necrotic infarct core (for reviews, see Schwartz-Bloom and Sah 2001; Doyle et al. 2008; Kostandy 2012). The bordering region, the peri-infarct zone, also suffers from a reduced blood flow and is a functionally impaired, but structurally intact region (Lo 2008). This region is not only interesting in respect to neuroprotection in the acute phase (<3 days in mice), as the tissue can be rescued upon reperfusion, but also in recovery in the sub-acute and chronic phase (>3 days in mice), as neurons and glial cells in this region can partly recover and facilitate remapping (Brown et al. 2009; Murphy and Corbett 2009; Clarkson et al. 2010; Lim et al. 2014). The glial cells deprived of blood flow become reactive, thus forming a persisting glial scar surrounding the infarct (Sofroniew and Vinters 2010; Li et al. 2014), which limits the infarct size and plays a role in the inflammatory response central to stroke pathogenesis (Burda and Sofroniew 2014; Sofroniew 2015). Although the glial scar has for long been regarded as a barrier for remapping and plasticity, aiding effects of glial scar on axon regeneration have also been reported (Anderson et al. 2016).

3.2 *GABA Signalling Disturbances After Ischemic Stroke*

During a stroke, numerous neurotransmitters are being released in the affected region setting the excitotoxic index critical for stroke outcome (Li et al. 2010a, b; Bhattacharya et al. 2014). Direct measurements of extracellular GABA have revealed an acute increase in the affected region after an ischemic episode in animal stroke models (Baldwin et al. 1994; Li et al. 2010a, b; Bhattacharya et al. 2014) (for review, see Green et al. 2000) and humans (Kanthan et al. 1995; Hutchinson et al. 2002; Kett-White et al. 2005). GABA is released not only in a vesicular manner as a consequence of depolarization but presumably also through reversal of GATs where GABA is released into the extracellular space (for review, see Schwartz-Bloom and Sah 2001). The reversal potential of glial GATs after ischemia, however, has not been addressed, but astrocytes from rats and humans can convert glutamatergic excitability to tonic GABA inhibition through reversal of GAT3 (Héja et al. 2009, 2012; Lee et al. 2011), but not BGT1 (Lee et al. 2011). Accordingly, GAT3 reversal is reported in reactive astrocytes in several animal disease models (Wu et al. 2014; Hernandez-Rabaza et al. 2016), and it is therefore likely that GAT3 reverses after ischemia. This is something to be aware of when using GAT inhibitors, as this could potentially result in further damage or development of rebound seizures, which have been noted using tiagabine (Knake et al. 1999).

The increase in extracellular GABA levels after stroke could be an endogenous protective mechanism meant to counterbalance glutamatergic excitotoxicity and minimize the extent of cell death (Bhattacharya et al. 2014). However, the elevation in GABAergic signalling and tonic inhibition appears to occur with such a delay that the benefits are limited (Chen et al. 2007; Clarkson et al. 2010; Li et al. 2014). Accordingly, if GABAergic signalling is enhanced through pharmacological intervention just hours after a stroke, neuroprotection can be achieved (for reviews, see Green et al. 2000; Schwartz-Bloom and Sah 2001; Trojnar et al. 2002) by lowering the excitotoxic index (Bhattacharya et al. 2014) and lowering the release of glutamate (Nelson et al. 2000; Ouyang et al. 2007). By contrast, continuous enhanced GABAergic signalling inhibits plasticity and remapping necessary for functional recovery on a longer time scale (Schallert et al. 1986; Lazar et al. 2002; Clarkson et al. 2010; Carmichael 2012). Thus, like it is reported for NMDA receptor signalling (Lo 2008), GABA signalling appears to have a biphasic role after an ischemic stroke, where GABA possesses neuroprotective effects in the acute phase but limits functional recovery by hindering plasticity in the sub-acute and chronic phase (Clarkson et al. 2010; Carmichael 2012).

3.3 *Regulation of GATs, GABA$_A$ Receptors and Tonic Inhibition After Stroke*

The GABAergic system is affected at many different levels after an ischemic stroke including the GATs. Several studies have shown that GAT3 levels are downregulated in the peri-infarct zone (Melone et al. 2003; Clarkson et al. 2010; Jaenisch et al. 2016), whereas the GAT1 expression level is unchanged (Melone et al. 2003; Frahm et al. 2004; Clarkson et al. 2010). No reports on BGT1 expression levels after stroke exist, but a downregulation of BGT1 is reported in a mouse model of reduced neuronal plasticity (Liu et al. 2015). Also changes in the expression of the extrasynaptic $GABA_A$ receptor subunits are reported with different outcomes depending on, among other factors, the stroke model and the age of the animals (Redecker et al. 2002; Jaenisch et al. 2016) (for review, see Schwartz-Bloom and Sah 2001; Hines et al. 2012). In general, $GABA_A$ receptor levels appear to be decreased after stroke in humans (Kim et al. 2014). Also the synthesis and synaptic release of GABA are most likely affected (Green et al. 2000), adding an extra layer of complexity to GABAergic signalling disturbances after an ischemic stroke.

Only few studies exist on tonic inhibition in ischemic stroke and how interventions affect neuroprotection and recovery (Johansen and Diemer 1991; Clarkson et al. 2010; Lake et al. 2015; Jaenisch et al. 2016). However, it is known that tonic inhibition mediated by $GABA_A$ receptors is increased in the peri-infarct region 3 to 14 days after a focal ischemic stroke in mice, coinciding with decreased levels and impaired function of GAT3 at day 7 days post-stroke, while GAT1 is unaffected in terms of level and function (Fig. 3) (Melone et al. 2003; Clarkson et al. 2010). No changes in phasic inhibition, the resting membrane potential or the GABA reversal

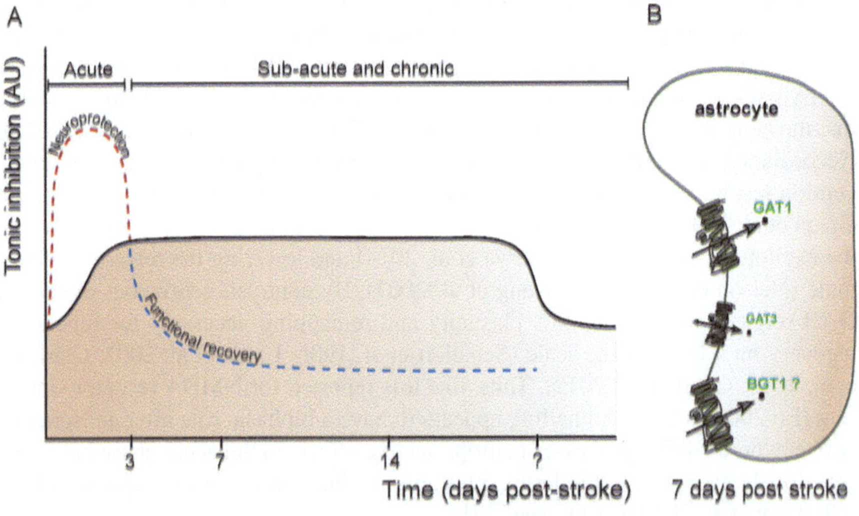

Fig. 3 Changes and modulation of GABAergic tonic inhibition after ischemic stroke. (**a**) Tonic inhibition mediated by $GABA_A$ receptors is increased in the peri-infarct zone 3–14 days after a focal photothrombotic stroke to the motor cortex (*full black line*). It is unknown when the increase in tonic inhibition turns back to baseline levels. Enhanced tonic inhibition in the acute phase (<3 days) promotes neuroprotection (*dashed red line*), whereas a decrease in tonic inhibition in the sub-acute and chronic phase (>3 days) promotes functional recovery (*dashed blue line*). (**b**) The increase in tonic inhibition is correlated with a decrease in GAT3 protein level 7 days post stroke (as illustrated by the smaller depiction of GAT3), whereas GAT1 is unchanged. It is unknown if BGT1 is affected by an ischemic stroke (See Sects. 3.3 and 3.4 for references)

potential were reported, supporting that the increase in tonic inhibition after a focal ischemic stroke is caused by a lowered GAT3-mediated GABA uptake (Clarkson et al. 2010).

A decrease in tonic inhibition in the peri-infarct region due to a lowered expression of the δ $GABA_A$ receptor subunit is reported in a more invasive stroke model, without any concomitant changes in GAT3 (Jaenisch et al. 2016). The same study reported a downregulation of GAT3 in the peri-infarct region after a focal ischemic stroke like Clarkson et al. and no change in δ and α5 $GABA_A$ receptor subunit expression (Clarkson et al. 2010; Jaenisch et al. 2016). Despite the discrepancy concerning the magnitude of tonic inhibition, both studies report changes in tonic inhibition after ischemic stroke and a downregulation of GAT3 after a focal ischemic stroke (Clarkson et al. 2010; Jaenisch et al. 2016). This points to a causal role for GAT3 in regulating tonic inhibition and GABAergic signalling, which is crucial for stroke outcome in rodents (Clarkson et al. 2010; Lake et al. 2015; Jaenisch et al. 2016) as well as humans (Blicher et al. 2015).

3.4 Pharmacological Inhibition of Glial GATs in Ischemic Stroke

3.4.1 Neuroprotection

Several studies on the neuroprotective role of GABA in ischemic stroke have been published, most of which employ $GABA_A$ receptor agonists, which are also used as antiepileptic drugs (for review, see Green et al. 2000; Schwartz-Bloom and Sah 2001; Trojnar et al. 2002). The use of $GABA_A$ receptor agonists has shown proof of concept in animal studies, but failed clinical trials (Ginsberg 2008), hence translation into humans has proven extremely difficult. This is partly because of the narrow time window during which the drugs can be administered successfully.

Newer studies have focused on the modulation of tonic inhibition for achieving neuroprotection. For instance, if tonic GABAergic signalling is enhanced by gaboxadol, a decrease in infarct volume is observed when given in combination with diazepam shortly after the stroke incident (Johansen and Diemer 1991). Accordingly, NMDA-induced cell death in vitro is minimized by gaboxadol (Kristensen et al. 2003) and muscimol at doses selective for the δ-containing $GABA_A$ receptors (Santhakumar et al. 2010). By contrast, if tonic inhibition is lowered acutely by the α5 $GABA_A$ receptor inverse agonist (L655,708), an increase in infarct volume results (Clarkson et al. 2010), illustrating the importance of enhancing tonic GABA currents to achieve neuroprotection in the acute phase (Fig. 3).

Despite the principle in targeting glial GATs for increasing tonic GABA signalling (Fig. 1), only GAT1 ligands have so far been investigated in ischemic stroke. Good evidence exists for promoting neuroprotection via inhibition of GAT1 with CI-966 or tiagabine in animal stroke models (Inglefield et al. 1995; Phillis 1995; Chen Xu et al. 2000; Yang et al. 2000; O'Connell et al. 2001; Iqbal et al. 2002) and in in vitro models where ischemia is induced by oxygen and glucose deprivation (Costa et al. 2004). GAT1 inhibition increases tonic inhibition under normal (Wu et al. 2014) and ischemic conditions (Clarkson et al. 2010) and hence highly involved in tonic inhibition, which is also reported in GAT1 KO mice (Jensen et al. 2003; Chiu et al. 2005). This illustrates the importance of controlling tonic inhibition in respect to neuroprotection and the beneficial use of targeting GAT1, preferably glial GAT1. Despite being used as an add-on therapy for partial epileptic seizures (Schousboe et al. 2011), tiagabine has not been tested for its neuroprotective effects in ischemic stroke patients, most likely due to the adverse effects of the drug (Kälviäinen 2002).

Another approach would be to block GAT3 and BGT1 to promote neuroprotection after ischemia, which, due to their presumed peri-/extrasynaptic localization, may selectively regulate tonic inhibition more isolated than GAT1 inhibition as it is reported for GAT3 (Song et al. 2013). Whether GAT3- and BGT1-selective inhibitors possess neuroprotective potential have not been studied directly, most probably due to the lack of selective brain-permeable compounds. (*S*)-SNAP-5114, the only available GAT3 compound, increases tonic inhibition under normal conditions (Clarkson et al. 2010; Song et al. 2013) or only in combination with a GAT1 blocker according to some studies (Keros and Hablitz 2005; Kersanté et al. 2013; Wu et al. 2014). Thus, it could be relevant to test its neuroprotective potential in animal stroke models. BGT1 is proposed to govern an extrasynaptic protecting mechanism following excitotoxic brain injury (Zhu and Ong 2004a, b), but ligands for BGT1 have not been studied in ischemic stroke models either.

3.4.2 Functional Recovery

The peri-infarct region has proven critical for functional recovery after stroke as surviving neurons can undergo remapping and plastic connectivity during the subacute and chronic phase (for review, see Murphy and Corbett 2009). If tonic GABA inhibition is enhanced during the recovery phase, detrimental outcomes have been reported in terms of re-emergence of stroke deficits in stroke patients (Lazar et al. 2002) and impaired recovery in rats (Schallert et al. 1986, 1992), underlining the negative impact that an increase in extracellular GABA has on plasticity (Hernandez and Schallert 1988; Collinson et al. 2002; Atack et al. 2006). In contrast, decreasing GABAergic inhibition is associated with functional recovery in humans and rodents after stroke (Fig. 3) (Clarkson et al. 2010; Kim et al. 2014; Lake et al. 2015; Jaenisch et al. 2016). Thus, the observed increased GABAergic signalling in the chronic phase is a putative target for recovery, where the modulation of tonic inhibition could be relevant in a much broader time window than thrombolytics.

Dampening tonic inhibition at a delayed onset pharmacologically via $\alpha5$-$GABA_A$ receptors by L655,708 or by genetic deletion of the extrasynaptic $\alpha5$ or δ-containing $GABA_A$ receptors has been demonstrated to promote functional recovery in stroke models (Clarkson et al. 2010; Lake et al. 2015). The improved recovery occurs even if the compounds were given as long as 7 days after the stroke, clearly showing an impressively broad therapeutic window. By contrast, if tonic inhibition was lowered acutely, within hours, no functional recovery was gained, and the infarct volume increased (Clarkson et al. 2010). Accordingly, an enhanced functional recovery in humans following stroke correlates with reduced $GABA_A$ receptor expression and GABAergic signalling in the chronic phase (Kim et al. 2014). This clearly illustrates the biphasic role of tonic GABA-mediated inhibition and the detrimental outcome on neuroprotection if excitability is not counterbalanced acutely. Further, it highlights the importance of timing the intervention. Unfortunately, no pharmacological studies on the role of glial GATs in stroke recovery exist. Nevertheless, it should be noted that (*S*)-SNAP-5114 enhanced plasticity and rescued long-term potentiation and working memory in a mouse model of Alzheimer's disease by inhibiting reversed GAT3 function, thus lowering tonic inhibition (Wu et al. 2014).

4 Perspectives and Concluding Remarks

Within recent years, tonic inhibition has emerged as an important player in the pathophysiology of epilepsy and ischemic stroke. Tonic inhibition can be regulated by indirect means via glial GATs that control ambient GABA levels, and GAT3 and BGT1 are therefore putative targets in the treatment of epilepsy and ischemic stroke. However, to truly understand the function of these transporters in each pathology, we must (1) characterize the expression profile across all brain regions, which is why validated antibodies are needed, and (2) characterize function also in respect to reversal of the transporters, which appears to be occurring in epilepsy (Rassner et al. 2016) and which becomes critical when considering the best therapeutic paradigm. Furthermore, we must (3) characterize changes in the temporal expression profile following various pathological stressors in order to time the pharmacological intervention properly to avoid detrimental outcomes. This will shed light on the function of GAT3 and BGT1 in epilepsy and ischemic stroke. To probe the therapeutic relevance of the two transporters, selective and potent ligands for the transporter subtypes are needed. During the last years, attempts have been made to create such pharmacological tools and potential drug-lead compounds.

To address the role of GAT3, more stable and selective compounds than (*S*)-SNAP-5114 are needed. The isatin-derived GAT3-selective compounds reported by Damgaard et al. could provide for a new type of GAT3 inhibitor (Damgaard et al. 2015). However, the specificity and usability of these compounds need to be further investigated. In the BGT1 field, the non-competitive inhibitor, *N*-(1-benzyl-4-piperidinyl)-2,4-dichlorobenzamide (BPDBA) (Kragholm et al. 2013); the cyclopropyl analogue of GABA, (1*S*,2*S*,5*R*)-5-aminobicyclo[3.1.0]hexane-2-carboxylic acid (Kobayashi et al. 2014); and the guanidine analogue of β-alanine, 2-amino-1,4,5,6-tetrahydropyrimidine-5-carboxylic acid (ATPCA) (Al-Khawaja et al. 2014), have been developed, all showing selectivity for BGT1 and potencies in the low micromolar range. Interestingly, the latter compound was also reported to have agonistic activity at synaptic but also extrasynaptic $GABA_A$ receptor isoforms

(Petersen et al. 2014). While this could prove to be an interesting multimodal profile in disorders such as epilepsy and ischemic stroke, the in vivo utility of this inhibitor awaits further investigation. In this regard, the physiochemical stability and the ability of the compounds to pass the blood-brain barrier are essential to assess.

While most studies have shown beneficial antiepileptic effects and neuroprotection in preclinical stroke models, some have also shown that GAT1 intervention with tiagabine can induce seizures (Knake et al. 1999; Vinton et al. 2005). Therefore, it could be speculated that glial GATs would be more suitable drug targets and give rise to less side effects. Altogether, it is important to systematically investigate the effects of GAT3- and BGT1-selective compounds for both epilepsy and ischemic stroke to optimize treatment strategies for each pathology.

Conflict of Interest The author declares no conflicts of interest.

References

Akbar MT, Rattray M, Williams RJ, Chong NWS, Meldrum BS. Reduction of GABA and glutamate transporter messenger RNAs in the severe-seizure genetically epilepsy-prone rat. Neuroscience. 1998;85(4):1235–51.

Al-Khawaja A, Petersen JG, Damgaard M, Jensen MH, Vogensen SB, Lie MEK, et al. Pharmacological identification of a guanidine-containing β-alanine analogue with low micromolar potency and selectivity for the betaine/GABA transporter 1 (BGT1). Neurochem Res. 2014;39(10):1988–96.

Anderson M, Burda J, Ren Y, Ao Y, O'Shea T, Kawaguchi R, et al. Astrocyte scar formation aids central nervous system axon regeneration. Nature. 2016;532(7598):195–200.

André V, Marescaux C, Nehlig A, Fritschy JM. Alterations of hippocampal GABAergic system contribute to development of spontaneous recurrent seizures in the rat lithium-pilocarpine model of temporal lobe epilepsy. Hippocampus. 2001;11(4):452–68.

Arellano JI, Muñoz A, Ballesteros-Yáñez I, Sola RG, DeFelipe J. Histopathology and reorganization of chandelier cells in the human epileptic sclerotic hippocampus. Brain. 2004;127(1):45–64.

Aronica E, Redeker S, Boer K, Spliet WGM, van Rijen PC, Gorter JA, et al. Inhibitory networks in epilepsy-associated gangliogliomas and in the perilesional epileptic cortex. Epilepsy Res. 2007;74(1):33–44.

Atack J, Bayley P, Seabrook G, Wafford K, McKernan R, Dawson G. L-655,708 enhances cognition in rats but is not proconvulsant at a dose selective for α5-containing $GABA_A$ receptors. Neuropharmacology. 2006;51(6):1023–9.

Baldwin HA, Williams JL, Snares M, Ferreira T, Cross AJ, Green AR. Attenuation by chlormethiazole administration of the rise in extracellular amino acids following focal ischaemia in the cerebral cortex of the rat. Br J Pharmacol. 1994 May;112(1):188–94.

Belelli D, Harrison NL, Maguire J, Macdonald RL, Walker MC, Cope DW. Extrasynaptic $GABA_A$ receptors: form, pharmacology, and function. J Neurosci. 2009;29(41):12757–63.

Berkovic SF, Mulley JC, Scheffer IE, Petrou S. Human epilepsies: interaction of genetic and acquired factors. Trends Neurosci. 2006;29(7):391–7.

Bettler B, Kaupmann K, Mosbacher J, Gassmann M. Molecular structure and physiological functions of $GABA_B$ receptors. Physiol Rev. 2004;84(3):835–67.

Bhattacharya P, Pandey AK, Paul S, Patnaik R. Alleviation of glutamate mediated neuronal insult by piroxicam in rodent model of focal cerebral ischemia: a possible mechanism of GABA agonism. J Physiol Biochem. 2014;70(4):901–13.

Blicher JU, Near J, Næss-Schmidt E, Stagg CJ, Johansen-Berg H, Nielsen JF, et al. GABA levels are decreased after stroke and GABA changes during rehabilitation correlate with motor improvement. Neurorehabil Neural Repair. 2015;29(3):278–86.

Bolvig T, Larsson OM, Pickering DS, Nelson N, Falch E, Krogsgaard-Larsen P, et al. Action of bicyclic isoxazole GABA analogues on GABA transporters and its relation to anticonvulsant activity. Eur J Pharmacol. 1999;375(1–3):367–74.

Borden LA, Caplan MJ. GABA transporter heterogeneity: pharmacology and cellular localization. Neurochem Int. 1996;29(4):335–56.

Borden LA, Smith KE, Gustafson EL, Branchek TA, Weinshank RL. Cloning and expression of a betaine/GABA transporter from human brain. J Neurochem. 1995;64(3):977–84.

Bouilleret V, Loup F, Kiener T, Marescaux C, Fritschy JM. Early loss of interneurons and delayed subunit-specific changes in $GABA_A$-receptor expression in a mouse model of mesial temporal lobe epilepsy. Hippocampus. 2000;10(3):305–24.

Brickley SG, Mody I. Extrasynaptic $GABA_A$ receptors: their function in the CNS and implications for disease. Neuron. 2012;73(1):23–34.

Brown CE, Aminoltejari K, Erb H, Winship IR, Murphy TH. In vivo voltage-sensitive dye imaging in adult mice reveals that somatosensory maps lost to stroke are replaced over weeks by new structural and functional circuits with prolonged modes of activation within both the peri-infarct zone and distant sites. J Neurosci. 2009;29(6):1719–34.

Burda JE, Sofroniew MV. Reactive gliosis and the multicellular response to CNS damage and disease. Neuron. 2014;81(2):229–48.

Calcagnotto ME, Paredes MF, Baraban SC. Heterotopic neurons with altered inhibitory synaptic function in an animal model of malformation-associated epilepsy. J Neurosci. 2002;22(17): 7596–605.

Calcagnotto ME, Paredes MF, Tihan T, Barbaro NM, Baraban SC. Dysfunction of synaptic inhibition in epilepsy associated with focal cortical dysplasia. J Neurosci. 2005;25(42):9649–57.

Campbell AM, Holmes O. Bicuculline epileptogenesis in the rat. Brain Res. 1984;323(2):239–46.

Carmichael ST. Brain excitability in stroke: the yin and yang of stroke progression. Arch Neurol. 2012;69(2):161–7.

Chen F, Suzuki Y, Nagai N, Jin L, Yu J, Wang H, et al. Rodent stroke induced by photochemical occlusion of proximal middle cerebral artery: evolution monitored with MR imaging and histopathology. Eur J Radiol. 2007;63(1):68–75.

Chen Xu W, Yi Y, Qiu L, Shuaib A. Neuroprotective activity of tiagabine in a focal embolic model of cerebral ischemia. Brain Res. 2000;874(1):75–7.

Chiu C-S, Brickley S, Jensen K, Southwell A, Mckinney S, Cull-Candy S, et al. GABA transporter deficiency causes tremor, ataxia, nervousness, and increased GABA-induced tonic conductance in cerebellum. J Neurosci. 2005;25(12):3234–45.

Clarkson AN. Perisynaptic GABA receptors the overzealous protector. Adv Pharmacol Sci. 2012;2012:708428.

Clarkson AN, Huang BS, Macisaac SE, Mody I, Carmichael ST. Reducing excessive GABA-mediated tonic inhibition promotes functional recovery after stroke. Nature. 2010;468(7321): 305–9.

Clausen RP, Moltzen EK, Perregaard J, Lenz SM, Sanchez C, Falch E, et al. Selective inhibitors of GABA uptake: synthesis and molecular pharmacology of 4-N-methylamino-4,5,6,7-tetrahydrobenzo[d]isoxazol-3-ol analogues. Bioorg Med Chem. 2005;13(3):895–908.

Collinson N, Kuenzi FM, Jarolimek W, Maubach KA, Cothliff R, Sur C, et al. Enhanced learning and memory and altered GABAergic synaptic transmission in mice lacking the α5 subunit of the $GABA_A$ receptor. J Neurosci. 2002;22(13):5572–80.

Conti F, Melone M, De Biasi S, Minelli A, Brecha NC, Ducati A. Neuronal and glial localization of GAT-1, a high-affinity γ-aminobutyric acid plasma membrane transporter, in human cerebral cortex: with a note on its distribution in monkey cortex. J Comp Neurol. 1998;396(1):51–63.

Conti F, Minelli A, Melone M. GABA transporters in the mammalian cerebral cortex: localization, development and pathological implications. Brain Res Rev. 2004;45(3):196–212.

Conti F, Zuccarello LV, Barbaresi P, Minelli A, Brecha NC, Melone M. Neuronal, glial, and epithelial localization of gamma-aminobutyric acid transporter 2, a high-affinity gamma-aminobutyric acid plasma membrane transporter, in the cerebral cortex and neighboring structures. J Comp Neurol. 1999;409(3):482–94.

Cope DW, Di Giovanni G, Fyson SJ, Orbán G, Errington AC, Lörincz ML, et al. Enhanced tonic $GABA_A$ inhibition in typical absence epilepsy. Nat Med. 2009;15(12):1392–8.

Costa C, Leone G, Saulle E, Pisani F, Bernardi G, Calabresi P. Coactivation of $GABA_A$ and $GABA_B$ receptor results in neuroprotection during in vitro ischemia. Stroke. 2004;35(2):596–600.

Curtis DR, Duggan AW, Felix D, Johnston G. GABA, bicuculline and central inhibition. Nature. 1970;226(5252):1222–4.

Dalby NO. GABA-level increasing and anticonvulsant effects of three different GABA uptake inhibitors. Neuropharmacology. 2000;39(12):2399–407.

Dalby NO. Inhibition of γ-aminobutyric acid uptake: anatomy, physiology and effects against epileptic seizures. Eur J Pharmacol. 2003;479(1–3):127–37.

Dalby NO, Thomsen C, Fink-Jensen A, Lundbeck J, Søkilde B, Man CM, et al. Anticonvulsant properties of two GABA uptake inhibitors NNC 05-2045 and NNC 05-2090, not acting preferentially on GAT-1. Epilepsy Res. 1997;28(1):51–61.

Damgaard M, Haugaard AS, Al-Khawaja A, Lie MEK, Wellendorph P, Clausen RP, et al. Development of non-GAT1 selective GABA uptake inhibitors. In: Ortega A, Schousboe A, editors. Glial amino acid transporters. Cham: Springer; in press.

Damgaard M, Al-Khawaja A, Vogensen SB, Jurik A, Sijm M, Lie MEK, et al. Identification of the first highly subtype-selective inhibitor of human GABA transporter GAT3. ACS Chem Neurosci. 2015;6(9):1591–9.

Das N, Dhanawat M, Shrivastava SK. An overview on antiepileptic drugs. Drug Discov Ther. 2012;6(4):178–93.

Dhar TGM, Borden LA, Tyagarajan S, Smith KE, Branchek TA, Weinshank RL, et al. Design, synthesis and evaluation of substituted triarylnipecotic acid derivatives as GABA uptake inhibitors: identification of a ligand with moderate affinity and selectivity for the cloned human GABA transporter GAT- 3. J Med Chem. 1994;37(15):2334–42.

Dobkin BH. Training and exercise to drive poststroke recovery. Nat Clin Pract Neurol. 2008;4(2):76–85.

Dobkin BH, Dorsch A. New evidence for therapies in stroke rehabilitation. Curr Atheroscler Rep. 2013;15(6):331.

Doi T, Ueda Y, Nagatomo K, Willmore LJ. Role of glutamate and GABA transporters in development of pentylenetetrazol-kindling. Neurochem Res. 2009;34(7):1324–31.

Doi T, Ueda Y, Tokumaru J, Willmore LJ. Molecular regulation of glutamate and GABA transporter proteins by clobazam during epileptogenesis in Fe^{+++}-induced epileptic rats. Mol Brain Res. 2005;142(2):91–6.

Doyle KP, Simon RP, Stenzel-Poore MP. Mechanisms of ischemic brain damage. Neuropharmacology. 2008 Sep;55(3):310–8.

During MJ, Ryder KM, Spencer DD. Hippocampal GABA transporter function in temporal-lobe epilepsy. Nature. 1995;376(6536):174–7.

Emberson J, Lees K, Lyden P, Blackwell L, Albers G, Bluhmki E, et al. Effect of treatment delay, age, and stroke severity on the effects of intravenous thrombolysis with alteplase for acute ischaemic stroke: a meta-analysis of individual patient data from randomised trials. Lancet. 2014;385(9958):1929–35.

Engelborghs S, D'Hooge R, De Deyn PP. Pathophysiology of epilepsy. Acta Neurol Belg. 2000;100(4):201–13.

Errington AC, Cope DW, Crunelli V. Augmentation of tonic $GABA_A$ inhibition in absence epilepsy: therapeutic value of inverse agonists at extrasynaptic $GABA_A$ receptors. Adv Pharmacol Sci. 2011;2011:1–12.

Esclapez M, Houser CR. Up-regulation of GAD65 and GAD67 in remaining hippocampal GABA neurons in a model of temporal lobe epilepsy. J Comp Neurol. 1999;412(3):488–505.

Farrant M, Nusser Z. Variations on an inhibitory theme: phasic and tonic activation of $GABA_A$ receptors. Nat Rev Neurosci. 2005;6(3):215–29.
Fink-Jensen A, Suzdak PD, Swedberg MDB, Judge ME, Hansen L, Nielsen PG. The γ-aminobutyric acid (GABA) uptake inhibitor, tiagabine, increases extracellular brain levels of GABA in awake rats. Eur J Pharmacol. 1992;220(2–3):197–201.
Frahm C, Haupt C, Weinandy F, Siegel G, Bruehl C, Witte OW. Regulation of GABA transporter mRNA and protein after photothrombotic infarct in rat brain. J Comp Neurol. 2004;478(2):176–88.
Freichel C, Potschka H, Ebert U, Brandt C, Löscher W. Acute changes in the neuronal expression of GABA and glutamate decarboxylase isoforms in the rat piriform cortex following status epilepticus. Neuroscience. 2006;141(4):2177–94.
Fu CY, He XY, Li XF, Zhang X, Huang ZW, Li J, et al. Nefiracetam attenuates pro-inflammatory cytokines and GABA transporter in specific brain regions of rats with post-ischemic seizures. Cell Physiol Biochem. 2015;37(5):2023–31.
Gastaut H, Naquet R, Poire R, Tassinari C. Treatment of status epilepticus with diazepam (Valium). Epilepsia. 1965;6:167–82.
Ginsberg MD. Neuroprotection for ischemic stroke: past, present and future. Neuropharmacology. 2008;55(3):363–89.
Gonsalves SF, Twitchell B, Harbaugh RE, Krogsgaard-Larsen P, Schousboe A. Anticonvulsant activity of the glial GABA uptake inhibitor, THAO, in chemical seizures. Eur J Pharmacol. 1989;168(2):265–8.
Green AR, Hainsworth AH, Jackson DM. GABA potentiation: a logical pharmacological approach for the treatment of acute ischaemic stroke. Neuropharmacology. 2000;39(9):1483–94.
Héja L, Barabas P, Nyitrai G, Kekesi KA, Lasztoczi B, Toke O, et al. Glutamate uptake triggers transporter-mediated GABA release from astrocytes. PLoS One. 2009;4(9):e7153.
Héja L, Nyitrai G, Kékesi O, Dobolyi Á, Szabó P, Fiáth R, et al. Astrocytes convert network excitation to tonic inhibition of neurons. BMC Biol. 2012;10(26):1–22.
Hernandez TD, Schallert T. Seizures and recovery from experimental brain damage. Exp Neurol. 1988;102(3):318–24.
Hernandez-Rabaza V, Cabrera-Pastor A, Taoro-Gonzalez L, Gonzalez-Usano A, Agusti A, Balzano T, et al. Neuroinflammation increases GABAergic tone and impairs cognitive and motor function in hyperammonemia by increasing GAT-3 membrane expression. Reversal by sulforaphane by promoting M2 polarization of microglia. J Neuroinflammation. 2016;13(1):83.
Hill RG, Simmonds MA, Straughan DW. A comparative study of some convulsant substances as γ-aminobutyric acid antagonists in the feline cerebral cortex. Br J Pharmacol. 1973;49(1):37–51.
Hines RM, Davies PA, Moss SJ, Maguire J. Functional regulation of $GABA_A$ receptors in nervous system pathologies. Curr Opin Neurobiol. 2012;22(3):552–8.
Hirao T, Morimoto K, Yamamoto Y, Watanabe T, Sato H, Sato K, et al. Time-dependent and regional expression of GABA transporter mRNAs following amygdala-kindled seizures in rats. Mol Brain Res. 1998;54(1):49–55.
Hirose S, Mitsudome A, Okada M, Kaneko S. Genetics of idiopathic epilepsies. Epilepsia. 2005;46(Suppl. 1):38–43.
Hutchinson PJ, O'Connell MT, Al-Rawi PG, Kett-White CR, Gupta AK, Maskell LB, et al. Increases in GABA concentrations during cerebral ischaemia: a microdialysis study of extracellular amino acids. J Neurol Neurosurg Psychiatry. 2002;72(1):99–105.
Høg S, Greenwood JR, Madsen KB, Larsson OM, Frølund B, Schousboe A, et al. Structure-activity relationships of selective GABA uptake inhibitors. Curr Top Med Chem. 2006;6(17):1861–82.
Inglefield JR, Perry JM, Schwartz RD. Postischemic inhibition of GABA reuptake by tiagabine slows neuronal death in the gerbil hippocampus. Hippocampus. 1995;5(5):460–8.
Iqbal S, Baziany A, Gordon S, Wright S, Hussain M, Miyashita H, et al. Neuroprotective effect of tiagabine in transient forebrain global ischemia: an in vivo microdialysis, behavioral, and histological study. Brain Res. 2002;946(2):162–70.
Jaenisch N, Liebmann L, Guenther M, Hübner CA, Frahm C, Witte OW. Reduced tonic inhibition after stroke promotes motor performance and epileptic seizures. Sci Report. 2016;6:26173.
Jardetzky O. Simple allosteric model for membrane pumps. Nature. 1966;211(5052):969–70.

Jensen K, Chiu CS, Sokolova I, Lester HA, Mody I. GABA transporter-1 (GAT1)-deficient mice: differential tonic activation of $GABA_A$ versus $GABA_B$ receptors in the hippocampus. J Neurophysiol. 2003;90(4):2690–701.

Johansen FF, Diemer NH. Enhancement of GABA neurotransmission after cerebral ischemia in the rat reduces loss of hippocampal CA1 pyramidal cells. Acta Neurol Scand. 1991;84(1):1–6.

Juhász G, Kékesi KA, Nyitrai G, Dobolyi A, Krogsgaard-Larsen P, Schousboe A. Differential effects of nipecotic acid and 4,5,6,7-tetrahydroisoxazolo[4,5-c]pyridin-3-ol on extracellular α-aminobutyrate levels in rat thalamus. Eur J Pharmacol. 1997;331(2–3):139–44.

Kaim S, Rosenstein I. Anticonvulsant properties of a new psychotherapeutic drug. Dis Nerv Syst. 1960;21(3):46–8.

Kang TC, Kim HS, Seo MO, Park SK, Kwon HY, Kang JH, et al. The changes in the expressions of γ-aminobutyric acid transporters in the gerbil hippocampal complex following spontaneous seizure. Neurosci Lett. 2001;310(1):29–32.

Kanthan R, Shuaib A, Griebel R, Miyashita H. Intracerebral human microdialysis. In vivo study of an acute focal ischemic model of the human brain. Stroke. 1995;26(5):870–3.

Keros S, Hablitz JJ. Subtype-specific GABA transporter antagonists synergistically modulate phasic and tonic $GABA_A$ conductances in rat neocortex. J Neurophysiol. 2005;94(3):2073–85.

Kersanté F, Rowley SCS, Pavlov I, Gutièrrez-Mecinas M, Semyanov A, Reul JMHM, et al. A functional role for both γ-aminobutyric acid (GABA) transporter-1 and GABA transporter-3 in the modulation of extracellular GABA and GABAergic tonic conductances in the rat hippocampus. J Physiol. 2013;591(10):2429–41.

Kett-White R, O'Connell MT, Hutchinson PJA, Al-Rawi PG, Gupta AK, Pickard JD, et al. Extracellular amino acid changes in patients during reversible cerebral ischaemia. Acta Neurochir Suppl. 2005;95:83–8.

Kim YK, Yang EJ, Cho K, Lim JY, Paik N-J. Functional recovery after ischemic stroke is associated with reduced GABAergic inhibition in the cerebral cortex: a GABA PET study. Neurorehabil Neural Repair. 2014;28(6):576–83.

Knake S, Hamer H, Schomburg U, Oertel W, Rosenow F. Tiagabine-induced absence status in idiopathic generalized epilepsy. Seizure. 1999;8:314–7.

Kobayashi T, Suemasa A, Igawa A, Ide S, Fukuda H, Abe H, et al. Conformationally restricted GABA with bicyclo[3.1.0]hexane backbone as the first highly selective BGT-1 inhibitor. ACS Med Chem Lett. 2014;5(8):889–93.

Kong S, Cheng Z, Liu J, Wang Y. Downregulated GABA and BDNF-TrkB pathway in chronic cyclothiazide seizure model. Neural Plast. 2014;2014:9–11.

Kostandy BB. The role of glutamate in neuronal ischemic injury: the role of spark in fire. Neurol Sci. 2012;33(2):223–37.

Kragholm B, Kvist T, Madsen KK, Jørgensen L, Vogensen SB, Schousboe A, et al. Discovery of a subtype selective inhibitor of the human betaine/GABA transporter 1 (BGT-1) with a non-competitive pharmacological profile. Biochem Pharmacol. 2013;86(4):521–8.

Kristensen AS, Andersen J, Jørgensen TN, Sørensen L, Eriksen J, Loland CJ, et al. SLC6 neurotransmitter transporters: structure, function, and regulation. Pharmacol Rev. 2011;63(3):585–640.

Kristensen BW, Noraberg J, Zimmer J. The $GABA_A$ receptor agonist THIP is neuroprotective in organotypic hippocampal slice cultures. Brain Res. 2003;973(2):303–6.

Kälviäinen R. Long-term safety of tiagabine. Epilepsia. 2001;42(Suppl 1):46–9.

Kälviäinen R. Tiagabine. Clinical efficacy and use in epilepsy. In: Antiepileptic drugs. 5th ed. New York: Raven Press; 2002. p. 698–704.

Lake EMR, Chaudhuri J, Thomason L, Janik R, Ganguly M, Brown M, et al. The effects of delayed reduction of tonic inhibition on ischemic lesion and sensorimotor function. J Cereb Blood Flow Metab. 2015;35(10):1601–9.

Lamar CD, Hurley RA, Rowland JA, Taber KH. Post-traumatic epilepsy: review of risks, pathophysiology, and potential biomarkers. J Neuropsychiatr Clin Neurosci. 2014;26(2):108–13.

Lazar RM, Fitzsimmons B-F, Marshall RS, Berman MF, Bustillo MA, Young WL, et al. Reemergence of stroke deficits with midazolam challenge. Stroke. 2002;33(1):283–5.

Lee M, McGeer E, McGeer P. Mechanisms of GABA release from human astrocytes. Glia. 2011;59(11):1600–11.

Lee TS, Bjørnsen LP, Paz C, Kim JH, Spencer SS, Spencer DD, et al. GAT1 and GAT3 expression are differently localized in the human epileptogenic hippocampus. Acta Neuropathol. 2006;111(4):351–63.

Lee V, Maguire J. The impact of tonic $GABA_A$ receptor-mediated inhibition on neuronal excitability varies across brain region and cell type. Front Neural Circuits. 2014;8(3):1–27.

Lehre AC, Rowley NM, Zhou Y, Holmseth S, Guo C, Holen T, et al. Deletion of the betaine-GABA transporter (BGT1; slc6a12) gene does not affect seizure thresholds of adult mice. Epilepsy Res. 2011;95(1–2):70–81.

Lerma J, Herreras O, Herranz AS, Munoz D, del Rio RM. In vivo effects of nipecotic acid on levels of extracellular gaba and taurine, and hippocampal excitability. Neuropharmacology. 1984;23(5):595–8.

Li C, Yan Z, Yang J, Chen H, Li H, Jiang Y, et al. Neuroprotective effects of resveratrol on ischemic injury mediated by modulating the release of neurotransmitter and neuromodulator in rats. Neurochem Int. 2010a;56(3):495–500.

Li H, Li C, Yan Z-Y, Yang J, Chen H. Simultaneous monitoring multiple neurotransmitters and neuromodulators during cerebral ischemia/reperfusion in rats by microdialysis and capillary electrophoresis. J Neurosci Methods. 2010b;189(2):162–8.

Li H, Zhang N, Lin HY, Yu Y, Cai QY, Ma L, et al. Histological, cellular and behavioral assessments of stroke outcomes after photothrombosis-induced ischemia in adult mice. BMC Neurosci. 2014;15(58):1–13.

Lim DH, LeDue JM, Mohajerani MH, Murphy TH. Optogenetic mapping after stroke reveals network-wide scaling of functional connections and heterogeneous recovery of the peri-infarct. J Neurosci. 2014;34(49):16455–66.

Liu S, Wang Y, Worley PF, Mattson MP, Gaiano N. The canonical Notch pathway effector RBP-J regulates neuronal plasticity and expression of GABA transporters in hippocampal networks. Hippocampus. 2015;25(5):670–8.

Lo EH. A new penumbra: transitioning from injury into repair after stroke. Nat Med. 2008;14(5): 497–500.

Macdonald R, Barker JL. Benzodiazepines specifically modulate GABA-mediated postsynaptic inhibition in cultured mammalian neurones. Nature. 1978;271(5645):563–4.

Madsen KK, Clausen RP, Larsson OM, Krogsgaard-Larsen P, Schousboe A, Steve WH. Synaptic and extrasynaptic GABA transporters as targets for anti-epileptic drugs. J Neurochem. 2009;109(Suppl. 1):139–44.

Madsen KK, Ebert B, Clausen RP, Krogsgaard-Larsen P, Schousboe A, White HS. Selective GABA transporter inhibitors tiagabine and EF1502 exhibit mechanistic differences in their ability to modulate the ataxia and anticonvulsant action of the extrasynaptic $GABA_A$ receptor agonist gaboxadol. J Pharmacol Exp Ther. 2011;338(1):214–9.

Madsen KK, White HS, Schousboe A. Neuronal and non-neuronal GABA transporters as targets for antiepileptic drugs. Pharmacol Ther. 2010;125(3):394–401.

Mao X, Guo F, Yu J, Min D, Wang Z, Xie N, et al. Up-regulation of GABA transporters and $GABA_A$ receptor α1 subunit in tremor rat hippocampus. Neurosci Lett. 2010;486(3):150–5.

Mathern GW, Mendoza D, Lozada A, Pretorius JK, Dehnes Y, Danbolt NC, et al. Hippocampal GABA and glutamate transporter immunoreactivity in patients with temporal lobe epilepsy. Neurology. 1999;52(3):453–72.

Medina-Ceja L, Sandoval-García F, Morales-Villagrán A, López-Pérez SJ. Rapid compensatory changes in the expression of EAAT-3 and GAT-1 transporters during seizures in cells of the CA1 and dentate gyrus. J Biomed Sci. 2012;19(78):1–11.

Meldrum BS, Horton RW. Convulsive effects of 4-deoxypyridoxine and of bicuculline in photosensitive baboons (*Papio papio*) and in rhesus monkeys (*Macaca mulatta*). Brain Res. 1971;35(2):419–36.

Melone M, Barbaresi P, Fattorini G, Conti F. Neuronal localization of the GABA transporter GAT-3 in human cerebral cortex: a procedural artifact? J Chem Neuroanat. 2005;30(1):45–54.

Melone M, Cozzi A, Pellegrini-Giampietro DE, Conti F. Transient focal ischemia triggers neuronal expression of GAT-3 in the rat perilesional cortex. Neurobiol Dis. 2003;14(1):120–32.

Murphy TH, Corbett D. Plasticity during stroke recovery: from synapse to behaviour. Nat Rev Neurosci. 2009;10(12):861–72.

Nagatomo K, Ueda Y, Doi T, Takaki M, Tsuru N. Functional role of GABA transporters for kindling development in GLAST KO mice. Neurosci Res. 2007;57(2):319–21.

Naquet R, Soulayrol R, Dolce G, Tassinari C, Broughton R, Loeb H. First attempt at treatment of experimental status epilepticus in animals and spontaneous status epilepticus in man with diazepam. Electroencephalogr Clin Neurophysiol. 1965;18:424–7.

Nelson RM, Green AR, Lambert DG, Hainsworth AH. On the regulation of ischaemia-induced glutamate efflux from rat cortex by GABA; in vitro studies with GABA, clomethiazole and pentobarbitone. Br J Pharmacol. 2000;130(5):1124–30.

Nielsen EB, Suzdak PD, Andersen KE, Knutsen LJS, Sonnewald U, Braestrup C. Characterization of tiagabine (NO-328), a new potent and selective GABA uptake inhibitor. Eur J Pharmacol. 1991;196(3):257–66.

O'Connell AW, Fox GB, Kjøller C, Gallagher HC, Murphy KJ, Kelly J, et al. Anti-ischemic and cognition-enhancing properties of NNC-711, γ-aminobutyric acid reuptake inhibitor. Eur J Pharmacol. 2001;424(1):37–44.

Olsen M, Sarup A, Larsson OMOM, Schousboe A. Effect of hyperosmotic conditions on the expression of the betaine-GABA-transporter (BGT-1) in cultured mouse astrocytes. Neurochem Res. 2005;30(6–7):855–65.

Olsen RW, Avoli M. GABA and epileptogenesis. Epilepsia. 1997;38(4):399–407.

Orozco-Suarez S, Brunson KL, Feria-Velasco A, Ribak CE. Increased expression of γ-aminobutyric acid transporter-1 in the forebrain of infant rats with corticotropin-releasing hormone-induced seizures but not in those with hyperthermia-induced seizures. Epilepsy Res. 2000;42(2–3):141–57.

Orser BA. Extrasynaptic $GABA_A$ receptors are critical targets for sedative-hypnotic drugs. J Clin Sleep Med. 2006;2(2):S12–8.

Ouyang C, Guo L, Lu Q, Xu X, Wang H. Enhanced activity of GABA receptors inhibits glutamate release induced by focal cerebral ischemia in rat striatum. Neurosci Lett. 2007;420(2):174–8.

Pabel J, Faust M, Prehn C, Wörlein B, Allmendinger L, Höfner G, et al. Development of an (*S*)-1-{2-[tris(4-methoxyphenyl)methoxy]ethyl}piperidine-3-carboxylic acid [(*S*)-SNAP-5114] carba analogue inhibitor for murine γ-aminobutyric acid transporter type 4. ChemMedChem. 2012;7(7):1245–55.

Patrylo PR, Spencer DD, Williamson A. GABA uptake and heterotransport are impaired in the dentate gyrus of epileptic rats and humans with temporal lobe sclerosis. J Neurophysiol. 2001;85(4):1533–42.

Pavlov I, Walker MC. Tonic GABAA receptor-mediated signalling in temporal lobe epilepsy. Neuropharmacology. 2013;69:55–61.

Petersen JG, Sørensen T, Damgaard M, Nielsen B, Jensen AA, Balle T, et al. Synthesis and pharmacological evaluation of 6-aminonicotinic acid analogues as novel $GABA_A$ receptor agonists. Eur J Med Chem. 2014;84:404–16.

Phillis JW. CI-966, a GABA uptake inhibitor, antagonizes ischemia-induced neuronal degeneration in the gerbil. Gen Pharmacol. 1995;26(5):1061–4.

Pirttimaki T, Parri HR, Crunelli V. Astrocytic GABA transporter GAT-1 dysfunction in experimental absence seizures. J Physiol. 2013;591(4):823–33.

Pitkänen A, Löscher W, Vezzani A, Becker AJ, Simonato M, Lukasiuk K, et al. Advances in the development of biomarkers for epilepsy. Lancet Neurol. 2016;15(8):843–56.

Princivalle AP, Duncan JS, Thom M, Bowery NG. $GABA_{B1a}$, $GABA_{B1b}$ and $GABA_{B2}$ mRNA variants expression in hippocampus resected from patients with temporal lobe epilepsy. Neuroscience. 2003a;122(4):975–84.

Princivalle AP, Richards DA, Duncan JS, Spreafico R, Bowery NG. Modification of $GABA_{B1}$ and $GABA_{B2}$ receptor subunits in the somatosensory cerebral cortex and thalamus of rats with absence seizures (GAERS). Epilepsy Res. 2003b;55(1–2):39–51.

Rassner MP, Moser A, Follo M, Joseph K, van Velthoven-Wurster V, Feuerstein TJ. Neocortical GABA release at high intracellular sodium and low extracellular calcium: an anti-seizure mechanism. J Neurochem. 2016;137(2):177–89.
Redecker C, Wang W, Fritschy J-M, Witte OW. Widespread and long-lasting alterations in $GABA_A$-receptor subtypes after focal cortical infarcts in rats: mediation by NMDA-dependent processes. J Cereb Blood Flow Metab. 2002;22(12):1463–75.
Richerson GB, Wu Y. Dynamic equilibrium of neurotransmitter transporters: not just for reuptake anymore. J Neurophysiol. 2003;90(3):1363–74.
Rowley NM, Smith MD, Lamb JG, Schousboe A, White HS. Hippocampal betaine/GABA transporter mRNA expression is not regulated by inflammation or dehydration post-status epilepticus. J Neurochem. 2011;117(1):82–90.
Santhakumar V, Jones RT, Mody I. Developmental regulation and neuroprotective effects of striatal tonic $GABA_A$ currents. Neuroscience. 2010;167(3):644–55.
Schallert T, Hernandez TD, Barth TM. Recovery of function after brain damage: severe and chronic disruption by diazepam. Brain Res. 1986;379(1):104–11.
Schallert T, Jones T, Weaver M, Shapiro L, Crippens D, Fulton R. Pharmacologic and anatomic considerations in recovery of function. Phys Med Rehabil. 1992;6(3):375–89.
Scharfman HE. The neurobiology of epilepsy. Curr Neuro Neurosci Rep. 2007;7(4):348–54.
Schijns O, Karaca Ü, Andrade P, de Nijs L, Küsters B, Peeters A, et al. Hippocampal GABA transporter distribution in patients with temporal lobe epilepsy and hippocampal sclerosis. J Chem Neuroanat. 2015;68:39–44.
Schipper S, Aalbers MW, Rijkers K, Swijsen A, Rigo JM, Hoogland G, et al. Tonic $GABA_A$ receptors as potential target for the treatment of temporal lobe epilepsy. Mol Neurobiol. 2016;53(8):5252–65.
Schousboe A. Role of astrocytes in the maintenance and modulation of glutamatergic and GABAergic neurotransmission. Neurochem Res. 2003;28(2):347–52.
Schousboe A, Larsson OM, Wood JD, Krogsgaard-Larsen P. Transport and metabolism of γ-aminobutyric acid in neurons and glia: implications for epilepsy. Epilepsia. 1983;24(5):531–8.
Schousboe A, Madsen KK, White HS. GABA transport inhibitors and seizure protection: the past and future. Future Med Chem. 2011;3(2):183–7.
Schousboe A, Wellendorph P, Frølund B, Clausen RP, Krogsgaard-Larsen P. Astrocytic GABA-transporters: pharmacological properties and targets for antiepileptic drugs. In: Ortega A, Schousboe A, editors. Glial amino acid transporters. Cham: Springer; in press.
Schwartz-Bloom RD, Sah R. γ-Aminobutyric $acid_A$ neurotransmission and cerebral ischemia. J Neurochem. 2001;77(2):353–71.
Shaju M, Abraham S. Innovations in epilepsy management - an overview. J Pharm Pharm Sci. 2013;16(4):564–76.
Smith MD, Saunders GW, Clausen RP, Frølund B, Krogsgaard-Larsen P, Larsson OM, et al. Inhibition of the betaine-GABA transporter (mGAT2/BGT-1) modulates spontaneous electrographic bursting in the medial entorhinal cortex (mEC). Epilepsy Res. 2008;79(1):6–13.
Sofroniew MV. Astrocyte barriers to neurotoxic inflammation. Nat Rev Neurosci. 2015;16(5):249–63.
Sofroniew MV, Vinters HV. Astrocytes: biology and pathology. Acta Neuropathol. 2010;119(1):7–35.
Song I, Volynski K, Brenner T, Ushkaryov Y, Walker M, Semyanov A. Different transporter systems regulate extracellular GABA from vesicular and non-vesicular sources. Front Cell Neurosci. 2013;7(23):1–7.
Southan C, Sharman JL, Benson HE, Faccenda E, Pawson AJ, Alexander SPH, et al. The IUPHAR/BPS Guide to Pharmacology in 2016: towards curated quantitative interactions between 1300 protein targets and 6000 ligands. Nucleic Acids Res. 2016;44(D1):D1054–68.
Sperk G, Schwarzer C, Heilman J, Furtinger S, Reimer RJ, Edwards RH, et al. Expression of plasma membrane GABA transporters but not of the vesicular GABA transporter in dentate granule cells after kainic acid seizures. Hippocampus. 2003;13(7):806–15.

Spreafico R, Tassi L, Colombo N, Bramerio M, Galli C, Garbelli R, et al. Inhibitory circuits in human dysplastic tissue. Epilepsia. 2000;41(Suppl. 6):S168–73.

Staley K. Molecular mechanisms of epilepsy. Nat Neurosci. 2015;18(3):367–72.

Su J, Yin J, Qin W, Sha S, Xu J, Jiang C. Role for pro-inflammatory cytokines in regulating expression of GABA transporter type 1 and 3 in specific brain regions of kainic acid-induced status epilepticus. Neurochem Res. 2015;40(3):621–7.

Sutch RJ, Davies CC, Bowery NG. GABA release and uptake measured in crude synaptosomes from Genetic Absence Epilepsy Rats from Strasbourg (GAERS). Neurochem Int. 1999;34(5):415–25.

Thomsen C, Sørensen PO, Egebjerg J. 1-(3-(9H-carbazol-9-yl)-1-propyl)-4-(2-methoxyphenyl)-4-piperidinol, a novel subtype selective inhibitor of the mouse type II GABA-transporter. Br J Pharmacol. 1997;120(6):983–5.

Treiman DM. GABAergic mechanisms in epilepsy. Epilepsia. 2001;42(Suppl. 3):8–12.

Trojnar MK, Małek R, Chrościńska M, Nowak S, Błaszczyk B, Czuczwar SJ. Neuroprotective effects of antiepileptic drugs. Pol J Pharmacol. 2002;54(6):557–66.

Ueda Y, Willmore LJ. Hippocampal γ-aminobutyric acid transporter alterations following focal epileptogenesis induced in rat amygdala. Brain Res Bull. 2000;52(5):357–61.

Usunoff G, Atsev E, Tchavdarov D. On the mechanisms of picrotoxin epileptic seizure (macro- and micro-electrode investigations). Electroencephalogr Clin Neurophysiol. 1969;27(4):444.

Vinton A, Kornberg A, Cowley M, Matkovic Z, Kilpatrick C, O'Brien T. Tiagabine-induced generalised non convulsive status epilepticus in patients with lesional focal epilepsy. J Clin Neurosci. 2005;12(2):128–33.

Vogensen SB, Jørgensen L, Madsen KK, Borkar N, Wellendorph P, Skovgaard-Petersen J, et al. Selective mGAT2 (BGT-1) GABA uptake inhibitors: design, synthesis, and pharmacological characterization. J Med Chem. 2013;56(5):2160–4.

Wang J-G, Cai Q, Zheng J, Dong Y-S, Li J-J, Li J-C, et al. Epigenetic suppression of GADs expression is involved in temporal lobe epilepsy and pilocarpine-induced mice epilepsy. Neurochem Res. 2016;41(7):1751–60.

Wellendorph P, Jacobsen J, Skovgaard-Petersen J, Jurik A, Vogensen SB, Ecker G, Schousboe A, Krogsgaard-Larsen P, Clausen RP. γ-Aminobutyric acid and glycine neurotransmitter transporters. In: Sitte HH, Ecker GF, Folkers G, Mannhold R, Buschmann H, Clausen RP, editors. Transporters as drug targets. Weinheim: Wiley-VCH Verlag GmbH & Co. KGaA; 2017. doi:10.1002/9783527679430.ch4.

White HS, Hunt J, Wolf HH, Swinyard EA, Falch E, Krogsgaard-Larsen P, et al. Anticonvulsant activity of the γ-aminobutyric acid uptake inhibitor N-4,4-diphenyl-3-butenyl-4,5,6,7-tetrahydroisoxazolo[4,5-c]pyridin-3-ol. Eur J Pharmacol. 1993;236(1):147–9.

White HS, Sarup A, Bolvig T, Kristensen AS, Petersen G, Nelson N, et al. Correlation between anticonvulsant activity and inhibitory action on glial γ-aminobutyric acid uptake of the highly selective mouse γ-aminobutyric acid transporter 1 inhibitor 3-hydroxy-4-amino-4,5,6,7-tetrahydro-1,2-benzisoxazole and its N-alkylated anal. J Pharmacol Exp Ther. 2002;302(2):636–44.

White HS, Watson WP, Hansen SL, Slough S, Perregaard J, Sarup A, et al. First demonstration of a functional role for central nervous system betaine/γ-aminobutyric acid transporter (mGAT2) based on synergistic anticonvulsant action among inhibitors of mGAT1 and mGAT2. J Pharmacol Exp Ther. 2005;312(2):866–74.

World Health Federation. Stroke [Internet]. 2016 [cited 2016 Sept 1]. Available from: www.world-heart-federation.org/cardiovascular-health/stroke/

WHO. Epilepsy Fact Sheet [Internet]. 2016a [cited 2016 Jun 30]. Available from: www.who.int/mediacentre/factsheets/fs999/en/

WHO. The top 10 causes of death Fact Sheet [Internet]. 2016b [cited 2016 Jun 30]. Available from: www.who.int/mediacentre/factsheets/fs310/en/

Wood JD, Johnson DD, Krogsgaard-Larsen P, Schousboe A. Anticonvulsant activity of the glial-selective GABA uptake inhibitor, THPO. Neuropharmacology. 1983;22(1):139–42.

Wu Z, Guo Z, Gearing M, Chen G. Tonic inhibition in dentate gyrus impairs long-term potentiation and memory in an Alzhiemer's disease model. Nat Commun. 2014;5(4159):1–13.

Yang Y, Li Q, Wang CX, Jeerakathil T, Shuaib A. Dose-dependent neuroprotection with tiagabine in a focal cerebral ischemia model in rat. Neuroreport. 2000;11(10):2307–11.

Yasiry Z, Shorvon SD. How phenobarbital revolutionized epilepsy therapy: the story of phenobarbital therapy in epilepsy in the last 100 years. Epilepsia. 2012;53(Suppl 8):26–39.

Yunger LM, Fowler PJ, Zarevics P, Setler PE. Novel inhibitors of γ-aminobutyric acid (GABA) uptake: anticonvulsant actions in rats and mice. J Pharmacol Exp Ther. 1984;228(1):109–15.

Zhou Y, Holmseth S, Hua R, Lehre AC, Olofsson AM, Poblete-Naredo I, et al. The betaine-GABA transporter (BGT1, slc6a12) is predominantly expressed in the liver and at lower levels in the kidneys and at the brain surface. AJP Ren Physiol. 2012;302(3):316–28.

Zhu XM, Ong WY. Changes in GABA transporters in the rat hippocampus after kainate-induced neuronal injury: decrease in GAT-1 and GAT-3 but upregulation of betaine/GABA transporter BGT-1. J Neurosci Res. 2004a;77(3):402–9.

Zhu XM, Ong WY. A light and electron microscopic study of betaine/GABA transporter distribution in the monkey cerebral neocortex and hippocampus. J Neurocytol. 2004b;33(2):233–40.

Glutamine/Glutamate Transporters in Glial Cells: Much More Than Participants of a Metabolic Shuttle

Angelina Rodríguez and Arturo Ortega

Abstract Glial glutamine and glutamate transporters play an important role in glial/neuronal interactions. An excellent model to establish the role of these membrane proteins is the cerebellum. The most abundant glutamatergic synapse in the central nervous system is present in the molecular layer of the cerebellar cortex, and it is entirely wrapped by Bergmann glial cells. The recycling of glutamate involves glutamate and glutamine transporters enriched in these radial glial processes. The functional properties of amino acid glial transporters allow, in an activity-dependent manner, the conformation of protein complexes important for the adequate support of glutamatergic neurotransmission. A detailed description of the most important features of glial glutamate and glutamine transporters follows, and a working model of the molecular mechanisms by which these glutamate and glutamine binding proteins interact, and by these means might modulate cerebellar glutamatergic transactions, is presented.

Keywords Glial cells • Glutamate transporters • Glutamine transporters • Glial/neuronal coupling

List of Abbreviations

AMPA	Alpha-amino-3-hydroxy-5-methylisoxazole-4-propionate
ASC	Alanine/serine/cysteine transporter subfamily
ATP	Adenosine triphosphate
BCH	2-Aminobicyclo-(2,2,1)-heptane-2-carboxylic acid
CATs	Cationic amino acid transporters

A. Rodríguez
Facultad de Ciencias Naturales, Universidad Autónoma de Querétaro,
Centro Universitario, Querétaro, México

A. Ortega (✉)
Departamento de Toxicología, Cinvestav-IPN,
Apartado Postal 14-740, México DF 07360, México
e-mail: arortega@cinvestav.mx

A. Ortega, A. Schousboe (eds.), *Glial Amino Acid Transporters*,
Advances in Neurobiology 16, DOI 10.1007/978-3-319-55769-4_8

CNS	Central nervous system
E15	Embryonic day 15
E19	Embryonic day 19
EAAT	Excitatory amino acid transporter
EAAT1	Excitatory amino acid transporter 1 (GLAST, SLC1a3)
EAAT2	Excitatory amino acid transporter 2 (GLT-1, SLC1a2)
EAAT3	Excitatory amino acid transporter 3
EAAT4	Excitatory amino acid transporter 4
EAAT5	Excitatory amino acid transporter 5
GABA	Gamma-aminobutyric acid
Glut1	Glucose transporter 1
GRIs	Ionotropic glutamate receptors
GRIA2	Glutamate ionotropic receptor AMPA type subunit 2
GRMs	Metabotropic glutamate receptors
GSH	Glutathione
KA	Kainate
L-AP-4	L-2-amino-phosphonobutanoate
LATs	Light subunits of amino acid transporters
MeAIB	2-Methylaminoisobutyric acid
mRNA	Messenger ribonucleic acid
mTOR	Mechanistic target of rapamycin
NMDA	N-Methyl-d-aspartate
PI3K	Phosphatidylinositol 3-kinase
PKB	Protein kinase B
PNS	Peripheral nervous system
Quis	Quisqualate
Rho	Ras homolog gene family
ROS	Reactive oxygen species
SA	System A
SLC1	Solute carrier family 1
SLC14A	Solute carrier family 1 member 4 (ASCT1)
SLC15A	Solute carrier family 1 member 5 (ASCT2)
SLC38	Solute carrier family 38
SLC38A1	Solute carrier family 38 member 1 (SNAT1)
SLC38A2	Solute carrier family 38 member 2 (SNAT2)
SLC38A3	Solute carrier family 38 member 3 (SNAT3)
SLC38A4	Solute carrier family 38 member 4 (SNAT4)
SLC38A5	Solute carrier family 38 member 5 (SNAT5)
SLC38A7	Solute carrier family 38 member 7 (SNAT7)
SLC38A8	Solute carrier family 38 member 8 (SNAT8)
SLC6	Solute carrier family 6
SLC6A14	Solute carrier family 6 member 14 (ATB°)
SLC6A15	Solute carrier family 6 member 15 (B°AT2)
SLC6A19	Solute carrier family 6 member 19 (B°AT1)
SLC7	Solute carrier family 7

SLC7A5	Solute carrier family 7 member 5 (LAT1)
SLC7A8	Solute carrier family 7 member 8 (LAT2)
SN	System N
SNAT	Sodium-coupled neutral amino acid transporter
tACPD	1-Amino-4,5-cyclopentane-trans-1,3-dicarboxylate
X_c^-	Anionic amino acid transporter light chain, Xc-system, member 11 (SLC7a11)
xCT	Functional subunit of anionic amino acid transporter light chain, Xc-system, member 11

1 Introduction

Once described as passive, non-excitable elements, glial cells have gained attraction in the past years. The expression of most of neurotransmitter receptors described thus far has attracted the attention of a plethora of researchers which have contributed to the study of glial physiology in the context of neurotransmission (Parpura and Verkhratsky 2012). In fact, the concept of a three-component (tripartite) synapse, in which radial glia was included, was described as early as 1991 (Teichberg 1991). Thereafter, the plausible role of glial cells in glutamatergic signaling in the brain was continuously challenged (Gallo and Ghiani 2000). The molecular characterization of glutamate uptake systems and the glial enrichment of some of these transporters provided a major input to the tripartite synapse concept (Gallo and Ghiani 2000; Storck et al. 1992). Likewise, the characterization of glial glutamine transporters further supported the important role of glia in glutamatergic transmission (Albrecht et al. 2010b). More recently, the description of activity-dependent recruitment of proteins and even organelles to glial glutamate transporters has apparently settled the critical participation of glial amino acid transporters in synaptic activity (Danbolt et al. 2016).

2 Glial Cells

It has been generally accepted that glial cells are more abundant than neurons (von Bartheld et al. 2016). Nevertheless, it has been demonstrated that although differences in the ratio of neurons to glial cells are present within particular brain structures, an average of ratio of one neuron per glial cell is obtained considering the brain as a whole (Azevedo et al. 2009). Different types of glial cells regulate aspects like architecture, function, and plasticity in the CNS as well as in the peripheral nervous system (PNS) (BrosiusLutz and Barres 2014). Glial cells are divided into two main types: microglia and macroglia. Microglia has been regarded as the immune cells of the CNS, although their involvement in other functional aspects of the brain has begun to emerge (Reemst et al. 2016). Ependymoglia, myelinating glia, and astrocytes constitute the macroglia. Ependymoglia includes radial glia from the retina and the cerebellum (Müller and Bergmann cells). Myelin-forming cells are the oligodendroglia in the CNS and the Schwann cells in the PNS. Astrocytes participate in brain development and neuronal migration (Nulty et al. 2015; Perea and Araque 2010). For example, astrocytes release heparan sulfate proteoglycans inducing excitatory synapses formation (Allen et al. 2012). In this context, macroglia is fundamental for neuronal survival and is associated with support and replenishment of metabolic substrates (Hertz et al. 2015). Despite this, recent findings have called the attention to the involvement of glia in synaptic transactions throughout the CNS (Martínez-Lozada and Ortega 2015). Astrocytes, through a battery of neurotransmitter receptors and transporters present in their plasma membrane, are capable to release neuroactive molecules (glutamate, D-serine, ATP, glutamine, GABA,) that bind to pre- and postsynaptic receptors. Although astrocytes are non-excitable cells, classical transmitters evoke transient increases in [Ca^{2+}] intracellular levels in cultured astrocytes or in brain slices (Robel and Sontheimer 2016). The synaptic control of the astrocyte Ca^{2+} signal is based in spatially restricted areas called "microdomains" of the astrocytic processes (Grosche et al. 1999). Moreover, ultrastructural studies have shown the presence of small synaptic-like vesicles located in close proximity to synapses, apposed either to presynaptic and postsynaptic elements that are thought to contain neuroactive substances (Jourdain et al. 2007). Glutamate was one of the first neuroactive molecules known to be released by astrocytes that exert an effect on neural excitability (Levi et al. 1982).

3 Glial Glutamate Receptors

Glutamate receptors have been classified in terms of their molecular structure and signaling mechanisms into ionotropic (GRIs) and metabotropic (GRMs) receptors. GRI are ligand-gated ion channels that are activated by the selective agonists – N-methyl-D-aspartate (NMDA), α-amino-3-hydroxy-5-methylisoxazole-4-propionate (AMPA), and kainate (KA) – each one of them representing a family of homo- or heteroligomer receptors (Gasic and Heinemann 1992). Metabotropic receptors are G-protein-coupled receptors that are divided based on their primary structure into group I, group II, and group III and are activated preferentially by quisqualate (quis), 1-amino-4,5-cyclopentane-trans-1,3-dicarboxylate (tACPD), and L-2-amino-phosphonobutanoate (L-AP-4) (Pin and Duvoisin 1995). Glial cells of different brain structures display both subtypes of receptors; these receptors have been extensively studied in glial cells that surround glutamatergic synapses like cerebellar Bergmann glia and retinal Müller glial cells (Bellamy and Ogden 2006). Bergmann glial cells display a glutamate-dependent continuous dialogue with Purkinje and granule cells, through Ca^{2+}-permeable AMPA receptors. A series of elegant experiments transducing the Na^{+}-determinant AMPA subunit, GRIA2 into Bergmann glial cells, modifies its architecture and its physical contacts with

Purkinje cells (Iino et al. 2001). It also should be noted that neuronal stimulation elicits glutamate-dependent changes in glial membrane potential in a number of preparations and that these electrical responses are carried out not only by glutamate receptors but also by the Na^+-dependent glutamate transporters (see below). In any event, glial glutamate receptors, like their neuronal counterpart, are linked to gene expression regulation both at the transcriptional and the translational level (Barrera et al. 2010; Gallo and Ghiani 2000; González-Mejia et al. 2006; Rosas et al. 2007; Zepeda et al. 2009). In this context, it is pertinent to emphasize that among the genes that are regulated by glial glutamate receptors are the ones that encode glutamate transporters. It is quite possible then that glutamate released from the synaptic terminal activates neuronal and glial receptors, modify gene expression patterns in both cell types, and that among the target genes those involved in glial/neuronal interactions are represented.

4 Glial Glutamate Transporters

Glial glutamate transporters are important for the removal of this neurotransmitter from the synaptic cleft. Five glutamate transporters have been characterized: the Na^+-dependent glutamate/aspartate transporter (GLAST/EAAT1), the glutamate transporter 1 (GLT-1/EAAT2), the excitatory amino acid carrier 1 (EAAC1/EAAT3), the excitatory amino acid transporter 4 (EAAT4), and the excitatory amino acid transporter 5 (EAAT5) (Table 1) (Danbolt et al. 2016). GLAST is present exclusively in glial cells, whereas GLT-1 is mainly expressed in glia but has also been found in hippocampal nerve terminals (Fig. 1). The other three transporters are expressed in neurons. The importance of glial glutamate transporters in pathological scenarios has been deduced from the knockout studies, in which an elevation of glutamate extracellular levels, neurodegeneration, and progressive paralysis have been recorded (Rothstein et al. 1996). The bulk of glutamate transport in the cerebellum is carried out by GLAST, whereas in the other brain areas, it is accomplished by GLT-1. Therefore, glial glutamate transporters are key elements in the prevention of overstimulation of glutamate receptors, a process that triggers plastic changes and excitotoxic cascades in several pathological conditions (Trotti et al. 2001). In this sense, it has been postulated that disruption of glial glutamate transport affects the time course, fidelity, and modulation of excitatory transmission. GLAST and GLT-1 mRNA levels have been investigated during development, and in pure glial preparations, at early stages of brain development, both mRNAs are present in significant amounts, especially at the time of gliogenesis (mouse E15-E19). At birth, GLAST is present in abundance, while GLT-1 is barely detectable. In fact, GLAST has been considered as a glial lineage marker (Kriegstein and Alvarez-Buylla 2009). In the adult organism, GLT-1 is an abundant protein (1 mg/g of tissue) and with the exemption of cerebellum and retina, the major glutamate uptake system. Glial glutamate transport is regulated in the short and long term. Short-term regulation includes cell surface expression and posttranslational modifications like phosphorylation,

Table 1 Glutamine/glutamate transporters

Transporter family	Number of integrands	Subgroup classification	Gene	Common protein name	Transport type	Brain expression
SLC1	7		SLC1A1	GLAST	Co-transport	Astrocyte
			SLC1A2	GLT-1	Co-transport	Astrocyte
			SLC1A3	EAAT3	Co-transport	Neurons
			SLC1A4	ASCT1	Antiport	Neurons
			SLC1A5	ASCT2	Antiport	Astrocyte
			SLC1A6	EAAT4	Co-transport	Neurons
			SLC1A7	EAAT5	Co-transport	Neurons
SLC6	21		SLC6A9	GlyT-1	Co-transport	Astrocyte
			SLC6A5	GlyT-2	Co-transport	–
			SLC6A14	ATB^{o}	Co-transport	Neurons
			SLC6A15	B^{o}AT2	Co-transport	Neurons
			SLC6A19	B^{o}AT1	Co-transport	–
SLC7	6	CAT	SLC7A6	y^{+}LAT2	Antiport	Astrocytes
			SLC7A7	y^{+}LAT1	Antiport	Astrocyte
			SLC7A9	B^{o+}AT1	Antiport	Astrocyte
		LAT	SLC7A5	LAT1	Antiport	Neurons
			SLC7A8	LAT2	Antiport	Astrocyte
SLC38	11	System A	SLC38A1	SNAT1	Co-transport	Neurons
			SLC38A2	SNAT2	Co-transport	Neurons
			SLC38A4	SNAT4	Co-transport	–
			SLC38A8	SNAT8	Co-transport	Neurons
		System N	SLC38A3	SNAT3	Antiport	Astrocyte
			SLC38A5	SNAT5	Antiport	Astrocyte
			SLC38A7	SNAT7	Antiport	Neurons

ubiquitination, and/or acetylation that in one way or another modify transporter expression at the plasma membrane (Robinson 2006).

Long-term regulation includes transcriptional as well as translational control (López-Bayghen and Ortega 2011; Martinez-Lozada et al. 2016). Diverse stimuli are known to affect glial glutamate transporters function, among them glutamate is the most important. It has been shown that glutamate regulates GLAST at the short and long term; in both cases the net result is a decrease in glutamate uptake activity, albeit the molecular mechanisms are different. In the short term, glutamate decreases the amount of plasma membrane transporters by interfering with the traffic of the protein to and from the membrane. This is a transporter-dependent effect (González and Ortega 2000). In contrast, in the long-term glutamate, acting through its receptors downregulates the transcription of the GLAST gene (Rosas et al. 2007). It should be mentioned that this regulation takes place in the cerebellum, while an opposite effect has been recorded for long-term effects of glutamate in the cerebral cortex, where an increase in GLAST has been detected (Gegelashvili et al. 2000). Recently, the role of transporters as signaling entities has begun to emerge, and glial glutamate transporters

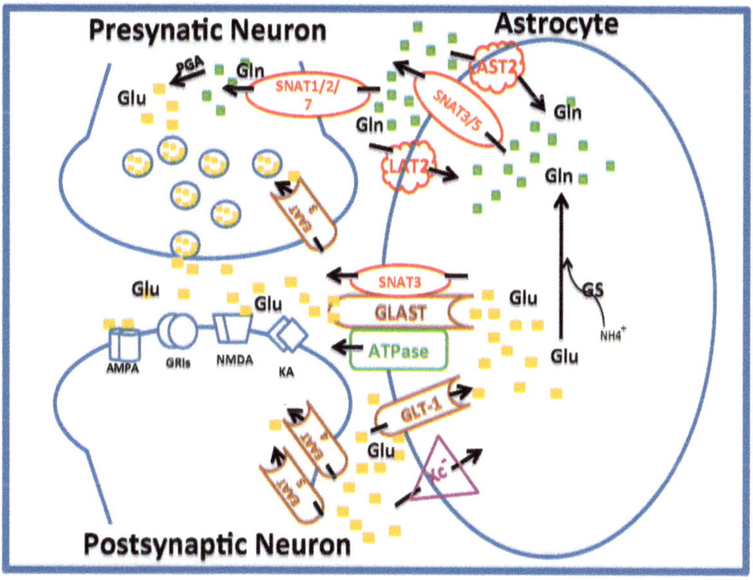

Fig. 1 Glutamate/glutamine transporters in the brain. Schematic localization of the amino acid transporters expression in neurons and astrocytes with their functional activity in different shuttles and metabolic pathways. Transporters are named by conventional names, and *arrows* indicate the most common flux direction

are no exception. A transporter-dependent increase in p42/44 mitogen kinase activity and in activity of the mammalian target of rapamycin (mTOR) has also been reported (Martínez-Lozada et al. 2011). Glutamate transporter signaling has also been demonstrated to participate in oligodendrocyte maturation (Martinez-Lozada et al. 2014). Furthermore, signaling complexes containing GLAST, the Na+/K+ ATPase, the Na^+-dependent neutral amino acid transporter 3 (SNAT 3), and the glucose transporter 1 (Glut 1) have been described (Flores-Mendez et al. 2016; Gegelashvili et al. 2007; Martínez-Lozada et al. 2013; Rose et al. 2009) .

4.1 *Cystine/Glutamate Antiporter*

Although it has been traditionally assumed that glutamate extracellular concentrations are dependent upon EAATs' function, glial uptake and its release are relevant for a proper glutamatergic transmission. In particular, the involvement of the cystine/glutamate antiporter x_c^- in glutamate homeostasis has been proposed (Fig. 1) (De Bundel et al. 2011). This exchanger protein takes up cystine and releases glutamate in a 1:1 ratio. It is formed by two subunits: xCT (functional subunit) and 4F2hc linked by a disulfide bridge. It participates in the synthesis of the antioxidant glutathione (GSH) (Albrecht et al. 2010a). Exposure to xenobiotics is linked to reactive oxygen species (ROS) production triggering x_c^- activity (Dal-Cim et al. 2016). Moreover, glutamate favors ROS synthesis and therefore x_c^- activity (Quincozes-Santos et al. 2014); this cellular mechanism is tightly controlled, since the glutamate released exacerbates its neurotoxic properties.

5 Glial Glutamine Transporters

Glutamine is the most abundant amino acid found in the human body (Labow and Souba 2000), accounting for more than 20% of the free amino acid pool in the plasma (Gerich et al. 2000); its main function is as a nitrogen carrier precursor for nucleotide and protein synthesis, but actually the attention is being paid to a plausible role as an intermediate in different metabolic pathways, as modulator of signal transduction, its probable involvement as maintainer of the cellular redox state, and also as an energy source (Deberardinis and Cheng 2009).

The historical controversy about the designation of glutamine as a nonessential amino acid due to the capacity to be synthesized de novo becomes cataloged as *conditionally essential* during specific illness conditions such as a trauma, sepsis, or cancer (Labow and Souba 2000) scenarios of the higher energy demand. The pool of circulating glutamine turnover is as rapid as that of circulating glucose (Gerich et al. 2000); therefore if the demand of glutamine is critical in stress or hypercatabolic states (Taylor and Curthoys 2004), the cell must take this amino acid from the diet or from cellular reservoirs like the muscle, where the cytosolic concentration ranges from 10 mM to 30 mM (Rennie et al. 1996).

The glutamine pool is maintained through de novo synthesis through the enzyme glutamine synthetase that attaches an ammonium ion to the glutamate. Therefore, the maintenance of plasma glutamine homeostasis is a balance between de novo synthesis, cell pool reservoir, and release and uptake of cell demand (Taylor and Curthoys 2004). Thus, to have a complete picture of the glutamine metabolism, it is necessary to consider the transport proteins responsible to regulate the glutamine concentration in a cell-specific manner. Four gene families of solute carrier transporters (SLC1, SLC6, SLC7, and SLC38) have been identified as glutamine transporters (Table 1) (Bhutia and Ganapathy 2016). The molecular and functional properties of amino acid transport are characterized by their overlapping substrate specificities, generally low substrate affinities, and widespread cellular distribution (Collarini and Oxender 1987).

5.1 SLC1

SLC1 are Na^+-dependent neutral amino acid exchange proteins known as the ASC family. Five of the seven members have high affinity for glutamate (Kanai et al. 2013), and SLC14A/ASCT1 and SLC15A/ASCT2 are neutral amino acid transporters. ASCT1 is the most abundant isoform in the brain, but glutamine is a poor substrate for this transporter (Zerangue and Kavanaugh 1996). ASCT2 functions as an exchanger, capable to uptake and release glutamine, serine, asparagine, and threonine. This process is electroneutral involving the influx of Na^+/amino acid coupled to the efflux of Na^+/amino acid; Na^+ cannot be substituted by Li^+ (Pingitore et al. 2013). Alanine, valine, and methionine are only inwardly transported (Scalise et al. 2016), and glutamate, lysine, arginine, 2-methylaminoisobutyric acid (MeAIB), and 2-aminobicyclo-(2,2,1)-heptane-2-carboxylic acid (BCH) are not transported (Pochini et al. 2014). ASCT2 transport is expressed mainly in primary cultures of rat astrocytes and also at lower levels in adult and embryonic brain (Table 1) (Bröer et al. 1999); evidence for a neuronal expression is weak (Su et al. 1997), although recent data demonstrates immunoreactivity in cerebellar Purkinje cell bodies and dendrites (Gliddon et al. 2009). It is interesting that ASCT2 is widely upregulated in different types of cancer-like neuroblastoma (Ren et al. 2015) or glioblastoma (Wise et al. 2008), and it is associated with alterations in signaling pathways and the potential coupling of this transporter to other amino acid transporters at a functional level (Nicklin et al. 2009). Recent studies have shown that ASCT2 is modulated by an epidermal growth factor (EGF)-triggered signaling pathway, involving phosphatidylinositol 3-kinase (PI3K)/protein kinase B (PKB) cascade (Avissar et al. 2008; Palmada et al. 2005). Moreover, silencing ASCT2 affects the mTOR pathway activity, modifying the protein repertoire (Fuchs et al. 2007).

5.2 *SLC6*

SLC6 is a Na^+/Cl^--coupled neurotransmitter transporter family. This family includes 21 transporters with substrate selectivity including molecules with neurotransmitter function (Broer 2006); SLC6A14/ATB^{0+}, SLC6A15/B^{0}AT2, and SLC6A19/B^{0}AT1 are able to transport glutamine (Pramod et al. 2013). In the brain, SLC6A14 is localized in the pituitary (Sloan and Mager 1999), and SLC6A15 is highly expressed in the cerebellum, olfactory bulb, pyramidal cells, dentate gyrus, thalamus, and hypothalamus. In all areas, the expression is confined to neurons (Table 1). Due to their localization, its physiological function is the transport of neurotransmitter precursors (Broer 2006). A correlation of the expression of these transports and tumor growth is based on their capacity to uptake essential and non-essential amino acids and their ability to regulate various metabolic pathways and inflammation (Babu et al. 2015; Bhutia and Ganapathy 2016).

5.3 *SLC7*

SLC7 is a glutamine transporter family of which its members are classified in two groups, cationic amino acid transporters (CATs) and the light subunits of amino acid transporters (LATs). In the human, six subunits have been identified (Pochini et al. 2014), but LAT1/SLC7A5 and LAT2/SLC7A8 are the most characterized.

LAT1 is expressed in the brain (Fig. 1), works as a heterodimer (Prasad et al. 1999), and is a Na^+-independent transporter, and its preference is mainly for leucine, giving rise to its cognate name as system L. A low-affinity, high-capacity glutamine uptake activity was described for this system in astrocytes and neurons (Nagaraja and Brookes 1996; Su et al. 1997). Functional studies suggest a minor role for system L-mediated glutamine transport in astrocytes, contributing with approximately only 10% of the total uptake (Broer and Brookes 2001; Sidoryk-Wmgrzynowicz et al. 2009). Probably the most interesting role of system L is the functional coupling between different transporters, like ASCT2 and LAT1 in cell proliferation (Fuchs and Bode 2005) or LAT1 and SNAT in activating mTOR (Taylor 2014).

5.4 SLC38

SCL38 is a family with 11 transporters divided in two different systems, named as system A (SA) and system N (SN), based on the ability to transport alanine (SA) or the capacity to transport amino acids with nitrogen in its R group (SN). These transmembrane proteins are Na^+-dependent neutral amino acid transporters and therefore are also known as SNATs. Both systems respond to hormonal regulation and are pH sensitive (Mackenzie and Erickson 2004). System A members include SNAT1, SNAT2, and SNAT4; transport small zwitterionic amino acids are inhibited by the substrate analogue MeAIB and have an adaptive regulation after amino acid depletion (Chaudhry et al. 2002; Tanaka et al. 2005). SNAT1 and SNAT2 are ubiquitously expressed in mammalian tissues. In the brain, both transporters are particularly associated with glutamatergic neurons (Conti and Melone 2006; Gonzalez-Gonzalez et al. 2005) with physiological relevance to the glutamate/glutamine cycle, transporting glutamine from extrasynaptic region, providing glutamine as a precursor for glutamate synthesis (Fig. 1). System N has two isoforms, SNAT3 and SNAT5, both coupled to Na^+ and H^+ gradients. These transporters are extremely important due to their capacity to uptake and release glutamine (Baird et al. 2006; Boulland et al. 2003; Bröer et al. 2002), and their almost exclusively expression in astrocytes (Fig. 1) (Boulland et al. 2003; Chaudhry et al. 1999). Therefore, are the major mediators of glutamine efflux once glial glutamate transporters (EAAT1 and EAAT2) take up glutamate from synaptic space and induce glutamine synthetase activity (Broer et al. 2004; Chaudhry et al. 1999). SNAT7 has been classified as a system N member, while SNAT8 as a system A protein; but the controversy is still maintained because these transporters share transport mechanisms and substrate preferences. SNAT7 was the first transporter reported to be expressed in GABAergic neurons and also in other type of neurons; later on, SNAT8 was described to be expressed in excitatory and inhibitory neurons, suggesting that both transporters participate in the glutamate/glutamine cycle in the brain (Hagglund et al. 2011, 2015).

6 Conclusion

Amino acid transporters belonging to the soluble carrier (SLC) family are classified according to their sequence homology (Table 1), affinity, and mechanism of substrate transport. Their expression is not symmetrical between species and tissues; some transporters are ubiquitous and others are cell specific. Moreover, they have different subcellular distributions in response to the microenvironmental conditions, transcriptional control, and functional co-regulation. Therefore this variability and complexity of response makes them critical proteins to maintain cell homeostasis. In particular, we focus on glutamate/glutamine transporters in the brain, not only on their role during neurotransmitter recycling but also in their involvement in biological processes such as signal transduction, metabolism control, amino acid availability, and protein translation that impact cell physiological conditions (Fig. 1). Therefore, transporters are a powerful alternative as drug targets. Nevertheless, a better understanding of the biochemical control mechanisms is critically needed.

Acknowledgments The work in our laboratories is supported by grants from Conacyt-México (255087) to A.O., *Soluciones para un México Verde* to A.O., and FOFI-UAQ2015 (FNN201608) to A.R.

Conflict of Interest The author declares no conflicts of interest.

References

Albrecht P, Lewerenz J, Dittmer S, Noack R, Maher P, Methner A. Mechanisms of oxidative glutamate toxicity: the glutamate/cystine antiporter system xc- as a neuroprotective drug target. CNS Neurol. Disord. Drug Targets. 2010a;9:373–82.

Albrecht J, Sidoryk-Węgrzynowicz M, Zielińska M, Aschner M. Roles of glutamine in neurotransmission. Neuron Glia Biol. 2010b;6:263–76. doi:10.1017/S1740925X11000093.

Allen NJ, Bennett ML, Foo LC, Wang GX, Chakraborty C, Smith SJ, Barres B a. Astrocyte glypicans 4 and 6 promote formation of excitatory synapses via GluA1 AMPA receptors. Nature. 2012;486:410–4. doi:10.1038/nature11059.

Avissar NE, Sax HC, Toia L. In human entrocytes, GLN transport and ASCT2 surface expression induced by short-term EGF are MAPK, PI3K, and rho-dependent. Dig Dis Sci. 2008; doi:10.1007/s10620-007-0120-y.

Azevedo FAC, Carvalho LRB, Grinberg LT, Farfel JM, Ferretti REL, Leite REP, Filho WJ, Lent R, Herculano-Houzel S. Equal numbers of neuronal and nonneuronal cells make the human brain an isometrically scaled-up primate brain. J Comp Neurol. 2009;513:532–41. doi:10.1002/cne.21974.

Babu E, Bhutia YD, Ramachandran S, Gnanaprakasam JP, Prasad PD, Thangaraju M, Ganapathy V. Deletion of the amino acid transporter Slc6a14 suppresses tumour growth in spontaneous mouse models of breast cancer. Biochem J. 2015;469:17–23. doi:10.1042/BJ20150437.

Baird FE, Pinilla-Tenas JJ, Ogilvie WLJ, Ganapathy V, Hundal HS, Taylor PM. Evidence for allosteric regulation of pH-sensitive System A (SNAT2) and System N (SNAT5) amino acid transporter activity involving a conserved histidine residue. Biochem J. 2006;397:369–75. doi:10.1042/BJ20060026.

Barrera I, Flores-Méndez M, Hernández-Kelly LC, Cid L, Huerta M, Zinker S, López-Bayghen E, Aguilera J, Ortega A. Glutamate regulates eEF1A phosphorylation and ribosomal transit time in Bergmann glial cells. Neurochem Int. 2010;57:795–803. doi:10.1016/j.neuint.2010.08.017.

von Bartheld CS, Bahney J, Herculano-Houzel S. The search for true numbers of neurons and glial cells in the human brain: A review of 150 years of cell counting. J Comp Neurol. 2016;524:3865–95. doi:10.1002/cne.24040.

Bellamy TC, Ogden D. Long-term depression of neuron to glial signalling in rat cerebellar cortex. Eur J Neurosci. 2006;23:581–6. doi:10.1111/j.1460-9568.2005.04588.x.

Bhutia YD, Ganapathy V. Glutamine transporters in mammalian cells and their functions in physiology and cancer. Biochim Biophys Acta. 2016;1863:2531–9. doi:10.1016/j.bbamcr.2015.12.017.

Boulland JL, Rafiki A, Levy LM, Storm-Mathisen J, Chaudhry FA. Highly differential expression of SN1, a bidirectional glutamine transporter, in astroglia and endothelium in the developing rat brain. Glia. 2003;41:260–75. doi:10.1002/glia.10188.

Broer S. The SLC6 orphans are forming a family of amino acid transporters. Neurochem Int. 2006; doi:10.1016/j.neuint.2005.11.021.

Broer S, Brookes N. Transfer of glutamine between astrocytes and neurons. J Neurochem. 2001;77:705–19. doi:10.1046/j.1471-4159.2001.00322.x.

Bröer a, Brookes N, Ganapathy V, Dimmer KS, Wagner C a, Lang F, Bröer S. The astroglial ASCT2 amino acid transporter as a mediator of glutamine efflux. J Neurochem. 1999;73:2184–94.

Bröer A, Albers A, Setiawan I, Edwards RH, Chaudhry F a, Wagner CA, Bröer S. Regulation of the glutamine transporter SN1 by extracellular pH and intracellular sodium ions. J Physiol. 2002;539:3–14. doi:10.1013/jphysiol.2001.013303.

Broer A, Deitmer JW, Broer S. Astroglial glutamine transport by system N is upregulated by glutamate. Glia. 2004;48:298–310. doi:10.1002/glia.20081.

BrosiusLutz A, Barres BA. Contrasting the glial response to axon injury in the central and peripheral nervous systems. Dev Cell. 2014;28:7–17. doi:10.1016/j.devcel.2013.12.002.

Chaudhry F a, Reimer RJ, Krizaj D, Barber D, Storm-Mathisen J, Copenhagen DR, Edwards RH. Molecular analysis of system N suggests novel physiological roles in nitrogen metabolism and synaptic transmission. Cell. 1999;99:769–80. doi:10.1016/S0092-8674(00)81674-8.

Chaudhry FA, Krizaj D, Larsson P, Reimer RJ, Wreden C, Storm-Mathisen J, Copenhagen D, Kavanaugh M, Edwards RH. Coupled and uncoupled proton movement by amino acid transport system N. EMBO J. 2002;20:7041–51. doi:10.1093/emboj/20.24.7041.

Collarini EJ, Oxender DL. Mechanisms of transport of amino acids across membranes. Annu Rev Nutr. 1987;7:75–90. doi:10.1146/annurev.nutr.7.1.75.

Conti F, Melone M. The glutamine commute: lost in the tube? Neurochem Int. 2006;48:459–64. doi:10.1016/j.neuint.2005.11.016.

Dal-Cim T, Martins WC, Thomaz DT, Coelho V, Poluceno GG, Lanznaster D, Vandresen-Filho S, Tasca CI. Neuroprotection promoted by guanosine depends on glutamine synthetase and glutamate transporters activity in hippocampal slices subjected to oxygen/glucose deprivation. Neurotox Res. 2016;29:460–8.

Danbolt NC, Furness DN, Zhou Y. Neuronal vs glial glutamate uptake: resolving the conundrum. Neurochem Int. 2016;98:29–45. doi:10.1016/j.neuint.2016.05.009.

De Bundel D, Schallier A, Loyens E, Fernando R, Miyashita H, Van Liefferinge J, Vermoesen K, Bannai S, Sato H, Michotte Y, Smolders I, Massie A. Loss of system x(c)- formula does not induce oxidative stress but decreases extracellular glutamate in hippocampus and influences spatial working memory and limbic seizure susceptibility. J Neurosci. 2011;31:5792–803.

Deberardinis R, Cheng T. Q's next: the diverse functions of glutamine in metabolism, cell biology and cancer. Oncogene. 2009;29:313–24. doi:10.1038/onc.2009.358.

Flores-M Endez M, Mendez-Flores OG, Ortega A. Glia plasma membrane transporters: key players in glutamatergic neurotransmission. Neurochem Int. 2016;98:46–55. doi:10.1016/j.neuint.2016.04.004.

Fuchs BC, Bode BP. Amino acid transporters ASCT2 and LAT1 in cancer: partners in crime? Semin Cancer Biol. 2005;15:254–66. doi:10.1016/j.semcancer.2005.04.005.

Fuchs BC, Finger RE, Onan MC, Bode BP. ASCT2 silencing regulates mammalian target-of-rapamycin growth and survival signaling in human hepatoma cells. Am J Physiol Cell Physiol. 2007;293:C55–63. doi:10.1152/ajpcell.00330.2006.

Gallo V, Ghiani CA. Glutamate receptors in glla: new cells, new inputs and new functions. Trends Pharmacol Sci. 2000;21:252–8.
Gasic GP, Heinemann S. Determinants of the calcium permeation of ligand-gated cation channels. Curr Opin Cell Biol. 1992;4:670–7. doi:10.1016/0955-0674(92)90088-T.
Gegelashvili G, Dehnes Y, Danbolt NC, Schousboe A. The high-affinity glutamate transporters GLT1, GLAST, and EAAT4 are regulated via different signalling mechanisms. Neurochem Int. 2000;37:163–70. doi:10.1016/S0197-0186(00)00019-X.
Gegelashvili M, Rodriguez-Kern A, Sung L, Shimamoto K, Gegelashvili G. Glutamate transporter GLAST/EAAT1 directs cell surface expression of FXYD2/γ subunit of Na, K-ATPase in human fetal astrocytes. Neurochem Int. 2007;50:916–20. doi:10.1016/j.neuint.2006.12.015.
Gerich JE, Meyer C, Stumvoll MW. Hormonal control of renal and systemic glutamine metabolism. J Nutr. 2000;130(4S Suppl):995S–1001S.
Gliddon CM, Shao Z, LeMaistre JL, Anderson CM. Cellular distribution of the neutral amino acid transporter subtype ASCT2 in mouse brain. J Neurochem. 2009;108:372–83. doi:10.1111/j.1471-4159.2008.05767.x.
González MI, Ortega A. Regulation of high-affinity glutamate uptake activity in Bergmann glia cells by glutamate. Brain Res. 2000;866:73–81. doi:10.1016/S0006-8993(00)02226-5.
Gonzalez-Gonzalez IM, Cubelos B, Gimenez C, Zafra F. Immunohistochemical localization of the amino acid transporter SNAT2 in the rat brain. Neuroscience. 2005;130:61–73. doi:10.1016/j.neuroscience.2004.09.023.
González-Mejia ME, Morales M, Hernández-Kelly LCR, Zepeda RC, Bernabé A, Ortega A. Glutamate-dependent translational regulation in cultured Bergmann glia cells: involvement of p70S6K. Neuroscience. 2006;141:1389–98. doi:10.1016/j.neuroscience.2006.04.076.
Grosche J, Matyash V, Möller T, Verkhratsky a, Reichenbach a, Kettenmann H. Microdomains for neuron-glia interaction: parallel fiber signaling to Bergmann glial cells. Nat Neurosci. 1999;2:139–43. doi:10.1038/5692.
Hagglund MGA, Sreedharan S, Nilsson VCO, Shaik JHA, Almkvist IM, Backlin S, Wrange O, Fredriksson R. Identification of SLC38A7 (SNAT7) protein as a glutamine transporter expressed in neurons. J Biol Chem. 2011; doi:10.1074/jbc.M110.162404.
Hagglund MGA, Hellsten SV, Bagchi S, Philippot G, Lofqvist E, Nilsson VCO, Almkvist I, Karlsson E, Sreedharan S, Tafreshiha A, Fredriksson R. Transport of l-glutamine, l-alanine, l-arginine and l-histidine by the neuron-specific slc38a8 (SNAT8) in cns. J Mol Biol. 2015; doi:10.1016/j.jmb.2014.10.016.
Hertz L, Gerkau NJ, Xu J, Durry S, Song D, Rose CR, Peng L. Roles of astrocytic Na(+) ,K(+) -ATPase and glycogenolysis for K(+) homeostasis in mammalian brain. J Neurosci Res. 2015;93:1019–30. doi:10.1002/jnr.23499.
Iino M, Goto K, Kakegawa W, Okado H, Sudo M, Ishiuchi S, Miwa a, Takayasu Y, Saito I, Tsuzuki K, Ozawa S. Glia-synapse interaction through Ca2+-permeable AMPA receptors in Bergmann glia. Science. 2001;292:926–9. doi:10.1126/science.1058827.
Jourdain P, Bergersen LH, Bhaukaurally K, Bezzi P, Santello M, Domercq M, Matute C, Tonello F, Gundersen V, Volterra A. Glutamate exocytosis from astrocytes controls synaptic strength. Nat Neurosci. 2007;10:331–9. doi:10.1038/nn1849.
Kanai Y, Clémençon B, Simonin A, Leuenberger M, Lochner M, Weisstanner M, Hediger MA. The SLC1 high-affinity glutamate and neutral amino acid transporter family. Mol Asp Med. 2013;34:108–20. doi:10.1016/j.mam.2013.01.001.
Kriegstein A, Alvarez-buylla A. The glial nature of embryonic and adult neural stem cells. Annu Rev Neurosci. 2009;32:149–84. doi:10.1146/annurev.neuro.051508.135600.
Labow BI, Souba WW. Glutamine. World J Surg. 2000; doi:10.1007/s002680010269.
Levi G, Gordon RD, Gallo V, Wilkin GP, Balàzs R. Putative acidic amino acid transmitters in the cerebellum I. Depolarization-induced release Brain Res. 1982;239:425–45. doi:10.1016/0006-8993(82)90520-0.
López-Bayghen E, Ortega A. Glial glutamate transporters: New actors in brain signaling. IUBMB Life. 2011;63:816–23. doi:10.1002/iub.536.

Mackenzie B, Erickson JD. Sodium-coupled neutral amino acid (System N/A) transporters of the SLC38 gene family. Pflugers Arch - Eur J Physiol. 2004;447:784–95. doi:10.1007/s00424-003-1117-9.

Martínez-Lozada Z, Ortega A. Glutamatergic transmission: A matter of three. Neural Plast. 2015;2015 doi:10.1155/2015/787396.

Martínez-Lozada Z, Hernández-Kelly LC, Aguilera J, López-Bayghen E, Ortega A. Signaling through EAAT-1/GLAST in cultured Bergmann glia cells. Neurochem Int. 2011;59:871–9. doi:10.1016/j.neuint.2011.07.015.

Martínez-Lozada Z, Guillem AM, Flores-Méndez M, Hernández-Kelly LC, Vela C, Meza E, Zepeda RC, Caba M, Rodríguez A, Ortega A. GLAST/EAAT1-induced Glutamine release via SNAT3 in Bergmann glial cells: Evidence of a functional and physical coupling. J Neurochem. 2013; doi:10.1111/jnc.12211.

Martinez-Lozada Z, Waggener CT, Kim K, Zou S, Knapp PE, Hayashi Y, Ortega A, Fuss B. Activation of sodium-dependent glutamate transporters regulates the morphological aspects of oligodendrocyte maturation via signaling through calcium/calmodulin-dependent kinase IIβ's actin-binding/-stabilizing domain. Glia. 2014;62:1543–58. doi:10.1002/glia.22699.

Martinez-Lozada, Z, Guillem, AM, Robinson, MB. Transcriptional regulation of glutamate transporters. From extracellular signals to transcription factors, 1st ed. Adv Pharmacol. Elsevier, 2016. doi:10.1016/bs.apha.2016.01.004

Nagaraja TN, Brookes N. Glutamine transport in mouse cerebral astrocytes. J Neurochem. 1996;66:1665–74.

Nicklin P, Bergman P, Zhang B, Triantafellow E, Wang H, Nyfeler B, Yang H, Hild M, Kung C, Wilson C, Myer VE, MacKeigan JP, Porter JA, Wang YK, Cantley LC, Finan PM, Murphy LO. Bidirectional Transport of Amino Acids Regulates mTOR and Autophagy. Cell. 2009; doi:10.1016/j.cell.2008.11.044.

Nulty J, Alsaffar M, Barry D. Radial glial cells organize the central nervous system via microtubule dependant processes. Brain Res. 2015;1625:171–9. doi:10.1016/j.brainres.2015.08.027.

Palmada M, Speil A, Jeyaraj S, Böhmer C, Lang F. The serine/threonine kinases SGK1, 3 and PKB stimulate the amino acid transporter ASCT2. Commun: Biochem. Biophys. Res; 2005. doi:10.1016/j.bbrc.2005.03.159.

Parpura V, Verkhratsky A. Astrocytes revisited: concise historic outlook on glutamate homeostasis and signaling. Croat Med J. 2012;53:518–28. doi:10.3325/cmj.2012.53.518.

Perea G, Araque A. GLIA modulates synaptic transmission. Brain Res Rev. 2010;63:93–102. doi:10.1016/j.brainresrev.2009.10.005.

Pin J-P, Duvoisin R. The metabotropic glutamate receptors: Structure and functions. Neuropharmacology. 1995;34:1–26. doi:10.1016/0028-3908(94)00129-G.

Pingitore P, Pochini L, Scalise M, Galluccio M, Hedfalk K, Indiveri C. Large scale production of the active human ASCT2 (SLC1A5) transporter in Pichia pastoris – Functional and kinetic asymmetry revealed in proteoliposomes. Biochim Biophys Acta Biomembr. 2013;1828:2238–46. doi:10.1016/j.bbamem.2013.05.034.

Pochini L, Scalise M, Galluccio M, Indiveri C, Giulivi C, Coe IR. Membrane transporters for the special amino acid glutamine: structure/function relationships and relevance to human health. Front Chem. 2014;11:61. doi:10.3389/fchem.2014.00061.

Pramod AB, Foster J, Carvelli L, Henry LK. SLC6 transporters: Structure, function, regulation, disease association and therapeutics. Mol Asp Med. 2013; doi:10.1016/j.mam.2012.07.002.

Prasad PD, Wang H, Huang W, Kekuda R, Rajan DP, Leibach FH, Ganapathy V. Human LAT1, a subunit of system L amino acid transporter: molecular cloning and transport function. Biochem Biophys Res Commun. 1999;255:283–8.

Quincozes-Santos A, Bobermin LD, Tramontina AC, Wartchow KM, Tagliari B, Souza DO, Wyse ATS, Gonçalves C. Toxicology in Vitro Oxidative stress mediated by NMDA, AMPA / KA channels in acute hippocampal slices: Neuroprotective effect of resveratrol. Toxicol in Vitro. 2014;28:544–51.

Reemst K, Noctor SC, Lucassen PJ, Hol EM. The Indispensable Roles of Microglia and Astrocytes during Brain Development. Front Hum Neurosci. 2016;10:1–28. doi:10.3389/fnhum.2016.00566.

Ren P, Yue M, Xiao D, Xiu R, Gan L, Liu H, Qing G. ATF4 and N-Myc coordinate glutamine metabolism in MYCN-amplified neuroblastoma cells through ASCT2 activation. J Pathol. 2015;235:90–100. doi:10.1002/path.4429.

Rennie MJ, Ahmed A, Khogali SE, Low SY, Hundal HS, Taylor PM. Glutamine metabolism and transport in skeletal muscle and heart and their clinical relevance. J Nutr. 1996;126:1142S–9S.

Robel S, Sontheimer H. Glia as drivers of abnormal neuronal activity. Nat Neurosci. 2016;19:28–33. doi:10.1038/nn.4184.

Robinson MB. Acute regulation of sodium-dependent glutamate transporters: A focus on constitutive and regulated trafficking. Handb Exp Pharmacol. 2006;175:251–75. doi:10.1007/3-540-29784-7-13.

Rosas S, Vargas MA, López-Bayghen E, Ortega A. Glutamate-dependent transcriptional regulation of GLAST/EAAT1: A role for YY1. J Neurochem. 2007;101:1134–44. doi:10.1111/j.1471-4159.2007.04517.x.

Rose EM, Koo JCP, Antflick JE, Ahmed SM, Angers S, Hampson DR. Glutamate transporter coupling to Na. K-ATPase J Neurosci. 2009;29:8143–55. doi:10.1523/JNEUROSCI.1081-09.2009.

Rothstein JD, Dykes-Hoberg M, Pardo CA, Bristol LA, Jin L, Kuncl RW, Kanai Y, Hediger MA, Wang Y, Schielke JP, Welty DF. Knockout of glutamate transporters reveals a major role for astroglial transport in excitotoxicity and clearance of glutamate. Neuron. 1996;16:675–86. doi:10.1016/S0896-6273(00)80086-0.

Scalise M, Pochini L, Galluccio M, Indiveri C, Bernardi P. Glutamine transport. From energy supply to sensing and beyond. Biochim Biophys Acta. 2016;1857:1147–57. doi:10.1016/j.bbabio.2016.03.006.

Sidoryk-Wmgrzynowicz M, Lee E, Albrecht J, Aschner M. Manganese disrupts astrocyte glutamine transporter expression and function. J Neurochem. 2009;110:822–30. doi:10.1111/j.1471-4159.2009.06172.x.

Sloan JL, Mager S. Cloning and functional expression of a human Na(+) and Cl(-)-dependent neutral and cationic amino acid transporter B(0+). J Biol Chem. 1999;274:23740–5. doi:10.1074/jbc.274.34.23740.

Storck T, Schulte S, Hofmann K, Stoffel W. Structure, expression, and functional analysis of a Na(+)-dependent glutamate/aspartate transporter from rat brain. Proc Natl Acad Sci U S A. 1992;89:10955–9. doi:10.1073/pnas.89.22.10955.

Su TZ, Campbell GW, Oxender DL. Glutamine transport in cerebellar granule cells in culture. Brain Res. 1997;757:69–78.

Tanaka K, Yamamoto A, Fujita T. Functional expression and adaptive regulation of Na+-dependent neutral amino acid transporter SNAT2/ATA2 in normal human astrocytes under amino acid starved condition. Neurosci Lett. 2005;378:70–5. doi:10.1016/j.neulet.2004.12.030.

Taylor PM. Role of amino acid transporters in amino acid sensing. Am J Clin Nutr. 2014; doi:10.3945/ajcn.113.070086.

Taylor L, Curthoys NP. Glutamine metabolism: Role in acid-base balance. Biochem Mol Biol Educ. 2004; doi:10.1002/bmb.2004.494032050388.

Teichberg VI. Glial glutamate receptors: likely actors in brain signaling. FASEB J. 1991;5:3086–91.

Trotti D, Aoki M, Pasinelli P, Berger UV, Danbolt NC, Brown RH, Hediger MA. Amyotrophic lateral sclerosis-linked glutamate transporter mutant has impaired glutamate clearance capacity. J Biol Chem. 2001;276:576–82. doi:10.1074/jbc.M003779200.

Wise DR, DeBerardinis RJ, Mancuso A, Sayed N, Zhang X-Y, Pfeiffer HK, Nissim I, Daikhin E, Yudkoff M, McMahon SB, Thompson CB. Myc regulates a transcriptional program that stimulates mitochondrial glutaminolysis and leads to glutamine addiction. Proc Natl Acad Sci U S A. 2008;105:18782–7. doi:10.1073/pnas.0810199105.

Zepeda RC, Barrera I, Castelán F, Suárez-Pozos E, Melgarejo Y, González-Mejia E, Hernández-Kelly LC, López-Bayghen E, Aguilera J, Ortega A. Glutamate-dependent phosphorylation of the mammalian target of rapamycin (mTOR) in Bergmann glial cells. Neurochem Int. 2009;55:282–7. doi:10.1016/j.neuint.2009.03.011.

Zerangue N, Kavanaugh MP. ASCT-1 is a neutral amino acid exchanger with chloride channel activity. J Biol Chem. 1996;271:27991–4. doi:10.1074/jbc.271.45.27991.

Glial Glutamate Transporters as Signaling Molecules

Edna Suárez-Pozos, Donají Chi-Castañeda, and Arturo Ortega

Abstract One of the most important processes of the synaptic transmission is neurotransmitter uptake, which is critical for the good performance of the nervous system by maintaining the neurotransmitter's baseline levels after its release. The major excitatory neurotransmitter in the central nervous system is glutamate; its extracellular levels are tightly regulated through high-affinity plasma membrane transporters. Most of the brain glutamate uptake activity is carried out by glial transporters that until recently have been regarded as important for the recycling of this excitatory amino acid. Besides, a role in the prevention of an overstimulation of neuronal glutamate receptors that would be linked to cell death has been established. Nevertheless, the Na^+ dependence of the uptake process paved the way to the plausible triggering of signal transduction cascades. Over the past decade a considerable amount of evidences suggesting an important intervention of these transporter proteins in glutamate signaling, mainly in glial cells, has been accumulated. Herein we provide a summary or the most important findings in this novel function of glial glutamate transporters as signal transduction entities, as the framework platform through which they may actively participate in glutamate-mediated transactions in the central nervous system.

Keywords EAATs • Gene expression regulation • Glutamate receptors • Glutamate transporters • Signal transduction

---

E. Suárez-Pozos • A. Ortega (✉)
Laboratorio de Neurotoxicología, Departamento de Toxicología,
Centro de Investigación y de Estudios Avanzados del Instituto Politécnico Nacional,
Apartado Postal 14-740, Ciudad de México 07000, Mexico
e-mail: arortega@cinvestav.mx

D. Chi-Castañeda
Laboratorio de Neurotoxicología, Departamento de Toxicología,
Centro de Investigación y de Estudios Avanzados del Instituto Politécnico Nacional,
Apartado Postal 14-740, Ciudad de México 07000, Mexico

Soluciones para un México Verde S.A. de C.V, Ciudad de México 01210, Mexico

A. Ortega, A. Schousboe (eds.), *Glial Amino Acid Transporters*,
Advances in Neurobiology 16, DOI 10.1007/978-3-319-55769-4_9

List of Abbreviations

AMPA	α-Amino-3-hydroxy-5-methyl-4-isoxazolepropionate
AP-1	Activator protein-1
BGC	Bergmann glial cells
DHPG	(RS)-3,5-Dihydroxyphenylglycine
EAAC1	Excitatory amino acid carrier
EAATs	Excitatory amino acid transporters
GLAST	Glutamate/aspartate transporter
Gln	Glutamine
GLT-1	Glutamate transporter 1
Glu	Glutamate
GluKs	Glutamate receptor ionotropic kainate
GluNs	Glutamate NMDA receptor subunit zeta
GPCR	G-protein-coupled receptor
GRIAs	Glutamate ionotropic receptors AMPA
GS	Glutamine synthetase
GTRAP	Glutamate transporter-4-associated protein
iGluRs	Ionotropic glutamate receptors
KA	Kainate
L-AP4	L-(+)-2-amino-4-phosphonobutyric acid
mGluRs	Metabotropic glutamate receptors
NMDA	N-Methyl-D-aspartate
PDC	L-trans-pyrrolidine-2,4-dicarboxylic acid
PI3K	Phosphoinositide 3-kinase
PKA	Cyclic adenosine monophosphate-dependent kinase
PKC	Diacylglycerol-dependent kinase
S-4C3HPG	(S)-4-Carboxy-3-hydroxy-phenylglycine
SNAT3	Na^+-dependent neutral amino acid transporter 3
THA	DL-threo-b-hyroxyaspartate
x_c^-	Cystine/glutamate antiporter

1 Introduction

For several decades, it has been assumed that glutamate (Glu) transactions are dependent on a wide range of signaling cascades triggered by specific Glu receptors present in the pre- and postsynaptic membrane, as well as in the plasma membrane of neighboring astrocytes. After neuronal stimulation, released Glu is removed by a family of Na^+-dependent Glu transporters, known as excitatory amino acid transporters (EAATs) (Martínez-Lozada and Ortega 2015). Inefficient clearance of Glu from the synaptic cleft overstimulates neuronal Glu receptors, triggering a massive Ca^{2+} influx that activates intracellular events that result in neuronal death.

This phenomenon is known as excitotoxicity and is the underlying mechanism of several neurodegenerative diseases, such as schizophrenia, Alzheimer's, and Parkinson's disease.

In this chapter, we focus on the glial Glu transporters, EAAT1/GLAST and EAAT2/GLT-1. Particularly, GLAST is highly expressed in the cerebellum, retina, and olfactory bulb, while GLT-1 is abundant in other brain areas. These glial transporters play a critical role in Glu turnover through the well-characterized Glu/glutamine (Gln) shuttle. The activation of the Glu/Gln shuttle is the result of an increase in *glutamine synthetase* (GS) activity and expression promoted by the response of glial cells to a synaptic activity (Lehmann et al. 2009). In other words, once Glu is removed from the synaptic cleft, Gln is produced as a result of the activation of GS (Tang et al. 2001). Within the cerebellum, an extraordinary association between neurons and their respective surrounding glial cells is essential to allow the excitatory transmission. In these terms, when cerebellar parallel fibers are depolarized, the activation of glial Glu transporters and receptors is observed, being a good example of a *tripartite synapse* (Balakrishnan and Bellamy 2009; López-Bayghen and Ortega 2011).

Over the last years, evidence suggesting a plausible role of Glu transporters as signaling transducer molecules has become stronger. It has been demonstrated that Glu triggers signaling transactions via Glu transporters in addition to established Glu receptors activation in GH3 cells (Fairman et al. 1995). It was also reported that Glu transporters are capable of regulating their translocation to the plasma membrane through receptor-like proteins (Abe and Saito 2001; Duan et al. 1999; González and Ortega 2000). Likewise, Glu transporters function as macromolecular complexes involved in the regulation of glutamatergic transmission (Martínez-Lozada and Ortega 2015).

2 Glial Glutamate Receptors

Two main subtypes of Glu receptors have been described thus far: metabotropic (GRMs) and ionotropic (GRIs) receptors. GRMs belong to the family of class C G-protein-coupled receptors (GPCRs), whose classification is based on the homology of their sequences, signal transduction mechanisms, and pharmacology. In these terms, group I includes GRM1 and GRM5. Specifically, this group is coupled to stimulation of phospholipase C with the release of intracellular Ca^{2+}. On the other hand, group II comprises GRM2 and GRM3, whereas GRM4, GRM6, GRM7, and GRM8 conform group III. Both groups are coupled to the inhibition of adenylate cyclase. Interestingly, GRMs are activated by the action of specific agonists, including (RS)-3,5-dihydroxyphenylglycine (DHPG, activates group I), (S)-4-carboxy-3-hydroxyphenylglycine ((S)-4C3HPG, acts on group II), and L-(+)-2-amino-4-phosphonobutyric acid (L-AP4, exerts its action on group III) (Kirischuk et al. 2007). In contrast, GRIs are classified mainly by pharmacological terms, in N-methyl-D-aspartate (NMDA), α-amino-3-hydroxy-5-methyl-4-isoxazolepropionate (AMPA), and kainate (KA) receptors. GRIs are composed by four subunits encoded by different genes. In this context, AMPA receptors are formed by the combination of the following subunits: GRIA1 (GluA1), GRIA2 (GluA2), GRIA3 (GluA3), and GRIA4 (GluA4) resulting in a wide range of cation channel properties; i.e., a Na^+ permeable channel requires the presence of at least one GRIA2 subunit. Ca^{2+} permeability is favored in the absence of GRIA2 subunits; these kinds of AMPA receptors are present in radial glial cells (Sommer et al. 1991; López et al. 1994; Müller et al. 1992). NMDA receptors are composed by one GluN1 subunit combined with GluN2A-D or GluN3A-B subunits to form functional channels. Finally, KA receptors are formed by the combination of four subunits of GluK1–5. It is important to mention that ionotropic receptors present a vast molecular diversity due to the fact that most of its subunits undergo RNA editing and splicing (Herguedas et al. 2013). In line with this information, it has been reported that Glu activates a wide range of signaling cascades through its receptors; these cascades are closely implicated in brain disorders and diseases, such as schizophrenia, Alzheimer's, Parkinson's disease, and multiple sclerosis, among other diseases.

3 Glutamate Transporters

Glu removal from the synaptic cleft is carried out by a family of Na^+-dependent transporters known as EAATs (Danbolt et al. 2016), which are widely expressed in neurons and glial cells. Excessive accumulation of extracellular Glu and over-activation of Glu receptors contribute to neuronal cell death. Therefore, an efficient Glu removal and regulation is compulsory to prevent excitotoxic events (Danbolt 2001; Gegelashvili et al. 1997). To date, five subtypes of high-affinity Glu transporters have been identified: EAAT1–5 in humans or GLAST, GLT-1, EAAC1, EAAT4, and EAAT5 in rodents (Fairman et al. 1995; Storck et al. 1992; Pines et al. 1992; Arriza et al. 1997). Studies to analyze the primary sequence homology between this group of transporters have shown that among the five of them, approximately 65% homology is shared. Nevertheless, a great variability is found between the transporters when several characteristics are compared, including structure, pharmacological properties, distribution, and expression patterns (López-Bayghen and Ortega 2011; Danbolt 2001). Sodium-dependent Glu transporters can be divided into two classes, astrocytic or neuronal. Accordingly, EAAT1 and EAAT2 are widely expressed in glial cells, although GLT-1 has also been described to be present in neuronal cells; whereas EAAT3–5 distribution is more restricted, EAAT3 is mainly present in neurons from the hippocampus, EAAT4 has been found in Purkinje cells, and EAAT5 has been reported in the retina (Danbolt 2001). Interestingly, within the brain, EAAT1 and EAAT2 are responsible for the 80–90% of the total Glu uptake (Danbolt 2001; Eulenburg and Gomeza 2010); in fact it has been calculated that EAAT2/GLT-1 represents approximately 2% of the total brain protein. Conversely, GLAST has been considered the most abundant glial Glu transporter during the early stages of development and therefore has been used as a glial marker (Danbolt 2001).

Its distribution in the adult organism is restricted to the cerebellum, retina, and olfactory bulb.

Glu transport begins after a Na^+-dependent Glu binding to the extracellular domain of the transporter, and the uptake is driven by the symport of Na^+ and antiport of K^+ (Zerangue and Kavanaugh 1996). First, Na^+, H^+, and K^+ are coupled to the transporter; then, ions are translocated and released into the cell cytosol. K^+ binding to the intracellular side promotes the substrate-free transport. Once the substrate-free faction is reoriented, K^+ molecule is released outside the cell. This uptake mechanism utilizes transmembrane ion gradients as a driving force to achieve the tight regulation of Glu concentrations inside and outside the cells (López-Bayghen and Ortega 2011; Zerangue and Kavanaugh 1996). The stoichiometry of the transport has been calculated: one molecule of Glu needs to be coupled to the symport of three Na^+ and one H^+ ions, as well as the antiport of one K^+ ion. Furthermore, under physiological conditions an anion conductance activation takes place, resulting on the influx of Cl^- (Jiang and Amara 2011).

The EAATs' general molecular structure includes eight to ten transmembranal domains (TM), of which six of them correspond to alpha helices in the N terminus portion of the protein and two to N-glycosylation sites, which have been localized in the extracellular loop between the III and IV TM domains. The structure also includes lower loops in the cell cytoplasm (López-Bayghen and Ortega 2011; Storck et al. 1992; Kanai and Hediger 1992). Specifically, EAAT1/GLAST and EAAT2/GLT-1 present a putative phosphorylation site for Ca^{2+}/diacylglycerol-dependent kinase (PKC) and the cAMP-dependent kinase (PKA), as well as AAXFIAQ and serine-rich motifs (Slotboom et al. 1999).

3.1 Glutamate Transporter Molecule Regulation through Glutamate Action

It has been widely reported that Glu acts as a modulator of its own transport through the regulation of its glial transporters at different levels: mRNA and protein quantities, transporter molecules in the plasma membrane, and modulation of the kinetic parameters of the transport (Poblete-Naredo et al. 2009). In these terms, several reports established that Glu self-regulates the expression of EAATs in the absence of neuronal factors. In addition, EAAT stimulation mediates the activation and release of astroglial factors. Particularly, Glu receptor stimulation in glial cells is related to the activation of a wide range of mechanisms (transcriptional and/or post-translational) that results in the activity regulation of the EAAT1/GLAST transporter. This favors the notion that the upregulation of Glu receptors mediates the activation of the Glu transporters in astrocyte cells, which resembles a system retribution (López-Bayghen and Ortega 2004).

Duan and Guillet, in 1999 and 2005, respectively, proposed that the regulation of Glu transporters is mediated by phosphorylation of both the transporter itself and accessory proteins (Duan et al. 1999; Guillet et al. 2005). Additionally, it has also been proposed that the traffic of the Glu transporters to and from the plasma membrane is regulated by signaling pathways that include PKA, PKC, and phosphoinositide 3-kinase (PI3K), resulting in the modulation of the total Glu uptake capacity of the system and the prevention of plausible excitotoxic episodes (Guillet et al. 2005). According to this finding, it has been described that the availability and functionality of EAAT1/GLAST is modified by Glu receptor-dependent and Glu receptor-independent signaling through metabotropic and ionotropic receptors involving PKC and PKA pathways (González and Ortega 1997; Rana and Hokin 1990).

3.2 Glial Glutamate Transporters as Signal Transducer Elements

Several years ago, it was demonstrated that Glu taken up by glial cells is rapidly converted to Gln to complete a Glu/Gln shuttle providing an evidence of the crucial role of glial cells in glutamatergic transmission (Shank and Campbell 1984). Later on the novel concept of *tripartite synapse* (pre- and postsynapse neurons and the glial cells) was revealed, based on how glial cells respond to neuronal activation modifying their intracellular Ca^{2+} concentrations and promoting the release of neuroactive substances endorsing a plausible feedback regulation of neuronal activity and synaptic strength (Shank and Campbell 1984; Coles and Abbott 1996; Araque et al. 1999). Moreover, Dityatev and Rusakov, in 2011, proposed the term of *tetrapartite synapse*, suggesting the involvement of the surrounding extracellular matrix (ECM) in neuronal communication (Dityatev and Rusakov 2011). Over the years, the regulation of Glu-mediated gene expression in glial cells has attracted the attention of several research groups (López-Bayghen and Ortega 2010; Gallo and Ghiani 2000; López-Bayghen et al. 2007; Balazs 2006). In this regard, the link between the transporter inward Na^+ current and a long-lasting response was analyzed. A clear Glu-dependent translational and transcriptional regulation, as well as the fact that GRIA receptor activation regulates both *glast* transcription and GLAST activity in a Na^+- and Ca^{2+}-dependent manner, was reported (Aguirre et al. 2002). Furthermore, it has also been suggested the involvement of EAAT1/GLAST in Glu signal transduction regulation (Zepeda et al. 2008). In consequence, several research groups have supported the role of EAATs as active and direct signaling molecules. Accordingly, Miller and colleagues (1994), using cortical astrocytes, demonstrated multiple glutamatergic signaling transduction pathways. The authors also determined that the stimulation of the same cellular model with L-trans-2,4-pyrrolidine dicarboxylic acid (L-trans-2,4-PDC; a potent inhibitor of Glu transporter) blocks cAMP accumulation, suggesting that this inhibitor may have effects via activation of metabotropic Glu receptors (López-Bayghen and Ortega 2011; Miller et al. 1994).

Later on, Fairman and colleagues (1995), using GH3 cells from the pituitary, proposed the existence of an excitatory amino acid transporter with the properties of a ligand-gated chloride channel. This type of transporter, once in the presence of Glu, triggers an increase of the Ca^{2+} cytosolic concentrations, which are mediated via Glu transporters rather than receptors' activation (Fairman et al. 1995). In this context, it has been reported that the C-termini fraction of the EAAT5 interacts with several PDZ (a molecular protein domain-biding motif) domains in PSD-95 protein (postsynaptic density-95 kDa protein). PSD-95 and related proteins function as multifunctional platform proteins that form signaling complexes. Therefore, the presence of a PDZ domain favors a plausible role of EAAT5 as a signaling entity in the retina (Sheng 1996). Gonzalez and Ortega reported that Glu regulates its uptake in a nondependent receptor manner and proposed a possible implication of modifications induced by the substrate translocation process of the transporter (González and Ortega 2000). In addition, Abe and colleagues (2001) showed ERK1/ERK2 phosphorylation induction in rat astrocyte cultures after incubating these cells with Glu in a receptor-independent manner (Abe and Saito 2001).

Interestingly enough, the effect of Glu was reproduced by Glu transporters substrates, such as aspartate (D and/or L) and by the action of transportable Glu uptake inhibitors, including DL-threo-β-hyroxyaspartate (THA) and PDC, suggesting that Glu induces ERK1/ERK2 phosphorylation by a mechanism related to Glu transporters (Abe and Saito 2001).

It is important to note that Na^+ influx associated with glia Glu uptake activates the sodium-calcium exchanger (NCX), which is a common characteristic of astroglia terminal processes contributing to Glu clearance from the synaptic cleft (Kirischuk et al. 2007). Moreover, Na^+ influx is linked to the activation of the Na^+/K^+-ATPase (Gegelashvili et al. 2007). Furthermore, it was confirmed that Glu transporters are in close association with the Na^+/K^+-ATPase pump in a common macromolecular complex (Rose et al. 2009). Interestingly, Pellerin and colleagues (2007) proposed that proteins that take part in either the Glu/Gln shuttle (such as GS), like the sodium-coupled neutral amino acid transporters (SNAT), or in the astrocyte/neuron lactate shuttle (as monocarboxylate transporters) might be under the control of the activity of the Glu transporters (Martínez-Lozada et al. 2011; Magistretti 2009). It has also been reported that Glu uptake activity prompts glucose influx (Magistretti 2009), as well as an increase in GS activity (Lehmann et al. 2009). Further reports pointed that Glu receptors and transporters regulate the initiation phase of the protein synthesis. Likewise, Glu modulates the phosphorylation of the eukaryotic initiation factor 2 (eIF2α) through GLAST/EAAT1 in a dose- and time-dependent manner (Flores-Méndez et al. 2013). Latest reports suggest that overstimulation of cerebellar parallel fibers would lead to a saturation of GLAST uptake activity, which favors the activation of AMPA and NMDA receptors, resulting in ROS production, cystine uptake, and Glu release via the cystine/Glu antiporter (x_c^-), exacerbating Purkinje cell death as has been reported for the oral administration of diphenylarsinic acid in rodent cerebellum (Kato et al. 2007) (Table 1).

Table 1 Glutamate transporters and protein interactions

Glutamate transporter	Protein interaction	Description	References
GLAST/EAAT1	Sept2	Sept2 bound directly to the carboxy-terminal region of GLAST in a GDP-form-dependent manner	Kinoshita et al. (2004)
	NHERF1	GLAST bound directly to the PDZ1 domain of NHERF1 via the C-terminal ETKM motif of GLAST	Lee et al. (2007),Sato et al. (2013)
	SNAT3	GLAST coupling between and SNAT3 in BGC	Martínez-Lozada et al. (2013)
	Mitochondria and proteins related to energetic metabolism	Compartmentalization of GLAST with mitochondria and proteins involved in energy metabolism could provide energetic support for Glu transport	Bauer et al. (2012)
GLT-1/EAAT2	LIM protein Ajuba	Cytoplasmic LIM protein, Ajuba, which interacts with the amino terminus of GLT-1	Marie et al. (2002)
	PDS-95	GLT-1b isoform contains a sequence motif that enables it to interact with proteins containing PDZ domains. Notably, it appears to interact with PSD-95, a scaffold protein that is involved in the organization of postsynaptic densities in glutamatergic synapses	González-González et al. (2008, 2009)
	PICK1	PICK1-GLT-1b interaction regulates the modulation of GLT-1 function by PKC	Bassan et al. (2008)
	MAGI-1	MAGI-1 regulates surface expression of GLT-1	Zou et al. (2011)
	Na^+/K^+ ATPase, glycolytic enzymes, and mitochondria	Na^+/K^+ ATPase, glycolytic enzymes, and mitochondria, providing a mechanism to spatially match the energy and buffering capacity to the demands imposed by transport	Genda et al. (2011)
EAAC1/EAAT3	GTRAP3–18	GTRAP3–18 interacts with the intracellular carboxy-terminal domain of EAAT3	Lin et al. (2001)
	PKCα	PKCα is implicated in EAAC1 redistribution to the plasma membrane and may be involved in the internalization of GLT-1a	González et al. (2003, 2005)
	Syntaxin 1A	Syntaxin 1A regulates the trafficking of EAAC1	Yu et al. (2006)
	α-Adducin	EAAC1 transporters associated with the actin-binding protein α-adducin in a PKC-insensitive manner	Bianchi et al. (2010)

(continued)

Table 1 (continued)

Glutamate transporter	Protein interaction	Description	References
	NCX	EAAT and NCX are both involved in Glu-induced ATP synthesis, with NCX playing a pivotal role	Magi et al. (2012, 2013)
EAAT4	GTRAP41 and GTRAP48	GTRAP41 and GTRAP48 interact with the intracellular carboxy-terminal domain of EAAT4 and modulate its Glu transport activity	Jackson et al. (2001)

Abbreviations: GTRAP Glu transporter-4-associated protein, *MAGI-1* membrane-associated guanylate kinase, *NCX* Na^+/Ca^{2+} exchange, *NHERF1* Na^+/H^+ exchanger regulatory factor, *PICK1* protein interacting with C kinase-1, *PKC* diacylglycerol-dependent kinase, *PSD-95* postsynaptic density protein-95, *SNAT3* Na^+-dependent neutral amino acid transporter 3; Sept2, septin 2 protein; x_c^-, cystine/glutamate antiporter

As pointed out, GLAST/EAAT1 activity results in Ca^{2+} influx, capable to induce mTOR phosphorylation and the DNA-binding activity of the transcription factor activator protein-1 (AP-1), suggesting GLAST/EAAT1 signaling could possibly participate in the transcriptional control gene expression in glial cells (López-Colomé et al. 2012). An activity-dependent differential distribution of the transporter molecules in lipid rafts has also been reported (Butchbach et al. 2004; Hou et al. 2008; González et al. 2007).

More recently, it has been reported that glial cells adjacent to glutamatergic synapses act as sensor of neuron-derived Glu; it was demonstrated that once the amino acid is internalized into the glial cells, the associated increase in Na^+ intracellular concentrations favors Gln release through the inverse mode of operation of the SNAT3. In fact, complexes between GLAST/EAAT1 and SNAT3 are regulated by Glu in a time-dependent manner (Martínez-Lozada et al. 2013).

The signaling properties of EAATs have also been documented in oligodendrocytes, in which the transporter-mediated Na^+ influx via the Na^+/Ca^{2+} exchanger activates the β isoform of the calcium/calmodulin-dependent kinase type II (CaMKII), accelerating the maturation of these glial cells (Bauer et al. 2009).

4 Conclusion

Glial glutamate transporters act as signaling transducer entities, and the regulation of not only glial function in the short and long term but most possibly through the modulation of the functionality and expression of proteins involved in glial/neuronal interactions plays an important role in neurotransmission. A schematic representation of the main findings described in this chapter is presented in Fig. 1.

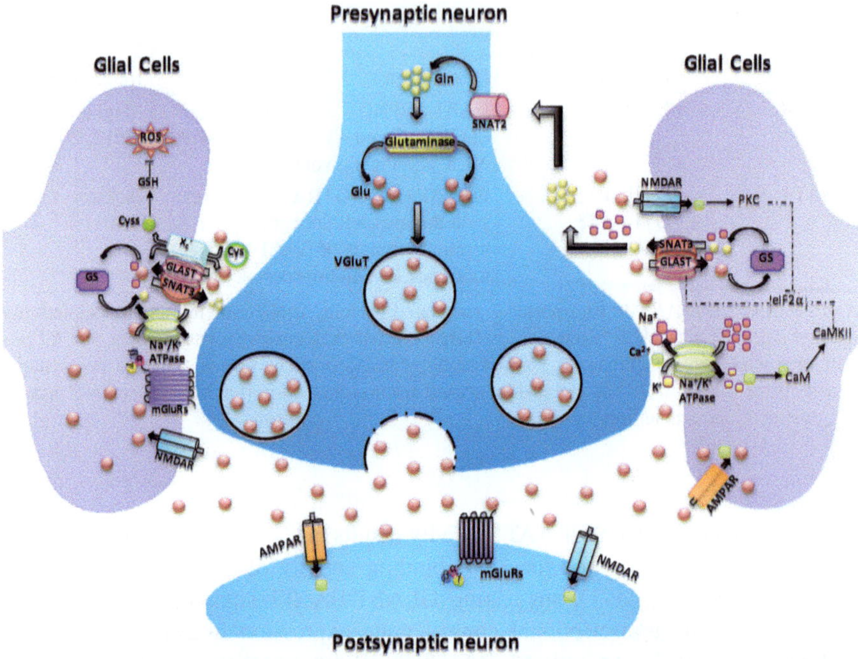

Fig. 1 Glial signaling through GLAST/EAAT1 (GLAST/SNAT3-x_c^- complex). Glu is released from the presynaptic neuron to the synaptic cleft and taken up into the glial cells by GLAST/EAAT1. An increase in intracellular Na^+ concentrations induces the operation, in reverse mode, of SNAT3 with consequent liberation of Gln, whereas the activation of the x_c^- system under stress conditions allows the release of Glu and the uptake of cystine to be transformed in the cysteine and finally produces glutathione (GSH)

Acknowledgments The work in the laboratory is supported by Conacyt-Mexico (255087) and *Soluciones para un México Verde S.A. de C.V.*; Edna Suárez-Pozos and Donají Chi-Castañeda are supported by Conacyt-Mexico fellowships (232180 and 376687, respectively).

Conflict of Interest The authors declare that they have no competing interests.

References

Abe K, Saito H. Possible linkage between glutamate transporter and mitogen-activated protein kinase cascade in cultured rat cortical astrocytes. J Neurochem. 2001;76(1):217–23.

Aguirre A, López-Bayghen E, Ortega A. Glutamate-dependent transcriptional regulation of the chkbp gene: signaling mechanisms. J Neurosci Res. 2002;70(1):117–27.

Araque A, Parpura V, Sanzgiri RP, Haydon PG. Tripartite synapses: glia, the unacknowledged partner. Trends Neurosci. 1999;22(5):208–15.

Arriza J, Eliasof S, Kavanaugh M, Amara S. Excitatory amino acid transporter 5, a retinal glutamate transporter coupled to a chloride conductance. PNAS. 1997;94(8):4155–60.

Balakrishnan S, Bellamy TC. Depression of parallel and climbing fiber transmission to Bergmann glia is input specific and correlates with increased precision of synaptic transmission. Glia. 2009;57(4):393–401.

Balazs R. Trophic effect of glutamate. Curr Top Med Chem. 2006;6(10):961–8.

Bassan M, Liu H, Madsen KL, Armsen W, Zhou J, DeSilva T, et al. Interaction between the glutamate transporter GLT1b and the synaptic PDZ domain protein PICK1. Eur J Neurosci. 2008;27(1):66–82.

Bauer NG, Richter-Landsberg C, Ffrench-Constant C. Role of the oligodendroglial cytoskeleton in differentiation and myelination. Glia. 2009;57(16):1691–705.

Bauer DE, Jackson JG, Genda EN, Montoya MM, Yudkoff M, Robinson MB. The glutamate transporter, GLAST, participates in a macromolecular complex that supports glutamate metabolism. Neurochem Int. 2012;61(4):566–74.

Bianchi MG, Gatti R, Torielli L, Padoani G, Gazzola GC, Bussolati O. The glutamate transporter excitatory amino acid carrier 1 associates with the actin-binding protein α-adducin. Neuroscience. 2010;169(2):584–95.

Butchbach MER, Tian G, Guo H, Lin CL. Association of Excitatory Amino Acid Transporters, especially EAAT2, with cholesterol-rich lipid raft Microdomains: importance for excitatory amino acid transporter localization and function. J Biol Chem. 2004;279(33):34388–96.

Coles JA, Abbott NJ. Signalling from neurones to glial cells in invertebrates. Trends Neurosci. 1996;19(8):358–62.

Danbolt NC. Glutamate uptake. Prog Neurobiol. 2001;65:1–105.

Danbolt NC, Furness DN, Zhou Y. Neuronal vs glial glutamate uptake: resolving the conundrum. Neurochem Int. 2016;98:29–45.

Dityatev A, Rusakov DA. Molecular signals of plasticity at the tetrapartite synapse. Curr Opin Neurobiol. 2011;21(2):353–9.

Duan S, Anderson CM, Stein BA, Swanson RA. Glutamate induces rapid upregulation of astrocyte glutamate transport and cell-surface expression of GLAST. J Neurosci. 1999;19(23):10193–200.

Eulenburg V, Gomeza J. Neurotransmitter transporters expressed in glial cells as regulators of synapse function. Brain Res Rev. 2010;63:103–12.

Fairman WA, Vandenberg RJ, Arriza JL, Kavanaugh MP, Amara SG. An excitatory amino-acid transporter with properties of a ligand-gated chloride channel. Nature. 1995;375(6532):599–603.

Flores-Méndez MA, Martínez-Lozada Z, Monroy HC, Hernández-Kelly LC, Barrera I, Ortega A. Glutamate-dependent translational control in cultured bergmann glia cells: EIF2α phosphorylation. Neurochem Res. 2013;38(7):1324–32.

Gallo V, Ghiani CA. Glutamate receptors in glia: new cells, new inputs and new functions. Trends Pharmacol Sci. 2000;21(7):252–8.

Gegelashvili G, Danbolt NC, Schousboe A. Neuronal soluble factors differentially regulate the expression of the GLT1 and GLAST glutamate transporters in cultured astroglia. J Neurochem. 1997;69(6):2612–5.

Gegelashvili M, Rodriguez-Kern A, Sung L, Shimamoto K, Gegelashvili G. Glutamate transporter GLAST/EAAT1 directs cell surface expression of FXYD2/gamma subunit of Na, K-ATPase in human fetal astrocytes. Neurochem Int. 2007;50(7–8):916–20.

Genda EN, Jackson JG, Sheldon AL, Locke SF, Greco TM, O'Donnell JC, et al. Co-compartmentalization of the astroglial glutamate transporter, GLT-1, with glycolytic enzymes and mitochondria. J Neurosci. 2011;31(50):18275–88.

González M, Ortega A. Regulation of the Na+–dependent high affinity glutamate/aspartate transporter in cultured Bergmann glia by phorbol esters. J Neurosci Res. 1997;50(4):585–90.

González MI, Ortega A. Regulation of high-affinity glutamate uptake activity in Bergmann glia cells by glutamate. Brain Res. 2000;866(1–2):73–81.

González MI, Bannerman PG, Robinson MB. Phorbol myristate acetate-dependent interaction of protein kinase Calpha and the neuronal glutamate transporter EAAC1. J Neurosci. 2003;23(13):5589–93.

González MI, Susarla BTS, Robinson MB. Evidence that protein kinase Cα interacts with and regulates the glial glutamate transporter GLT-1. J Neurochem. 2005;94(5):1180–8.
González MI, Susarla BTS, Fournier KM, Sheldon AL, Robinson MB. Constitutive endocytosis and recycling of the neuronal glutamate transporter, excitatory amino acid carrier 1. J Neurochem. 2007;103(5):1917–31.
González-González IM, García-Tardón N, Cubelos B, Giménez C, Zafra F. The glutamate transporter GLT1b interacts with the scaffold protein PSD-95. J Neurochem. 2008 Jun;105(5):1834–48.
González-González IM, García-Tardón N, Giménez C, Zafra F. Splice variants of the glutamate transporter GLT1 form hetero-oligomers that interact with PSD-95 and NMDA receptors. J Neurochem. 2009;110(1):264–74.
Guillet B, Velly L, Canolle B, Masmejean F, Nieoullon A, Pisano P. Differential regulation by protein kinases of activity and cell surface expression of glutamate transporters in neuron-enriched cultures. Neurochem Int. 2005;46(4):337–46.
Herguedas B, Krieger J, Greger IH. Receptor heteromeric assembly-how it works and why it matters: the case of ionotropic glutamate receptors. In:Molecular and cellular changes in the cancer cell; 2013. p. 361–86.
Hou Q, Huang Y, Amato S, Snyder SH, Huganir RL, Man HY. Regulation of AMPA receptor localization in lipid rafts. Mol Cell Neurosci. 2008;38(2):213–23.
Jackson M, Song W, Liu MY, Jin L, Dykes-Hoberg M, Lin CLG, et al. Modulation of the neuronal glutamate transporter EAAT4 by two interacting proteins. Nature. 2001;410(6824):89–93.
Jiang J, Amara SG. New views of glutamate transporter structure and function: advances and challenges. Neuropharmacology. 2011;60(1):172–81.
Kanai Y, Hediger MA. Primary structure and functional characterization of a high-affinity glutamate transporter. Nature. 1992;360(6403):467–71.
Kato K, Mizoi M, An Y, Nakano M, Wanibuchi H, Endo G, et al. Oral administration of diphenylarsinic acid, a degradation product of chemical warfare agents, induces oxidative and nitrosative stress in cerebellar Purkinje cells. Life Sci. 2007;81(21–22):1518–25.
Kinoshita N, Kimura K, Matsumoto N, Watanabe M, Fukaya M, Ide C. Mammalian septin Sept2 modulates the activity of GLAST, a glutamate transporter in astrocytes. Genes Cells. 2004;9(1):1–14.
Kirischuk S, Kettenmann H, Verkhratsky A. Membrane currents and cytoplasmic sodium transients generated by glutamate transport in Bergmann glial cells. Pflugers Arch. 2007;454(2):245–52.
Lee A, Rayfield A, Hryciw DH, Ma TA, Wang D, Pow D, et al. Na+ – H+ exchanger regulatory factor 1 is a PDZ scaffold for the astroglial glutamate transporter GLAST. Glia. 2007;55(2):119–29.
Lehmann C, Bette S, Engele J. High extracellular glutamate modulates expression of glutamate transporters and glutamine synthetase in cultured astrocytes. Brain Res. 2009;1297:1–8.
Lin CI, Orlov I, Ruggiero AM, Dykes-Hoberg M, Lee A, Jackson M, et al. Modulation of the neuronal glutamate transporter EAAC1 by the interacting protein GTRAP3-18. Nature. 2001;410(6824):84–8.
López T, López-Colomé A, Ortega A. AMPA/KA receptor expression in radial glia. Neuroreport. 1994;5(4):504–6.
López-Bayghen E, Ortega A. Glutamate-dependent transcriptional regulation of GLAST: role of PKC. J Neurochem. 2004;91(1):200–9.
López-Bayghen E, Ortega A. Glial cells and synaptic activity: translational control of metabolic coupling. Rev Neurol. 2010;50(10):607–15.
López-Bayghen E, Ortega A. Glial glutamate transporters: new actors in brain signaling. IUBMB Life. 2011;63(10):816–23.
López-Bayghen E, Rosas S, Castelán F, Ortega A. Cerebellar Bergmann glia: an important model to study neuron–glia interactions. Neuron Glia Biol. 2007;3(2):155–67.
López-Colomé A, Martínez-Lozada Z, Guillem A, López E, Ortega A. Glutamate transporter-dependent mTOR phosphorylation in Müller glia cells. ASN Neuro. 2012;4(5):AN20120022.
Magi S, Lariccia V, Castaldo P, Arcangeli S, Nasti AA, Giordano A, et al. Physical and functional interaction of NCX1 and EAAC1 transporters leading to glutamate-enhanced ATP production in brain mitochondria. PLoS One. 2012;7(3):e34015.

Magi S, Arcangeli S, Castaldo P, Nasti AA, Berrino L, Piegari E, et al. Glutamate-induced ATP synthesis: relationship between plasma membrane Na+/Ca2+ exchanger and excitatory amino acid transporters in brain and heart cell models. Mol Pharmacol. 2013;84(4):603–14.
Magistretti PJ. Role of glutamate in neuron-glia metabolic coupling. Am J Clin Nutr. 2009;90(3):875S–80S.
Marie H, Billups D, Bedford FK, Dumoulin A, Goyal RK, Longmore GD, et al. The amino terminus of the glial glutamate transporter GLT-1 interacts with the LIM protein Ajuba. Mol Cell Neurosci. 2002;19(2):152–64.
Martínez-Lozada Z, Ortega A. Glutamatergic transmission: a matter of three. Neural Plast. 2015;2015:1–11.
Martínez-Lozada Z, Hernández-Kelly LC, Aguilera J, López-Bayghen E, Ortega A. Signaling through EAAT-1/GLAST in cultured Bergmann glia cells. Neurochem Int. 2011;59(6):871–9.
Martínez-Lozada Z, Guillem AM, Flores-Méndez M, Hernández-Kelly LC, Vela C, Meza E, et al. GLAST/EAAT1-induced glutamine release via SNAT3 in Bergmann glial cells: evidence of a functional and physical coupling. J Neurochem. 2013;125(4):545–54.
Miller S, Bridges RJ, Chamberlin AR, Cotman CW. Pharmacological dissociation of glutamatergic metabotropic signal transduction pathways in cortical astrocytes. Eur J Pharmacol. 1994;269(2):235–41.
Müller T, Möller T, Berger T, Schnitzer J, Kettenmann H. Calcium entry through kainate receptors and resulting potassium-channel blockade in Bergmann glial cells. Science. 1992;256(5063):1563–6.
Pellerin L, Bouzier-Sore A-K, Aubert A, Serres S, Merle M, Costalat R, et al. Activity-dependent regulation of energy metabolism by astrocytes: An update. Glia. septiembre de 2007;55(12):1251–62.
Pines G, Danbolt N,C, Bjørås M, Zhang Y, Bendahan A, et al. Cloning and expression of a rat brain L-glutamate transporter. Nature. 1992;360(6403):464–7.
Poblete-Naredo I, Angulo C, Hernández-Kelly L, López-Bayghen E, Aguilera J, Ortega A. Insulin-dependent regulation of GLAST/EAAT1 in Bergmann glial cells. Neurosci Lett. 2009;451(2):134–8.
Rana RS, Hokin LE. Role of phosphoinositides in transmembrane signaling. Physiol Rev. 1990;70(1):115–64.
Rose EM, Koo JCP, Antflick JE, Ahmed SM, Angers S, Hampson DR. Glutamate transporter coupling to Na K-ATPase. J Neurosci. 2009;29(25):8143–55.
Sato K, Otsu W, Otsuka Y, Inaba M. Modulatory roles of NHERF1 and NHERF2 in cell surface expression of the glutamate transporter GLAST. Biochem Biophys Res Commun. 2013;430(2):839–45.
Shank RP, Campbell GL. Amino acid uptake, content, and metabolism by neuronal and glial enriched cellular fractions from mouse cerebellum. J Neurosci. 1984;4(1):58–69.
Sheng M. PDZs and receptor/channel clustering: rounding up the latest suspects. Neuron. 1996;17(4):575–8.
Slotboom DJ, Konings WN, Lolkema JS. Structural features of the glutamate transporter family. Microbiol Mol Biol Rev. 1999;63(2):293–307.
Sommer B, Köhler M, Sprengel R, Seeburg PH. RNA editing in brain controls a determinant of ion flow in glutamate-gated channels. Cell. 1991;67(1):11–9.
Storck T, Schulte S, Hofmann K, Stoffel W. Structure, expression, and functional analysis of a Na(+)-dependent glutamate/aspartate transporter from rat brain. Proc Natl Acad Sci U S A. 1992;89(22):10955–9.
Tang H, Hornstein E, Stolovich M, Levy G, Livingstone M, Templeton D, et al. Amino acid-induced translation of TOP mRNAs is fully dependent on phosphatidylinositol 3-kinase-mediated signaling, is partially inhibited by rapamycin, and is independent of S6K1 and rpS6 phosphorylation. Mol Cell Biol. 2001;21(24):8671–83.
Yu YX, Shen L, Xia P, Tang YW, Bao L, Pei G. Syntaxin 1A promotes the endocytic sorting of EAAC1 leading to inhibition of glutamate transport. J Cell Sci. 2006;119(Pt 18):3776–87.

Zepeda RC, Barrera I, Castelán F, Soto-Cid A, Hernández-Kelly LC, López-Bayghen E, et al. Glutamate-dependent transcriptional regulation in Bergmann glia cells: involvement of p38 MAP kinase. Neurochem Res. 2008;33(7):1277–85.

Zerangue N, Kavanaugh MP. Flux coupling in a neuronal glutamate transporter. Nature. 1996;383(6601):634–7.

Zou S, Pita-Almenar JD, Eskin A. Regulation of glutamate transporter GLT-1 by MAGI-1. J Neurochem. 2011;117(5):833–40.

Regulation of Glutamate Transporter Expression in Glial Cells

Donají Chi-Castañeda, Edna Suárez-Pozos, and Arturo Ortega

Abstract Glutamate (Glu) is the major excitatory neurotransmitter in the vertebrate central nervous system. During synaptic activity, Glu is released into the synaptic cleft and binds to Glu receptors activating a wide variety of signal transduction cascades. Extracellular Glu concentrations are maintained exclusively within physiological levels mainly by glial Glu transporters. Inefficient clearance of synaptic Glu may be neurotoxic owing to prolonged hyperactivation of postsynaptic Glu receptors, causing a multitude of intracellular events in the postsynaptic neuron, which ultimately results in neuronal cell death. This phenomenon is known as excitotoxicity and is the underlying mechanisms of a number of neurodegenerative diseases. Therefore, it is important to understand the regulation of Glu transporters' function. Transporter activity can be regulated in different ways, including gene expression, transporter protein targeting and trafficking, and posttranslational modifications of the transporter protein. The identification of these mechanisms has allowed to understand the role of Glu transporters during pathology and will aid in the development of therapeutic strategies for treating or preventing pathologies associated with excitotoxicity.

Keywords EAATs • Gene expression • Glutamate transporters • Growth factors • Protein kinases • Transcriptional factors

---

D. Chi-Castañeda
Laboratorio de Neurotoxicología, Departamento de Toxicología,
Centro de Investigación y de Estudios Avanzados del Instituto Politécnico Nacional,
Apartado Postal 14-740, Ciudad de México 07000, México

Soluciones para un México Verde S.A. de C.V., Ciudad de México 01210, México

E. Suárez-Pozos • A. Ortega (✉)
Laboratorio de Neurotoxicología, Departamento de Toxicología,
Centro de Investigación y de Estudios Avanzados del Instituto Politécnico Nacional,
Apartado Postal 14-740, Ciudad de México 07000, México
e-mail: arortega@cinvestav.mx

A. Ortega, A. Schousboe (eds.), *Glial Amino Acid Transporters*,
Advances in Neurobiology 16, DOI 10.1007/978-3-319-55769-4_10

List of Abbreviation

ATP	Adenosine triphosphate
dbcAMP	Dibutyryl cyclic adenosine monophosphate
CNS	Central nervous system
DNA	Deoxyribonucleic acid
EAAC1	Excitatory amino acid carrier 1
EAAT	Excitatory amino acid transporters
EGF	Epidermal growth factor
GLAST	Glutamate/aspartate transporter
GLT-1	Glutamate transporter 1
Glu	Glutamate
Gln	Glutamine
GSH	Glutathione
GTRP	Glutamate transporter-associated protein
mRNA	Messenger ribonucleic acid
NCM	Neuron-conditioned media
NF-κB	Nuclear factor-kappa B
NGF	Nerve growth factor
PACAP	Pituitary adenylate cyclase-activating peptide
PI3K	Phosphoinositide 3-kinase
PKA	Protein kinase A
PKC	Protein kinase C
SLC1	Solute carrier 1 gene
TGFα	Transforming growth factor alpha

1 Introduction

To guarantee efficient neurotransmission in the brain, it is critical that synaptic glutamate (Glu) is rapidly removed. High-affinity Glu transporters are responsible to maintain a low concentration of synaptic Glu and therefore to attenuate glutamatergic signaling. These uptake processes occur to a vast extent on surrounding astrocytes although it also occurs presynaptically. Aberrant expression and functioning of Glu transporters will easily lead to a disturbance of synaptic transmission, contributing thereby to the pathology of both acute neuronal injury and chronic neurodegenerative conditions. With L-glutamate as the main excitatory neurotransmitter in the brain, precise function and regulation of Glu transporters are crucial to maintein the synaptic concentration of this amino acid below excitotoxic levels. In the present chapter, we will focus mainly on the transcriptional and translational regulation of Glu transporters.

2 Glutamate

L-glutamate is the major excitatory neurotransmitter in the mammalian central nervous system (CNS). It is present in practically all areas of the brain, and its receptors are widely distributed and expressed on neuronal and non neuronal cells. This neurotransmitter plays an important role in processes of cognition, learning and memory formation (Fonnum 1984; Headley and Grillner 1990), as well as, in other plastic changes involved in the regulation of CNS development including synapse induction and elimination (Rabacchi et al. 1992; Durand et al. 1996), cellular migration and differentiation (Komuro and Rakic 1993; Rossi and Slater 1993). It is known that the concentration of Glu in the synaptic cleft is in the low millimolar range (Danbolt 2001). However, excessive accumulation of extracellular Glu and overactivation of Glu receptors contribute to neurotoxicity and neurodegenerative diseases (McEntee and Crook 1993; Lauriat and McInnes 2007; Domingues et al. 2010; Gegelashvili and Bjerrum 2014). To keep the levels of Glu in the synaptic cleft sufficiently low, Glu has to be removed from the extracellular space; the only way to remove Glu rapidly from the synaptic cleft is by cellular uptake. This uptake is carried out by sodium-dependent high-affinity transport systems located on the plasma membrane of perisynaptic astrocytes and, in a lower degree, to neurons themselves (Kanai et al. 1993). Under pathological conditions, Glu uptake becomes extremely important because of that; in the synaptic cleft this neurotransmitter rises to a great extent above normal levels, high enough to trigger excitotoxicity in the postsynaptic neuron (Maragakis and Rothstein 2004). Once Glu is taken up, it is used for metabolic purposes (protein synthesis, energetic metabolism) or recycled as transmitter by entering into the Glu/glutamine cycle, which as it is known; is the major pathway by which Glu is recycled (Danbolt 2001).

3 Characterization and Differential Localization of Glutamate Transporters

Since 1992, molecular cloning and molecular biology studies have identified five plasma membrane Glu transporter subtypes, known as excitatory amino acid transporters EAAT1-5 (human nomenclature) or GLAST/GLT-1/EAAC1/EAAT4/EAAT5 (rodent nomenclature for the five respective human transporters) (Storck et al. 1992; Pines et al. 1992; Kanai and Hediger 1992; Arriza et al. 1994; Fairman et al. 1995; Arriza et al. 1997). In addition to their main substrate Glu, these transporters mediate L- and D-aspartate uptake. However, EAAT3/EAAC1 also accepts L-cysteine as a substrate (Danbolt 2001). Glu transporters use predominantly the Na^+ gradient across the membrane as the main energy source for intracellular accumulation of Glu. The transport stoichiometry for EAATs corresponds to three Na^+

and one H^+ per transport cycle of one molecule of Glu, while one K^+ is concurrently released from the cell (Danbolt 2001; Zerangue and Kavanaugh 1996; Levy et al. 1998). This stoichiometry allows the transporters to generate up to a million-fold concentration gradient across the membrane (Zerangue and Kavanaugh 1996).

Glu transporters display distinct molecular and pharmacological characteristics, as well as discrete cellular and regional localizations (Table 1). EAAT1/GLAST (glutamate/aspartate transporter) is the major Glu transporter present in the cerebellar astrocytes (Lehre and Danbolt 1998; Takatsuru et al. 2007), the inner ear (Furness and Lehre 1997; Takumi et al. 1997), the circumventricular organs (Berger and Hediger 2000), and in the retina (Derouiche and Rauen 1995; Derouiche 1996; Rauen et al. 1996, 1998; Lehre et al. 1997; Pow and Barnett 1999). EAAT2/GLT-1 (glutamate transporter 1) is almost exclusively glial and is widespread and abundant throughout the forebrain, cerebellum, and spinal cord (Furuta et al. 1997). EAAT3/EAAC1 (excitatory amino acid carrier 1) is a neuronal transporter (Kanai and Hediger 1992; Rothstein et al. 1994; Bjørås et al. 1996) and is not expressed in glial cells (Holmseth et al. 2012; Shashidharan et al. 1997). It appears to be expressed in the majority if not all neurons throughout the CNS, but is selectively targeted to somata and dendrites avoiding axon terminals (Holmseth et al. 2012; Shashidharan et al. 1997). EAAT4 is predominantly found in cerebellar Purkinje cells (Fairman et al. 1995; Dehnes et al. 1998) where it is targeted to the dendrites, the spines in particular (Dehnes et al. 1998), but there is also EAAT4 expression in a subset of forebrain neurons (Dehnes et al. 1998; Massie et al. 2008; de Vivo et al. 2010). Finally, EAAT5 is preferentially expressed in rod photoreceptors and bipolar cells of the retina. EAAT5 expression in the brain is very low (Arriza et al. 1997; Pow and Barnett 2000). It is important to mention that approximately 80–90% of the total brain removal of Glu from the synaptic cleft is carried out by EAAT1/GLAST and/or EAAT2/GLT-1 (Lehre and Danbolt 1998; Eulenburg and Gomeza 2010), reflecting that these proteins are abundantly expressed in glial cells.

Furthermore, expression of these transporters is also differentially controlled during development. For example, GLAST is found at relatively high levels during the early stages of development, while GLT-1 levels increase dramatically during development (Furuta et al. 1997), suggesting that GLT-1 may be a marker of astrocyte maturation.

4 Regulation of Glutamate Transporters

Glutamatergic transmission participates in many important brain functions, while elevated concentrations of extracellular Glu can cause severe excitotoxic damage to the receiving neurons. Since no extracellular enzymes exist for Glu breakdown, it is critical to have an efficient Glu uptake activity. To date, little is known about the mechanisms of Glu transporter regulation, as well as the factors that are responsible for regulating its protein expression levels and transporter activity. However, it still seems interesting to consider the mechanisms that regulate Glu transporters as

Table 1 Main characteristics of the high-affinity glutamate transporters

Transporter	Human homolog	Gen	Major cell type	CNS distribution	Outside CNS distribution	Properties	References
GLAST	EAAT1	*Slc1a3*	Astrocytes, oligodendrocytes	Cerebellum, cortex, spinal cord	Testis, heart, kidney, lung, spleen, liver, muscles	Major subtype expressed during CNS development	Maragakis and Rothstein (2004), Storck et al. (1992), Lehre and Danbolt (1998), Takatsuru et al. (2007), Furuta et al. (1997)
GLT-1	EAAT2	*Slc1a2*	Astrocytes, neurons	Brain, spinal cord, retina, cerebellum	Liver	Responsible for >90% of total Glu uptake	Lauriat and McInnes (2007), Maragakis and Rothstein (2004), Furuta et al. (1997)
EAAC1	EAAT3	*Slc1a1*	Neurons	Hippocampus, cerebellum, striatum	Spinal cord, kidney, lung, muscle, heart, liver	Membrane expression under dynamic regulation by intracellular kinases and cholesterol. Possibly important in areas of higher neuron-glia ratio	Kanai and Hediger (1992), Rothstein et al. (1994), Bjørås et al. (1996), Conti et al. (1998), He et al. (2000), Nieoullon et al. (2006)
EAAT4	EAAT4	*Slc1a6*	Purkinje cells	Cerebellum	Heart	High Cl^- conductance. Regulator of neuronal excitability, counteracting depolarization of neurons	Storck et al. (1992), Pines et al. (1992), Kanai and Hediger (1992), Fairman et al. (1995), Dehnes et al. (1998), Massie et al. (2008)
EAAT5	EAAT5	*Slc1a7*	Photoreceptors and bipolar cells	Retina	Heart	High Cl^- conductance	Storck et al. (1992), Pines et al. (1992), Kanai and Hediger (1992), Fairman et al. (1995), Arriza et al. (1997), Pow and Barnett (2000)

potential drug targets. It is possible that a better understanding of the mechanisms involved in the alteration of the expression of Glu transporters may lead to a broader understanding of the pathogenesis of diverse neurologic and psychiatric diseases. Therefore, in this chapter, we will focus on the mechanisms responsible for Glu transporter regulation.

A variety of mechanisms regulate to the Glu transporters, including deoxyribonucleic acid (DNA) transcription, messenger ribonucleic acid (mRNA) maturation and stabilization, posttranslational modifications, trafficking to and from the plasma membrane (Robinson 2002, 2006), and diffusion in the membrane (Shin et al. 2009; Benediktsson et al. 2012; Murphy-Royal et al. 2015). While DNA transcription and protein expression events require a long time period (several hours) to reflect effects on the activity of the transporter, posttranslational modifications may occur within minutes. As with most biological events, it is likely that a combination of all of these mechanisms is essential for efficient Glu uptake to occur.

4.1 *Transcriptional and Translational Regulation of EAAT1/GLAST*

The human gene product, EAAT1 protein, has 542 amino acid residues in length. This protein has a molecular weight of 59.5 KDa and 96% amino acid sequence identity with the rat GLAST sequence (Storck et al. 1992). In contrast, EAAT1 has 52% identity with EAAT4 and 49% with EAAT3 (Arriza et al. 1997). Originally, the human EAAT1 gene (*SLC1A3*) was mapped to chromosome 5p13 (Takai et al. 1995); while the mouse GLAST gene (*Slc1a3*) was mapped to chromosome 15A2 (Hagiwara et al. 1996). Actually, *SLC1A3* is located on chromosome 5p11–p12, and it is organized in ten exons interposed by nine introns (Stoffel et al. 1996).

Early studies of the human and rodent EAAT1/GLAST promoter regions revealed that both species have a low homology among their sequences (Kim et al. 2003; Unger et al. 2012). Specifically, the full-length promoters of EAAT1/GLAST lack well-defined *cis*-acting elements such as TATA and CCAAT boxes. In contrast, both promoters contain a GC box, whereas the human gene also contains an E box. Interestingly, the absence of TATA box in the EAAT promoters identified so far is not unusual; therefore it has been suggested that TATA-less promoters regulate tissue-specific genes (Müller et al. 2007). Despite the lack of sequence homology, the EAAT1/GLAST promoters from rat and human share various common putative transcription factor binding sites, including nuclear factor-kappa B (NF-κB), cartilage homeoprotein 1 (CART), CCAAT/enhancer-binding protein (CEBP), GATA-binding factors (GATA), human acute myelogenous leukemia factors (HAML), myeloid zinc finger 1 factors (MZF1), NKX homeodomain factors (NKXH), octamer-binding protein (Oct1), specificity protein 1 (SP1), and yin yang 1 (YY1), implying that constitutive EAAT1/GLAST expression in both species is controlled by similar mechanism (Unger et al. 2012).

In 1996, Gegelashvili and coworkers reported that in cultured astroglia Glu itself regulates GLAST and induces an increase in Glu uptake and in the expression of the protein but not in its mRNA levels (Gegelashvili et al. 1996). The Glu-mediated increase in GLAST was repressed by inhibitors of α-amino-3-hydroxy-5-methyl-4-isoxazolepropionic acid (AMPA)/kainate receptors. Years later, Gegelashvili and colleagues also showed that activation of group II metabotropic Glu receptors (mGluRs) upregulates GLAST mRNA and protein levels (Gegelashvili et al. 2000; Aronica et al. 2003) via the extracellular signal-regulated kinase (ERK)/phosphoinositide 3-kinase (PI3K)/NF-κB pathway (Lin et al. 2014).

Simultaneously, several groups demonstrated in primary cultures of astrocytes that dibutyryl cyclic adenosine monophosphate (dbcAMP) also increases both GLAST mRNA and protein levels; the dbcAMP-mediated upregulation of the transporter is blocked by protein kinase A (PKA) inhibitors (Schlag et al. 1998). Later on it was revealed that, a neuron-derived peptide, the pituitary adenylate cyclase-activating polypeptide (PACAP) regulates the maximal velocity of GLAST-mediated Glu uptake and protein levels. PACAP-inactivating antibodies, PACAP receptor and PKA antagonists, and inhibitors of PACAP receptors (PAC1R) attenuate the stimulatory effect of neuron-conditioned medium (NCM) on GLAST expression (Figiel and Engele 2000). In this report, the authors also indicate that the increase in the expression of GLAST exerted by PACAP is dependent on the activation of PKA signaling pathway.

Other groups of activators of GLAST expression are steroids (Table 2). Specifically, estrogen and tamoxifen increase mRNA and protein levels of the transporter in primary astroglia cells (Karki et al. 2014a; Lee et al. 2009; Pawlak et al. 2005). The activation of both the G-protein-coupled receptor 30 (GPR30) and the estrogen nuclear receptors (ERα and β), particularly ERα, mediates the effects of estrogen and tamoxifen on GLAST regulation (Karki et al. 2014a; Lee et al. 2009; Sato et al. 2003). It should be noted that the transforming growth factor alpha (TGFα) is responsible for the effects of the estrogen on this glial transporter regulation (Karki et al. 2014a; Lee et al. 2009).

Other growth and neurotrophic factors that increase both GLAST mRNA and protein levels are epidermal growth factor (EGF), basic fibroblast growth factor (bFGF), insulin-like growth factor-1 (IGF-1), and glial cell line-derived neurotrophic factor (GDNF) (Bonde et al. 2003; Suzuki et al. 2001; Zelenaia et al. 2000). Specifically, growth factors exert their effect on GLAST through the Ras/mitogen extracellular signal-regulated kinase (Ras/MEK), PI3K/Akt, and PKA pathways (Lee et al. 2009; Dhandapani et al. 2005; Figiel et al. 2003), which activate NF-κB (Karki et al. 2014a; Figiel et al. 2003).

Contrary to mentioned above, it has been reported that activators of EAAT1/GLAST expression, such as EGF, TGFα, and PACAP, among others, failed to further increase the activity of a reporter gene carrying the full-length rat or human EAAT1/GLAST promoter. However, using reporter constructions containing the 3′-untranslated region (3′-UTR) of the human EAAT1 gene in addition to the human or rat full-length promoter instead of the factors mentioned above, it was restored the stimulatory effect of the activators of EAAT1/GLAST expression, revealing that

Table 2 Extracellular molecules that up- or downregulate transcription of EAAT1/GLAST

Stimulus	Cell type	Effect	Inhibitors of effect	References
Upregulation				
Glu	Astrocyte cultures	↑Protein levels and GLAST-mediated uptake	Inhibitors of AMPA/kainate receptors	Gegelashvili et al. (1996)
dbcAMP	Astrocyte cultures	↑mRNA, protein levels and GLAST-mediated uptake	Inhibitors of PKA	Gegelashvili et al. (1996), Schlag et al. (1998), Zelenaia et al. (2000), Swanson et al. (1997)
Group II mGluR	Astroglia cultures with NCM	↑mRNA and protein levels	Group II mGluR antagonist	Gegelashvili et al. (2000), Aronica et al. (2003)
TGFα	Astrocyte cultures	↑mRNA and protein levels	–	Zelenaia et al. (2000)
EGF	Astrocyte cultures	↑mRNA, protein levels and GLAST-mediated uptake	Inhibitors of PKC and tyrosine kinases	Suzuki et al. (2001)
Estrogen	Primary astroglia cells, human astrocyte cultures	↑mRNA, protein levels and GLAST-mediated uptake	Specific blocker of the TGF-β1 receptor	Karki et al. (2014a), Lee et al. (2009), Pawlak et al. (2005)
PACAP	Glia cultures	↑mRNA and protein levels	Antagonist for both PAC1 and VPAC1 receptors	Figiel and Engele (2000)
Tamoxifen	Primary astroglia cells	↑mRNA and protein levels	Specific blocker of the TGF-β1 receptor	Karki et al. (2014a), Lee et al. (2009) Pawlak et al. (2005)
bFGF	Astrocyte cultures	↑mRNA, protein levels and GLAST-mediated uptake	Inhibitors of tyrosine kinases	Suzuki et al. (2001)
IGF-1	Astrocyte cultures	↑mRNA, protein levels and GLAST-mediated uptake	Inhibitors of PKC	Suzuki et al. (2001)
GDNF	Hippocampal slice cultures	↑Expression	–	Bonde et al. (2003)
Downregulation				
3′-UTR	Astroglia cultures	↓Constitutive expression	Construct containing only the full-length promoter sequence	Unger et al. (2012)
Hypoxia	Astrocyte cultures	↓Basal expression	FGF2 and specific inhibitors of JAK/STAT	Raymond et al. (2011), Feng et al. (2015)

(continued)

Table 2 (continued)

Stimulus	Cell type	Effect	Inhibitors of effect	References
iGluR	BGC	↓*chglast* promoter activity	AMPA antagonist	López-Bayghen et al. (2003)
Insulin	BGC	↓*chglast* promoter activity	–	Poblete-Naredo et al. (2009)
Endothelins	Cortical astrocyte cultures	↓Protein levels	–	Rozyczka et al. (2004), Luo and Grammas (2010)
Arsenite	BGC	↓Expression and GLAST-mediated uptake	Inhibitors of PKC	Castro-Coronel et al. (2011)

Abbreviations: *3′-UTR* 3′-untranslated region, *AMPA* α-amino-3-hydroxy-5-methyl-4-isoxazolepropionic acid, *bFGF* basic fibroblast growth factor, *BGC* Bergmann glial cells, *dbcAMP* dibutyryl cyclic adenosine monophosphate, *EGF* epidermal growth factor, *FGF2* fibroblastic growth factor 2, *GDNF* glial cell line-derived neurotrophic factor, *Glu* glutamate, *IGF-1* insulin-like growth factor-1, *JAK* Janus kinase, *mGluR* metabotropic glutamate receptors, *NCM* neuron-conditioned media, *PACAP* pituitary adenylate cyclase-activating peptide, *PAC1* PACAP receptor, *PKA* protein kinase A, *PKC* protein kinase C, *STAT* signal transducer and activator of transcription, *TGFα* transforming growth factor alpha, *TNFα* tumor necrosis factor alpha, *VPAC1* vasoactive intestinal peptide receptor type 1

the human 3′-UTR acts as a transcriptional repressor of constitutive transcription of this transporter (Unger et al. 2012).

Curiously, it has been reported that under hypoxic conditions the expression of Glu transporters is reduced in astrocytes, unlike what happens under hypoxic conditions in which the Janus kinase (JAK)/signal transducer and activator of transcription (STAT) signaling pathway is inhibited, and the expression of GLAST is downregulated without affecting the expression of GLT-1 (Raymond et al. 2011; Feng et al. 2015). At this point, direct interaction between the GLAST promoter and STAT remains unclear. Taken together these findings, it is tempting to speculate the existence of different signaling cascades acting downstream of the growth factors that may regulate the transcription of GLAST. Other research groups have proposed diverse Glu receptors as downregulators of GLAST transcription. In this sense, Lyon and colleagues have shown that knocking down the expression of the group II mGluR subunit mGluR3 in mice, it was possible to reduce the GLAST mRNA and protein levels (Lyon et al. 2008). Furthermore, the use of a selective mGluR group I agonist downregulated GLAST protein expression in astrocytes cultured in the presence of growth factors (Aronica et al. 2003). Moreover, using cortical neuron cultures, it demonstrated a downregulation of GLAST expression in the presence of selective mGluR group I agonists (Gegelashvili et al. 2000). Further reports suggested that mGluRs regulate the expression of GLAST through the ERK/PI3K/NF-κB signaling pathway (Lin et al. 2014).

On the other hand, GLAST expression is also regulated by ionotropic Glu receptors (iGluR) as previously shown by López-Bayghen and collaborators (López-Bayghen et al. 2003). It was demonstrated that using an agonist of AMPA to stimulate iGluRs results in a substantial reduction of *chglast* promoter activity through the Ca^{2+}/PKC signaling pathway. In addition, it was described that exogenous expression of the subunits of the transcription factors, including protein-1 (AP-1), c-Jun, or c-Fos, reduces the *chglast* promoter activity (López-Bayghen et al. 2003; López-Bayghen and Ortega 2004). Further studies demonstrated that incubating BGC cultures with Glu and AMPA promotes in a dose- and time-dependent manner an increase of the YY1 transcription factor DNA binding to the GLAST promoter, whereas an overexpression of this transcription factor reduces GLAST mRNA levels, all related with transcriptional repression of the *chglast* promoter (Rosas et al. 2007). Moreover, Poblete-Naredo and coworkers showed an insulin-dependent GLAST downregulation, through an increase in YY1 binding to the *chglast* promoter (Poblete-Naredo et al. 2009). Additionally, it has been published that the incubation of cocultures of primary astrocytes with T cells incubated in the presence of TNFα reduces GLAST protein levels (Korn et al. 2005). Furthermore, using cortical astroglia cultures exposed to endothelins, it was observed a reduction of GLAST protein levels even in the presence of dbcAMP, PACAP, EGF, or TGFα (Rozyczka et al. 2004; Luo and Grammas 2010).

Recent findings have documented a GLAST expression downregulation after an exposure to environmental toxicants. Kwakye and coworkers suggested that exposing astrocytes to Mn^{2+} results in the reduction of GLAST expression; this is closely related to TNFα increase (Kwakye et al. 2015). Whereas, Lee and collaborators demonstrated that GLAST downregulation is reverted using a TNFα blocker (Lee et al. 2009). Moreover, exposure to arsenite seems to decrease GLAST expression and transport activity, as well as increase in transcription factors, such as Nrf2 and AP-1, binding to the GLAST promoter sequence (Castro-Coronel et al. 2011).

4.2 *Transcriptional and Translational Regulation of EAAT2/GLT-1*

The human gene product sequence, EAAT2 protein, has 574 amino acid residues in length. EAAT2 has a molecular weight of 62.1 KDa, and it has 95% identity with the corrected sequence (Kanner 1993) of rat GLT-1 (Pines et al. 1992). The EAAT2 gene (*SLC1A2*) was mapped to chromosome 11 bands p13-p12 (Li and Francke 1995); conversely, the mouse GLT-1 (*Slc1a2*) gene is allocated to the middle region of chromosome 2 (Kirschner et al. 1994). Similar to EAAT1, EAAT2 includes ten exons that range from 127 to 251 bp (Meyer et al. 1997); it does not have a TATA box in the proximal promoter (Su et al. 2003), lacks well-defined *cis* elements, and contains five specificity protein 1 (*Sp1*) sites and GC-rich repeats (Su et al. 2003), although it is unknown if these sites are necessary for GLT-1 expression. In accordance to Regan and coworkers, it has been described the existence of a close correlation between the expression of the reporter of the bacterial artificial chromosome (BAC), which contains the human *SLC1A2* gen, and GLT-1 expression in transgenic mice (Regan et al. 2007), suggesting that transcriptional control is similar between both species.

Pure astrocyte cultures express only GLAST; but in cocultures of astrocytes with neurons, expression of GLT-1 is induced in astrocytes, while expression of GLAST is slightly augmented (Gegelashvili et al. 1997; Swanson et al. 1997). Subsequently, several studies carried out in mouse astrocyte cultures have documented low levels of GLT-1; additionally, neurons induce GLT-1 expression (Apricò et al. 2004; Ghosh et al. 2016). This information strongly suggests the existence of neuronal soluble factors that increase Glu transporter protein and mRNA, although axonal interactions with astroglia also play a role in this effect (Zelenaia et al. 2000; Gegelashvili et al. 1997; Yang et al. 2009). In line with these reports, it has been proposed that upregulation of GLT-1 by NCM is dependent on the activation of the p42/44 MAP kinases via the tyrphostin-sensitive receptor tyrosine kinase (RTK) signaling pathway (Gegelashvili et al. 2000; Swanson et al. 1997). Together, NCM also phosphorylates to CREM-1 and ATF-1 transcription factors (Gegelashvili et al. 2000; Swanson et al. 1997); nonetheless, NCM-induced upregulation of GLT-1 is completely abolished by inhibitors of PI3K, tyrosine kinase, and NF-κB (Zelenaia et al. 2000; Swanson et al. 1997).

Fascinatingly, treatments with different growth factors in astrocyte/neuronal cocultures, and treating astrocytes with dbcAMP induce expression of GLT-1 (Swanson et al. 1997; Zelenaia et al. 2000; Schlag et al. 1998). All of these treatments are related with changes in astrocyte morphology from polygonal (fibroblast-like) shape to a process-bearing stellate shape that has been thought to reflect a more differentiated state. Continuing with dbcAMP, several research groups have reported that dbcAMP has a significant upregulation effect on GLT-1 (Swanson et al. 1997; Zelenaia et al. 2000; Schlag et al. 1998). Similar to what was found with NCM, the response caused by dbcAMP is blocked by inhibitors of PI3K and NF-κB (Zelenaia et al. 2000).

Among the soluble factors responsible for transporter upregulation, growth factors play an important role (Table 3). Zelenaia and colleagues demonstrated in cultured cortical astrocytes that both EGF and TGFα increase GLT-1 mRNA, protein levels, and GLT-1-mediated uptake. Inhibitors of NF-κB, PI3K, and receptor tyrosine kinase block the response of these growth factors (Zelenaia et al. 2000). Afterward, using a dominant-negative variant of Akt kinase, it was shown that the effect of EGF is reduced, while expression of constitutively active form of Akt mimics the effect of EGF (Li et al. 2006). These results are consistent with a growth factor/PI3K/Akt/NF-κB pathway regulating the transcription of GLT-1 transporter. Accordingly with this study, the laboratory group of Figiel reported similar stimulatory effects on GLT-1 expression with EGF and TGFα (Figiel et al. 2003). Conversely, no effect was observed in GLT-1 regulation with stimuli of insulin, bFGF, and nerve growth factor (NGF) (Zelenaia et al. 2000). There is controversy whether platelet-derived growth factor (PDGF), GDNF, and brain-derived neurotrophic factor (BDNF) regulate or not GLT-1 expression (Suzuki et al. 2001; Zelenaia et al. 2000; Dhandapani et al. 2005), although it has also been reported that neither of these factors have an effect on GLT-1 (Zelenaia et al. 2000; Figiel et al. 2003). Particularly, the effects of BDNF are blocked by pharmacological inhibition of the ERK/NF-κB signaling pathway (Rodriguez-Kern et al. 2003). These differ-

Table 3 Extracellular molecules that up- or downregulate transcription of EAAT2/GLT-1

Stimulus	Cell type	Effect	Inhibitors of effect	References
Upregulation				
NCM	Cortical astrocytes cocultured with neurons	↑Expression	Inhibitors of PI3K, tyrosine kinase, and NF-κB	Gegelashvili et al. (2000), Zelenaia et al. (2000), Swanson et al. (1997)
dbcAMP	dbcAMP-treated astrocyte cultures, cortical astrocytes cocultured with neurons	↑mRNA, protein levels and GLT-1-mediated uptake	Inhibitors of PI3K, PKA, and NF-κB	Schlag et al. (1998), Zelenaia et al. (2000), Figiel et al. (2003), Swanson et al. (1997)
EGF	Cortical astrocyte cultures, PHFA cultures	↑mRNA, protein levels and GLT-1-mediated uptake	Inhibitors of PI3K, tyrosine kinase, and NF-κB	Zelenaia et al. (2000), Figiel et al. (2003), Su et al. (2003), Li et al. (2006)
TGFα	Cortical astrocyte cultures, PHFA cultures	↑mRNA, protein levels and GLT-1-mediated uptake	Inhibitors of PI3K, tyrosine kinase, Akt, and NF-κB	Zelenaia et al. (2000), Figiel et al. (2003), Su et al. (2003)
PDGF	Primary astrocyte-enriched cultures, cortical astroglia cultures	↑mRNA and protein levels	Inhibitors of Akt	Zelenaia et al. (2000), Figiel et al. (2003)
GDNF	Hippocampal slice cultures	↑Expression	–	Bonde et al. (2003)
BDNF	Differentiated astrocyte cultures	↑Expression and GLT-1-mediated uptake	Inhibition of the ERK/NF-κB signaling pathway	Rodriguez-Kern et al. (2003)
PACAP	Cortical astroglia cultures, primary glia cultures	↑mRNA and protein levels	PACAP-38 antiserum, PACAP receptor antagonist, and PKC inhibitor	Figiel and Engele (2000), Figiel et al. (2003)
Estrogens	Primary astrocyte cultures	↑mRNA, protein levels, and GLT-1-mediated uptake	Inhibitors PKA and NF-κB	Karki et al. (2013, 2014a), Lee et al. (2012)
Glucocorticoids	Cortical astrocyte cultures	↑mRNA, protein levels, and GLT-1-mediated uptake	Antagonists of glucocorticoid or mineralocorticoid receptors	Zschocke et al. (2005)

(continued)

Table 3 (continued)

Stimulus	Cell type	Effect	Inhibitors of effect	References
ATP	Hippocampal slices	↑Expression and GLT-1-mediated uptake	MEK 1/2 inhibitors	Frizzo et al. (2007)
Adenosine	Astrocytic cell line C8-D1A	↑mRNA levels	A1 receptor-specific antagonists	Wu et al. (2011)
Downregulation				
TNFα	PHFA cultures	↓mRNA	Inhibitors of NF-κB activation	Su et al. (2003)
YY1	Primary astrocyte cultures	↓Basal expression	HDAC inhibitors	Karki et al. (2014b)
Endothelins	Cortical astrocyte cultures	↓mRNA and protein levels	Inhibitors of PKA	Rozyczka et al. (2004)
Dopamine	Striatal astrocyte cultures	↓mRNA and protein levels	D1 receptor antagonist	Brito et al. (2009)
Retinoic acid	Primary astrocyte cultures	↓mRNA and protein levels	RXR antagonists	Chan et al. (2012)
DHPG	Human glia cell cultures	↓Protein levels	Antagonists of mGluR1	Aronica et al. (2003)

Abbreviations: Akt a serine/threonine-protein kinase, *ATP* adenosine triphosphate, *BDNF* brain-derived neurotrophic factor, *dbcAMP* dibutyryl cyclic adenosine monophosphate, *DHPG* (S)-3,5-dihydroxyphenylglycine, *EGF* epidermal growth factor, *ERK* extracellular signal-regulated kinase, *GDNF* glial cell line-derived neurotrophic factor, *HDAC* histone deacetylase, *MEK* mitogen-activated protein kinase, *mGluR1* group I metabotropic glutamate receptors, *NCM* neuron-conditioned media, *NF-κB* nuclear factor-kappa B, *PACAP* pituitary adenylate cyclase-activating peptide, *PDGF* platelet-derived growth factor, *PHFA* primary human fetal astrocytes, *PI3K* phosphoinositide 3-kinase, *PKA* protein kinase A, *PKC* protein kinase C, *RXR* retinoic X receptor, *TGFα* transforming growth factor alpha, *TNFα* tumor necrosis factor alpha, *YY1* yin yang 1

ences could be explained by two possible reasons: first, different populations of astrocytes regulate differently to GLT-1, and second, the presence of neurons modifies the response of astrocytes to the growth factor stimulus.

Early studies showed the presence of the consensus binding sequence of NF-κB in the 5′-UTR region of the human EAAT2 gene (Meyer et al. 1996). To confirm these findings, years later, the EAAT2 promoter was cloned. Bioinformatic analysis of the EAAT2 promoter sequence established several regulatory transcription factor binding elements, including nuclear factor of activated T cells (NFAT), N-myc proto-oncogene protein (N-*myc*), and NF-κB, which contribute to expression and regulation of the transporter. At transcriptional level, TNFα factor decreases EAAT2 expression, whereas EGF and TGFα increase EAAT2 mRNA expression in primary human fetal astrocyte (PHFA) cultures (Su et al. 2003). Another important finding is the fact that overexpression of promoter-deleted constructions showed the involvement of NF-κB in EAAT2 transcriptional regulation (Su et al. 2003). It is clear a convergence between independent signaling pathways, similar to what was reported earlier in astrocyte cultures, is important in EAAT2 transcriptional control (Zelenaia et al. 2000).

Almost 17 years ago, a neuron-derived peptide, PACAP was determined to induce the expression of the transporter protein, as well as, glutamine synthetase (GS) in astroglia cultures. These effects were blocked by anti-PACAP-directed antibodies or with PAC1 antagonists (Figiel and Engele 2000). It should be noted that PACAP-dependent increase of GLT-1 is carried out through activation of PKA and PKC pathways (Figiel et al. 2003).

Recently, it has been proven that estrogens are involved in the GLT-1 regulation through ERα and β and GRP30. Particularly, when the estrogen receptors are activated, an increased binding to CREB (site -308) and NF-κB (sites -251, -272, and -583) in GLT-1 promoter is present (Karki et al. 2013, 2014a; Lee et al. 2012). According to these reports, it has been proposed that part of the response generated by estrogens is dependent on estrogen-dependent upregulation of TGFα, which functions as an autocrine factor capable of regulating GLT-1 expression via the MEK/ERK and PI3K/Akt signaling pathways (Karki et al. 2014a; Lee et al. 2012). With the studies carried out to date, the two mechanisms that have been proposed to explain the estrogen regulation of GLT-1 expression are GRP30/cAMP/PKA/CREB signaling pathway and pathways that activate TGFα, including TGFα/PI3K/Akt/NF-κB and MEK/ERK/NF-κB (Karki et al. 2013, 2014a; Lee et al. 2012).

Another group of hormones implicated in the EAAT2 regulation are the glucocorticoids. Although very little is known about this regulation, it has been suggested that glucocorticoids increase GLT-1 mRNA and protein levels (Zschocke et al. 2005). Specifically, antagonists of either the glucocorticoid or the mineralocorticoid receptors prevent the effect exerted by dexamethasone (Zschocke et al. 2005).

Interestingly, GLT-1 expression also increases both ATP and adenosine stimuli (Rosas et al. 2007; Poblete-Naredo et al. 2009) through P2Y (ATP receptors) and A1 (adenosine receptors) receptors. Probably the effect exerted by ATP could be explained via ATP/ERK/NF-κB signaling pathway, although this hypothesis needs to be demonstrated.

Concerning EAAT2 downregulation, it has been proposed that TNFα increases NF-κB binding to the GLT-1 promoter (Sitcheran et al. 2005). Contrary to what has been described above, when this transcription factor is activated by TNFα, it represses GLT-1. Together these results suggest a bidirectional regulation of GLT-1 via NF-κB. The response to this phenomenon was described using the EAAT2 promoter fragment constructs. Sitcheran and collaborators described that the TNFα-induced NF-κB activation is dependent on the recruitment of N-*myc*, while EGF-mediated activation is independent on IκB degradation (Sitcheran et al. 2005). This information allows us to consider that the interaction of NF-κB with other transcription factors may regulate the direction of NF-κB effect.

Another transcription factor regulated by TNFα is YY1, which represses basal GLT-1 expression (Karki et al. 2014b). As we know, NF-κB is a regulator itself of YY1 expression, since expression of p65 activates its promoter. In line with these findings, Karki and colleagues demonstrated that p65 is sufficient to activate GLT-1 promoter; nevertheless, p65 and YY1 co-expression inhibits the activation of this promoter (Karki et al. 2014b). Curiously, similar to N-*myc*, when YY1 is bound to the GLT-1 promoter, it changes the effect of NF-κB from activation to suppression.

Interestingly, endothelins and dopamine also decrease GLT-1 protein and mRNA levels. Particularly, an inhibitor of PKA represses the effect of the endothelins (Rozyczka et al. 2004), while D1 receptor mediates the effect of dopamine (Brito et al. 2009). Since D1 receptors are coupled to cAMP increase, probably both responses are mediated by the PKA signaling pathway.

Finally, several signals suppressing GLT-1 expression have been described. One of them is retinoic acid, which increases the binding of retinoid X receptor (RXR) to the retinoic acid response element (RARE) in the promoter of the transporter resulting in decreasing GLT-1 protein levels (Chan et al. 2012). On the other hand, DHPG also reduces GLT-1 protein levels, and its effects are blocked by antagonists of GRM1 (Aronica et al. 2003), suggesting metabotropic Glu receptors regulate GLT-1.

4.3 Regulation of EAAT3/EAAC1

The sequence of EAAT3 consists of 525 amino acids in length. This protein has a molecular weight of 57.2 KDa and 92% amino acid sequence identity with the rabbit sequence termed EAAC1 (Kanai and Hediger 1992). EAAC1 shares 66% amino acid sequence similarity with GLAST and 61% with GLT-1 (Malandro and Kilberg 1996). In situ fluorescence hybridization showed that the gene that encodes human EAAT3 (*SLC1A1*) is localized to chromosome 9 band p24 (Smith et al. 1994). Its mouse homolog is located in chromosome 19 at the centromere (Home – Gene – NCBI [Internet], https://www.ncbi.nlm.nih.gov/gene).

Numerous studies have proposed that EAAC1 may be more important for the synthesis of the glutathione (GSH, an important antioxidant to protect cells from oxidative damage) than for the clearance of neurotransmitter pools of Glu by importing cysteine (Aoyama and Nakaki 2013). According to this information, EAAC1 mediates 70–80% of neuronal cysteine uptake (Shanker et al. 2001), indicating that, in the brain, this transporter plays an essential role as a cysteine transporter, rather than a Glu transporter. Studies performed in mice lacking in EAAC1 have been used as a good model of chronic neuronal oxidative stress; these animals present diminished neuronal GSH levels, increased markers of neuronal oxidative stress, increased susceptibility to oxidant injury, and slow age-dependent reduction in overall brain size; these changes are blocked by N-acetylcysteine (NAC), a membrane-permeable cysteine precursor (Aoyama et al. 2006; Berman et al. 2011).

Therefore, EAAC1 deficiency leads to impaired GSH homeostasis and neurodegeneration. Accordingly, a conserved antioxidant responsive elements (AREs)-related sequence was found in the promoter of this transporter (Escartin et al. 2011). Nuclear factor erythroid 2-related factor 2 (Nrf2) is one of the transcription factors that bind to AREs. In C6 glioma cells, it has been described that both activators of Nrf2 (L-sulforaphane, SR; *tert*-butylhydroquinone, t-BHQ) and its overexpression resulted in EAAC1 transcriptional activation (Escartin et al. 2011). Escartin and colleagues also established that selective expression of Nrf2 in brain neurons increases both neuronal EAAC1 protein and GSH levels (Escartin et al. 2011).

In 2006, Ma and coworkers identified a binding sequence for the regulatory factor X1 (RFX1) in human EAAT3 promoter. Interestingly, RFX1 sequence was not present in the promoter regions of the genes encoding for EAAT1 and EAAT2. In C6 and SH-SY5Y cell lines, the authors demonstrated that transfection of RFX1 increases both, protein levels and the activity of the promoter of EAAT3 (Table 4). Knockout of RFX1 decreases EAAT3 expression in cultured rat cortical neurons (Ma et al. 2006). All these results suggest that the effects of RFX1 are specific on EAAT3.

Using C6 rat glioma cells, it was demonstrated that all-*trans*-retinoic acid (ATRA) treatment has a clear stimulatory effect on EAAC1 mRNA and protein levels (Bianchi et al. 2008). The effect is inhibited by LE540, a specific retinoic acid receptor (RAR) antagonist, while the specific RAR agonist, Am80, mimics the ATRA effect (Bianchi et al. 2009). Curiously, Am80 also causes an evident increase EAAC1 expression. This evidence suggests that ATRA-dependent stimulation of EAAC1 transport activity requires the interaction of the retinoid with a RAR receptor. It should be noted that this or similar sequences are not present in the human counterpart (Bianchi et al. 2009); therefore it is not clear if these effects are also present in humans. The β subtype of the RAR plays an important role in EAAC1 induction since its expression increases after ATRA treatment, suggesting that RARβ might be an intermediate of ATRA-dependent EAAC1 increase.

Finally, it is also interesting to note that GRM2 knockout mice shows a decrease in EAAC1 levels, pointing out a plausible role of GRM2 in EAAC1 gene expression regulation (Lyon et al. 2008).

For the regulation of the uptake activity, several signaling pathways modulate EAAC1 cell surface expression. In 1998, it was shown that EAAC1 is inserted in the plasma membrane surface in C6 glioma cells after PKC activation. This effect is associated with a redistribution of EAAC1 from subcellular compartments to the membrane (Davis et al. 1998). Using pharmacological approaches combined with downregulation of specific PKC subtypes, González and coworkers provided evidence to suggest that PCKα regulates EAAC1 redistribution (González et al. 2003); whereas PKCε regulates EAAC1 catalytic efficiency (González et al. 2002). Furthermore, activation of PKC rapidly increases (within minutes) EAAC1-mediated Glu uptake. PKC inhibitors block both PKC-dependent induction cell surface expression and uptake, while inhibition of PI3K abolishes the effect on cell surface expression but does not completely block the increase in uptake (Davis et al. 1998). All these observations indicate that PKC activation increases EAAC1-mediated uptake by augmenting the number of transporter molecules expressed at

Table 4 Extracellular molecules that up- or downregulate transcription of EAAT3/EAAC1

Stimulus	Cell type	Effect	Inhibitors of effect	References
Upregulation				
Nrf2	C6 glioma cells	↑Protein levels and EAAT3-mediated uptake	–	Escartin et al. (2011)
RFX1	C6 glioma cells, SH-SY5Y cell line	↑Protein levels and EAAT3-mediated uptake	RFX1 antisense oligonucleotides	Ma et al. (2006)
ATRA	C6 glioma cells	↑mRNA and protein levels	Specific RAR antagonist	Bianchi et al. (2009)
Downregulation				
Group II mGluR	*mGluR2*$^{-/-}$ mice	↓mRNA	–	Lyon et al. (2008)

Abbreviations: ATRA trans-retinoic acid, *mGluR* group II metabotropic glutamate receptors, *Nrf2* nuclear factor erythroid 2-related factor 2, *RAR* retinoic acid receptor, *RFX1* regulatory factor X1

the cell membrane and/or by altering the catalytic efficiency of the transporter. Similar effects in cell surface expression of EAAC1 were observed in C6 glioma cells when stimulated with PDGF (Sims et al. 2000). This growth factor stimulated PI3K activity and rapidly increased both, the activity and cell surface expression of EAAC1. The effects of PDGF were blocked by inhibitors of PI3K, but not by PKC inhibitors (Sims et al. 2000), suggesting that two independent but converging signaling pathways may be involved in the regulation of both the activity and cell surface expression of EAAC1. Sims and coworkers also evaluated other growth factors, including insulin, NGF, and EGF, which had no effect on Glu transport or effect as PI3K activity promoters (Sims et al. 2000). From these results it was concluded that these growth factors do not contribute to the regulation of EAAC1 activity.

In 2001, Lin and colleagues identified an EAAC1-associated protein, glutamate transporter-associated protein 3-18 (GTRAP3-18), which downregulates EAAC1 transport activity by lowering substrate affinity without altering EAAC1 trafficking (Lin et al. 2001). Later on, it was reported that another EAAC1-associated protein, syntaxin 1A, potentiates EAAC1 trafficking and inhibits EAAC1-mediated Glu transport without altering its substrate affinity (Yu et al. 2006). Finally, it has been confirmed that the glia-derived cholesterol regulates the activity and expression of EAAC1 through an increase of the total number of Glu transporter proteins and the number of Glu transport sites at the cell surface (Canolle et al. 2004).

4.4 Regulation of EAAT4

The amino acid sequence of EAAT4 exhibits 65%, 41%, and 48% amino acid identity to the human Glu transporters EAAT1, EAAT2, and EAAT3, respectively (Fairman et al. 1995; Malandro and Kilberg 1996). The human EAAT4 gene (*SLC1A6*) is located in chromosome 19 in the band 13.12, while the mouse gene is mapped in chromosome 10 in the centromeric region (Home – Gene – NCBI [Internet], https://www.ncbi.nlm.nih.gov/gene).

There are very few studies about the transcriptional regulation of the EAAT4 subtype of Glu transporter. To date, EAAT4 promoter studies are restricted to anatomical expression studies in vivo using EAAT4-eGFP BAC reporter transgenic mice (de Vivo et al. 2010) containing the human *SLC1A6* gene plus 107 kb of upstream sequence and 54 kb downstream of the last exon (Gincel et al. 2007), which indicate predominant expression of EAAT4 in neuronal cells with highest density of expression in cerebellar Purkinje cells. However, it is important to mention that the expression of the reporter protein correlates with EAAT4 protein, suggesting that there is evolutionary conservation of transcriptional regulation.

Additionally, EAAT4 and EAAT5 exhibit features of Glu-gated chloride channels (Storck et al. 1992; Pines et al. 1992; Kanai and Hediger 1992; Fairman et al. 1995; Arriza et al. 1997); nevertheless, chloride fluxes are not coupled with the translocation of Glu. In BT4C glioma cells, Glu promotes the redistribution of EAAT4 (Gegelashvili et al. 2000). This phenomenon can be explained by the substrate-dependent activation of the intrinsic chloride channels present in the transporter. These chloride channels are functionally associated with cytoskeletal elements (Ullrich and Sontheimer 1997), which may be involved in trafficking of EAAT4 to the cell surface. In line with these hypotheses, it has been mentioned that GTRAP41 and 48 positively modulate EAAT4, increasing uptake by enhancing the cell surface expression of the transporter. This modulation can be explained by two mechanisms: first, GTRAP41 is involved in the coupling of EAAT4 to the actin cytoskeleton, and second, GTRP48 may couple EAAT4 to the Rho GTPase signaling pathway (Jackson et al. 2001).

In 2001, Mitrovic and colleagues reported that Zn^{2+} inhibits the chloride conductance of EAAT4 without affecting the Glu transport rate, which indicates that Zn^{2+} binds in a different site than Glu (Mitrovic et al. 2001). In the same study, it was reported that the alignment of the amino acid sequences of the EAATs showed that the two histidine residues of EAAT1 that bind Zn^{2+} are conserved in EAAT4 sequence. Using site-directed mutagenesis, it was shown that mutation of either of the two histidine residues (H146 or H156) to alanine abolishes Zn^{2+} inhibition of the EAAT4 anion conductance (Mitrovic et al. 2001), demonstrating that Zn^{2+} interacts with EAAT4 at a similar site to that of EAAT1. It is interesting to note that, although the Zn^{2+} binding sites are similar on EAAT1 and EAAT4, the effects of Zn^{2+} are different. In the case of EAAT1, Zn^{2+} inhibits transport without affecting the chloride conductance (Vandenberg et al. 1998), whereas for EAAT4, Zn^{2+} has no effect on the rate of Glu transport but inhibits the chloride channel activity (Mitrovic et al. 2001).

It has been reported that EAAT4 mRNA expression is reduced in both, bipolar and major depressive disorders (McCullumsmith and Meador-Woodruff 2002), as well as a decrease in the expression levels of EAAT4 during chronic restraint stress (Zink et al. 2010), suggesting a role for the glutamatergic dysfunction in the pathophysiology of mood disorders.

4.5 Regulation of EAAT5

The EAAT5 subtype of Glu transporter is composed of 560 amino acid residues in length and has a molecular mass of 61 KDa. EAAT5 has 46% identity with EAAT1, 43% with EAAT4, 37% with EAAT3, and 36% identity with EAAT2 (Arriza et al. 1997). The human EAAT5 gene (*SLC1A7*) localizes to chromosome 1 band 32.3, and the mouse gene is mapped in the centromere of chromosome 4 (Home – Gene – NCBI [Internet], https://www.ncbi.nlm.nih.gov/gene).

Essentially, not much is known about the regulation of EAAT5; however, from the identification of this transporter, in 1997 it was reported that the amino acid residues at the EAAT5 C terminus correspond to a sequence motif found in synaptic membrane proteins: E-S or T-X-V-COOH (Arriza et al. 1997; Sheng 1996). Interestingly, the C termini interact with several PDZ (a molecular protein-binding motif) domains in PSD-95 protein (postsynaptic density-95 KDa) (Sheng 1996), which suggested that the transporter is a component of the signal transduction pathway. Its channel-like properties may indicate a role in retinal physiology different from neurotransmitter clearance.

Nowadays, the only report concerning EAAT5 regulation demonstrates that the co-expression of the transporter with the serum/glucocorticoid-regulated kinase 1 (SGK1, a serine/threonine kinase known to regulate Glu transport) and SGK3 increases EAAT5 activity by increasing cell surface abundance of the carrier in a *Xenopus laevis* (Boehmer et al. 2005), suggesting that these kinases are potent regulators of EAAT5.

5 Conclusion

Previously, astrocytes were recognized solely for their maintenance role in the regulation of brain homeostasis. Nowadays, that idea has changed, and we know that these cells play a crucial role in regulating neuronal activity and signal transmission. In particular, Glu transporters responsible for approximately 90% of the removal of this neurotransmitter from the synaptic cleft are astrocytic in nature, highlighting the importance of these cells. Currently, the number of diseases related to the imbalance of the glutamatergic system, which leads to serious pathophysiological scenarios, is increasing. Therefore, the interest and importance of the regulation of Glu transporters have also increased. The investigations carried out in relation to this regulation have allowed us to elucidate various mechanisms by which these transporters are regulated, including gene expression, transporter protein targeting and trafficking, as well as posttranslational modifications of the transporter protein. In this way, a better knowledge of the regulatory mechanisms of Glu transporters will allow us to develop new therapeutic strategies for a significant number of neurological diseases associated to glutamatergic neurotransmission.

Acknowledgments The work in the laboratory is supported by Conacyt-Mexico (255087) and *Soluciones para un México Verde S.A. de C.V.*; Donají Chi-Castañeda and Edna Suárez-Pozos are supported by Conacyt-México fellowships (376687 and 232180, respectively).

Conflict of Interest The author declares no conflicts of interest.

References

Aoyama K, Nakaki T. Neuroprotective properties of the excitatory amino acid carrier 1 (EAAC1). Amino Acids. 2013 Jul;45(1):133–42.

Aoyama K, Suh SW, Hamby AM, Liu J, Chan WY, Chen Y, et al. Neuronal glutathione deficiency and age-dependent neurodegeneration in the EAAC1 deficient mouse. Nat Neurosci. 2006 Jan;9(1):119–26.

Apricò K, Beart PM, Crawford D, O'Shea RD. Binding and transport of [3H](2S,4R)-4-methylglutamate, a new ligand for glutamate transporters, demonstrate labeling of EAAT1 in cultured murine astrocytes. J Neurosci Res. 2004 Mar 15;75(6):751–9.

Aronica E, Gorter JA, Ijlst-Keizers H, Rozemuller AJ, Yankaya B, Leenstra S, et al. Expression and functional role of mGluR3 and mGluR5 in human astrocytes and glioma cells: opposite regulation of glutamate transporter proteins. Eur J Neurosci. 2003 May;17(10):2106–18.

Arriza JL, Fairman WA, Wadiche JI, Murdoch GH, Kavanaugh MP, Amara SG. Functional comparisons of three glutamate transporter subtypes cloned from human motor cortex. J Neurosci. 1994 Sep;14(9):5559–69.

Arriza JL, Eliasof S, Kavanaugh MP, Amara SG. Excitatory amino acid transporter 5, a retinal glutamate transporter coupled to a chloride conductance. Proc Natl Acad Sci U S A. 1997 Apr 15;94(8):4155–60.

Benediktsson AM, Marrs GS, Tu JC, Worley PF, Rothstein JD, Bergles DE, et al. Neuronal activity regulates glutamate transporter dynamics in developing astrocytes. Glia. 2012 Feb;60(2):175–88.

Berger UV, Hediger MA. Distribution of the glutamate transporters GLAST and GLT-1 in rat circumventricular organs, meninges, and dorsal root ganglia. J Comp Neurol. 2000 Jun 5;421(3):385–99.

Berman AE, Chan WY, Brennan AM, Reyes RC, Adler BL, Suh SW, et al. N-acetylcysteine prevents loss of dopaminergic neurons in the EAAC1-/- mouse. Ann Neurol. 2011 Mar;69(3):509–20.

Bianchi MG, Gazzola GC, Tognazzi L, Bussolati O. C6 glioma cells differentiated by retinoic acid overexpress the glutamate transporter excitatory amino acid carrier 1 (EAAC1). Neuroscience. 2008 Feb;151(4):1042–52.

Bianchi MG, Gazzola GC, Cagnin S, Kagechika H, Bussolati O. The ATRA-dependent overexpression of the glutamate transporter EAAC1 requires RARbeta induction. Biochim Biophys Acta. 2009 Sep;1788(9):1861–8.

Bjørås M, Gjesdal O, Erickson JD, Torp R, Levy LM, Ottersen OP, et al. Cloning and expression of a neuronal rat brain glutamate transporter. Brain Res Mol Brain Res. 1996 Feb;36(1):163–8.

Boehmer C, Rajamanickam J, Schniepp R, Kohler K, Wulff P, Kuhl D, et al. Regulation of the excitatory amino acid transporter EAAT5 by the serum and glucocorticoid dependent kinases SGK1 and SGK3. Biochem Biophys Res Commun. 2005 Apr 8;329(2):738–42.

Bonde C, Sarup A, Schousboe A, Gegelashvili G, Noraberg J, Zimmer J. GDNF pre-treatment aggravates neuronal cell loss in oxygen-glucose deprived hippocampal slice cultures: a possible effect of glutamate transporter up-regulation. Neurochem Int. 2003;43(4-5):381–8.

Brito VI, Rozanski VE, Beyer C, Küppers E. Dopamine regulates the expression of the glutamate transporter GLT1 but not GLAST in developing striatal astrocytes. J Mol Neurosci. 2009 Nov;39(3):372–9.

Canolle B, Masmejean F, Melon C, Nieoullon A, Pisano P, Lortet S. Glial soluble factors regulate the activity and expression of the neuronal glutamate transporter EAAC1: implication of cholesterol. J Neurochem. 2004 Mar;88(6):1521–32.

Castro-Coronel Y, Del Razo LM, Huerta M, Hernandez-Lopez A, Ortega A, Lopez-Bayghen E. Arsenite exposure downregulates EAAT1/GLAST transporter expression in glial cells. Toxicol Sci. 2011 Aug 1;122(2):539–50.

Chan TJ, Her LS, Liaw HJ, Chen MC, Tzeng SF. Retinoic acid mediates the expression of glutamate transporter-1 in rat astrocytes through genomic RXR action and non-genomic protein kinase C signaling pathway. J Neurochem. 2012 May;121(4):537–50.

Conti F, DeBiasi S, Minelli A, Rothstein JD, Melone M. EAAC1, a high-affinity glutamate tranporter, is localized to astrocytes and gabaergic neurons besides pyramidal cells in the rat cerebral cortex. Cereb Cortex. 1998 Mar;8(2):108–16.

Danbolt NC. Glutamate uptake. Prog Neurobiol. 2001;65(1):1–105.

Davis KE, Straff DJ, Weinstein EA, Bannerman PG, Correale DM, Rothstein JD, et al. Multiple signaling pathways regulate cell surface expression and activity of the excitatory amino acid carrier 1 subtype of Glu transporter in C6 glioma. J Neurosci. 1998 Apr 1;18(7):2475–85.

Dehnes Y, Chaudhry FA, Ullensvang K, Lehre KP, Storm-Mathisen J, Danbolt NC. The glutamate transporter EAAT4 in rat cerebellar Purkinje cells: a glutamate-gated chloride channel concentrated near the synapse in parts of the dendritic membrane facing astroglia. J Neurosci. 1998 May 15;18(10):3606–19.

Derouiche A. Possible role of the Müller cell in uptake and metabolism of glutamate in the mammalian outer retina. Vis Res. 1996 Dec;36(24):3875–8.

Derouiche A, Rauen T. Coincidence of L-glutamate/L-aspartate transporter (GLAST) and glutamine synthetase (GS) immunoreactions in retinal glia: evidence for coupling of GLAST and GS in transmitter clearance. J Neurosci Res. 1995 Sep 1;42(1):131–43.

Dhandapani KM, Wade FM, Mahesh VB, Brann DW. Astrocyte-derived transforming growth factor-β mediates the neuroprotective effects of 17β-estradiol: involvement of nonclassical genomic signaling pathways. Endocrinology. 2005 Jun;146(6):2749–59.

Domingues AM de J, Taylor M, Fern R. Glia as transmitter sources and sensors in health and disease. Neurochem Int. 2010 Nov;57(4):359–66.

Durand GM, Kovalchuk Y, Konnerth A. Long-term potentiation and functional synapse induction in developing hippocampus. Nature. 1996 May 2;381(6577):71–5.

Escartin C, Joon Won S, Malgorn C, Auregan G, Berman AE, Chen P-C, et al. Nuclear factor erythroid 2-related factor 2 facilitates neuronal glutathione synthesis by upregulating neuronal excitatory amino acid transporter 3 expression. J Neurosci. 2011 May 18;31(20):7392–401.

Eulenburg V, Gomeza J. Neurotransmitter transporters expressed in glial cells as regulators of synapse function. Brain Res Rev. 2010;63(1):103–12.

Fairman WA, Vandenberg RJ, Arriza JL, Kavanaugh MP, Amara SG. An excitatory amino-acid transporter with properties of a ligand-gated chloride channel. Nature. 1995 Jun 15;375(6532):599–603.

Feng D, Guo B, Liu G, Wang B, Wang W, Gao G, et al. FGF2 alleviates PTSD symptoms in rats by restoring GLAST function in astrocytes via the JAK/STAT pathway. Eur Neuropsychopharmacol. 2015 Aug;25(8):1287–99.

Figiel M, Engele J. Pituitary adenylate cyclase-activating polypeptide (PACAP), a neuron-derived peptide regulating glial glutamate transport and metabolism. J Neurosci. 2000 May 15;20(10):3596–605.

Figiel M, Maucher T, Rozyczka J, Bayatti N, Engele J. Regulation of glial glutamate transporter expression by growth factors. Exp Neurol. 2003 Sep;183(1):124–35.

Fonnum F. Glutamate: a neurotransmitter in mammalian brain. J Neurochem. 1984 Jan;42(1):1–11.

Frizzo ME, Frizzo JK, Amadio S, Rodrigues JM, Perry ML, Bernardi G, et al. Extracellular adenosine triphosphate induces glutamate transporter-1 expression in hippocampus. Hippocampus. 2007 Apr;17(4):305–15.

Furness DN, Lehre KP. Immunocytochemical localization of a high-affinity glutamate-aspartate transporter, GLAST, in the rat and guinea-pig cochlea. Eur J Neurosci. 1997 Sep;9(9):1961–9.

Furuta A, Rothstein JD, Martin LJ. Glutamate transporter protein subtypes are expressed differentially during rat CNS development. J Neurosci. 1997 Nov 1;17(21):8363–75.

Gegelashvili G, Bjerrum OJ. High-affinity glutamate transporters in chronic pain: an emerging therapeutic target. J Neurochem. 2014 Dec;131(6):712–30.

Gegelashvili G, Civenni G, Racagni G, Danbolt NC, Schousboe I, Schousboe A. Glutamate receptor agonists up-regulate glutamate transporter GLAST in astrocytes. Neuroreport. 1996 Dec 20;8(1):261–5.

Gegelashvili G, Danbolt NC, Schousboe A. Neuronal soluble factors differentially regulate the expression of the GLT1 and GLAST glutamate transporters in cultured astroglia. J Neurochem. 1997 Dec;69(6):2612–5.

Gegelashvili G, Dehnes Y, Danbolt NC, Schousboe A. The high-affinity glutamate transporters GLT1, GLAST, and EAAT4 are regulated via different signalling mechanisms. Neurochem Int. 2000;37(2-3):163–70.

Ghosh M, Lane M, Krizman E, Sattler R, Rothstein JD, Robinson MB. The transcription factor Pax6 contributes to the induction of GLT-1 expression in astrocytes through an interaction with a distal enhancer element. J Neurochem. 2016 Jan;136(2):262–75.

Gincel D, Regan MR, Jin L, Watkins AM, Bergles DE, Rothstein JD. Analysis of cerebellar Purkinje cells using EAAT4 glutamate transporter promoter reporter in mice generated via bacterial artificial chromosome-mediated transgenesis. Exp Neurol. 2007 Jan;203(1):205–12.

González MI, Kazanietz MG, Robinson MB. Regulation of the neuronal glutamate transporter excitatory amino acid carrier-1 (EAAC1) by different protein kinase C subtypes. Mol Pharmacol. 2002 Oct;62(4):901–10.

González MI, Bannerman PG, Robinson MB. Phorbol myristate acetate-dependent interaction of protein kinase Calpha and the neuronal glutamate transporter EAAC1. J Neurosci. 2003 Jul 2;23(13):5589–93.

Hagiwara T, Tanaka K, Takai S, Maeno-Hikichi Y, Mukainaka Y, Wada K. Genomic organization, promoter analysis, and chromosomal localization of the gene for the mouse glial high-affinity glutamate transporter Slc1a3. Genomics. 1996;33(3):508–15.

He Y, Janssen WG, Rothstein JD, Morrison JH. Differential synaptic localization of the glutamate transporter EAAC1 and glutamate receptor subunit GluR2 in the rat hippocampus. J Comp Neurol. 2000 Mar 13;418(3):255–69.

Headley PM, Grillner S. Excitatory amino acids and synaptic transmission: the evidence for a physiological function. Trends Pharmacol Sci. 1990 May;11(5):205–11.

Holmseth S, Dehnes Y, Huang YH, Follin-Arbelet VV, Grutle NJ, Mylonakou MN, et al. The density of EAAC1 (EAAT3) glutamate transporters expressed by neurons in the mammalian CNS. J Neurosci. 2012 Apr 25;32(17):6000–13.

Home – Gene – NCBI [Internet]. Available from: https://www.ncbi.nlm.nih.gov/gene.

Jackson M, Song W, Liu MY, Jin L, Dykes-Hoberg M, Lin CG, et al. Modulation of the neuronal glutamate transporter EAAT4 by two interacting proteins. Nature. 2001 Mar 1;410(6824):89–93.

Kanai Y, Hediger MA. Primary structure and functional characterization of a high-affinity glutamate transporter. Nature. 1992 Dec 3;360(6403):467–71.

Kanai Y, Smith CP, Hediger MA. A new family of neurotransmitter transporters: the high-affinity glutamate transporters. FASEB J. 1993 Dec;7(15):1450–9.

Kanner BI. Glutamate transporters from brain. A novel neurotransmitter transporter family. FEBS Lett. 1993 Jun 28;325(1-2):95–9.

Karki P, Webb A, Smith K, Lee K, Son DS, Aschner M, et al. cAMP response element-binding protein (CREB) and nuclear factor κB mediate the tamoxifen-induced up-regulation of glutamate transporter 1 (GLT-1) in rat astrocytes. J Biol Chem. 2013 Oct 4;288(40):28975–86.

Karki P, Webb A, Zerguine A, Choi J, Son DS, Lee E. Mechanism of raloxifene-induced upregulation of glutamate transporters in rat primary astrocytes. Glia. 2014a Aug;62(8):1270–83.

Karki P, Webb A, Smith K, Johnson J, Lee K, Son D-S, et al. Yin Yang 1 is a repressor of glutamate transporter EAAT2, and it mediates manganese-induced decrease of EAAT2 expression in astrocytes. Mol Cell Biol. 2014b Apr;34(7):1280–9.

Kim SY, Choi SY, Chao W, Volsky DJ. Transcriptional regulation of human excitatory amino acid transporter 1 (EAAT1): cloning of the EAAT1 promoter and characterization of its basal and inducible activity in human astrocytes. J Neurochem. 2003 Dec;87(6):1485–98.

Kirschner MA, Copeland NG, Gilbert DJ, Jenkins NA, Amara SG. Mouse excitatory amino acid transporter EAAT2: isolation, characterization, and proximity to neuroexcitability loci on mouse chromosome 2. Genomics. 1994 Nov;24(2):218–24.

Komuro H, Rakic P. Modulation of neuronal migration by NMDA receptors. Science. 1993 Apr 2;260(5104):95–7.

Korn T, Magnus T, Jung S. Autoantigen specific T cells inhibit glutamate uptake in astrocytes by decreasing expression of astrocytic glutamate transporter GLAST: a mechanism mediated by tumor necrosis factor-alpha. FASEB J. 2005 Nov;19(13):1878–80.

Kwakye G, Paoliello M, Mukhopadhyay S, Bowman A, Aschner M. Manganese-induced parkinsonism and Parkinson's disease: shared and distinguishable features. Int J Environ Res Public Health. 2015 Jul 6;12(7):7519–40.

Lauriat TL, McInnes LA. EAAT2 regulation and splicing: relevance to psychiatric and neurological disorders. Mol Psychiatry. 2007 Dec;12(12):1065–78.

Lee E-SY, Sidoryk M, Jiang H, Yin Z, Aschner M. Estrogen and tamoxifen reverse manganese-induced glutamate transporter impairment in astrocytes. J Neurochem. 2009 Jul;110(2):530–44.

Lee E, Sidoryk-Wêgrzynowicz M, Wang N, Webb A, Son DS, Lee K, et al. GPR30 regulates glutamate transporter GLT-1 expression in rat primary astrocytes. J Biol Chem. 2012 Aug 3;287(32):26817–28.

Lehre KP, Danbolt NC. The number of glutamate transporter subtype molecules at glutamatergic synapses: chemical and stereological quantification in young adult rat brain. J Neurosci. 1998 Nov 1;18(21):8751–7.

Lehre KP, Davanger S, Danbolt NC. Localization of the glutamate transporter protein GLAST in rat retina. Brain Res. 1997 Jan 2;744(1):129–37.

Levy LM, Warr O, Attwell D. Stoichiometry of the glial glutamate transporter GLT-1 expressed inducibly in a Chinese hamster ovary cell line selected for low endogenous Na^+-dependent glutamate uptake. J Neurosci. 1998 Dec 1;18(23):9620–8.

Li X, Francke U. Assignment of the gene SLC1A2 coding for the human glutamate transporter EAAT2 to human chromosome 11 bands p13-p12. Cytogenet Cell Genet. 1995;71(3):212–3.

Li L-B, Toan SV, Zelenaia O, Watson DJ, Wolfe JH, Rothstein JD, et al. Regulation of astrocytic glutamate transporter expression by Akt: evidence for a selective transcriptional effect on the GLT-1/EAAT2 subtype. J Neurochem. 2006 May;97(3):759–71.

Lin CI, Orlov I, Ruggiero AM, Dykes-Hoberg M, Lee A, Jackson M, et al. Modulation of the neuronal glutamate transporter EAAC1 by the interacting protein GTRAP3-18. Nature. 2001 Mar 1;410(6824):84–8.

Lin CH, You JR, Wei KC, Gean PW. Stimulating ERK/PI3K/NFκB signaling pathways upon activation of mGluR2/3 restores OGD-induced impairment in glutamate clearance in astrocytes. Eur J Neurosci. 2014 Jan;39(1):83–96.

López-Bayghen E, Ortega A. Glutamate-dependent transcriptional regulation of GLAST: role of PKC. J Neurochem. 2004 Oct;91(1):200–9.

López-Bayghen E, Espinoza-Rojo M, Ortega A. Glutamate down-regulates GLAST expression through AMPA receptors in Bergmann glial cells. Brain Res Mol Brain Res. 2003 Jul 4;115(1):1–9.

Luo J, Grammas P. Endothelin-1 is elevated in Alzheimer's disease brain microvessels and is neuroprotective. J Alzheimers Dis. 2010;21(3):887–96.

Lyon L, Kew JNC, Corti C, Harrison PJ, Burnet PWJ. Altered hippocampal expression of glutamate receptors and transporters in GRM2 and GRM3 knockout mice. Synapse. 2008 Nov;62(11):842–50.

Ma K, Zheng S, Zuo Z. The transcription factor regulatory factor X1 increases the expression of neuronal glutamate transporter type 3. J Biol Chem. 2006 Jul 28;281(30):21250–5.

Malandro MS, Kilberg MS. Molecular biology of mammalian amino acid transporters. Annu Rev Biochem. 1996;65:305–36.

Maragakis NJ, Rothstein JD. Glutamate transporters: animal models to neurologic disease. Neurobiol Dis. 2004 Apr;15(3):461–73.
Massie A, Cnops L, Smolders I, McCullumsmith R, Kooijman R, Kwak S, et al. High-affinity Na^+/K^+-dependent glutamate transporter EAAT4 is expressed throughout the rat fore- and midbrain. J Comp Neurol. 2008 Nov 10;511(2):155–72.
McCullumsmith RE, Meador-Woodruff JH. Striatal excitatory amino acid transporter transcript expression in schizophrenia, bipolar disorder, and major depressive disorder. Neuropsychopharmacology. 2002 Mar;26(3):368–75.
McEntee WJ, Crook TH. Glutamate: its role in learning, memory, and the aging brain. Psychopharmacology. 1993;111(4):391–401.
Meyer T, Meyer B, Sitte W, Ludolph AC, Speer A, Küther G. The glial glutamate transporter complementary DNA in patients with amyotrophic lateral sclerosis. Ann Neurol. 1996 Sep;40(3):456–9.
Meyer T, Ludolph AC, Morkel M, Hagemeier C, Speer A. Genomic organization of the human excitatory amino acid transporter gene GLT-1. Neuroreport. 1997 Feb 10;8(3):775–7.
Mitrovic AD, Plesko F, Vandenberg RJ. Zn(2+) inhibits the anion conductance of the glutamate transporter EEAT4. J Biol Chem. 2001 Jul 13;276(28):26071–6.
Müller F, Demény MA, Tora L. New problems in RNA polymerase II transcription initiation: matching the diversity of core promoters with a variety of promoter recognition factors. J Biol Chem. 2007 May 18;282(20):14685–9.
Murphy-Royal C, Dupuis JP, Varela JA, Panatier A, Pinson B, Baufreton J, et al. Surface diffusion of astrocytic glutamate transporters shapes synaptic transmission. Nat Neurosci. 2015 Jan 12;18(2):219–26.
Nieoullon A, Canolle B, Masmejean F, Guillet B, Pisano P, Lortet S. The neuronal excitatory amino acid transporter EAAC1/EAAT3: does it represent a major actor at the brain excitatory synapse? J Neurochem. 2006 Aug;98(4):1007–18.
Pawlak J, Brito V, Küppers E, Beyer C. Regulation of glutamate transporter GLAST and GLT-1 expression in astrocytes by estrogen. Mol Brain Res. 2005 Jul 29;138(1):1–7.
Pines G, Danbolt NC, Bjørås M, Zhang Y, Bendahan A, Eide L, et al. Cloning and expression of a rat brain L-glutamate transporter. Nature. 1992 Dec 3;360(6403):464–7.
Poblete-Naredo I, Angulo C, Hernández-Kelly L, López-Bayghen E, Aguilera J, Ortega A. Insulin-dependent regulation of GLAST/EAAT1 in Bergmann glial cells. Neurosci Lett. 2009 Feb 20;451(2):134–8.
Pow DV, Barnett NL. Changing patterns of spatial buffering of glutamate in developing rat retinae are mediated by the Müller cell glutamate transporter GLAST. Cell Tissue Res. 1999 Jul;297(1):57–66.
Pow DV, Barnett NL. Developmental expression of excitatory amino acid transporter 5: a photoreceptor and bipolar cell glutamate transporter in rat retina. Neurosci Lett. 2000 Feb 11;280(1):21–4.
Rabacchi S, Bailly Y, Delhaye-Bouchaud N, Mariani J. Involvement of the N-methyl D-aspartate (NMDA) receptor in synapse elimination during cerebellar development. Science. 1992 Jun 26;256(5065):1823–5.
Rauen T, Rothstein JD, Wässle H. Differential expression of three glutamate transporter subtypes in the rat retina. Cell Tissue Res. 1996 Dec;286(3):325–36.
Rauen T, Taylor WR, Kuhlbrodt K, Wiessner M. High-affinity glutamate transporters in the rat retina: a major role of the glial glutamate transporter GLAST-1 in transmitter clearance. Cell Tissue Res. 1998 Jan;291(1):19–31.
Raymond M, Li P, Mangin J-M, Huntsman M, Gallo V. Chronic perinatal hypoxia reduces glutamate–aspartate transporter function in astrocytes through the Janus Kinase/signal transducer and activator of transcription pathway. J Neurosci. 2011;31(49):17864–71.
Regan MR, Huang YH, Kim YS, Dykes-Hoberg MI, Jin L, Watkins AM, et al. Variations in promoter activity reveal a differential expression and physiology of glutamate transporters by glia in the developing and mature CNS. J Neurosci. 2007 Jun 20;27(25):6607–19.

Robinson MB. Regulated trafficking of neurotransmitter transporters: common notes but different melodies. J Neurochem. 2002 Jan;80(1):1–11.
Robinson MB. Acute regulation of sodium-dependent glutamate transporters: a focus on constitutive and regulated trafficking. In:Neurotransmitter transporters. Berlin/Heidelberg: Springer-Verlag; 2006. p. 251–75.
Rodriguez-Kern A, Gegelashvili M, Schousboe A, Zhang J, Sung L, Gegelashvili G. Beta-amyloid and brain-derived neurotrophic factor, BDNF, up-regulate the expression of glutamate transporter GLT-1/EAAT2 via different signaling pathways utilizing transcription factor NF-kappaB. Neurochem Int. 2003;43(4-5):363–70.
Rosas S, Vargas MA, López-Bayghen E, Ortega A. Glutamate-dependent transcriptional regulation of GLAST/EAAT1: a role for YY1. J Neurochem. 2007 May;101(4):1134–44.
Rossi DJ, Slater NT. The developmental onset of NMDA receptor-channel activity during neuronal migration. Neuropharmacology. 1993 Nov;32(11):1239–48.
Rothstein JD, Martin L, Levey AI, Dykes-Hoberg M, Jin L, Wu D, et al. Localization of neuronal and glial glutamate transporters. Neuron. 1994 Sep;13(3):713–25.
Rozyczka J, Figiel M, Engele J. Endothelins negatively regulate glial glutamate transporter expression. Brain Pathol. 2004 Oct;14(4):406–14.
Sato K, Matsuki N, Ohno Y, Nakazawa K. Estrogens inhibit l-glutamate uptake activity of astrocytes via membrane estrogen receptor alpha. J Neurochem. 2003 Sep;86(6):1498–505.
Schlag BD, Vondrasek JR, Munir M, Kalandadze A, Zelenaia OA, Rothstein JD, et al. Regulation of the glial Na^+-dependent glutamate transporters by cyclic AMP analogs and neurons. Mol Pharmacol. 1998 Mar;53(3):355–69.
Shanker G, Allen JW, Mutkus LA, Aschner M. The uptake of cysteine in cultured primary astrocytes and neurons. Brain Res. 2001 Jun 1;902(2):156–63.
Shashidharan P, Huntley GW, Murray JM, Buku A, Moran T, Walsh MJ, et al. Immunohistochemical localization of the neuron-specific glutamate transporter EAAC1 (EAAT3) in rat brain and spinal cord revealed by a novel monoclonal antibody. Brain Res. 1997 Oct 31;773(1–2):139–48.
Sheng M. PDZs and receptor/channel clustering: rounding up the latest suspects. Neuron. 1996 Oct;17(4):575–8.
Shin JW, Nguyen KTD, Pow DV, Knight T, Buljan V, Bennett MR, et al. Distribution of glutamate transporter GLAST in membranes of cultured astrocytes in the presence of glutamate transport substrates and ATP. Neurochem Res. 2009 Oct 8;34(10):1758–66.
Sims KD, Straff DJ, Robinson MB. Platelet-derived growth factor rapidly increases activity and cell surface expression of the EAAC1 subtype of glutamate transporter through activation of phosphatidylinositol 3-kinase. J Biol Chem. 2000 Feb 18;275(7):5228–37.
Sitcheran R, Gupta P, Fisher PB, Baldwin AS. Positive and negative regulation of EAAT2 by NF-κB: a role for N-myc in TNFα-controlled repression. EMBO J. 2005 Feb 9;24(3):510–20.
Smith CP, Weremowicz S, Kanai Y, Stelzner M, Morton CC, Hediger MA. Assignment of the gene coding for the human high-affinity glutamate transporter EAAC1 to 9p24: potential role in dicarboxylic aminoaciduria and neurodegenerative disorders. Genomics. 1994 Mar 15;20(2):335–6.
Stoffel W, Sasse J, Düker M, Müller R, Hofmann K, Fink T, et al. Human high affinity, Na^+-dependent L-glutamate/L-aspartate transporter GLAST-1 (EAAT-1): gene structure and localization to chromosome 5p11-p12. FEBS Lett. 1996;386(2):189–93.
Storck T, Schulte S, Hofmann K, Stoffel W. Structure, expression, and functional analysis of a Na(+)-dependent glutamate/aspartate transporter from rat brain. Proc Natl Acad Sci U S A. 1992 Nov 15;89(22):10955–9.
Su ZZ, Leszczyniecka M, Kang DC, Sarkar D, Chao W, Volsky DJ, et al. Insights into glutamate transport regulation in human astrocytes: cloning of the promoter for excitatory amino acid transporter 2 (EAAT2). Proc Natl Acad Sci. 2003 Feb 18;100(4):1955–60.
Suzuki K, Ikegaya Y, Matsuura S, Kanai Y, Endou H, Matsuki N. Transient upregulation of the glial glutamate transporter GLAST in response to fibroblast growth factor, insulin-like growth factor and epidermal growth factor in cultured astrocytes. J Cell Sci. 2001 Oct;114(Pt 20):3717–25.

Swanson RA, Liu J, Miller JW, Rothstein JD, Farrell K, Stein BA, et al. Neuronal regulation of glutamate transporter subtype expression in astrocytes. J Neurosci. 1997 Feb 1;17(3):932–40.
Takai S, Yamada K, Kawakami H, Tanaka K, Nakamura S. Localization of the gene (SLC1A3) encoding human glutamate transporter (GluT-1) to 5p13 by fluorescence in situ hybridization. Cytogenet Cell Genet. 1995;69(3-4):209–10.
Takatsuru Y, Iino M, Tanaka K, Ozawa S. Contribution of glutamate transporter GLT-1 to removal of synaptically released glutamate at climbing fiber-Purkinje cell synapses. Neurosci Lett. 2007 Jun 8;420(1):85–9.
Takumi Y, Matsubara A, Danbolt NC, Laake JH, Storm-Mathisen J, Usami S, et al. Discrete cellular and subcellular localization of glutamine synthetase and the glutamate transporter GLAST in the rat vestibular end organ. Neuroscience. 1997 Aug;79(4):1137–44.
Ullrich N, Sontheimer H. Cell cycle-dependent expression of a glioma-specific chloride current: proposed link to cytoskeletal changes. Am J Phys. 1997 Oct;273(4 Pt 1):C1290–7.
Unger T, Lakowa N, Bette S, Engele J. Transcriptional regulation of the GLAST/EAAT-1 gene in rat and man. Cell Mol Neurobiol. 2012 May 18;32(4):539–47.
Vandenberg RJ, Mitrovic AD, Johnston GA. Molecular basis for differential inhibition of glutamate transporter subtypes by zinc ions. Mol Pharmacol. 1998 Jul;54(1):189–96.
de Vivo L, Melone M, Bucci G, Rothstein JD, Conti F. Quantitative analysis of EAAT4 promoter activity in neurons and astrocytes of mouse somatic sensory cortex. Neurosci Lett. 2010 Apr 19;474(1):42–5.
Wu J, Lee MR, Kim T, Johng S, Rohrback S, Kang N, et al. Regulation of ethanol-sensitive EAAT2 expression through adenosine A1 receptor in astrocytes. Biochem Biophys Res Commun. 2011 Mar 4;406(1):47–52.
Yang Y, Gozen O, Watkins A, Lorenzini I, Lepore A, Gao Y, et al. Presynaptic regulation of astroglial excitatory neurotransmitter transporter GLT1. Neuron. 2009 Mar;61(6):880–94.
Yu Y-X, Shen L, Xia P, Tang Y-W, Bao L, Pei G. Syntaxin 1A promotes the endocytic sorting of EAAC1 leading to inhibition of glutamate transport. J Cell Sci. 2006 Sep 15;119(Pt 18):3776–87.
Zelenaia O, Schlag BD, Gochenauer GE, Ganel R, Song W, Beesley JS, et al. Epidermal growth factor receptor agonists increase expression of glutamate transporter GLT-1 in astrocytes through pathways dependent on phosphatidylinositol 3-kinase and transcription factor NF-kappaB. Mol Pharmacol. 2000 Apr;57(4):667–78.
Zerangue N, Kavanaugh MP. Flux coupling in a neuronal glutamate transporter. Nature. 1996 Oct 17;383(6601):634–7.
Zink M, Vollmayr B, Gebicke-Haerter PJ, Henn FA. Reduced expression of glutamate transporters vGluT1, EAAT2 and EAAT4 in learned helpless rats, an animal model of depression. Neuropharmacology. 2010 Feb;58(2):465–73.
Zschocke J, Bayatti N, Clement AM, Witan H, Figiel M, Engele J, et al. Differential promotion of glutamate transporter expression and function by glucocorticoids in astrocytes from various brain regions. J Biol Chem. 2005 Oct 14;280(41):34924–32.

Glutamate Transport System as a Novel Therapeutic Target in Chronic Pain: Molecular Mechanisms and Pharmacology

Georgi Gegelashvili and Ole Jannik Bjerrum

Abstract The vast majority of peripheral neurons sensing noxious stimuli and conducting pain signals to the dorsal horn of the spinal cord utilize glutamate as a chemical transmitter of excitation. High-affinity glutamate transporter subtypes GLAST/EAAT1, GLT1/EAAT2, EAAC1/EAAT3, and EAAT4, differentially expressed on sensory neurons, postsynaptic spinal interneurons, and neighboring glia, ensure fine modulation of glutamate neurotransmission in the spinal cord. The glutamate transport system seems to play important roles in molecular mechanisms underlying chronic pain and analgesia. Downregulation of glutamate transporters (GluTs) often precedes or occurs simultaneously with development of hypersensitivity to thermal or tactile stimuli in various models of chronic pain. Moreover, antisense knockdown or pharmacological inhibition of these membrane proteins can induce or aggravate pain. In contrast, upregulation of GluTs by positive pharmacological modulators or by viral gene transfer to the spinal cord can reverse the development of such pathological hypersensitivity. Furthermore, some multi-target drugs displaying analgesic properties (e.g., tricyclic antidepressant amitriptyline, riluzole, anticonvulsant valproate, tetracycline antibiotic minocycline, β-lactam antibiotic ceftriaxone and its structural analog devoid of antibacterial activity, clavulanic acid) can significantly increase the spinal glutamate uptake. Thus, mounting evidence points at GluTs as prospective therapeutic target for chronic pain treatment. However, design and development of new analgesics based on the modulation of glutamate uptake will require more precise knowledge of molecular mechanisms underlying physiological or aberrant functioning of this transport system in the spinal cord.

G. Gegelashvili (✉)
Institute of Chemical Biology, Ilia State University,
K. Cholokashvili Ave 3/5, 0162 Tbilisi, Georgia

Department of Drug Design and Pharmacology, Faculty of Health and Medical Sciences,
University of Copenhagen, Copenhagen 2100, Denmark
e-mail: ggegelas@neuregeron.com

O.J. Bjerrum
Department of Drug Design and Pharmacology, Faculty of Health and Medical Sciences,
University of Copenhagen, Copenhagen 2100, Denmark

A. Ortega, A. Schousboe (eds.), *Glial Amino Acid Transporters*,
Advances in Neurobiology 16, DOI 10.1007/978-3-319-55769-4_11

Keywords Glutamate • EAAT • Pain • Amino acid transporter • Opioid tolerance • Gene transfer • β-lactam antibiotic • Ceftriaxone • Clavulanic acid • Riluzole • Tricyclic antidepressant

List of Abbreviations

AA	Arachidonic acid
ALS	Amyotrophic lateral sclerosis
AMPA	α-Amino-3-hydroxy-5-methyl-4-isoxazolepropionic acid
ASCT	Alanine/serine/cysteine transporter
BBB	Blood-brain barrier
BDNF	Brain-derived neurotrophic factor
CCI	Chronic constriction injury;
DHK	Dihydrokainate
DL-THA	DL-threo-beta-hydroxyaspartate
DRG	Dorsal root ganglion
EAAC1	Excitatory amino acid carrier 1
EAAT	Excitatory amino acid transporter
EAE	Experimental autoimmune encephalitis
FDA	Food and Drug Administration
FGF	Fibroblast growth factor
GLAST	Glutamate/aspartate transporter
GLT1	Glutamate transporter 1
GluT	Glutamate transporter
HSD	Hypoxic spreading depression
HSF-1	Heat-shock factor 1
IL	Interleukin
INFy	Interferon γ
L-*t*2,4-PDC	L-*trans*-Pyrrolidine-2,4-dicarboxylic acid
MAP kinase	Mitogen-activated protein kinase
MG-132	Z-Leu-Leu-Leu-CHO peptide
MND	Motor neuron disease; MS, multiple sclerosis
MS-153	(*R*)-(−)-5-methyl-1-nicotinoyl-2-pyrazoline
NF-κB	Nuclear factor κB
NG108-15	Neuroblastoma-glioma hybrid cell line clone 108–15
NGF	Nerve growth factor
NMDA	N-methyl-D-aspartic acid
Nrf2	Nuclear factor erythroid 2-related factor 2
NSAID	Nonsteroidal anti-inflammatory drug
OGD	Oxygen-glucose deprivation
PGE	Prostaglandin E
PI3 kinase	Phosphoinisitide-3 kinase
PKC	Protein kinase C

pSNL	Partial sciatic nerve ligation
SIN	Sciatic inflammatory neuropathy
SMA	Spinal muscular atrophy
SNI	Spared nerve injury
SNL	Spinal nerve ligation
SRI6	So-sparing peroxynitrite decomposition catalyst
STZ	Streptozotocin
TBOA	Threo-beta-benzyloxyaspartate
TG	Trigeminal ganglion
TNF	Tumor necrosis factor
TRPV-1	Transient receptor potential cation channel subfamily V member 1
WB	Western blotting
xCT	Cystine/glutamate exchange transporter

1 Introduction

1.1 Glutamatergic Mechanisms of Sensory Signal Processing in the Spinal Cord

Noxious signals from periphery are transmitted to the dorsal horn of the spinal cord via primary afferent neurons. These neurons, as well as the majority of secondary interneurons in the dorsal horn, are glutamatergic (Todd et al. 2003). However, primary afferent neurons are heterogeneous with regard to co-transmitter profiles and myelination levels, hence conductive properties (Millan 1999; Miller et al. 2011). The fine modulation and integration of the incoming sensory signals in the dorsal horn are ensured by various elements of the glutamate-sensing and -transporting systems expressed both in neurons and glial cells. These include mainly different types/splice variants of metabotropic and ionotropic glutamate receptors (NMDA-, kainate-, and AMPA-preferring subtypes) and various transport systems for moving glutamate across the plasma membrane in both directions (e.g., glutamate uptake and exocytosis), as well as for refilling synaptic vesicles (Larsson and Broman 2011; Osikowicz et al. 2013). The efficacy of glutamatergic neurotransmission is dramatically increased under the so-called central sensitization occurring in the spinal dorsal horn (Ji et al. 2003). The development of this phenomenon requires some complex spatiotemporal changes in the glutamatergic machinery and its supporting systems that could be triggered, e.g., by damage to peripheral sensory neurons and/or persistent noxious stimuli. The central sensitization is thought to be involved in the development of typical features of chronic pain: Hyperalgesia (amplified pain perception to noxious stimuli) and allodynia (painful response to otherwise innocuous stimuli) (Sandkuhler 2009).

Physiological or aberrant pain signaling in the spinal dorsal horn is finely tuned by the high-affinity glutamate uptake system that precisely regulates levels of this excitatory neurotransmitter in the vicinity of glutamate receptors. However, the roles of this transport system in chronic pain mechanisms is poorly elucidated (for recent review, see (Gegelashvili and Bjerrum 2014). The present review will cover several mechanistic aspects of pharmacological- and epigenetic regulation of glutamate transporters in chronic pain models and discuss perspectives of this system as of prospective therapeutic target for chronic pain treatment.

2 High-Affinity Glutamate Transporters

2.1 SLC1A Family of Solute Transporters

High-affinity uptake of neurotransmitter glutamate both in CNS and peripheral sensory path is ensured by the members of the SLC1A family of solute transporters: GLT1 (or EAAT2) (Pines et al. 1992), GLAST (or, according to alternative classification, EAAT1) (Storck et al. 1992), EAAC1 (or EAAT3) (Kanai and Hediger 1992), EAAT4 (Fairman et al. 1995), and EAAT5 (Arriza et al. 1997). In certain conditions (e.g., lowered pH), glutamate can be transported by another member of this family, ASCT2 (Gegelashvili et al. 2006), that otherwise operates as a neutral amino acid carrier (Arriza et al. 1993; Utsunomiya-Tate et al. 1996) (for comprehensive reviews, see (Danbolt 2001; Gegelashvili and Schousboe 1997, 1998; Gegelashvili et al. 2001). EAAT4 and EAAT5, in addition to their transporter properties, exhibit chloride-channel activities as well (Fairman et al. 1995; Arriza et al. 1997).

The EAAT4 and EAAC1 subtypes are predominantly expressed in neurons, though EAAC1 was occasionally detected in astrocytes (Conti et al. 1998), as well as in oligodendrocytes (Domercq and Matute 1999).

Glutamate transporters GLT1 and GLAST are typically expressed in astroglia but are also found in activated microglia (Xin et al. 2009) and oligodendrocytes (Domercq and Matute 1999). These two glutamate transporter subtypes ensure absolute majority (>95%) of glutamate uptake (Danbolt 2001). GLT1 and GLAST exist in several functional splice variants, some of those expressed in minor quantities, often at atypical locations (Chen et al. 2002; Rauen et al. 2004; Schmitt et al. 2002; Sullivan et al. 2007; Holmseth et al. 2009). For example, small amounts of originally cloned GLT1 or its GLT1b splice variant were found in oxygen-glucose-deprived neurons (Bonde et al. 2003) or in some hippocampal neurons under physiological conditions (Furness et al. 2008; Chen et al. 2002; Schmitt et al. 2002; Holmseth et al. 2009). Hypoxic neurons seem to express the GLAST1b splice variant as well (Sullivan et al. 2007). Atypical expression and distribution of glutamate transporter subtypes were also shown in retinal cells (Maenpaa et al. 2004).

Glutamate transporter subtypes and their splice variants carry glutamate across the plasma membrane with different affinities and velocities. While the basic driving force is transmembrane gradients of Na^+ and K^+, kinetic parameters of glutamate transporters (K_m and V_{max}) are significantly affected by pH, posttranslational modifications, association with intracellular scaffolding proteins, composition of the lipid bilayer, or direct interaction with a number of chemical compounds (Gegelashvili and Schousboe 1997; 1998; Beart and O'Shea 2007).

2.2 *High-Affinity Glutamate Transporters in the Spinal Cord and Primary Afferents*

In the spinal cord, more pronounced expression of GLT1, GLAST, and EAAC1 was found in the dorsal horn as compared to ventral horn. Among these subtypes, GLT1 was responsible for up to 40% of total transport of glutamate. At the same time, moderate quantitative differences among the spinal levels were observed. For example, EAAC1 expression is more pronounced at the cervical and thoracic levels than at the lumbar level, whereas GLT1 is most abundant in gray matter of lumbar dorsal horn (Queen et al. 2007). EAAC1 and lesser amounts of GLAST are also present in the ventral horn, while GLT1 is undetectable (Rothstein et al. 1994; Vera-Portocarrero et al. 2002; Queen et al. 2007).

Concerning cell type- and subcellular distribution, both GLT1 and GLAST are most abundantly expressed in spinal astrocytes, predominantly on their thin processes (Vera-Portocarrero et al. 2002; Rothstein et al. 1994, 1995). However, transient production of small quantities of GLT1 in growing spinal axons was also detected (Yamada et al. 1998).

Pronounced expression of GLAST is detected in satellite glial cells surrounding neuronal cell bodies in DRG, whereas lesser amounts of this subtype are found in rat dorsal root ganglion (DRG) neurons and myelin of sciatic nerve (Berger and Hediger 2000; Carozzi et al. 2008). GLT1 co-localizes with GLAST in the cytoplasmic compartment of satellite glial cells, but unlike GLAST, is absent in DRG neuronal cell bodies. GLT1 is expressed in axons located in the sciatic nerve. In the same nerve segment, GLT1 is also found in the cytoplasmic compartment of Schwann cells surrounding axons (Carozzi et al. 2008). EAAC1 is predominantly expressed in cytoplasmic compartments of axons and central terminals of DRG neurons and in the myelin layer ensheathing axons of sciatic nerve (Tao et al. 2004; Carozzi et al. 2008). In a single study, expression of EAAT4 in spinal astrocytes has been reported (Hu et al. 2003).

Spinal glutamate transporters by regulating glutamate concentrations in the synaptic cleft and in its vicinity not only modulate physiological parameters of direct excitatory neurotransmission but also control extra-synaptic leak of glutamate to neighboring neurons as well. Thus they provide an effective rescue mechanism for neurons against toxicity caused by glutamate. In contrast, aberrant functioning of the glutamate transport system – e.g., deficit of uptake capacity due to the downregulation of glutamate transporters or reversed operation of these carrier proteins in case of energy deprivation – may lead to sustained increase in extracellular glutamate concentrations and excitotoxic neuronal death. This is a likely mechanism contributing, for example, to the degeneration of motor neurons in amyotrophic lateral sclerosis (ALS) (Rothstein et al. 1992, 1995; Schousboe et al. 1997; Gegelashvili and Schousboe 1997, 1998; Gegelashvili et al. 2001; Bonde et al. 2003).

Since aberrant glutamate signaling in the sensory circuitry of the spinal cord is believed to be involved in chronic pain mechanisms, the glutamate transport system has recently come under scrutiny as potential culprit. For example, spinal ischemia that can develop under spinal cord injury causes reversed operation of glutamate transporters. This leads to excessive accumulation of glutamate that in turn activates dorsal horn neurons involved in the processing of nociceptive signal (Chen et al. 2007). Not only the reversed operation but also pronounced changes in the expression and function of glutamate transporters could be the consequence of spinal cord injury or spinal nerve damage (Vera-Portocarrero et al. 2002). It is noteworthy that well before cloning and pharmacological characterization of high-affinity glutamate transporter subtypes, partial injury to the spinal cord or dissection of spinal nerves has been known to induce dramatic deficit of transport of glutamate in the spinal cord (Potashner and Tran 1984, 1985). It is also noteworthy that neuropathic pain, one of the widespread and least treatable forms of chronic pain, often develops following nerve injury. Thus, significant interest has emerged toward actual involvement of high-affinity glutamate transporters in the mechanisms of chronic pain.

3 Regulation of Glutamate Transporters in Animal Models of Chronic Pain

3.1 Animal Models of Neuropathic Pain

Various animal models of neuropathic pain have been utilized to study expression and activity of glutamate transporters. The common paradigm for these models is/are surgically introduced injury(ies) to peripheral sensory nerves. However, precise anatomical site and severity of the nerve damage define individual characteristics of various models. The most versatile models of neuropathic pain utilize the procedures of partial sciatic nerve ligation (pSNL model), spinal nerve ligation (Honore et al. 2011), spared nerve injury (Honore et al. 2011), and chronic constriction injury of the main branch of sciatic nerve (Honore et al. 2011). Reproducible development of hypersensitivity to mechanical and/or thermal stimuli in these animal models of neuropathic pain has been reported in numerous studies (Sorkin and Yaksh 2009; Barrot 2012; Honore et al. 2011).

In these experimental models, frequently observed postoperative increase in glutamate transporter expression (Cavaliere et al. 2007; Sung et al. 2003) was followed by sustained downregulation of these proteins or decreased net glutamate uptake. In most cases, the downregulation preceded or occurred concurrently to the development of hypersensitivity to thermal, mechanical, or chemical stimuli. In most cases, glutamate transporter GLT1 was selectively downregulated, although in some models, expression levels of other subtypes were also affected, sometimes in opposite directions (Cirillo et al. 2011). The observed changes occur at different cell types and/or subcellular sites in the spinal cord. In some studies, total amount of GLT1 was found to be only slightly affected, while major changes occurred in distribution of transporter molecules between the plasma membrane and cytosol (Maeda et al. 2008; Ramos et al. 2010). GLT1 levels are predominantly reduced in spinal astrocytes, while in some cases, concomitant increase in the expression of this subtype could be detected in spinal microglia activated following injury (Berger et al. 2011). Only minor downregulation of total GLT1 and GLAST pools was found in some studies. However, significant changes in glutamate transporter expression at strategically located small "hot spots" were suggested in these cases (Napier et al. 2012).

3.2 Molecular Mechanisms of GluT Regulation in Neuropathic Pain Models

Molecular mechanisms of the nerve injury-induced alterations in the expression and functional activity of glutamate transporters in the spinal cord are largely obscure. Initial upregulation, occurring immediately after surgical operation of peripheral nerve, might serve a purpose of counteracting increased glutamate signaling (Sung et al. 2003; Cavaliere et al. 2007). This compensatory mechanism involves changes in tyrosine kinase and MAP kinase signaling (Sung et al. 2003). Analogous transient upregulation of glutamate transporter expression was detected after acute formalin injection (Niederberger et al. 2003, 2006) or acute optic nerve injury (Mawrin et al. 2003).

One of the likely causes of the sustained downregulation of glutamate transporters in spinal astrocytes after nerve injury could be a deficit of various neuronally released factors (Thone-Reineke et al. 2008) essential for the induction and/or maintenance of GLT1 and other membrane transport systems in astrocytes (Gegelashvili et al. 1997, 2000; Rodriguez-Kern et al. 2003; Figiel et al. 2003; Gegelashvili et al. 1993). For example, increased levels of NGF in the spinal cord both can both upregulate GLT1 expression and mitigate neuropathic pain in rats (Cirillo et al. 2011). Increased production of another signaling molecule – arachidonic acid, in CCI model (Sung et al. 2007) – was suggested to contribute to both deficit of glutamate uptake and development of neuropathic pain. Arachidonic acid is an effective inhibitor of glutamate transporters (Zerangue et al. 1995; Trotti et al. 1995), and it can thus trigger sustained increase in glutamate concentrations in and around the synaptic cleft (Potashner and Tran 1984, 1985; Binns et al. 2005; Hu et al. 2010). Excessive glutamate can also diffuse to more remote sites, normally not directly targeted by primary sensory neurons in the spinal cord and elicit aberrant signaling in both distantly located spinal astrocytes and neurons (Nie and Weng 2010; Nie et al. 2010a). In targeted astrocytes, the leaked glutamate can via different pathways trigger release of additional glutamate stored in astroglial secretory vesicles (Rojas et al. 2007; Yoshizumi et al. 2012). This turns spinal astrocytes into aberrant magnifiers of glutamate signaling, chronically disturbing normal nociceptive circuitry in the dorsal horn (Nie et al. 2010a).

4 Pharmacological and Epigenetic Modulation of Glutamate Transporters in Pain

4.1 Direct Pharmacological Inhibitors of Glutamate Transporters

One of the most convincing indications of the involvement of the glutamate transport system in pain mechanisms is the effect of specific inhibitors of high-affinity glutamate transporters on nociceptive behaviors in various in vivo models of pain.

Thus, increased extracellular glutamate concentrations in the spinal cord, as well as neuropathic or visceral pain develop in normal rats after infusion of glutamate transporter inhibitors dihydrokainate (DHK, specific inhibitor of GLT1), L-*trans*-2,4-PDC (general transportable inhibitor of glutamate uptake), or DL-TBOA (blocker of all glutamate transporter subtypes) (Gosselin et al. 2010). All these effects, including pain behaviors, could be mitigated by administration of NMDA- or non-NMDA receptor antagonists (Liaw et al. 2005).

Furthermore, in various chronic pain models, DHK can reverse analgesic effects produced by ceftriaxone (Gunduz et al. 2011), valproate (Hobo et al. 2011), or by overexpression of GLT1 via viral gene transfer (Lin et al. 2011).

However, not all studies have confirmed the pain-inducing effects of glutamate transporter inhibitors. In contrast to abovementioned studies, blocking of glutamate transport by either of the four inhibitors, DL-TBOA, DHK, DL-THA, or L-*trans*-2,4-PDC, markedly reduces inflammatory pain behaviors developing after injection of formalin (Niederberger et al. 2003; Yaster et al. 2011) and thermal hyperalgesia induced by complete Freund's adjuvant (Yaster et al. 2011). Furthermore, spinal infusion of DL-TBOA in mice prevents mechanical allodynia inducible by NMDA, AMPA, or prostaglandins PGE_2 and PGE_{2a} (Minami et al. 2001).

Thus, despite the clear pharmacological evidence for involvement of glutamate transporters in spinal mechanisms of pain provided by many studies, the inhibition of glutamate transport seems to paradoxically produce opposite effects on pain behaviors. The mechanism of mitigating effect of glutamate transporter inhibitors on pain is particularly unclear and may involve several pathways:

1. Prolonged elevation of extracellular glutamate concentrations may cause desensitization of postsynaptic ionotropic glutamate receptors and, thus, block transmission of excitatory signal from primary sensory neurons.

2. Chronically increased glutamate concentrations can also activate presynaptic inhibitory metabotropic glutamate receptors and, thus, diminish exocytotic release of glutamate from primary sensory neurons.
3. Higher glutamate concentrations may stimulate descending locus coeruleus neurons that provide noradrenergic inhibition in the spinal cord (Hayashida et al. 2010).
4. Blocking of glutamate transporters in spinal astrocytes would abolish co-entry of Na^+, hence, slow down Na^+/Ca^{2+} exchanger. This would, consequently, reduce the entry of Ca^{2+} and, as a result, prevent exocytotic release of vesicular glutamate in the vicinity of postsynaptic glutamate receptors (Rojas et al. 2007; Yoshizumi et al. 2012).
5. Inhibitors of glutamate carriers may block excessive release of glutamate by transporters operating in the reverse mode, i.e., carrying glutamate from the cytosol to extracellular space. The reversal of glutamate transport may typically occur in case of depletion of energy substrates, a situation that can emerge under pathologically increased metabolic activity within the sensory path (for review, see (Tao et al. 2005).

All the above-listed mechanisms underlying pain-inducing or mitigating activities of glutamate transport inhibitors may take place across the pain transduction pathways. However, it seems likely that the type of effect is determined by the actual site (e.g., by fine compartmentalization) and time frame of inhibitor action. It is also noteworthy that the possible concomitant action of some of the glutamate transporter inhibitors/ligands on glutamate receptors (Shimamoto 2008) may not be overlooked when critically evaluating the above findings.

4.2 Anticancer Drug Taxol

The commonly used anticancer drug, taxol (paclitaxel), induces peripheral sensory nerve damage and neuropathic pain (Kaplan et al. 1993; Chaudhry et al. 1994; Rowinsky et al. 1993). It was also found to downregulate glutamate transporters GLAST and GLT1 in activated astrocytes of rat spinal dorsal horn (Weng et al. 2005; Cata et al. 2006; Zhang et al. 2012). These observations are in line with the decreased expression of GLT1 in activated spinal astroglia in a mice model of neuropathic pain (Cavaliere et al. 2007), although, there is a principal discrepancy among these studies concerning production/involvement of pro-inflammatory cytokines in these processes. The mechanism of taxol-induced downregulation of astroglial GluTs is yet unexplained, especially, in the light of taxol's ability to increase cell surface expression and activity of neuronal glutamate transporter EAAC1 through stabilization of microtubules (Bianchi et al. 2006). However, among possible mechanisms of downregulation of spinal glutamate uptake in taxol-induced neuropathies (Doyle et al. 2012) is the increased production of some pro-inflammatory cytokines, e.g., TNFα and IL-1β, known to inhibit/downregulate expression of astroglial glutamate transporters GLT1 and GLAST (Fine et al. 1996; Prow and Irani 2008; Wang et al. 2003). These cytokines increase also production of peroxynitrite that can directly impair glutamate transporter subtypes (Trotti et al. 1996). In support of the latter mechanism, two orally active peroxynitrite decomposition catalysts, SRI6 and SRI110, could effectively prevent taxol-induced neuropathic pain (Doyle et al. 2012).

4.3 *Multi-target Anticonvulsants with Analgesic Properties*

4.3.1 Valproate

Valproic acid, an effective anticonvulsant noted for its analgesic properties (Gill et al. 2011; Waszkielewicz et al. 2011), was found to induce glutamate transporter GLAST in cerebellar astrocytes in vitro (Aguirre et al. 2008). At molecular level, valproate induces chromatin modifications expressed in demethylation of certain parts of GLT1 promoter and its enrichment in acetylated histone H4, leading to enhanced GLT1 transcription (Perisic et al. 2010). In SNL model of neuropathic pain, oral administration of valproate appeared to increase expression of both GLAST and GLT1 in the spinal dorsal horn and, at the same time, to significantly reduce mechanical allodynia. Reversal of the valproate-induced analgesia by GLT1 inhibitor, DHK, strongly supports the presence of causal relationship between the valproate-dependent upregulation of GluTs and its mitigating effect on tactile hypersensitivity developing in this model. These data highlight also a predominant role of GLT1 in the observed effect, despite the concomitant upregulation of GLAST by valproate (Hobo et al. 2011). However, involvement of complementary pathways, other than induction of GluTs, triggered by valproic acid (Fukuchi et al. 2009), could not be ruled out.

It is noteworthy that the analgesic effect of valproate could be enhanced by riluzole, another modulator of glutamate transport (Hobo et al. 2011). This study, thus, pointed at convenient pair of compounds for combination therapy, both with an effect on glutamate transporters, but also affecting other pathways.

4.3.2 Riluzole

Riluzole, initially introduced as a compound with anticonvulsive properties and currently approved for treatment of motor neuron disease (MND), has rapidly emerged as a wide-range neuroprotective agent with multiple cellular mechanisms of action, including those modulating glutamatergic circuits relevant for nociception. Indeed, this compound can produce components of general anesthesia, including analgesia in mice (Irifune et al. 2007). Riluzole appeared to attenuate and reverse neuropathic behaviors in various models of neuropathic pain (Sung et al. 2003; Hama and Sagen 2011; Coderre et al. 2007). Riluzole can also specifically augment visceral (gastrointestinal) nociceptive hypersensitivity induced by early-life stress in rats but not nociceptive responses to acute somatic pain stimulation (Gosselin et al. 2010). It also effectively reverses thermal hypersensitivity induced by chronic morphine treatment (Mao et al. 2002). It is noteworthy that some limited clinical studies could not confirm efficacy of riluzole in the treatment of peripheral neuropathies (Galer et al. 2000) or acute inflammatory pain (Hammer et al. 1999).

From the mechanistic point of view, strong correlation between neuromodulatory actions of riluzole, including analgesia, and its capability to positively modulate high-affinity glutamate transport system has been noted in a number of recent studies; thus, chronic treatment with riluzole prevents the pronounced loss of all major spinal glutamate transporter subtypes (EAAC1, GLAST, and GLT1) in CCI model of neuropathic pain (Sung et al. 2003) and restores normal levels of GLT1 in the hippocampus of aged rats, with no effect on younger animals (Brothers et al. 2013). It is noteworthy that no influence of riluzole on GLT1 levels was found in rat model of Parkinson's disease (Carbone et al. 2012). In contrast, in vivo administration of riluzole in young rats appeared to increase affinity and V_{max} for high-affinity glutamate uptake in spinal cord synaptosomes (Azbill et al. 2000). This effect could be recapitulated in synaptosomes from untreated animals, directly incubated with riluzole in vitro (Azbill et al. 2000; Dunlop et al. 2003). Riluzole was found to increase expression and activity of GLT1 in mouse striatal astrocytes (Carbone et al. 2012) and in NG108-15 neuroprogenitor cells (Liu et al. 2011), to elevate levels of another glutamate transporter, EAAC1, and glutamate uptake in C6 astroglial cells (Dall'igna et al. 2013). Riluzole seems to be capable of enhancing the activity of all three major glutamate transporters, GLT1, EAAC1, and GLAST, in clonal cell lines stably expressing these carrier proteins, primarily by increasing their affinity to glutamate (Fumagalli et al. 2008). In cerebral astroglial cultures, riluzole stimulates glutamate uptake, but at the same time, paradoxically triggers glutamate release (Hayashida et al. 2010) through a chain of consequential events: (i) initially, a massive cotransport of glutamate and Na+ into astrocytes through glutamate transporters, (ii) subsequent activation of Na^+/Ca^{2+} exchanger, (iii) elevation of cytosolic Ca^{2+} and, finally, (iv) Ca-dependent release of astroglial vesicular glutamate (Yoshizumi et al. 2012). The increased extracellular glutamate seems to provide analgesic effect in SNL model, most likely via stimulation of descending locus coeruleus neurons that provide noradrenergic inhibition in the spinal cord (Hayashida et al. 2010).

The actual mechanism of enhancement of glutamate transporter expression and/or activity by riluzole is poorly understood. In neuroprogenitor cells, the observed upregulation of GLT1 probably occurs through riluzole-induced heat shock factor 1 (HSF-1) (Liu et al. 2011). In spinal synaptosomes, stimulatory effect of riluzole apparently involves cholera toxin-sensitive G protein signaling mechanism (Azbill et al. 2000). The increase of EAAC1 expression in C6 cells by riluzole is most likely mediated through FGF receptor signaling but may also utilize at least two other independent intracellular pathways involving protein kinase C (Bianchi et al. 2006) and PI3 kinase (Dall'igna et al. 2013). In contrast, riluzole decreases EAAC1 activity in oocytes through inhibition of PKC (Choi et al. 2013), whereas pharmacological inhibition of PKC has no effect on riluzole-stimulated glutamate uptake in spinal synaptosomes that typically contain neuronal glutamate transporter EAAC1 (Azbill et al. 2000).

At systemic level, the site of analgesic action of riluzole may be species, model, and administration route dependent. For example, in neuropathic spinal cord injury (SCI) model, analgesic effect of riluzole is achieved only upon intracerebroventricular injection of the drug but not following intrathecal administration (Hama and Sagen 2011). In contrast, chronic intrathecal injections of riluzole are effective to reverse neuropathic behaviors and loss of glutamate transporters in CCI model of neuropathic pain (Sung et al. 2003). Furthermore, opposite modes of action of riluzole on astroglial glutamate transport across the plasma membrane (i.e., enhanced uptake vs. transporter-dependent release of vesicular glutamate), both resulting in analgesia, may take place in the spinal cord. It seems likely that either mechanism may dominate in restricted anatomical subcompartments and target distinct pain-processing circuits. For example, one mechanism may limit excitatory neurotransmission to ascending neurons, while another may provide glutamate-dependent activation of descending inhibitory signaling.

Thus, positive modulation of spinal glutamate transporters might be one of the key underlying mechanistic targets of analgesia provided by riluzole, although its actions on neurotransmission through other multiple signaling pathways (for review, see Cifra et al. 2013) may not be underscored.

4.4 Tetracycline Antibiotics with Analgesic Properties That Affect Glutamate Transporter Function

Minocycline, a tetracycline antibiotic, with wide-range neuroprotective actions beyond its core antibacterial activity, has been shown to be a quite effective analgesic in a number of models of chronic and acute pain. Thus, systemic administration of minocycline, depending on time frame and duration of the treatment, prevents thermal hypersensitivity (Vanelderen et al. 2013) or both mechanical allodynia and thermal hyperalgesia (Padi and Kulkarni 2008; Amin et al. 2012) in CCI-operated rats; it attenuates mechanical allodynia in SNL (Pu et al. 2013) and in sciatic inflammatory neuropathy (SIN) models (Ledeboer et al. 2005) and inhibits second phase of formalin-induced inflammatory pain (Cho et al. 2006); it also attenuates experimental diabetic neuropathy (Pabreja et al. 2011) and reduces mechanical allodynia in rat bone cancer-induced pain model (Wang et al. 2012).

It is noteworthy that minocycline can potentiate mitigating effects of another antibiotic, β-lactam ceftriaxone, on thermal hyperalgesia and mechanical allodynia (Amin et al. 2012).

Actual mechanism underlying analgesic properties of minocycline is not known. However, glutamate transporters have recently been suggested as possible target. Several studies point at the capability of minocycline to positively regulate glutamate transporter expression and activity in the spinal cord and neural cells.

Minocycline was found to markedly increase glutamate uptake in mice spinal cord and, thus, to rescue motor neurons from neurotropic virus-induced degeneration (Darman et al. 2004). Recently, it has been demonstrated that minocycline can prevent downregulation of astroglial transporters GLT1 and GLAST in the spinal dorsal horn of neuropathic rats, normalize NMDA receptor signaling, and prevent the development of mechanical allodynia following partial sciatic nerve ligation (pSNL) (Nie et al. 2010b). It is noteworthy that minocycline do not affect expression of glutamate transporters in normal (Mimura et al. 2011) or sham-operated rats (Nie et al. 2010b).

Molecular mechanism of effect of minocycline on glutamate transport has yet to be clarified. Genetic polymorphism, presence of supporting factors, and signaling pathways may be essential. For example, minocycline may either increase or decrease glutamate uptake executed by GLAST and EAAC1 in SY5Y neuroblastoma cells, depending on the expression or absence of HFE protein in these cells (Mitchell et al. 2011). Minocycline reduces production of certain cytokines, such as TNFα and IL1β (Ledeboer et al. 2005), that can impair glutamate transporters. It also elicits several intracellular signaling pathways, e.g., MAP kinase and NF-κB pathways that are known to be essential for the expression and maintenance of glutamate transporter GLT1 in astroglia (Gegelashvili et al. 2000; Rodriguez-Kern et al. 2003). Minocycline can activate Nrf2 factor (Sakata et al. 2012) that triggers expression of genes supporting antioxidant protection, among those glutamate transporter EAAC1 (Escartin et al. 2011) and glutamate /cysteine exchanger xCT (Sasaki et al. 2002). Furthermore, minocycline-induced protective mechanisms can prevent oxidative damage and functional impairment of glutamate transporter subtypes (Volterra et al. 1994). It cannot be ruled out that chelating properties of minocycline ensure scavenging of divalent ions, including Zn^{+2} ions, that are known to directly inhibit glutamate transporters GLAST and GLT1 (Vandenberg et al. 1998) or decrease production of GLT1 in spinal cultures (Yao 2009). This can, in turn, enhance glutamate transport in the spinal cord.

5 β-Lactam Antibiotic Ceftriaxone and Structurally Related Compounds

5.1 Influence on GLT1 Expression

In a search for pharmacological agents capable of increasing expression of GLT1 / EAAT2, J.D. Rothstein's group (Rothstein et al. 2005) screened a broad panel of FDA-approved drugs and nutritional substances and discovered that several β-lactam antibiotics, including ceftriaxone, penicillin G, and ampicillin, but not other antibiotics without β-lactam rings, could activate the promoter of GLT1 gene in cultured human fetal astroglia and actually increase its functional expression in rat spinal cord organotypic slice cultures, neuron-glial cocultures, as well as in chronically treated animals. While ampicillin, penicillin, as well as ceftriaxone and three other tested cephalosporins displayed comparable potencies in vitro, ceftriaxone appeared to be most effective in inducing GLT1 in vivo (Rothstein et al. 2005). Both GLT1a and GLT1b splice variants were almost equally upregulated, whereas no effect on other glutamate transporter subtypes has been detected. Thus, the induction of GLT1 has been linked to a pronounced neuroprotective effect of ceftriaxone observed both in two different in vitro models of excitotoxic neurodegeneration, in an in vivo model of ALS, as well as in some limited clinical trials (Rothstein et al. 2005; Berry et al. 2013).

5.2 *Pharmacological Effects in Pain Models*

Because downregulation of glutamate uptake is believed to contribute to the spinal mechanisms of pain, ceftriaxone as a proposed enhancer of glutamate transport has been increasingly tested in different animal models of acute or chronic pain.

Thus, chronic administration of ceftriaxone prevented, reduced, or fully reversed tactile and/or thermal hypersensitivity, as well as pain behaviors in rat CCI model (Yan et al. 2009; Ramos et al. 2010), mice models of inflammatory colitis and colorectal and bladder distension (Yang et al. 2011; Lin et al. 2011), streptozotocin-induced diabetes model (Gunduz et al. 2011), experimental autoimmune encephalitis model (Ramos et al. 2010), and cervical nerve root injury-based model of radicular pain (Nicholson et al. 2014). Ceftriaxone could also attenuate development of tolerance to acute analgesia offered by opioids (Rawls et al. 2010b) or nicotine (Schroeder et al. 2011). Ceftriaxone is effective in counteracting development of pain hypersensitivity typically occurring in result of chronic administration of opioids (Chen et al. 2012). Furthermore, ceftriaxone is capable of enhancing analgesic effect of nicotine (Schroeder et al. 2011).

In some models, analgesic effects provided by chronic administration of ceftriaxone were paralleled by enhanced glutamate uptake and/or increased expression of total GLT1 (Nicholson et al. 2014), GLT1 monomers (Hu et al. 2010), or only membrane-bound, dimerized pools of this transporter(Ramos et al. 2010) in lumbar sections of the spinal cord. Furthermore, the reversal of the preventive or therapeutic effects of ceftriaxone by pharmacological or antisense downregulation of GLT1 provided additional arguments in favor of the hypothesis that the analgesic potency of this β-lactam compound is based on its glutamate transport-enhancing properties.

However, the exact mechanism of β-lactam antibiotic-sensitive modulation of glutamate transport, as well as the actual nature of its analgesic or neuroprotective activities, is still unclear. Moreover, ceftriaxone and its structural analogs might utilize pathways beyond increases in glutamate uptake. These issues need further in-depth elucidation, due to some paradoxical and/or yet unexplained data reviewed below.

5.3 *Effects on Signaling Pathways and Transcriptional Mechanisms*

In human astroglial cultures, ceftriaxone via a yet unidentified pathway stimulate proteosomal degradation of an inhibitory protein, IκB, a process that liberates the active NF-κB complex, thus supports translocation and binding of this transcription factor to one of the NF-κB sites identified on the target promoter, and, finally leads to increased synthesis of GLT1 mRNA and its protein product (Lee et al. 2008). Furthermore, in the spinal cords of ceftriaxone-treated mice, 2.7-fold and 6-fold upregulation of GLT1 mRNA and protein, correspondingly, were observed (Nizzardo et al. 2011). It is noteworthy that at higher concentrations of ceftriaxone (Klein et al. 2006), significantly exceeding those measured in humans in CSF under clinical treatment (Nau et al. 1993), this β-lactam acts as a proteasome inhibitor and, actually, suppresses NF-κB activity (Bachetti et al. 2010).

Of note, in some studies, ceftriaxone treatment could not increase GLT1 promoter activity in cultured rat astrocytes and also failed to elevate GLT1 mRNA and protein levels in rat brain (Thone-Reineke et al. 2008; Hu et al. 2010) or in various CNS regions in C57BL/6 mice, incl., cortex, hippocampus, optic nerve, spinal cord (Melzer et al. 2008).

5.4 *Translational/Posttranslational Mechanisms of Regulation*

Translational/posttranslational mechanisms, rather than transcriptional activation, seem to play predominant roles in the control of GLT1 levels by β-lactam compounds in some models of CNS pathology. At least a part of β-lactam-sensitive upregulation of GLT1 was due to enhanced translation of one of the extended GLT1 RNA species in vivo, in primary astroglial cultures, and in GLT1-transfected astrocytes (Tian et al. 2007).

Increase in GLT1 dimerization and its targeting to the cell surface might represent one of the modes of enhancement of glutamate transport activity by ceftriaxone (Ramos et al. 2010).

Interestingly, in the CCI model of chronic pain, relative loss of both membrane-associated dimeric and cytoplasmic pools of GLT1 has been detected. However, ceftriaxone was only capable of selectively restoring the levels of the membrane-bound, dimeric, i.e., functionally active, GLT1 pool, but not of the cytoplasmic depot of this protein (Ramos et al. 2010).

5.5 *Direct Effects on Glutamate Transport Activity*

In in vitro model of oxygen-glucose deprivation (OGD) using organotypic slice cultures, ceftriaxone was incapable of increasing GLT1 expression but still provided neuroprotection by enhancing glutamate transport activity (Lipski et al.

2007). Similarly, while strongly neuroprotective in a rat model of transient focal ischemia, ceftriaxone could not affect GLT1 levels but strongly stimulated glutamate uptake in different brain regions, as well as in cultured astrocytes (Thone-Reineke et al. 2008).

Mechanism of ceftriaxone-dependent increase in transport activity occurring independently from transcriptional or translational upregulation of glutamate transporters remains unresolved. Direct, i.e., cell signaling-independent influence of ceftriaxone on glutamate transporter proteins has been suggested (Lipski et al. 2007). Indeed, contribution of pronounced metal-chelating properties of β-lactams (Ji et al. 2005) or chemical modification of GLT1 through direct binding of ceftriaxone may not be a priori ruled out. In fact, under physiological conditions, β-lactam ring of this class of antibiotics opens spontaneously, and the formed penicilloyl can readily react with lysine residues of proteins (Schneider and de Weck 1966), including GLT1, thus modifying its structure/conformation. Such chemical mechanisms may, in particular, reduce inhibition of transporters by some polyvalent metal ions (Mafra et al. 2001; Mutkus et al. 2005) or, for example, trigger trafficking of GLT1 from cytosol to the plasma membrane and/or stimulate dimerization of GLT1.

Contrary to the above findings, no effect of ceftriaxone on glutamate uptake, even at very high concentrations (0.5–1 mM, under both acute and chronic administration), was detected in mammalian cell line tsA201 ectopically expressing human EAAT2 or in rat neuron-glial cocultures, where glutamate uptake was predominantly executed by GLT1 (Melzer et al. 2008). This observation somewhat refutes the hypothetical mechanism based on direct, covalent, or non-covalent modulation of glutamate transporter proteins by β-lactams, mentioned above. Of note, cephalosporin B, a prototype compound with β-lactam ring, i.e., possessing chemical reactivity similar to that of ceftriaxone, was incapable of affecting glutamate transport system in the original study (Rothstein et al. 2005).

5.6 A Novel Molecular Target for Ceftriaxone, Directly Affecting Glutamate Homeostasis

It has recently been shown that in HT22 cell line, expression of glutamate/cystine exchanger, particularly, of its light chain, xCT, could be significantly increased by ceftriaxone through induction of nuclear factor Nrf2 (Lewerenz et al. 2009). Ceftriaxone can also upregulate both xCT and Nrf2 in vivo, as demonstrated in a murine model of spinal muscular atrophy (Nizzardo et al. 2011). In this regard, inhibition of xCT-mediated glutamate release was shown to mitigate cancer-induced bone pain (Ungard et al. 2014).

5.7 Protective Mechanisms Beyond Upregulation of Glutamate Transport System

Interestingly, chronic treatment with ceftriaxone significantly dampened excitotoxic brain damage in a mouse model of multiple sclerosis (MS), while the observed therapeutic effect of ceftriaxone was preserved even if glutamate transport was largely blocked by a selective GLT1 inhibitor – dihydrokainate (DHK) (Melzer et al. 2008). Similarly, another inhibitor of glutamate transport, DL-TBOA, failed to reverse neuroprotective effects of ceftriaxone in HT22 cell cultures where this β-lactam also strongly upregulated GLT1 (Lewerenz et al. 2009).

Thus, despite the therapeutic importance of hampering aberrant glutamate homeostasis in MS, it has been suggested (Melzer et al. 2008) that, at least in this particular MS model, ceftriaxone targeted molecular mechanisms other than upregulation of the glutamate transport system. In other words, ceftriaxone exerted protective immunological effects, such as impairment of T cell proliferation and secretion of pro-inflammatory cytokines, INFγ and IL17, and reduction of T cell invasion into the CNS.

It cannot be ruled out that ceftriaxone exerts its effect on the glutamate transport system, as well as on other neuroprotective pathways in the CNS via affecting downstream signaling cascades and subsequently elicited molecular/cellular events. Ceftriaxone appeared to be capable of reducing glial activation (Ramos et al. 2010), thus, likely interfering with production and secretion of factors affecting both glutamate transporters and other functional systems. For example, this β-lactam compound can reduce production/secretion of some cytokines altered in experimental meningitis (Klein et al. 2006), incl. pro-inflammatory cytokines INFγ and IL17 (Melzer et al. 2008). On the other hand, ceftriaxone can enhance production of both BDNF and its receptor trkB (Thone-Reineke et al. 2008). In this regard, it is noteworthy that expression of GLT1 can be upregulated by BDNF, in combination with other factors, such, as for example, β-amyloid peptide (Rodriguez-Kern et al. 2003).

5.8 Gene Expression Profile and Possible Cross Talk of Multiple Molecular Pathways and Factors

Recent study employing the global gene expression profiling approach in murine model of SMA supported the notion that the basis of neuroprotective efficacy of ceftriaxone might extend well beyond the upregulation of glutamate transporter GLT1 or cystine/glutamate exchanger: Chronic administration of ceftriaxone, exerting marked neuroprotection in this murine model, resulted in upregulation of 82 genes and downregulation of 147 genes in the spinal cord. It is noteworthy that gene expression patterns for many solute transporters, other than xCT and GLT1, were found to be significantly altered in this study (Nizzardo et al. 2011). In light of these findings, the analgesic or neuroprotective effects of ceftriaxone may not be founded solely on the upregulation of glutamate transport but may involve synergistic action of multiple factors altered by this compound.

5.9 Concluding Remarks on Ceftriaxone Effects

Based on the accumulated data, it seems that in different model systems, but sometimes in almost similar models/experimental conditions, ceftriaxone modulates glutamate transport at distinct molecular levels – transcriptional, translational, or posttranslational. Furthermore, in some cases, this compound completely fails to influence glutamate uptake, while still offering neuroprotection. Thus, under certain stressors or modalities (that may suppress β-lactam-dependent intracellular pathways leading to upregulation of GLT1), ceftriaxone might not be effective at all. It seems likely that analgesic or neuroprotective effects of ceftriaxone may not be solely founded on the upregulation of glutamate transport but may involve synergistic action of multiple factors altered by this compound (including other solute transporters, neurotrophic factors, and cytokines). The character and extent of ceftriaxone effects in experimental animals might also be shaped by the genetic background and origin of individual strain (Rode et al. 2007; Kristensen et al. 2016).

6 Glutamate Transport and Analgesic Mechanisms of Tricyclic Antidepressants

Amitriptyline, a tricyclic antidepressant, provides strong analgesia, especially, in neuropathic pain patients (Moore et al. 2012). In CCI model of neuropathic pain, amitriptyline was found to mitigate hypersensitivity to thermal stimuli (Huang et al. 2013).

Mechanisms underlying pain-relieving effects of tricyclic antidepressants are still enigmatic. However, recent data point at capability of amitriptyline to influence expression and activity of glutamate transporters in general and, particularly, to modulate spinal pain mechanisms through this system. Thus, administration of amitriptyline for up to 24 h can significantly increase levels of mRNA encoding neuronal glutamate transporter EAAC1 (Andin et al. 2004), although, functional significance of this observation is not clear, especially, in the light of negative effect of amitriptyline on the maximal velocity (V_{max}) of EAAC1 in vitro (Baik et al. 2009). In SNI model, chronic treatment with amitriptyline not only prevents loss of astroglial glutamate transporters GLAST and GLT1 in the spinal dorsal horn but also elevates its levels beyond initial, preoperative levels. Interestingly, the upregulation of these transporters occurs also in amitriptyline-treated sham-operated rats. However, in this study, amitriptyline attenuated mechanical allodynia only in SNI rats, without affecting initial levels of sensitivity to mechanical stimuli in sham-operated rats (Mao and Yang 2010). Chronic administration of amitriptyline alone would not affect levels of glutamate transporter expression or exert mitigating effect on pain hypersensitivity. However, amitryptyline attenuated development of tolerance to morphine. Furthermore, such treatment prevented dramatic downregulation of spinal glutamate transporters EAAC1, GLT1, and GLAST, occurring in morphine-tolerant rats, and even increased production of these proteins over the control levels (Tai et al. 2006). Acute treatment with amitriptyline could also restore/potentiate analgesic effect of opioids in morphine-tolerant rats by enhancing rapid translocation of glutamate transporters GLAST and GLT1 from the cytosol to the plasma membrane, thus enhancing glutamate uptake (Tai et al. 2007). In a model of neuropathic pain induced by pertussis toxin, morphine was incapable of provide pain relief, whereas acute pretreatment with amitriptyline could induce analgesic effect of morphine. This was matched by a relatively fast (within 2 h) restorative increase in glutamate transporter expression to control levels. It is noteworthy that the sequential combination of amitriptyline and morphine was essential for the expression of these effects (Lin et al. 2008). Very little is known about signaling pathways underlying chronic or acute effects of amitriptyline on the glutamate transport system. In morphine-tolerant rats, the increased relocation of GLAST and GLT1 from intracellular depots to the cell surface requires suppression of a phosphorylated form of protein kinase A, as well as activation of three distinct subtypes of protein kinase C (Tai et al. 2007). Furthermore, amitriptyline-induced overexpression of all three major glutamate transporter subtypes, GLT1, GLAST, and EAAC1, is mediated through the activation of transcription factor NF-*k*B (Tai et al. 2008). It is noteworthy that the essential requirement for this factor has previously been demonstrated only with regard to GLT1 expression (Rodriguez-Kern et al. 2003). Amitriptyline seems to predominantly regulate posttranscriptional pathways enhancing synthesis and/or stability of GLT1, rather than gene promoter activity (Perisic et al. 2010).

7 Concluding Remarks and Outlook

The glutamate transport system – a less explored component of the glutamatergic machinery – is obviously emerging as a prospective therapeutic target in chronic pain. A bulk of evidence indicate that malfunctioning of glutamate transporters significantly contribute to aberrant glutamate signaling involved in the development of central sensitization in chronic pain. Thus, downregulation of glutamate transporters precedes or occurs concurrently to the development of pathological hypersensitivity to thermal or tactile stimuli in different models of chronic pain. Moreover, antisense knockdown or pharmacological inhibition of these membrane proteins appeared to induce or aggravate pain sensitivity. In contrast, upregulation of GluTs by positive pharmacological modulators or by direct viral gene transfer to the spinal cord could reverse the development of such pathological hypersensitivity. On the other hand, some multi-target drugs displaying also analgesic properties (e.g., tricyclic antidepressant amitriptyline, riluzole, anticonvulsant valproate, tetracycline antibiotic minocycline, β-lactam antibiotic ceftriaxone or its structural analog devoid of antibacterial activity, clavulanic acid) can significantly increase spinal glutamate uptake.

Thus, a novel mechanistic concept for the development of new analgesics has emerged: It might be possible to correct aberrant glutamate signaling in the spinal cord by increasing glutamate uptake either via genetic upregulation of glutamate transport or its direct activation by specific enhancers.

In this regard, one of the prospective search directions would be rational design, syntheses, and testing of analgesic properties of structural analogs of the abovementioned multi-target drugs (i.e., anticonvulsants, antibiotics, antidepressants), "stripped off" of their core therapeutic efficacies but retaining glutamate transport-enhancing activities.

Versatility of such approach could be illustrated by most recent data on analgesic effects of clavulanic acid (CA), a structural analog of β-lactam ceftriaxone, devoid of antibacterial activity and displaying BBB-permeability superior to that of ceftriaxone. CA was shown to mimic some anti-opioid and antiseizure effects of ceftriaxone (Rawls et al. 2010a; Schroeder et al. 2014), to mitigate formalin-induced pain in mice (Hajhashemi and Dehdashti 2014), to alleviate tactile and thermal hypersensitivity in CCI model of neuropathic pain (Kristensen et al. 2016, unpublished observation), and to induce glutamate transporter GLT1 in mice (Kim et al. 2016), its splice variant GLT1b in rat spinal cord, as well as in astroglial cultures of rodent and human origins (unpublished observation from Pain Research Unit, Bjerrum et al., under review).

Testing of chemically modified tetracyclines-minocycline analogues (e.g., incyclinide), devoid of antibacterial activity (Lokeshwar 2011), would be another prospective avenue in the search for new analgesics. Yet, no information is available on possible effects of these compounds on glutamatergic spinal mechanisms underlying chronic pain.

However, the rational modification of these drugs will apparently require better understanding of the structure-function relations. For example, just the presence of a β-lactam ring in chemical structure is probably not enough (e.g., for a prototype β-lactam compound – cephalosporin C) to mimic effects of ceftriaxone or some other closely related β-lactam drugs on the expression of glutamate transporters (Rothstein et al. 2005).

It is noteworthy that such modified drugs may not necessarily be able to provide full alleviation of pain but may in combination with other analgesics generate pronounced synergistic pain-relieving effects. This notion applies, for example, to minocycline that can potentiate mitigating effects of another antibiotic, β-lactam ceftriaxone, on thermal hyperalgesia and mechanical allodynia (Amin et al. 2012). Another example is synergistic anti-nociceptive effect produced by the combination of NSAIDs with ceftriaxone in inflammatory pain models (Stepanovic-Petrovic et al. 2013).

However, as reviewed in the present chapter, the role of the glutamate transport system in chronic pain mechanisms is more sophisticated than just general decrease in glutamate uptake. It seems that relatively minor but sustained changes in localized areas rather than global increases in glutamate concentrations might underlie aberrant pain signaling in the spinal cord. Thus, in-depth elucidation of molecular and cellular processes, including signal transduction pathways, responsible for fine tuning of individual glutamate transporter subtypes in the spinal cord will be essential for more precise therapeutic targeting of this system in chronic pain. In this regard, gene therapy based on targeting glutamate transporter expression to specific sites in the spinal cord can develop into another promising strategy supplementing the pharmacotherapeutical approaches (Lin et al. 2011; Maeda et al. 2008).

Acknowledgments This work was supported by grants from the Lundbeck Foundation (R88-A9077-B1012), Augustinus Fonden (2014–2016), Oda and Hans Henningsen's Foundation (2015), and Ilia State University. Personal assistance by Mr. Alexandre Gegelashvili is cordially acknowledged.

Conflict of Interest The author declares no conflicts of interest.

References

Aguirre G, Rosas S, Lopez-Bayghen E, Ortega A. Valproate-dependent transcriptional regulation of GLAST/EAAT1 expression: involvement of Ying-Yang 1. Neurochem Int. 2008;52:1322–31.

Amin B, Hajhashemi V, Hosseinzadeh H, Abnous K. Antinociceptive evaluation of ceftriaxone and minocycline alone and in combination in a neuropathic pain model in rat. Neuroscience. 2012;224:15–25.

Andin J, Stenfors C, Ross SB, Marcusson J. Modulation of neuronal glutamate transporter rEAAC1 mRNA expression in rat brain by amitriptyline. Brain Res Mol Brain Res. 2004;126:74–7.

Arriza JL, Kavanaugh MP, Fairman WA, Wu YN, Murdoch GH, North RA, Amara SG. Cloning and expression of a human neutral amino acid transporter with structural similarity to the glutamate transporter gene family. J Biol Chem. 1993;268:15329–32.

Arriza JL, Eliasof S, Kavanaugh MP, Amara SG. Excitatory amino acid transporter 5, a retinal glutamate transporter coupled to a chloride conductance. Proc Natl Acad Sci U S A. 1997;94:4155–60.

Azbill RD, Mu X, Springer JE. Riluzole increases high-affinity glutamate uptake in rat spinal cord synaptosomes. Brain Res. 2000;871:175–80.

Bachetti T, Di ZE, Balbi P, Bocca P, Prigione I, Deiana GA, Rezzani A, Ceccherini I, Sechi G. In vitro treatments with ceftriaxone promote elimination of mutant glial fibrillary acidic protein and transcription down-regulation. Exp Cell Res. 2010;316:2152–65.

Baik HJ, Lee SA, Washington JM, Zuo ZY. Amitriptyline inhibits the activity of the rat glutamate transporter EAAT3 expressed in Xenopus oocytes. J Pharm Pharmacol. 2009;61:577–81.

Barrot M. Tests and models of nociception and pain in rodents. Neuroscience. 2012;211:39–50.

Beart PM, O'Shea RD. Transporters for L-glutamate: an update on their molecular pharmacology and pathological involvement. Br J Pharmacol. 2007;150:5–17.

Berger UV, Hediger MA. Distribution of the glutamate transporters GLAST and GLT-1 in rat circumventricular organs, meninges, and dorsal root ganglia. J Comp Neurol. 2000;421:385–99.

Berger JV, Deumens R, Goursaud S, Schafer S, Lavand'homme P, Joosten EA, Hermans E. Enhanced neuroinflammation and pain hypersensitivity after peripheral nerve injury in rats expressing mutated superoxide dismutase 1. J Neuroinflammation. 2011;8:33.

Berry JD, Shefner JM, Conwit R, Schoenfeld D, Keroack M, Felsenstein D, Krivickas L, David WS, Vriesendorp F, Pestronk A, Caress JB, Katz J, Simpson E, Rosenfeld J, Pascuzzi R, Glass J, Rezania K, Rothstein JD, Greenblatt DJ, Cudkowicz ME. Design and initial results of a multi-phase randomized trial of ceftriaxone in amyotrophic lateral sclerosis. PLoS One. 2013;8:e61177.

Bianchi MG, Rotoli BM, Dall'Asta V, Gazzola GC, Gatti R, Bussolati O. PKC-dependent stimulation of EAAT3 glutamate transporter does not require the integrity of actin cytoskeleton. Neurochem Int. 2006;48:341–9.

Binns BC, Huang Y, Goettl VM, Hackshaw KV, Stephens RL Jr. Glutamate uptake is attenuated in spinal deep dorsal and ventral horn in the rat spinal nerve ligation model. Brain Res. 2005;1041:38–47.

Bonde C, Sarup A, Schousboe A, Gegelashvili G, Zimmer J, Noraberg J. Neurotoxic and neuroprotective effects of the glutamate transporter inhibitor DL-threo-beta-benzyloxyaspartate (DL-TBOA) during physiological and ischemia-like conditions. Neurochem Int. 2003; 43:371–80.

Brothers HM, Bardou I, Hopp SC, Kaercher RM, Corona AW, Fenn AM, Godbout JP, Wenk GL. Riluzole partially rescues age-associated, but not LPS-induced, loss of glutamate transporters and spatial memory. J Neuroimmune Pharmacol. 2013;8:1098–105.

Carbone M, Duty S, Rattray M. Riluzole elevates GLT-1 activity and levels in striatal astrocytes. Neurochem Int. 2012;60:31–8.

Carozzi VA, Canta A, Oggioni N, Ceresa C, Marmiroli P, Konvalinka J, Zoia C, Bossi M, Ferrarese C, Tredici G, Cavaletti G. Expression and distribution of 'high affinity' glutamate transporters GLT1, GLAST, EAAC1 and of GCPII in the rat peripheral nervous system. J Anat. 2008;213:539–46.

Cata JP, Weng HR, Chen JH, Dougherty PM. Altered discharges of spinal wide dynamic range neurons and down-regulation of glutamate transporter expression in rats with paclitaxel-induced hyperalgesia. Neuroscience. 2006;138:329–38.

Cavaliere C, Cirillo G, Bianco MR, Rossi F, De Novellis V, Maione S, Papa M. Gliosis alters expression and uptake of spinal glial amino acid transporters in a mouse neuropathic pain model. Neuron Glia Biol. 2007;3:141–53.

Chaudhry V, Rowinsky EK, Sartorius SE, Donehower RC, Cornblath DR. Peripheral neuropathy from taxol and cisplatin combination chemotherapy: clinical and electrophysiological studies. Ann Neurol. 1994;35:304–11.

Chen W, Aoki C, Mahadomrongkul V, Gruber CE, Wang GJ, Blitzblau R, Irwin N, Rosenberg PA. Expression of a variant form of the glutamate transporter GLT1 in neuronal cultures and in neurons and astrocytes in the rat brain. J Neurosci. 2002;22:2142–52.

Chen M, Tao YX, Gu JG. Inward currents induced by ischemia in rat spinal cord dorsal horn neurons. Mol Pain. 2007;3:10.

Chen Z, He Y, Wang ZJ. The beta-lactam antibiotic, ceftriaxone, inhibits the development of opioid-induced hyperalgesia in mice. Neurosci Lett. 2012;509:69–71.

Cho IH, Chung YM, Park CK, Park SH, Lee H, Kim D, Piao ZG, Choi SY, Lee SJ, Park K, Kim JS, Jung SJ, Oh SB. Systemic administration of minocycline inhibits formalin-induced inflammatory pain in rat. Brain Res. 2006;1072:208–14.

Choi JS, Ryu JH, Zuo Z, Yang SM, Chang HW, Do SH. Riluzole attenuates excitatory amino acid transporter type 3 activity in Xenopus oocytes via protein kinase C inhibition. Eur J Pharmacol. 2013;713:39–43.

Cifra A, Mazzone GL, Nistri A. Riluzole: what it does to spinal and brainstem neurons and how it does it. Neuroscientist. 2013;19:137–44.

Cirillo G, Bianco MR, Colangelo AM, Cavaliere C, Daniele dL, Zaccaro L, Alberghina L, Papa M. Reactive astrocytosis-induced perturbation of synaptic homeostasis is restored by nerve growth factor. Neurobiol Dis. 2011;41:630–9.

Coderre TJ, Kumar N, Lefebvre CD, Yu JS. A comparison of the glutamate release inhibition and anti-allodynic effects of gabapentin, lamotrigine, and riluzole in a model of neuropathic pain. J Neurochem. 2007;100:1289–99.

Conti F, DeBiasi S, Minelli A, Rothstein JD, Melone M. EAAC1, a high-affinity glutamate tranporter, is localized to astrocytes and gabaergic neurons besides pyramidal cells in the rat cerebral cortex. Cereb Cortex. 1998;8:108–16.

Dall'igna OP, Bobermin LD, Souza DO, Quincozes-Santos A. Riluzole increases glutamate uptake by cultured C6 astroglial cells. Int J Dev Neurosci. 2013;31:482–6.

Danbolt NC. Glutamate uptake. Prog Neurobiol. 2001;65:1–105.
Darman J, Backovic S, Dike S, Maragakis NJ, Krishnan C, Rothstein JD, Irani DN, Kerr DA. Viral-induced spinal motor neuron death is non-cell-autonomous and involves glutamate excitotoxicity. J Neurosci. 2004;24:7566–75.
Domercq M, Matute C. Expression of glutamate transporters in the adult bovine corpus callosum. Brain Res Mol Brain Res. 1999;67:296–302.
Doyle T, Chen Z, Muscoli C, Bryant L, Esposito E, Cuzzocrea S, Dagostino C, Ryerse J, Rausaria S, Kamadulski A, Neumann WL, Salvemini D. Targeting the overproduction of peroxynitrite for the prevention and reversal of paclitaxel-induced neuropathic pain. J Neurosci. 2012;32:6149–60.
Dunlop J, Beal MH, She Y, Howland DS. Impaired spinal cord glutamate transport capacity and reduced sensitivity to riluzole in a transgenic superoxide dismutase mutant rat model of amyotrophic lateral sclerosis. J Neurosci. 2003;23:1688–96.
Escartin C, Won SJ, Malgorn C, Auregan G, Berman AE, Chen PC, Deglon N, Johnson JA, Suh SW, Swanson RA. Nuclear factor erythroid 2-related factor 2 facilitates neuronal glutathione synthesis by upregulating neuronal excitatory amino acid transporter 3 expression. J Neurosci. 2011;31:7392–401.
Fairman WA, Vandenberg RJ, Arriza JL, Kavanaugh MP, Amara SG. An excitatory amino-acid transporter with properties of a ligand-gated chloride channel. Nature. 1995;375:599–603.
Figiel M, Maucher T, Rozyczka J, Bayatti N, Engele J. Regulation of glial glutamate transporter expression by growth factors. Exp Neurol. 2003;183:124–35.
Fine SM, Angel RA, Perry SW, Epstein LG, Rothstein JD, Dewhurst S, Gelbard HA. Tumor necrosis factor alpha inhibits glutamate uptake by primary human astrocytes. Implications for pathogenesis of HIV-1 dementia. J Biol Chem. 1996;271:15303–6.
Fukuchi M, Nii T, Ishimaru N, Minamino A, Hara D, Takasaki I, Tabuchi A, Tsuda M. Valproic acid induces up- or down-regulation of gene expression responsible for the neuronal excitation and inhibition in rat cortical neurons through its epigenetic actions. Neurosci Res. 2009;65:35–43.
Fumagalli E, Funicello M, Rauen T, Gobbi M, Mennini T. Riluzole enhances the activity of glutamate transporters GLAST, GLT1 and EAAC1. Eur J Pharmacol. 2008;578:171–6.
Furness DN, Dehnes Y, Akhtar AQ, Rossi DJ, Hamann M, Grutle NJ, Gundersen V, Holmseth S, Lehre KP, Ullensvang K, Wojewodzic M, Zhou Y, Attwell D, Danbolt NC. A quantitative assessment of glutamate uptake into hippocampal synaptic terminals and astrocytes: new insights into a neuronal role for excitatory amino acid transporter 2 (EAAT2). Neuroscience. 2008;157:80–94.
Galer BS, Twilling LL, Harle J, Cluff RS, Friedman E, Rowbotham MC. Lack of efficacy of riluzole in the treatment of peripheral neuropathic pain conditions. Neurology. 2000;55:971–5.
Gegelashvili G, Bjerrum OJ. High-affinity glutamate transporters in chronic pain: an emerging therapeutic target. J Neurochem. 2014;131:712–30.
Gegelashvili G, Schousboe A. High affinity glutamate transporters: regulation of expression and activity. Mol Pharmacol. 1997;52:6–15.
Gegelashvili G, Schousboe A. Cellular distribution and kinetic properties of high-affinity glutamate transporters. Brain Res Bull. 1998;45:233–8.
Gegelashvili G, Andersson AM, Schousboe A, Bock E. Cyclic AMP regulates NCAM expression and phosphorylation in cultured mouse astrocytes. Eur J Cell Biol. 1993;62:343–51.
Gegelashvili G, Danbolt NC, Schousboe A. Neuronal soluble factors differentially regulate the expression of the GLT1 and GLAST glutamate transporters in cultured astroglia. J Neurochem. 1997;69:2612–5.
Gegelashvili G, Dehnes Y, Danbolt NC, Schousboe A. The high-affinity glutamate transporters GLT1, GLAST, and EAAT4 are regulated via different signalling mechanisms. Neurochem Int. 2000;37:163–70.
Gegelashvili G, Robinson MB, Trotti D, Rauen T. Regulation of glutamate transporters in health and disease. Prog Brain Res. 2001;132:267–86.
Gegelashvili M, Rodriguez-Kern A, Pirozhkova I, Zhang J, Sung L, Gegelashvili G. High-affinity glutamate transporter GLAST/EAAT1 regulates cell surface expression of glutamine/neutral amino acid transporter ASCT2 in human fetal astrocytes. Neurochem Int. 2006;48:611–5.

Gill D, Derry S, Wiffen PJ, Moore RA. Valproic acid and sodium valproate for neuropathic pain and fibromyalgia in adults. Cochrane Database Syst Rev. 2011:CD009183

Gosselin RD, O'Connor RM, Tramullas M, Julio-Pieper M, Dinan TG, Cryan JF. Riluzole normalizes early-life stress-induced visceral hypersensitivity in rats: role of spinal glutamate reuptake mechanisms. Gastroenterology. 2010;138:2418–25.

Gunduz O, Oltulu C, Buldum D, Guven R, Ulugol A. Anti-allodynic and anti-hyperalgesic effects of ceftriaxone in streptozocin-induced diabetic rats. Neurosci Lett. 2011;491:23–5.

Hajhashemi V, Dehdashti K. Antinociceptive effect of clavulanic acid and its preventive activity against development of morphine tolerance and dependence in animal models. Res Pharm Sci. 2014;9:315–21.

Hama A, Sagen J. Antinociceptive effect of riluzole in rats with neuropathic spinal cord injury pain. J Neurotrauma. 2011;28:127–34.

Hammer NA, Lilleso J, Pedersen JL, Kehlet H. Effect of riluzole on acute pain and hyperalgesia in humans. Br J Anaesth. 1999;82:718–22.

Hayashida K, Parker RA, Eisenach JC. Activation of glutamate transporters in the locus coeruleus paradoxically activates descending inhibition in rats. Brain Res. 2010;1317:80–6.

Hobo S, Eisenach JC, Hayashida K. Up-regulation of spinal glutamate transporters contributes to anti-hypersensitive effects of valproate in rats after peripheral nerve injury. Neurosci Lett. 2011;502:52–5.

Holmseth S, Scott HA, Real K, Lehre KP, Leergaard TB, Bjaalie JG, Danbolt NC. The concentrations and distributions of three C-terminal variants of the GLT1 (EAAT2; slc1a2) glutamate transporter protein in rat brain tissue suggest differential regulation. Neuroscience. 2009;162:1055–71.

Honore PH, Basnet A, Kristensen P, Andersen LM, Neustrup S, Mollgaard P, Eljaja L, Bjerrum OJ. Predictive validity of pharmacologic interventions in animal models of neuropathic pain. Scand J Pain. 2011;2:178–84. Ref Type: Abstract

Hu WH, Walters WM, Xia XM, Karmally SA, Bethea JR. Neuronal glutamate transporter EAAT4 is expressed in astrocytes. Glia. 2003;44:13–25.

Hu Y, Li W, Lu L, Cai J, Xian X, Zhang M, Li Q, Li L. An anti-nociceptive role for ceftriaxone in chronic neuropathic pain in rats. Pain. 2010;148:284–301.

Huang KL, Shieh JP, Chu CC, Cheng KI, Wang JJ, Lin MT, Yeh MY. Prolonged analgesic effect of amitriptyline base on thermal hyperalgesia in an animal model of neuropathic pain. Eur J Pharmacol. 2013;702:20–4.

Irifune M, Kikuchi N, Saida T, Takarada T, Shimizu Y, Endo C, Morita K, Dohi T, Sato T, Kawahara M. Riluzole, a glutamate release inhibitor, induces loss of righting reflex, antinociception, and immobility in response to noxious stimulation in mice. Anesth Analg. 2007;104:1415–21. table

Ji RR, Kohno T, Moore KA, Woolf CJ. Central sensitization and LTP: do pain and memory share similar mechanisms? Trends Neurosci. 2003;26:696–705.

Ji HF, Shen L, Zhang HY. Beta-lactam antibiotics are multipotent agents to combat neurological diseases. Biochem Biophys Res Commun. 2005;333:661–3.

Kanai Y, Hediger MA. Primary structure and functional characterization of a high-affinity glutamate transporter. Nature. 1992;360:467–71.

Kaplan JG, Einzig AI, Schaumburg HH. Taxol causes permanent large fiber peripheral nerve dysfunction: a lesson for preventative strategies. J Neuro-Oncol. 1993;16:105–7.

Kim J, John J, Langford D, Walker E, Ward S, Rawls SM. Clavulanic acid enhances glutamate transporter subtype I (GLT-1) expression and decreases reinforcing efficacy of cocaine in mice. Amino Acids. 2016;48:689–96.

Klein M, Paul R, Angele B, Popp B, Pfister HW, Koedel U. Protein expression pattern in experimental pneumococcal meningitis. Microbes Infect. 2006;8:974–83.

Kristensen PJ, Heegaard AM, Hestehave S, Jeggo RD, Bjerrum OJ, Munro G. Vendor-derived differences in injury-induced pain phenotype and pharmacology of Sprague-Dawley rats: does it matter? Eur J Pain. 2016; 4:692–704.

Larsson M, Broman J. Synaptic plasticity and pain: role of ionotropic glutamate receptors. Neuroscientist. 2011;17:256–73.

Ledeboer A, Sloane EM, Milligan ED, Frank MG, Mahony JH, Maier SF, Watkins LR. Minocycline attenuates mechanical allodynia and proinflammatory cytokine expression in rat models of pain facilitation. Pain. 2005;115:71–83.

Lee SG, Su ZZ, Emdad L, Gupta P, Sarkar D, Borjabad A, Volsky DJ, Fisher PB. Mechanism of ceftriaxone induction of excitatory amino acid transporter-2 expression and glutamate uptake in primary human astrocytes. J Biol Chem. 2008;283:13116–23.

Lewerenz J, Albrecht P, Tien ML, Henke N, Karumbayaram S, Kornblum HI, Wiedau-Pazos M, Schubert D, Maher P, Methner A. Induction of Nrf2 and xCT are involved in the action of the neuroprotective antibiotic ceftriaxone in vitro. J Neurochem. 2009;111:332–43.

Liaw WJ, Stephens RL Jr, Binns BC, Chu Y, Sepkuty JP, Johns RA, Rothstein JD, Tao YX. Spinal glutamate uptake is critical for maintaining normal sensory transmission in rat spinal cord. Pain. 2005;115:60–70.

Lin JA, Tsai RY, Lin YT, Lee MS, Cherng CH, Wong CS, Tzeng JI. Amitriptyline pretreatment preserves the antinociceptive effect of morphine in pertussis toxin-treated rats by lowering CSF excitatory amino acid concentrations and reversing the downregulation of glutamate transporters. Brain Res. 2008;1232:61–9.

Lin Y, Roman K, Foust KD, Kaspar BK, Bailey MT, Stephens RL. Glutamate transporter GLT-1 upregulation attenuates visceral nociception and hyperalgesia via spinal mechanisms not related to anti-inflammatory or probiotic effects. Pain Res Treat. 2011;2011:507029.

Lipski J, Wan CK, Bai JZ, Pi R, Li D, Donnelly D. Neuroprotective potential of ceftriaxone in in vitro models of stroke. Neuroscience. 2007;146:617–29.

Liu AY, Mathur R, Mei N, Langhammer CG, Babiarz B, Firestein BL. Neuroprotective drug riluzole amplifies the heat shock factor 1 (HSF1)- and glutamate transporter 1 (GLT1)-dependent cytoprotective mechanisms for neuronal survival. J Biol Chem. 2011;286:2785–94.

Lokeshwar BL. Chemically modified non-antimicrobial tetracyclines are multifunctional drugs against advanced cancers. Pharmacol Res. 2011;63:146–50.

Maeda S, Kawamoto A, Yatani Y, Shirakawa H, Nakagawa T, Kaneko S. Gene transfer of GLT-1, a glial glutamate transporter, into the spinal cord by recombinant adenovirus attenuates inflammatory and neuropathic pain in rats. Mol Pain. 2008;4:65.

Maenpaa H, Gegelashvili G, Tahti H. Expression of glutamate transporter subtypes in cultured retinal pigment epithelial and retinoblastoma cells. Curr Eye Res. 2004;28:159–65.

Mafra RA, Araujo DA, Beirao PS, Cruz JS. Glutamate transport in rat cerebellar granule cells is impaired by inorganic epileptogenic agents. Neurosci Lett. 2001;310:85–8.

Mao QX, Yang TD. Amitriptyline upregulates EAAT1 and EAAT2 in neuropathic pain rats. Brain Res Bull. 2010;81:424–7.

Mao J, Sung B, Ji RR, Lim G. Chronic morphine induces downregulation of spinal glutamate transporters: implications in morphine tolerance and abnormal pain sensitivity. J Neurosci. 2002;22:8312–23.

Mawrin C, Pap T, Pallas M, Dietzmann K, Behrens-Baumann W, Vorwerk CK. Changes of retinal glutamate transporter GLT-1 mRNA levels following optic nerve damage. Mol Vis. 2003;9:10–3.

Melzer N, Meuth SG, Torres-Salazar D, Bittner S, Zozulya AL, Weidenfeller C, Kotsiari A, Stangel M, Fahlke C, Wiendl H. A beta-lactam antibiotic dampens excitotoxic inflammatory CNS damage in a mouse model of multiple sclerosis. PLoS One. 2008;3:e3149.

Millan MJ. The induction of pain: an integrative review. Prog Neurobiol. 1999;57:1–164.

Miller KE, Hoffman EM, Sutharshan M, Schechter R. Glutamate pharmacology and metabolism in peripheral primary afferents: physiological and pathophysiological mechanisms. Pharmacol Ther. 2011;130:283–309.

Mimura K, Tomimatsu T, Minato K, Jugder O, Kinugasa-Taniguchi Y, Kanagawa T, Nozaki M, Yanagihara I, Kimura T. Ceftriaxone preconditioning confers neuroprotection in neonatal rats through glutamate transporter 1 upregulation. Reprod Sci. 2011;18:1193–201.

Minami T, Matsumura S, Okuda-Ashitaka E, Shimamoto K, Sakimura K, Mishina M, Mori H, Ito S. Characterization of the glutamatergic system for induction and maintenance of allodynia. Brain Res. 2001;895:178–85.

Mitchell RM, Lee SY, Simmons Z, Connor JR. HFE polymorphisms affect cellular glutamate regulation. Neurobiol Aging. 2011;32:1114–23.

Moore RA, Derry S, Aldington D, Cole P, Wiffen PJ. Amitriptyline for neuropathic pain and fibromyalgia in adults. Cochrane Database Syst Rev. 2012;12:CD008242.

Mutkus L, Aschner JL, Fitsanakis V, Aschner M. The in vitro uptake of glutamate in GLAST and GLT-1 transfected mutant CHO-K1 cells is inhibited by manganese. Biol Trace Elem Res. 2005;107:221–30.

Napier IA, Mohammadi SA, Christie MJ. Glutamate transporter dysfunction associated with nerve injury-induced pain in mice. J Neurophysiol. 2012;107:649–57.

Nau R, Prange HW, Muth P, Mahr G, Menck S, Kolenda H, Sorgel F. Passage of cefotaxime and ceftriaxone into cerebrospinal fluid of patients with uninflamed meninges. Antimicrob Agents Chemother. 1993;37:1518–24.

Nicholson KJ, Gilliland TM, Winkelstein BA. Upregulation of GLT-1 by treatment with ceftriaxone alleviates radicular pain by reducing spinal astrocyte activation and neuronal hyperexcitability. J Neurosci Res. 2014;92:116–29.

Nie H, Weng HR. Impaired glial glutamate uptake induces extrasynaptic glutamate spillover in the spinal sensory synapses of neuropathic rats. J Neurophysiol. 2010;103:2570–80.

Nie H, Zhang H, Weng HR. Bidirectional neuron-glia interactions triggered by deficiency of glutamate uptake at spinal sensory synapses. J Neurophysiol. 2010a;104:713–25.

Nie H, Zhang H, Weng HR. Minocycline prevents impaired glial glutamate uptake in the spinal sensory synapses of neuropathic rats. Neuroscience. 2010b;170:901–12.

Niederberger E, Schmidtko A, Rothstein JD, Geisslinger G, Tegeder I. Modulation of spinal nociceptive processing through the glutamate transporter GLT-1. Neuroscience. 2003;116:81–7.

Niederberger E, Schmidtko A, Coste O, Marian C, Ehnert C, Geisslinger G. The glutamate transporter GLAST is involved in spinal nociceptive processing. Biochem Biophys Res Commun. 2006;346:393–9.

Nizzardo M, Nardini M, Ronchi D, Salani S, Donadoni C, Fortunato F, Colciago G, Falcone M, Simone C, Riboldi G, Govoni A, Bresolin N, Comi GP, Corti S. Beta-lactam antibiotic offers neuroprotection in a spinal muscular atrophy model by multiple mechanisms. Exp Neurol. 2011;229:214–25.

Osikowicz M, Mika J, Przewlocka B. The glutamatergic system as a target for neuropathic pain relief. Exp Physiol. 2013;98:372–84.

Pabreja K, Dua K, Sharma S, Padi SS, Kulkarni SK. Minocycline attenuates the development of diabetic neuropathic pain: possible anti-inflammatory and anti-oxidant mechanisms. Eur J Pharmacol. 2011;661:15–21.

Padi SS, Kulkarni SK. Minocycline prevents the development of neuropathic pain, but not acute pain: possible anti-inflammatory and antioxidant mechanisms. Eur J Pharmacol. 2008;601:79–87.

Perisic T, Zimmermann N, Kirmeier T, Asmus M, Tuorto F, Uhr M, Holsboer F, Rein T, Zschocke J. Valproate and amitriptyline exert common and divergent influences on global and gene promoter-specific chromatin modifications in rat primary astrocytes. Neuropsychopharmacology. 2010;35:792–805.

Pines G, Danbolt NC, Bjoras M, Zhang Y, Bendahan A, Eide L, Koepsell H, Storm-Mathisen J, Seeberg E, Kanner BI. Cloning and expression of a rat brain L-glutamate transporter. Nature. 1992;360:464–7.

Potashner SJ, Tran PL. Decreased uptake and release of D-aspartate in the guinea pig spinal cord after dorsal root section. J Neurochem. 1984;42:1135–44.

Potashner SJ, Tran PL. Decreased uptake and release of D-aspartate in the guinea pig spinal cord after partial cordotomy. J Neurochem. 1985;44:1511–9.

Prow NA, Irani DN. The inflammatory cytokine, interleukin-1 beta, mediates loss of astroglial glutamate transport and drives excitotoxic motor neuron injury in the spinal cord during acute viral encephalomyelitis. J Neurochem. 2008;105:1276–86.

Pu S, Xu Y, Du D, Yang M, Zhang X, Wu J, Jiang W. Minocycline attenuates mechanical allodynia and expression of spinal NMDA receptor 1 subunit in rat neuropathic pain model. J Physiol Biochem. 2013; 69:349–57.
Queen SA, Kesslak JP, Bridges RJ. Regional distribution of sodium-dependent excitatory amino acid transporters in rat spinal cord. J Spinal Cord Med. 2007;30:263–71.
Ramos KM, Lewis MT, Morgan KN, Crysdale NY, Kroll JL, Taylor FR, Harrison JA, Sloane EM, Maier SF, Watkins LR. Spinal upregulation of glutamate transporter GLT-1 by ceftriaxone: therapeutic efficacy in a range of experimental nervous system disorders. Neuroscience. 2010;169:1888–900.
Rauen T, Wiessner M, Sullivan R, Lee A, Pow DV. A new GLT1 splice variant: cloning and immunolocalization of GLT1c in the mammalian retina and brain. Neurochem Int. 2004;45:1095–106.
Rawls SM, Karaca F, Madhani I, Bhojani V, Martinez RL, Abou-Gharbia M, Raffa RB. Beta-lactamase inhibitors display anti-seizure properties in an invertebrate assay. Neuroscience. 2010a;169:1800–4.
Rawls SM, Zielinski M, Patel H, Sacavage S, Baron DA, Patel D. Beta-lactam antibiotic reduces morphine analgesic tolerance in rats through GLT-1 transporter activation. Drug Alcohol Depend. 2010b;107:261–3.
Rode F, Thomsen M, Brolos T, Jensen DG, Blackburn-Munro G, Bjerrum OJ. The importance of genetic background on pain behaviours and pharmacological sensitivity in the rat spared serve injury model of peripheral neuropathic pain. Eur J Pharmacol. 2007;564:103–11.
Rodriguez-Kern A, Gegelashvili M, Schousboe A, Zhang J, Sung L, Gegelashvili G. Beta-amyloid and brain-derived neurotrophic factor, BDNF, up-regulate the expression of glutamate transporter GLT-1/EAAT2 via different signaling pathways utilizing transcription factor NF-kappaB. Neurochem Int. 2003;43:363–70.
Rojas H, Colina C, Ramos M, Benaim G, Jaffe EH, Caputo C, DiPolo R. Na+ entry via glutamate transporter activates the reverse Na+/Ca2+ exchange and triggers Ca(i)2+-induced Ca2+ release in rat cerebellar Type-1 astrocytes. J Neurochem. 2007;100:1188–202.
Rothstein JD, Martin LJ, Kuncl RW. Decreased glutamate transport by the brain and spinal cord in amyotrophic lateral sclerosis. N Engl J Med. 1992;326:1464–8.
Rothstein JD, Martin L, Levey AI, Dykes-Hoberg M, Jin L, Wu D, Nash N, Kuncl RW. Localization of neuronal and glial glutamate transporters. Neuron. 1994;13:713–25.
Rothstein JD, Van KM, Levey AI, Martin LJ, Kuncl RW. Selective loss of glial glutamate transporter GLT-1 in amyotrophic lateral sclerosis. Ann Neurol. 1995;38:73–84.
Rothstein JD, Patel S, Regan MR, Haenggeli C, Huang YH, Bergles DE, Jin L, Dykes HM, Vidensky S, Chung DS, Toan SV, Bruijn LI, Su ZZ, Gupta P, Fisher PB. Beta-lactam antibiotics offer neuroprotection by increasing glutamate transporter expression. Nature. 2005;433:73–7.
Rowinsky EK, Chaudhry V, Cornblath DR, Donehower RC. Neurotoxicity of taxol. J Natl Cancer Inst Monogr. 1993;1993:107–15.
Sakata H, Niizuma K, Yoshioka H, Kim GS, Jung JE, Katsu M, Narasimhan P, Maier CM, Nishiyama Y, Chan PH. Minocycline-preconditioned neural stem cells enhance neuroprotection after ischemic stroke in rats. J Neurosci. 2012;32:3462–73.
Sandkuhler J. Models and mechanisms of hyperalgesia and allodynia. Physiol Rev. 2009;89:707–58.
Sasaki H, Sato H, Kuriyama-Matsumura K, Sato K, Maebara K, Wang H, Tamba M, Itoh K, Yamamoto M, Bannai S. Electrophile response element-mediated induction of the cystine/glutamate exchange transporter gene expression. J Biol Chem. 2002;277:44765–71.
Schmitt A, Asan E, Lesch KP, Kugler P. A splice variant of glutamate transporter GLT1/EAAT2 expressed in neurons: cloning and localization in rat nervous system. Neuroscience. 2002;109:45–61.
Schneider CH, de Weck AL. Chemical aspects of penicillin allergy. The direct penicilloylation of epsilon-amino groups by penicillin at pH 7.4. Helv Chim Acta. 1966;49:1695–706.
Schousboe A, Sonnewald U, Civenni G, Gegelashvili G. Role of astrocytes in glutamate homeostasis. Implications for excitotoxicity. Adv Exp Med Biol. 1997;429:195–206.

Schroeder JA, Quick KF, Landry PM, Rawls SM. Glutamate transporter activation enhances nicotine antinociception and attenuates nicotine analgesic tolerance. Neuroreport. 2011;22:970–3.

Schroeder JA, Tolman NG, McKenna FF, Watkins KL, Passeri SM, Hsu AH, Shinn BR, Rawls SM. Clavulanic acid reduces rewarding, hyperthermic and locomotor-sensitizing effects of morphine in rats: a new indication for an old drug? Drug Alcohol Depend. 2014;142:41–5.

Shimamoto K. Glutamate transporter blockers for elucidation of the function of excitatory neurotransmission systems. Chem Rec. 2008;8:182–99.

Sorkin LS, Yaksh TL. Behavioral models of pain states evoked by physical injury to the peripheral nerve. Neurotherapeutics. 2009;6:609–19.

Stepanovic-Petrovic RM, Micov AM, Tomic MA, Kovacevic JM, Boskovic BD. Antihyperalgesic/antinociceptive effects of ceftriaxone and its synergistic interactions with different analgesics in inflammatory pain in rodents. Anesthesiology. 2013;120(3):737–50.

Storck T, Schulte S, Hofmann K, Stoffel W. Structure, expression, and functional analysis of a Na(+)-dependent glutamate/aspartate transporter from rat brain. Proc Natl Acad Sci U S A. 1992;89:10955–9.

Sullivan SM, Macnab LT, Bjorkman ST, Colditz PB, Pow DV. GLAST1b, the exon-9 skipping form of the glutamate-aspartate transporter EAAT1 is a sensitive marker of neuronal dysfunction in the hypoxic brain. Neuroscience. 2007;149:434–45.

Sung B, Lim G, Mao J. Altered expression and uptake activity of spinal glutamate transporters after nerve injury contribute to the pathogenesis of neuropathic pain in rats. J Neurosci. 2003;23:2899–910.

Sung B, Wang S, Zhou B, Lim G, Yang L, Zeng Q, Lim JA, Wang JD, Kang JX, Mao J. Altered spinal arachidonic acid turnover after peripheral nerve injury regulates regional glutamate concentration and neuropathic pain behaviors in rats. Pain. 2007;131:121–31.

Tai YH, Wang YH, Wang JJ, Tao PL, Tung CS, Wong CS. Amitriptyline suppresses neuroinflammation and up-regulates glutamate transporters in morphine-tolerant rats. Pain. 2006;124:77–86.

Tai YH, Wang YH, Tsai RY, Wang JJ, Tao PL, Liu TM, Wang YC, Wong CS. Amitriptyline preserves morphine's antinociceptive effect by regulating the glutamate transporter GLAST and GLT-1 trafficking and excitatory amino acids concentration in morphine-tolerant rats. Pain. 2007;129:343–54.

Tai YH, Tsai RY, Wang YH, Cherng CH, Tao PL, Liu TM, Wong CS. Amitriptyline induces nuclear transcription factor-kappaB-dependent glutamate transporter upregulation in chronic morphine-infused rats. Neuroscience. 2008;153:823–31.

Tao F, Liaw WJ, Zhang B, Yaster M, Rothstein JD, Johns RA, Tao YX. Evidence of neuronal excitatory amino acid carrier 1 expression in rat dorsal root ganglion neurons and their central terminals. Neuroscience. 2004;123:1045–51.

Tao YX, Gu J, Stephens RL Jr. Role of spinal cord glutamate transporter during normal sensory transmission and pathological pain states. Mol Pain. 2005;1:30.

Thone-Reineke C, Neumann C, Namsolleck P, Schmerbach K, Krikov M, Schefe JH, Lucht K, Hortnagl H, Godes M, Muller S, Rumschussel K, Funke-Kaiser H, Villringer A, Steckelings UM, Unger T. The beta-lactam antibiotic, ceftriaxone, dramatically improves survival, increases glutamate uptake and induces neurotrophins in stroke. J Hypertens. 2008;26:2426–35.

Tian G, Lai L, Guo H, Lin Y, Butchbach ME, Chang Y, Lin CL. Translational control of glial glutamate transporter EAAT2 expression. J Biol Chem. 2007;282:1727–37.

Todd AJ, Hughes DI, Polgar E, Nagy GG, Mackie M, Ottersen OP, Maxwell DJ. The expression of vesicular glutamate transporters VGLUT1 and VGLUT2 in neurochemically defined axonal populations in the rat spinal cord with emphasis on the dorsal horn. Eur J Neurosci. 2003;17:13–27.

Trotti D, Volterra A, Lehre KP, Rossi D, Gjesdal O, Racagni G, Danbolt NC. Arachidonic acid inhibits a purified and reconstituted glutamate transporter directly from the water phase and not via the phospholipid membrane. J Biol Chem. 1995;270:9890–5.

Trotti D, Rossi D, Gjesdal O, Levy LM, Racagni G, Danbolt NC, Volterra A. Peroxynitrite inhibits glutamate transporter subtypes. J Biol Chem. 1996;271:5976–9.

Ungard RG, Seidlitz EP, Singh G. Inhibition of breast cancer-cell glutamate release with sulfasalazine limits cancer-induced bone pain. Pain. 2014;155:28–36.

Utsunomiya-Tate N, Endou H, Kanai Y. Cloning and functional characterization of a system ASC-like Na+-dependent neutral amino acid transporter. J Biol Chem. 1996;271:14883–90.

Vandenberg RJ, Mitrovic AD, Johnston GA. Molecular basis for differential inhibition of glutamate transporter subtypes by zinc ions. Mol Pharmacol. 1998;54:189–96.

Vanelderen P, Rouwette T, Kozicz T, Heylen R, Van ZJ, Roubos EW, Vissers K. Effects of chronic administration of amitriptyline, gabapentin and minocycline on spinal brain-derived neurotrophic factor expression and neuropathic pain behavior in a rat chronic constriction injury model. Reg Anesth Pain Med. 2013;38:124–30.

Vera-Portocarrero LP, Mills CD, Ye Z, Fullwood SD, McAdoo DJ, Hulsebosch CE, Westlund KN. Rapid changes in expression of glutamate transporters after spinal cord injury. Brain Res. 2002;927:104–10.

Volterra A, Trotti D, Tromba C, Floridi S, Racagni G. Glutamate uptake inhibition by oxygen free radicals in rat cortical astrocytes. J Neurosci. 1994;14:2924–32.

Wang Z, Pekarskaya O, Bencheikh M, Chao W, Gelbard HA, Ghorpade A, Rothstein JD, Volsky DJ. Reduced expression of glutamate transporter EAAT2 and impaired glutamate transport in human primary astrocytes exposed to HIV-1 or gp120. Virology. 2003;312:60–73.

Wang LN, Yang JP, Zhan Y, Ji FH, Wang XY, Zuo JL, Xu QN. Minocycline-induced reduction of brain-derived neurotrophic factor expression in relation to cancer-induced bone pain in rats. J Neurosci Res. 2012;90:672–81.

Waszkielewicz AM, Gunia A, Sloczynska K, Marona H. Evaluation of anticonvulsants for possible use in neuropathic pain. Curr Med Chem. 2011;18:4344–58.

Weng HR, Aravindan N, Cata JP, Chen JH, Shaw AD, Dougherty PM. Spinal glial glutamate transporters downregulate in rats with taxol-induced hyperalgesia. Neurosci Lett. 2005;386:18–22.

Xin WJ, Weng HR, Dougherty PM. Plasticity in expression of the glutamate transporters GLT-1 and GLAST in spinal dorsal horn glial cells following partial sciatic nerve ligation. Mol Pain. 2009;5:15.

Yamada K, Watanabe M, Shibata T, Nagashima M, Tanaka K, Inoue Y. Glutamate transporter GLT-1 is transiently localized on growing axons of the mouse spinal cord before establishing astrocytic expression. J Neurosci. 1998;18:5706–13.

Yan H, Li CM, Li YL, Gong ZH. Effect of spinal glutamate transporter 1 on chronic constriction injury of sciatic nerve and morphine tolerance of rats. Yao Xue Xue Bao. 2009;44:581–5.

Yang M, Roman K, Chen DF, Wang ZG, Lin Y, Stephens RL Jr. GLT-1 overexpression attenuates bladder nociception and local/cross-organ sensitization of bladder nociception. Am J Physiol Renal Physiol. 2011;300:F1353–9.

Yao X. Effect of zinc exposure on HNE and GLT-1 in spinal cord culture. Neurotoxicology. 2009;30:121–6.

Yaster M, Guan X, Petralia RS, Rothstein JD, Lu W, Tao YX. Effect of inhibition of spinal cord glutamate transporters on inflammatory pain induced by formalin and complete Freund's adjuvant. Anesthesiology. 2011;114:412–23.

Yoshizumi M, Eisenach JC, Hayashida K. Riluzole and gabapentinoids activate glutamate transporters to facilitate glutamate-induced glutamate release from cultured astrocytes. Eur J Pharmacol. 2012;677:87–92.

Zerangue N, Arriza JL, Amara SG, Kavanaugh MP. Differential modulation of human glutamate transporter subtypes by arachidonic acid. J Biol Chem. 1995;270:6433–5.

Zhang H, Yoon SY, Zhang H, Dougherty PM. Evidence that spinal astrocytes but not microglia contribute to the pathogenesis of Paclitaxel-induced painful neuropathy. J Pain. 2012;13:293–303.

Molecular Characteristics, Regulation, and Function of Monocarboxylate Transporters

Roberto Elizondo-Vega and María Angeles García-Robles

Abstract Lactate transporters play an important role in the glutamate recycling. Here their kinetics and tissue distribution with emphasis on the brain are addressed. Recent evidence shows their participation in important brain functions that involve intercellular communication, such as hypothalamic glucose sensing. Furthermore, we describe the regulation of their expression and some animal models that have allowed clarification of their functions.

Keywords Lactate transporters • Glutamate transporters • Glucosensing • Lactate • Ketone bodies • Tanycytes • Astrocytes • Oligodendrocytes

List of Abbreviations

$[Ca^{2+}]_i$	Intracellular calcium concentrations
3V	Third ventricle
4-CIN	4-hydroxycinnamate
ALS	Amyotrophic lateral sclerosis
AMP	Adenosine monophosphate
AMPK	Protein kinase activated by AMP
AN	Arcuate nucleus
ATP	Adenosine triphosphate
DEPC	Diethyl pyrocarbonate
DIDS	4,4-di-isotiocianoestibeno acid -2,2-disulfonic
DMN	Dorsomedial nucleus
DNA	Deoxyribonucleic acid
GFAP	Glial fibrillary acidic protein
GLAST	Glutamate aspartate transporter
GLT-1	Glutamate transporter 1 (GLT-1)

R. Elizondo-Vega • M.A. García-Robles (✉)
Departamento de Biología Celular, Universidad de Concepción, Concepción, Chile
e-mail: mgarcia@udec.cl

A. Ortega, A. Schousboe (eds.), *Glial Amino Acid Transporters*,
Advances in Neurobiology 16, DOI 10.1007/978-3-319-55769-4_12

GLUT4	Glucose transporter type 4
HIF-1α	Hypoxia-inducible factor 1-α
HUGO	Human Genome Organization
IGF-1	Insulin-like growth factor 1
MCTs	Monocarboxylate transporters
mTOR	AKT kinase and mechanistic target of rapamycin
NFAT	Nuclear factor of activated T cells
NPY	Neuropeptide Y
pCMB	P-chloromercuribenzene sulfonate
PEPCK	Phosphoenolpyruvate carboxykinase
PGC-1α	Peroxisome proliferator-activated receptor gamma coactivator 1-alpha
PVN	Paraventricular nucleus
T3	Triiodothyronine
TM	Transmembrane

1 Introduction

MCTs comprise a family of 14 members (MCT1-14) distributed in a wide variety of tissues (Halestrap 2012, 2013a, 2013b); they use the electrochemical gradient of protons to translocate monocarboxylates and protons at a 1:1 ratio in the same direction (Gladden 2004; Halestrap and Price 1999). MCT1 has the widest selectivity and may transport several short chain fatty acids, such as acetate, butyrate, lactate, or pyruvate (Carneiro and Pellerin 2015). In the brain, lactate is an important oxidative energy substrate (Magistretti and Pellerin 1999), and its intracerebroventricular (icv) administration decreases both food intake and blood glucose levels in rats (Lam et al. 2008). Ketone bodies also affect food intake (Carneiro et al. 2016; Hawkins et al. 1986) and body weight in an opposing manner (Iwata et al. 2011). There is evidence that lactic acid produced by astrocytes can be exported for use in neurons as respiratory fuel or for glucose synthesis. Such is the case in a model that proposes a metabolic coupling between glutamatergic neurons and astrocytes (Pellerin and Magistretti 1994). In the glutamatergic synapse, astrocytes located between pre- and postsynaptic neurons remove 95% of the glutamate released through the glutamate transporters, glutamate aspartate transporter (GLAST) and glutamate transporter 1 (GLT-1) (Anderson and Swanson 2000). Sodium glutamate-coupled transport results in the accumulation of intracellular sodium and consequent activation of the Na^+/K^+ pump, causing a decrease in intracellular ATP content, which is enhanced by metabolizing glutamate to glutamine. Thus, glutamate released into the synaptic cleft triggers glucose uptake by astrocytes, after which glucose is converted to lactate. Astrocyte modulation of glucose uptake in response to glutamate has recently been demonstrated in a human stem cell-derived coculture system (NT2-derived neurons and astrocytes) (Tarczyluk et al. 2015). Neuronal lactate can be used as a fuel. On the other hand, it has been postulated that in steady-state conditions, glutamate release from neurons and lactate generation in astrocytes from Krebs cycle intermediates could result in lactate export into the blood (Sonnewald 2014). Unlike neurons, astrocytes express malate dehydrogenase (oxaloacetate-decarboxylating) and phosphoenolpyruvate carboxykinase (PEPCK); therefore, both malate and oxaloacetate can be converted into lactate (Sonnewald 2014), which can leave the brain via the blood or go to the periventricular system. However, during brain activation, when glycolysis is strongly upregulated, aerobic glycolysis releases glucose-derived lactate into the brain (Sonnewald 2014). Increased oxidation of glutamate after its uptake into astrocytes may also contribute to increased lactate efflux (Pellerin and Magistretti 1994). Both options may be occurring in varying proportions, probably depending on the energy state of the organism. Regardless of that, lactate transporters undoubtedly play an important function in glutamate recycling.

2 Monocarboxylate Transporters: Distribution and Specific Functions

The MCTs belong to the solute carrier 16 (SLC16) gene family according to the Human Genome Organization (HUGO) nomenclature. MCT (SLC16) family is composed of 14 isoforms (MCT1–MCT14). They are predicted to contain 12 transmembrane helices with intracellular C- and N-termini and a large cytosolic loop between TM6 and TM7. All members contain two highly conserved sequences located in TM1 and TM5 and are not glycosylated (Halestrap and Price 1999). Only MCT1, MCT2, MCT3, and MCT4 have been described as symporters of monocarboxylates and protons, such as L-lactate, L-acetoacetate, and DL-β-hydroxybutyrate, in a stoichiometric ratio of 1:1 (Poole and Halestrap 1993). However, MCT7 has been recently described to be associated with β-hydroxybutyrate transport in the plasma membrane of Zebrafish hepatocytes (Hugo et al. 2012). MCT6 has been described as a transporter of the diuretic drug, bumetanide; however, its endogenous substrate is unknown (Murakami et al. 2005). MCT8 has been described as a high-affinity transporter of thyroid hormones (Friesema et al. 2003), and it has been suggested that MCT9 facilitates the transport of carnitine (Suhre et al. 2011). MCT10 has been described as an aromatic amino acid transporter (Kim et al. 2002). The other members of the family of MCTs are recognized as orphan transporters given that their substrates remain unknown (Halestrap 2013a).

Of the 14 known isoforms, only MCTs 1–4 have been characterized kinetically for transporting monocarboxylates (Table 1). The transport capacity of MCT1, MCT2, and MCT4 is reversibly affected by a series of competitive inhibitors, such as 4-hydroxycinnamate (4-CIN); inhibitors of general anionic transport, such as 4,4-di-isotiocianoestibeno acid -2,2-disulfonic (DIDS); and the flavonoids, quercetin and phloretin (Broer et al. 1997, 1998, 1999; Dimmer et al. 2000; Lin et al. 1998). Transport can also be irreversibly affected by organomercury compounds, such as

Table 1 Km values of MCT isoforms expressed in the brain

Protein name	MCT1	MCT2	MCT3	MCT4
Gene name	SLC16A1	SLC16A7	SLC16A8	SLC16A3
Substrates	Lactate Pyruvate Ketones bodies	Lactate Pyruvate Ketones bodies	Lactate	Lactate Pyruvate Ketones bodies
Km for L-lactate (mM)	7.7	0.74	5.8	34
Km for pyruvate (mM)	1.0	0.08	–	153
Km for D-ß-hydroxybutyrate (mM)	12.5	1.2	–	64
Km for Acetoacetate (mM)	5.5	0.8	–	31
Expression in the brain	Cortical hippocampal and supraoptic nucleus astrocytes; choroid plexus; ependymal cells; endothelial cell; pericytes; α- and β-tanycytes	Neurons of cerebral cortex; Purkinje cells; ependymal cells; subependymal astrocytes; hypothalamic orexigenic and anorexigenic neurons	Basolateral membrane of choroid plexus	Bergmann glia; cerebellum, hippocampus and corpus callosum astrocytes; cerebral cortex; ependymal cells; α- and β-tanycytes
References	Broer et al. (1997), Broer et al. (1998), Broer et al. (1999), Dimmer et al. (2000), Carpenter (1994), Cortes-Campos et al. (2011), Gerhart et al. (1997), Gerhart et al. (1998) Hanu et al. (2000), Leino et al. (1999), Pierre et al. (2000), Pellerin et al. (2005), Rafiki et al. (2003), Koehler-Stec et al. (1998)	Broer et al. (1998), Lin et al. (1998), Pierre and Pellerin (2005), Bergersen et al. (2001), Bergersen (2007), Gerhart et al. (1997), Pierre et al. (2000), Cortes-Campos et al. (2013)	Grollman et al. (2000), Philp et al. (2003)	Broer et al. (1997), Broer et al. (1998), Broer et al. (1999), Dimmer et al. (2000), Carpenter (1994), Rafiki et al. (2003), Bergersen et al. (2001), Cortes-Campos et al. (2011)

p-chloromercuribenzene sulfonate (pCMBS), which is capable of inhibiting the transport of MCT1 and MCT4 by eliminating disulfide bridges present in the chaperone protein, basigin, or CD147 (Wilson et al. 2005). Thus, their association with MCT1 and MCT4 is inhibited. However, pCMBS does not alter MCT2, a transporter that interacts with the chaperone protein, embigin (also known as GP-70), which is insensitive to pCMBS (Broer et al. 1999; Garcia et al. 1995; Wilson et al. 2005). The MCT3 transporter is only inhibited by diethyl pyrocarbonate (DEPC); however, the effect of the inhibitors described above has not been reported (Grollman et al. 2000).

Specific inhibitors for MCT1, AR-C117977, and AR-C155858 have recently been generated. However, AR-C155858 also inhibits MCT2 when it interacts with basigin but not when it is associated with embigin (Ovens et al. 2010a, 2010b). In contrast, AR-C117977 inhibits both transporters with similar affinity (Murray et al. 2005; Ovens et al. 2010a).

Localization studies have determined that MCT1 is ubiquitously expressed both in tissues producing monocarboxylates (e.g., erythrocytes) and in those consuming monocarboxylates (e.g., striated muscle skeletal and cardiac striated muscle) (Broer et al. 1997; Halestrap 2013a; Halestrap and Price 1999; Jackson and Halestrap 1996). Its peripheral location has been reported mainly in myocytes (Halestrap et al. 1997), striated skeletal muscle (Garcia et al. 1995; Wilson et al. 1998), basolateral membranes of renal cortex proximal tubules, choroid plexus, blood vessels, and ependyma (Koehler-Stec et al. 1998; Pierre et al. 2000). In the CNS, MCT1 has been detected in astrocytes (Gerhart et al. 1997; Hanu et al. 2000; Pierre et al. 2000), endothelial cells, pericytes (Gerhart et al. 1997; Leino et al. 1999), primary glial cultures, and hypothalamic neurons (Ainscow et al. 2002). In the hypothalamus, MCT1 is expressed in α- and β-tanycytes lining the ventricular walls and floor of the third ventricle (3V), as well as in endothelial cells of blood vessels (Cortes-Campos et al. 2011). In α-tanycytes, MCT1 is polarized in ventricular cell membranes and feet terminals contacting endothelial cells of blood vessels (Cortes-Campos et al. 2011; Elizondo-Vega et al. 2016; Elizondo-Vega et al. 2015). In the ventricular wall, β1v-tanycytes possess an intense immunoreaction for MCT1 in the apical membrane, which is also observed in their cellular processes that contact orexigenic neurons in the arcuate nucleus (AN), blood vessels, and the outer region of the brain (Cortes-Campos et al. 2011). A weak immunoreaction has also been detected in β2-tanycytes located in the median eminence. Importantly, there is intense reactivity for MCT1 in astrocytes and marginal glia in this region. By immunohistochemical analysis, MCT1 expression has recently been reported in neurons that express NPY (neuropeptide Y) (Carneiro et al. 2016), which is coincident with the capacity of this carrier to allow both the influx and release of monocarboxylates.

MCT4 has been located only in tissues that produce lactate (Manning Fox et al. 2000; Rafiki et al. 2003) and possess high glycolytic capacity, including the retinal pigment epithelium of rats (Bergersen et al. 1999), white skeletal muscle fibers (Garcia et al. 1995; Wilson et al. 1998), chondrocytes (Meredith et al. 2002), leukocytes (Wilson et al. 1998), and striated skeletal muscle of neonatal rats (Hatta et al. 2001; Wilson et al. 1998). In the brain, MCT4 has been detected in the astrocytes of the cortex, hippocampus, cerebellum, corpus callosum, and internal capsule (Bergersen et al. 2001; Pellerin et al. 2005; Rafiki et al. 2003). MCT4 has also been detected in hypothalamic astrocytes and ependymal cells close to the paraventricular nucleus (PVN) (Pellerin et al. 2005). Similar to MCT1, MCT4 was also identified in the lateral region of the AN, particularly in dorsal β1-tanycyte processes, which contact anorexigenic neurons. These processes are positive for GFAP but have low immunoreactivity for vimentin, compared to ventral β1-tanycyte processes. In the median eminence, MCT4 is expressed in subependymal astrocytes, astrocytes, and some blood vessels (Cortes-Campos et al. 2011).

Unlike MCT1 and MCT4, MCT2 is differentially expressed in various tissues depending on the species; however, it is usually found in tissues incorporating monocarboxylates, such as the testis, kidney, stomach, liver, lung, brain, and epididymis (Koehler-Stec et al. 1998). In the brain, MCT2 expression is confined especially in the postsynaptic density of neurons, a region rich in mitochondria, and is believed capable of preferentially oxidizing lactate as an energy source (Bergersen et al. 2001; Chiry et al. 2008; Pierre and Pellerin 2005). MCT2 expression is restricted to neurons (Pierre et al. 2000), glia *limitans*, ependymal cells (Gerhart et al. 1997), and astrocyte vascular processes (Gerhart et al. 1997; Pierre et al. 2000). In the hypothalamus, MCT2 has been detected in orexigenic and anorexigenic neurons of the AN (Cortes-Campos et al. 2013). The presence of MCT1 and MCT4 in tanycyte processes and MCT2 in hypothalamic neurons strongly suggests that there may be a transfer of lactate as a signal derived from glucose from tanycytes to the neurons to support the hypothalamic glucosensing. In addition, MCT2 expression increases in the dorsomedial nucleus (DMN) and anterior hypothalamic nucleus in rats fed with a high-fat diet (Pierre et al. 2007). The MCT3 isoform has a more restricted location, in the basolateral membranes of the retinal pigment epithelium and choroid plexus cells (Philp et al. 2003; Yoon et al. 1999).

3 MCT Regulation

The expression of the MCT isoforms depends on the metabolic state of the cell, and the regulation may be mediated by both transcriptional and posttranscriptional mechanisms. In the brain, an increase in MCT1 expression was reported in obese mice or mice subjected to ketosis and chronic hyperglycemia (Canis et al. 2009; Leino et al. 2001; Pierre et al. 2007).

It has been demonstrated that MCT4 is greatly increased by hypoxia (Perez de Heredia et al. 2010; Ullah et al. 2006), which is consistent with its proposed role in exporting lactic acid produced by glycolysis. This effect can be mediated by hypoxia-inducible factor 1-α (HIF-1α) (Perez de Heredia et al. 2010; Ullah et al. 2006), as well as the hypoxia response elements found in the *MCT4* promoter (Ullah et al. 2006).

Similarly, it has been proposed that during intense aerobic exercise, the increase in intracellular calcium concentrations ($[Ca^{2+}]_i$) and AMP levels can induce MCT1 and MCT4 expression in a mechanism dependent on calcineurin and protein kinase activated by AMP (AMPK) (Halestrap 2012). Calcineurin dephosphorylation and activation induce NFAT transcription factor activation (Bigard et al. 2000), resulting in its interaction with DNA-binding sequences located in the promoter region of *MCT1* (Halestrap 2012). In addition, AMPK activation in conjunction with increased $[Ca^{2+}]_i$ activates the transcriptional coactivator, PGC1α (Lee et al. 2006; Ojuka 2004), which increases MCT1 expression during muscle activity (Benton et al. 2008).

It has also been reported that the thyroid hormone, T3, is capable of increasing the transcription of MCT1 and MCT4 in skeletal muscle; however, only increased protein expression levels of MCT4 were detected (Wang et al. 2003). Downregulation of MCT1 expression has also been reported in pancreatic β cells (Zhao et al. 2001), via a posttranscriptional mechanism that includes the expression of microRNA 29 (miRNA 29) that prevents MCT1 mRNA translation (Pullen et al. 2011).

Literature describing the transcriptional regulation of MCT2 is more limited. However, following a 48-h fast, an increase in β-hydroxybutyrate concentration and increased MCT2 expression were detected in the brainstem of female rats (Matsuyama et al. 2009). In this context, it has been reported that expression of MCT2 mRNA is greater in the neonatal brain than in the adult brain, which may be associated with the high concentration of ketone bodies produced by hepatic lipid oxidation from breast milk, the main food component in the neonatal stage (Cremer 1982; Dombrowski et al. 1989; Hawkins et al. 1971). In addition, there is evidence that the expression of MCT2 increases in response to norepinephrine (Chenal and Pellerin 2007), insulin, and insulin-like growth factor 1 (IGF-1), a signaling mechanism mediated by AKT kinase and mechanistic target of rapamycin (mTOR) (Chenal et al. 2008). Another potential mechanism of MCT regulation includes transporter translocation involving intracellular storing compartments of the plasma membrane, as has been reported for the GLUT4 transporter (Bogan 2012). In this context, it has been described that hypertrophy of the left ventricle of the heart, in response to ischemia, occurs in parallel with an increase in MCT1 levels at the plasma membrane, which occurs without changes in its mRNA expression (Johannsson et al. 2001). Recently, it has been reported that the expression of MCTs in the hypothalamus and in different brain regions may be affected by prolonged fasting in rats, with upregulated levels of MCT1 and MCT2 in the cerebral cortex, hippocampus, and hypothalamus. However, increased MCT4 was only detected in the hypothalamus, suggesting that the expression of MCTs is dependent on specific brain functions, including control of food intake in the hypothalamus (Takimoto and Hamada 2014).

The expression of MCT1 is also essential in development, as MCT1 knockout is lethal; however, MCT1 +/− mice survive, developing axonal degeneration observed in the spinal cord and brain (Lee et al. 2012). In this context, it is interesting that in patients suffering from amyotrophic lateral sclerosis (ALS), there is a reduction in the expression of MCT1 by oligodendrocytes (Lee et al. 2012), suggesting that it has a role in the pathogenesis of the disease. This would be consistent with the role of oligodendrocytes in metabolically supporting axonal and neuronal cells to deliver lactate as an energy supply, as has been proposed in the glia-neuron model of metabolic coupling (Bergersen 2007; Pierre and Pellerin 2005). Finally, it has recently been shown that MCT1 +/− mice exhibit resistance to obesity induced by a high-fat diet, as well as reduced insulin resistance and decreased hepatic steatosis, compared to MCT1 +/+ mice (Lengacher et al. 2013). Interestingly, MCT1 +/− animals have normal levels of lactate and ketone bodies in the blood but have reduced levels of insulin and leptin. These findings uncover the important role of MCT1 in the regulation of energy balance in animals exposed to an obesogenic diet (Lengacher et al. 2013).

4 Conclusion

The different studies associated with MCTs have focused on describing their location or transport capacity, as well as their alterations in diverse pathologies, either at the CNS level or in peripheral tissues. The inhibition of these transporters has been proposed to solve problems related to cancer or brain bioenergetics; however, the lack of studies focused on elucidating the physiological role of these carriers makes it difficult to fully understand their effects and extrapolate these results to a pharmacological therapy. On the other hand, it is important to emphasize that many members of the MCT family remain as orphan transporters, needing to deepen the functional studies and the physiological or pathophysiological importance of these transporters.

In this context, it has been determined that some of these transporters, such as MCT5, MCT6, MCT9, MCT12, and MCT14, lack important residues for the transport of protons, which has led to propose that these mutations would affect its functionality (Halestrap 2013b). However, one cannot rule out the possibility that these proteins can transport molecules in a non-proton-associated way or even that they are not functioning as transporters but as receptors. On the other hand, studies on the regulation of expression of MCTs have focused mainly on MCT1, partly because of its greater expression and ubiquity. A closer look at the regulatory aspects of expression for MCT2, MCT3, MCT4, MCT8, and MCT10 would allow us to design new and better treatments for pathologies where these transporters are involved, allowing the development of more efficient therapeutic strategies.

Conflict of Interest The author declares no conflicts of interest.

References

Ainscow EK, Mirshamsi S, Tang T, Ashford ML, Rutter GA. Dynamic imaging of free cytosolic ATP concentration during fuel sensing by rat hypothalamic neurones: evidence for ATP-independent control of ATP-sensitive K(+) channels. J Physiol. 2002;544(Pt 2):429–45.

Anderson CM, Swanson RA. Astrocyte glutamate transport: review of properties, regulation, and physiological functions. Glia. 2000;32(1):1–14.

Benton CR, Yoshida Y, Lally J, Han X-X, Hatta H, Bonen A. PGC-1α increases skeletal muscle lactate uptake by increasing the expression of MCT1 but not MCT2 or MCT4. Physiol Genomics. 2008;35(1):45–54.

Bergersen LH. Is lactate food for neurons? Comparison of monocarboxylate transporter subtypes in brain and muscle. Neuroscience. 2007;145(1):11–9.

Bergersen L, Johannsson E, Veruki ML, Nagelhus EA, Halestrap A, Sejersted OM, Ottersen OP. Cellular and subcellular expression of monocarboxylate transporters in the pigment epithelium and retina of the rat. Neuroscience. 1999;90(1):319–31.

Bergersen L, Waerhaug O, Helm J, Thomas M, Laake P, Davies AJ, Wilson MC, Halestrap AP, Ottersen OP. A novel postsynaptic density protein: the monocarboxylate transporter MCT2 is co-localized with delta-glutamate receptors in postsynaptic densities of parallel fiber-Purkinje cell synapses. Exp Brain Res. 2001;136(4):523–34.

Bigard X, Sanchez H, Zoll J, Mateo P, Rousseau V, Veksler V, Ventura-Clapier R. Calcineurin co-regulates contractile and metabolic components of slow muscle phenotype. J Biol Chem. 2000;275(26):19653–60.

Bogan JS. Regulation of glucose transporter translocation in health and diabetes. Annu Rev Biochem. 2012;81:507–32.

Broer S, Rahman B, Pellegri G, Pellerin L, Martin JL, Verleysdonk S, Hamprecht B, Magistretti PJ. Comparison of lactate transport in astroglial cells and monocarboxylate transporter 1 (MCT 1) expressing Xenopus laevis oocytes. Expression of two different monocarboxylate transporters in astroglial cells and neurons. J Biol Chem. 1997;272(48):30096–102.

Broer S, Schneider HP, Broer A, Rahman B, Hamprecht B, Deitmer JW. Characterization of the monocarboxylate transporter 1 expressed in Xenopus laevis oocytes by changes in cytosolic pH. Biochem J. 1998;333(Pt 1):167–74.

Broer S, Broer A, Schneider HP, Stegen C, Halestrap AP, Deitmer JW. Characterization of the high-affinity monocarboxylate transporter MCT2 in Xenopus laevis oocytes. Biochem J. 1999;341(Pt 3):529–35.

Canis M, Maurer MH, Kuschinsky W, Duembgen L, Duelli R. Increased densities of monocarboxylate transporter MCT1 after chronic hyperglycemia in rat brain. Brain Res. 2009;1257:32–9.

Carneiro L, Pellerin L. Monocarboxylate transporters: new players in body weight regulation. Obes Rev. 2015;16(Suppl 1):55–66.

Carneiro L, Geller S, Fioramonti X, Hebert A, Repond C, Leloup C, Pellerin L. Evidence for hypothalamic ketone body sensing: impact on food intake and peripheral metabolic responses in mice. Am J Physiol Endocrinol Metab. 2016;310(2):E103–15.

Carpenter L, Halestrap AP. The kinetics, substrate and inhibitor specificity of the lactate transporter of Ehrlich-Lettre tumour cells studied with the intracellular pH indicator BCECF. Biochemical Journal. 1994;304(3):751–760.

Cortes-Campos C, Elizondo R, Llanos P, Uranga RM, Nualart F, Garcia MA. MCT expression and lactate influx/efflux in tanycytes involved in glia-neuron metabolic interaction. PLoS One. 2011;6(1):e16411.

Cortes-Campos C, Elizondo R, Carril C, Martinez F, Boric K, Nualart F, Garcia-Robles MA. MCT2 expression and lactate influx in anorexigenic and orexigenic neurons of the arcuate nucleus. PLoS One. 2013;8(4):e62532.

Cremer JE. Substrate utilization and brain development. J Cereb Blood Flow Metab. 1982;2(4):394–407.

Chenal J, Pellerin L. Noradrenaline enhances the expression of the neuronal monocarboxylate transporter MCT2 by translational activation via stimulation of PI3K/Akt and the mTOR/S6K pathway. J Neurochem. 2007;102(2):389–97.

Chenal J, Pierre K, Pellerin L. Insulin and IGF-1 enhance the expression of the neuronal monocarboxylate transporter MCT2 by translational activation via stimulation of the phosphoinositide 3-kinase-Akt-mammalian target of rapamycin pathway. Eur J Neurosci. 2008;27(1):53–65.

Chiry O, Fishbein WN, Merezhinskaya N, Clarke S, Galuske R, Magistretti PJ, Pellerin L. Distribution of the monocarboxylate transporter MCT2 in human cerebral cortex: an immunohistochemical study. Brain Res. 2008;1226:61–9.

Dimmer KS, Friedrich B, Lang F, Deitmer JW, Broer S. The low-affinity monocarboxylate transporter MCT4 is adapted to the export of lactate in highly glycolytic cells. Biochem J. 2000;350(Pt 1):219–27.

Dombrowski GJ Jr, Swiatek KR, Chao KL. Lactate, 3-hydroxybutyrate, and glucose as substrates for the early postnatal rat brain. Neurochem Res. 1989;14(7):667–75.

Elizondo-Vega R, Cortes-Campos C, Barahona MJ, Oyarce KA, Carril CA, Garcia-Robles MA. The role of tanycytes in hypothalamic glucosensing. J Cell Mol Med. 2015;19(7):1471–82.

Elizondo-Vega R, Cortes-Campos C, Barahona MJ, Carril C, Ordenes P, Salgado M, Oyarce K, Garcia-Robles ML. Inhibition of hypothalamic MCT1 expression increases food intake and alters orexigenic and anorexigenic neuropeptide expression. Sci Rep. 2016;6:33606.

Friesema EC, Ganguly S, Abdalla A, Manning Fox JE, Halestrap AP, Visser TJ. Identification of monocarboxylate transporter 8 as a specific thyroid hormone transporter. J Biol Chem. 2003;278(41):40128–35.

Garcia CK, Brown MS, Pathak RK, Goldstein JL. cDNA cloning of MCT2, a second monocarboxylate transporter expressed in different cells than MCT1. J Biol Chem. 1995;270(4):1843–9.

Gerhart DZ, Enerson BE, Zhdankina OY, Leino RL, Drewes LR. Expression of monocarboxylate transporter MCT1 by brain endothelium and glia in adult and suckling rats. Am J Phys. 1997;273(1 Pt 1):E207–13.

Gerhart DZ, Enerson BE, Zhdankina OY, Leino RL, Drewes LR. Expression of the monocarboxylate transporter MCT2 by rat brain glia. Glia. 1998;22(3):272–281.

Gladden LB. Lactate metabolism: a new paradigm for the third millennium. J Physiol. 2004;558(Pt 1):5–30.

Grollman EF, Philp NJ, McPhie P, Ward RD, Sauer B. Determination of transport kinetics of chick MCT3 monocarboxylate transporter from retinal pigment epithelium by expression in genetically modified yeast. Biochemistry. 2000;39(31):9351–7.

Halestrap AP. The monocarboxylate transporter family – structure and functional characterization. IUBMB Life. 2012;64(1):1–9.

Halestrap AP. Monocarboxylic acid transport. Compr Physiol. 2013a;3(4):1611–43.

Halestrap AP. The SLC16 gene family – structure, role and regulation in health and disease. Mol Asp Med. 2013b;34(2–3):337–49.

Halestrap AP, Price NT. The proton-linked monocarboxylate transporter (MCT) family: structure, function and regulation. Biochem J. 1999;343(Pt 2):281–99.

Halestrap AP, Wang X, Poole RC, Jackson VN, Price NT. Lactate transport in heart in relation to myocardial ischemia. Am J Cardiol. 1997;80(3A):17A–25A.

Hanu R, McKenna M, O'Neill A, Resneck WG, Bloch RJ. Monocarboxylic acid transporters, MCT1 and MCT2, in cortical astrocytes in vitro and in vivo. Am J Physiol Cell Physiol. 2000;278(5):C921–30.

Hatta H, Tonouchi M, Miskovic D, Wang Y, Heikkila JJ, Bonen A. Tissue-specific and isoform-specific changes in MCT1 and MCT4 in heart and soleus muscle during a 1-yr period. Am J Physiol Endocrinol Metab. 2001;281(4):E749–56.

Hawkins RA, Williamson DH, Krebs HA. Ketone-body utilization by adult and suckling rat brain in vivo. Biochem J. 1971;122(1):13–8.

Hawkins RA, Mans AM, Davis DW. Regional ketone body utilization by rat brain in starvation and diabetes. Am J Phys. 1986;250(2 Pt 1):E169–78.

Hugo SE, Cruz-Garcia L, Karanth S, Anderson RM, Stainier DY, Schlegel A. A monocarboxylate transporter required for hepatocyte secretion of ketone bodies during fasting. Genes Dev. 2012;26(3):282–93.

Iwata K, Kinoshita M, Yamada S, Imamura T, Uenoyama Y, Tsukamura H, Maeda K. Involvement of brain ketone bodies and the noradrenergic pathway in diabetic hyperphagia in rats. J Physiol Sci. 2011;61(2):103–13.

Jackson VN, Halestrap AP. The kinetics, substrate, and inhibitor specificity of the monocarboxylate (lactate) transporter of rat liver cells determined using the fluorescent intracellular pH indicator, 2',7'-bis(carboxyethyl)-5(6)-carboxyfluorescein. J Biol Chem. 1996;271(2):861–8.

Johannsson E, Lunde PK, Heddle C, Sjaastad I, Thomas MJ, Bergersen L, Halestrap AP, Blackstad TW, Ottersen OP, Sejersted OM. Upregulation of the cardiac monocarboxylate transporter MCT1 in a rat model of congestive heart failure. Circulation. 2001;104(6):729–34.

Kim DK, Kanai Y, Matsuo H, Kim JY, Chairoungdua A, Kobayashi Y, Enomoto A, Cha SH, Goya T, Endou H. The human T-type amino acid transporter-1: characterization, gene organization, and chromosomal location. Genomics. 2002;79(1):95–103.

Koehler-Stec EM, Simpson IA, Vannucci SJ, Landschulz KT, Landschulz WH. Monocarboxylate transporter expression in mouse brain. Am J Phys. 1998;275(3 Pt 1):E516–24.

Lam CK, Chari M, Wang PY, Lam TK. Central lactate metabolism regulates food intake. Am J Physiol Endocrinol Metab. 2008;295(2):E491–6.

Lee WJ, Kim M, Park HS, Kim HS, Jeon MJ, Oh KS, Koh EH, Won JC, Kim MS, Oh GT, Yoon M, Lee KU, Park JY. AMPK activation increases fatty acid oxidation in skeletal muscle by activating PPARalpha and PGC-1. Biochem Biophys Res Commun. 2006;340(1):291–5.

Lee Y, Morrison BM, Li Y, Lengacher S, Farah MH, Hoffman PN, Liu Y, Tsingalia A, Jin L, Zhang PW, Pellerin L, Magistretti PJ, Rothstein JD. Oligodendroglia metabolically support axons and contribute to neurodegeneration. Nature. 2012;487(7408):443–8.

Leino RL, Gerhart DZ, Drewes LR. Monocarboxylate transporter (MCT1) abundance in brains of suckling and adult rats: a quantitative electron microscopic immunogold study. Brain Res Dev Brain Res. 1999;113(1–2):47–54.

Leino RL, Gerhart DZ, Duelli R, Enerson BE, Drewes LR. Diet-induced ketosis increases monocarboxylate transporter (MCT1) levels in rat brain. Neurochem Int. 2001;38(6):519–27.

Lengacher S, Nehiri-Sitayeb T, Steiner N, Carneiro L, Favrod C, Preitner F, Thorens B, Stehle JC, Dix L, Pralong F, Magistretti PJ, Pellerin L. Resistance to diet-induced obesity and associated metabolic perturbations in haploinsufficient monocarboxylate transporter 1 mice. PLoS One. 2013;8(12):e82505.

Lin RY, Vera JC, Chaganti RS, Golde DW. Human monocarboxylate transporter 2 (MCT2) is a high affinity pyruvate transporter. J Biol Chem. 1998;273(44):28959–65.

Magistretti PJ, Pellerin L. Cellular mechanisms of brain energy metabolism and their relevance to functional brain imaging. Philos Trans R Soc Lond Ser B Biol Sci. 1999;354(1387):1155–63.

Manning Fox JE, Meredith D, Halestrap AP. Characterisation of human monocarboxylate transporter 4 substantiates its role in lactic acid efflux from skeletal muscle. J Physiol. 2000;529(Pt 2):285–93.

Matsuyama S, Ohkura S, Iwata K, Uenoyama Y, Tsukamura H, Maeda K, Kimura K. Food deprivation induces monocarboxylate transporter 2 expression in the brainstem of female rat. J Reprod Dev. 2009;55(3):256–61.

Meredith D, Bell P, McClure B, Wilkins R. Functional and molecular characterisation of lactic acid transport in bovine articular chondrocytes. Cell Physiol Biochem. 2002;12(4):227–34.

Murakami Y, Kohyama N, Kobayashi Y, Ohbayashi M, Ohtani H, Sawada Y, Yamamoto T. Functional characterization of human monocarboxylate transporter 6 (SLC16A5). Drug Metab Dispos. 2005;33(12):1845–51.

Murray CM, Hutchinson R, Bantick JR, Belfield GP, Benjamin AD, Brazma D, Bundick RV, Cook ID, Craggs RI, Edwards S, Evans LR, Harrison R, Holness E, Jackson AP, Jackson CG, Kingston LP, Perry MW, Ross AR, Rugman PA, Sidhu SS, Sullivan M, Taylor-Fishwick DA, Walker PC, Whitehead YM, Wilkinson DJ, Wright A, Donald DK. Monocarboxylate transporter MCT1 is a target for immunosuppression. Nat Chem Biol. 2005;1(7):371–6.

Ojuka EO. Role of calcium and AMP kinase in the regulation of mitochondrial biogenesis and GLUT4 levels in muscle. Proc Nutr Soc. 2004;63(2):275–8.

Ovens MJ, Davies AJ, Wilson MC, Murray CM, Halestrap AP. AR-C155858 is a potent inhibitor of monocarboxylate transporters MCT1 and MCT2 that binds to an intracellular site involving transmembrane helices 7–10. Biochem J. 2010a;425(3):523–30.

Ovens MJ, Manoharan C, Wilson MC, Murray CM, Halestrap AP. The inhibition of monocarboxylate transporter 2 (MCT2) by AR-C155858 is modulated by the associated ancillary protein. Biochem J. 2010b;431(2):217–25.

Pellerin L, Magistretti PJ. Glutamate uptake into astrocytes stimulates aerobic glycolysis: a mechanism coupling neuronal activity to glucose utilization. Proc Natl Acad Sci U S A. 1994;91(22):10625–9.

Pellerin L, Bergersen LH, Halestrap AP, Pierre K. Cellular and subcellular distribution of monocarboxylate transporters in cultured brain cells and in the adult brain. J Neurosci Res. 2005;79(1–2):55–64.

Perez de Heredia F, Wood IS, Trayhurn P. Hypoxia stimulates lactate release and modulates monocarboxylate transporter (MCT1, MCT2, and MCT4) expression in human adipocytes. Pflugers Arch. 2010;459(3):509–18.

Philp NJ, Ochrietor JD, Rudoy C, Muramatsu T, Linser PJ. Loss of MCT1, MCT3, and MCT4 expression in the retinal pigment epithelium and neural retina of the 5A11/basigin-null mouse. Invest Ophthalmol Vis Sci. 2003;44(3):1305–11.

Pierre K, Pellerin L. Monocarboxylate transporters in the central nervous system: distribution, regulation and function. J Neurochem. 2005;94(1):1–14.

Pierre K, Pellerin L, Debernardi R, Riederer BM, Magistretti PJ. Cell-specific localization of monocarboxylate transporters, MCT1 and MCT2, in the adult mouse brain revealed by double immunohistochemical labeling and confocal microscopy. Neuroscience. 2000;100(3):617–27.

Pierre K, Parent A, Jayet PY, Halestrap AP, Scherrer U, Pellerin L. Enhanced expression of three monocarboxylate transporter isoforms in the brain of obese mice. J Physiol. 2007;583(Pt 2):469–86.

Poole RC, Halestrap AP. Transport of lactate and other monocarboxylates across mammalian plasma membranes. Am J Phys. 1993;264(4 Pt 1):C761–82.

Pullen TJ, da Silva Xavier G, Kelsey G, Rutter GA. miR-29a and miR-29b contribute to pancreatic beta-cell-specific silencing of monocarboxylate transporter 1 (Mct1). Mol Cell Biol. 2011;31(15):3182–94.

Rafiki A, Boulland JL, Halestrap AP, Ottersen OP, Bergersen L. Highly differential expression of the monocarboxylate transporters MCT2 and MCT4 in the developing rat brain. Neuroscience. 2003;122(3):677–88.

Sonnewald U. Glutamate synthesis has to be matched by its degradation – where do all the carbons go? J Neurochem. 2014;131(4):399–406.

Suhre K, Shin SY, Petersen AK, Mohney RP, Meredith D, Wagele B, Altmaier E, CardioGram, Deloukas P, Erdmann J, Grundberg E, Hammond CJ, de Angelis MH, Kastenmuller G, Kottgen A, Kronenberg F, Mangino M, Meisinger C, Meitinger T, Mewes HW, Milburn MV, Prehn C, Raffler J, Ried JS, Romisch-Margl W, Samani NJ, Small KS, Wichmann HE, Zhai G, Illig T, Spector TD, Adamski J, Soranzo N, Gieger C. Human metabolic individuality in biomedical and pharmaceutical research. Nature. 2011;477(7362):54–60.

Takimoto M, Hamada T. Acute exercise increases brain region-specific expression of MCT1, MCT2, MCT4, GLUT1, and COX IV proteins. J Appl Physiol (1985). 2014;116(9):1238–50.

Tarczyluk MA, Nagel DA, Rhein Parri H, Tse EH, Brown JE, Coleman MD, Hill EJ. Amyloid beta 1-42 induces hypometabolism in human stem cell-derived neuron and astrocyte networks. J Cereb Blood Flow Metab. 2015;35(8):1348–57.

Ullah MS, Davies AJ, Halestrap AP. The plasma membrane lactate transporter MCT4, but not MCT1, is up-regulated by hypoxia through a HIF-1alpha-dependent mechanism. J Biol Chem. 2006;281(14):9030–7.

Wang Y, Tonouchi M, Miskovic D, Hatta H, Bonen A. T3 increases lactate transport and the expression of MCT4, but not MCT1, in rat skeletal muscle. Am J Physiol Endocrinol Metab. 2003;285(3):E622–8.
Wilson MC, Jackson VN, Heddle C, Price NT, Pilegaard H, Juel C, Bonen A, Montgomery I, Hutter OF, Halestrap AP. Lactic acid efflux from white skeletal muscle is catalyzed by the monocarboxylate transporter isoform MCT3. J Biol Chem. 1998;273(26):15920–6.
Wilson MC, Meredith D, Fox JE, Manoharan C, Davies AJ, Halestrap AP. Basigin (CD147) is the target for organomercurial inhibition of monocarboxylate transporter isoforms 1 and 4: the ancillary protein for the insensitive MCT2 is EMBIGIN (gp70). J Biol Chem. 2005;280(29):27213–21.
Yoon H, Donoso LA, Philp NJ. Cloning of the human monocarboxylate transporter MCT3 gene: localization to chromosome 22q12.3-q13.2. Genomics. 1999;60(3):366–70.
Zhao C, Wilson MC, Schuit F, Halestrap AP, Rutter GA. Expression and distribution of lactate/monocarboxylate transporter isoforms in pancreatic islets and the exocrine pancreas. Diabetes. 2001;50(2):361–6.

Glial Excitatory Amino Acid Transporters and Glucose Incorporation

Germán Fernando Gutiérrez Aguilar, Ivan Alquisiras-Burgos, Mónica Espinoza-Rojo, and Penélope Aguilera

Abstract Excitatory amino acid transporters (EAATs) expressed in astrocytes remove the glutamate released by neurons in and around the synaptic cleft. In this manner, astrocytes preserve the signaling functions mediated by glutamate on synapses and prevent excitotoxicity. Additionally, EAAT activation stimulates glucose utilization in astrocytes, linking neuronal activity with astrocyte metabolism. In this chapter, we briefly review the characteristics of the EAATs and the glucose transporters (GLUTs) expressed in the brain. Thereafter, we focus on the effect of EAATs activation and its association with glucose utilization in astrocytes, specifically addressing the role played by Na^+ and Ca^{2+} ions. Next, we analyze evidence that proposes mechanisms by which the activity of GLUTs could be modulated after EAAT activation (e.g., kinases altering GLUTs traffic to cell membrane). Finally, we analyzed the current knowledge on EAAT function during energy deficiency as a possible inducer of GLUT expression to prevent neuronal damage.

Keywords Glutamate transporters • EAATs • GLT-1 • Glucose transporters • GLUT1 • Glucose metabolism • Astrocytes

List of Abbreviations

AICAR	5-Aminoimidazole-4-carboxamide 1-β-D-ribofuranoside
AMPA	DL-α-Amino-3-hydroxy-5-methyl-4-isoxazolepropionic acid
AMPK	AMP-activated protein kinase

G.F. Gutiérrez Aguilar • I. Alquisiras-Burgos • P. Aguilera (✉)
Laboratorio de Patología Vascular Cerebral, Instituto Nacional de Neurología y Neurocirugía "Manuel Velasco Suárez", Insurgentes Sur #3877, Col. La Fama, Tlalpan, Ciudad de México 14269, México
e-mail: gutierrezref@gmail.com; burgos_inc@hotmail.com; penelope.aguilera@gmail.com

M. Espinoza-Rojo
Laboratorio de Biología Molecular y Genómica, Universidad Autónoma de Guerrero, Chilpancingo, Guerrero 39087, México
e-mail: moniespinoza@yahoo.com

A. Ortega, A. Schousboe (eds.), *Glial Amino Acid Transporters*,
Advances in Neurobiology 16, DOI 10.1007/978-3-319-55769-4_13

APV	D-(−)-2-Amino-5-phosphonopentanoic acid
BAPTA	1,2-Bis(2-aminophenoxy) ethane-N,N,N′,N′-tetraacetic acid
CNQX	6-Cyano-7-nitroquinoxaline-2,3-dione
EAAT	Excitatory amino acid transporter
ETDG	4,6-Ethylidine-D-glucose
GLUT	Glucose transporter
L-AP3	L-(+)-2-Amino-3-phosphonopropionic acid
L-CCG II	L-threo-β-Hydroxyaspartate
NMDA	N-Methyl-D-aspartate
NMDG+	N-Methyl-D-glucamine
OGD	Oxygen-glucose deprivation
PKC	Protein kinase C
TBOA	DL-threo-β-Benzyloxyaspartate
THA	DL-threo-β-Hydroxyaspartate
THA	threo-β-Hydroxyaspartate
TPA	2-O-Tetradecanoylphorbol-13-acetate
trans-ACPD	(±)-1-Aminocyclopentane-*trans*-1,3-dicarboxylic acid

1 Introduction

Astrocytes are the most numerous glial cells in the brain; they display a remarkable heterogeneity in their morphology and functions that theoretically are almost as complex as those observed in neurons (Hu et al. 2016). This type of glia deals with numerous tasks: builds up the general architecture of the brain, maintains brain homeostasis, controls the development of other cells, maintains synaptic function, and favors synaptogenesis (Verkhratsky and Nedergaard 2016). Noticeably, astrocytes fulfill a very important role on storage and distribution of the energy sources. In this line, evidence suggests that astrocytes are the major providers of energetic substrates to neurons; however, in recent decades an intense debate has continued without unequivocal demonstration how this process is controlled.

Glutamatergic neurotransmission in the brain requires continual resetting of the ionic gradients across membrane. In astrocytes, a considerable amount of energy is expended in clearing synaptically released glutamate in order to limit the excitatory signaling in neurons. The extracellular glutamate concentration is finely regulated through the Na^+-dependent glutamate transporter [also called excitatory amino acid transporter (EAAT)], the operation of which depends on the Na^+/K^+ gradient generated across the cell membrane by the Na^+/K^+ pump. Therefore, the preservation of the transmembrane Na^+ gradient signifies an enormous energy cost for the astrocyte (Robinson and Jackson 2016). Some arguments presume that glutamate uptake may not induce an increase on glucose uptake in astrocytes since total glutamate oxidation can produce even more ATP than the amount spent by the EAAT (McKenna 2013). However, new concepts support the idea that glycogenolysis is needed because it may maintain a high concentration of glucose-6-phosphate while simultaneously provides a substrate to sustain both the astrocytic and metabolic pathway and, additionally, to free up extracellular glucose and lactate to satisfy immediate neuronal energy demands at stimulus onset (Dienel and Cruz 2015). Consequently, as glycogen storage has to be replaced, glucose transporter (GLUT) might respond to EAAT activation.

Interestingly, glutamate (as part of the glutamatergic transmission) might activate a signal in the astrocyte that will advise about neuronal energetic requirements. Recent findings suggest that glucose metabolism and glutamate transport in astrocytes are highly interconnected; nevertheless, how glucose and glutamate transport and their metabolism alter astroglial functional properties still remains elusive. The aim of this review was to analyze the current knowledge on the relations existing between glutamate and GLUTs under normal and pathological conditions and describe the studies that demonstrate the mechanism activated in astrocytes in order to fulfill neuronal energetic demands.

2 Glucose Transporter Expression in the Brain

To obtain its energetic fuel, neurons require the delivery of glucose from blood. Although, not completely understood, it has been suggested that glucose has to transit across a plasmatic membrane of endothelial cells, which constitute the blood-brain barrier; then, glucose must enter glial cells where it is metabolized to lactate and, finally, reach neuronal cells where it is completely oxidized. This sequence of events is recognized as "the Astrocyte-to-Neuron Lactate Shuttle Hypothesis" (Pellerin and Magistretti 1994) and might occur under intense neuronal activity. Supporting the hypothesis, lactate was monitored in vivo demonstrating its flux from astrocytes to neurons (Mächler et al. 2016). Alternatively, blood-derived glucose might enter directly into neurons to be metabolized.

A pivotal point in this process is the passage of glucose through GLUT, which belongs to the sodium-independent facilitated hexose transporters, family of integral membrane proteins that includes 13 members (Deng and Yan 2016). The principal GLUTs expressed in rodent and the human brain are GLUT1 and GLUT3 (Maher et al. 1991; Maher et al. 1992). Even though at very low levels, the presence of other isoforms has also been reported throughout the brain, but its function, localization, and kinetics have not been completely characterized (Maher et al. 1994; Simpson et al. 2008). GLUT3 frequently received its designation as "the neuronal glucose transporter" because when it was cloned from a mouse library, showed a restricted distribution in the brain (Simpson et al. 2008), but it is also expressed in some peripheral tissues in humans (Maher et al. 1994). GLUT1 is a ubiquitous protein which is expressed in the brain as two molecular mass forms: 40–45 and 50–60 kDa which are distinguished in their degree of glycosylation. The heavier form is expressed in the endothelial cells of the blood-brain barrier, and it is also detected in choroid plexus and ependymal cells. The 40–45 kDa form of GLUT1 has been identified in neuronal and glial membranes with the most important location in glia. GLUT1 is detectable in all primary cultures of brain cells (including neurons, astrocytes, and microglia) and in most cells in culture. In the brain, GLUT1/GLUT3 ratio is 1:1, an amount associated to the high level of expression of each transporter seen in astrocytes and neurons, respectively (Maher et al. 1994). Analysis of glucose transporter kinetic parameters has revealed that GLUT3 has a higher affinity for glucose and greater transport capacity than GLUT1 (GLUT3, Km = 1.4 mM, Kcat = 6500/s; GLUT1, Km = 6.9 mM, Kcat = 1200/s). Since glucose concentration in rat and human brains has been estimated to be around 1–2 mM, while it is between 5–6 mM in serum, it has been suggested that cells expressing GLUT3 (i.e., neurons) have a preferential access to glucose (Maher et al. 1994; Simpson et al. 2008). Evidence supports that neurons consume more energy than astrocytes, although they are slower metabolizers of glucose; additionally, astrocytes have a higher glucose transport capacity (Jakoby et al. 2014). Data confirm that under resting conditions neurons and astrocytes will consume similar amounts of glucose; nevertheless, during neurotransmission astrocytes will sustain energetic requirements of neurons (Chuquet et al. 2010). It is possible that astrocytes adjust their metabolism in order to support neuronal functioning during stress and pathological conditions, a situation that favors the use of lactate over glucose (Wyss et al. 2011).

3 EAAT Function in the Brain

Glutamate is the main excitatory neurotransmitter in the central nervous system. During neuronal activity, glutamate is released to the synaptic cleft, and, then, it is rapidly removed by the EAATs. This process is indispensable to assure the reliability of synaptic transmission, but also to sustain low and nontoxic extracellular glutamate levels at the synapsis.

EAATs belong to solute carrier family 1, a class of integral membrane proteins. This family includes five EAAT members characterized in mammalian central nervous system: GLAST/EAAT1, GLT-1/EAAT2, EAAC1/EAAT3, EAAT4, and EAAT5. GLAST and GLT-1 were isolated from rats (Danbolt et al. 1992). GLAST is specifically expressed in the brain, in both neurons and astroglia; it has a prominent expression in the cerebellar cortex, it is also present in the cortex and hippocampus, but no mRNA is found in structures known to contain high density of glutamatergic synapses such as the cerebellar granular layer (Storck et al. 1992; Rothstein et al. 1994). GLT-1 is the most important transporter in the brain, representing around 1% of total brain protein. It is localized almost exclusively to astrocytic process throughout the brain and spinal cord (Rothstein et al. 1994). GLT-1 is detected both in the immediate vicinity of the synaptic cleft (60%) and in extrasynaptic locations (40%). About 35% of synaptic GLT-1 is closely related to excitatory terminals and 25% with inhibitory ones (e.g., GABAergic terminals) (Minelli et al. 2001). GLAST and GLT-1 are expressed through the brain with a specific pattern during development and with regional, cellular, and subcellular distribution. Remarkably, astrocytes in culture showed different expression patterns dependenting on if they were grown alone or co-cultured with neurons. The latter displays more complex morphologies (has a highly branched and polygonal stellate shape) and EAAT expression patterns (express both GLT-1 and GLAST) (Swanson et al. 1997). On the other hand, EAAC1 was first isolated from rabbits (Kanai and Hediger 1992). It appears to be expressed in specific neurons, such as Purkinje cells and large pyramidal cortical neurons, but does not seem to be selective for glutamatergic ones. It is highly enriched in the cortex, hippocampus, and caudate-putamen and is restricted to pre- and postsynaptic elements (Rothstein et al. 1994). EAAT4 is expressed mostly in the cerebellum at the Purkinje cell bodies but not at granule cells. It has a lower expression in the forebrain, present mainly in small caliber dendrites in neocortex and hippocampus, and is rarely found in forebrain astrocytes (Fairman et al. 1995; Furuta et al. 1997). Finally, EAAT5 is expressed predominantly in the retina (Arriza et al. 1997).

4 EAAT and Glycolytic Activity on Astrocytes

The stoichiometry of the glutamate transport has been determined to be as follows: cotransports of three Na^+ or two Na^+ and one H^+ and counter-transports of one K^+ per transported glutamate. It is an electrogenic transport process that depends on the electrochemical gradient maintained by the Na^+/K^+-ATPase. Under physiological conditions, it provides the driving force to accumulate glutamate into the cell against its concentration gradient (Focke et al. 2013). The increment on extracellular glutamate during intense neuronal activity provokes the overload of Na^+ ions in the astrocyte. Intracellular increase of Na^+ due to activation of EAATs has the ability to diffuse through gap junctions and propagate as a Na^+ wave. The Na^+ signal reaches values of 41 ± 6.5 mM that decrease along the spreading axes of the wave to 10.8 ± 4.1 mM at a distance of 165 μm from the stimulated cells. The Na^+ wave initiated by the glutamate transport serves to mediate the spreading of the cellular metabolic response within groups of astrocytes, a response that is probably associated to neuronal activity (Bernardinelli et al. 2004). In response to glutamate application, mitochondrial Na^+ concentration can pass from 11–12 mM to over 30 mM. Na^+ changes resulting from Na^+/glutamate transport activity appears to be transmitted to the mitochondrial matrix. This increase on internal mitochondrial Na^+ concentration impacts the oxidative energy production (Chatton et al. 2016; Fernández-Moncada and Barros 2014). Interestingly, there is also a quasi-perfect match between the spread of the Na^+ wave and the shape of increased glucose uptake (Bernardinelli et al. 2004). It can be deduced that the connectivity of astrocytes allows them to coordinate the glutamatergic activity to the induction of glucose metabolism in astrocytes (Fig. 1).

In intact brains, glucose utilization during and after neuronal firing occurs principally in the neuropil, region rich in synapses and astrocytes (Chuquet et al. 2010). Supporting this data, glycolysis is rapidly induced by glutamate in astrocytes in

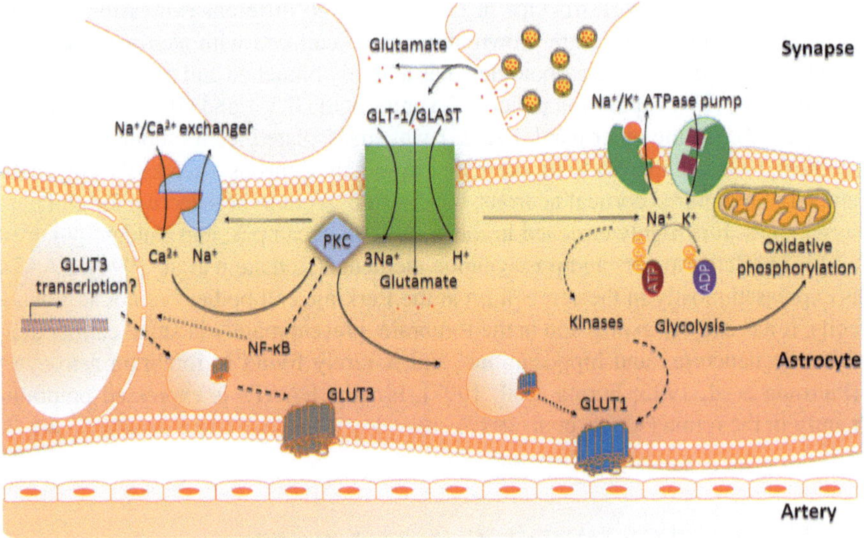

Fig. 1 Functional coupling in glutamate capture by astrocytes. GLT-1/GLAST function in coordination with the Na^+/Ca^{2+} exchanger and Na^+/K^+ pump to maintain glutamate capture. The Na^+/K^+ pump uses energy from glycolysis and oxidative phosphorylation to maintain ionic homeostasis, while the exchanger works in reverse. The increase of intracellular Ca^{2+} promotes PKC activation and the following translocation of GLUT1 to the cellular membrane. Activation of the NF-κB pathway promotes transcription and translocation of GLUT3 to the membrane in conditions of oxygen and glucose deprivation

culture. Short stimulation with glutamate (<5 min) elicits a minor glycolytic inhibition, but glycolysis rises when extracellular K^+ is applied together with glutamate. Additionally, exposure to glutamate (5–50 μM) for longer periods (5–20 min) provokes robust, late, and persistent stimulation of astrocytic glycolysis (Bittner et al. 2011). This response suggests that intense neurotransmission (or even pathological conditions in which massive concentration of glutamate is released to the synapsis) might activate a signal to increase glucose metabolism in astrocytes to sustain neuronal energetics.

Experiments on astrocytes in culture have revealed that glutamate is indispensable to activate glucose metabolism because glycolysis is not stimulated in the presence of EAAT inhibitors [DL-*threo*-*β*-benzyloxyaspartate (TBOA), DL-*threo*-*β*-hydroxyaspartate (THA), or L-*threo*-*β*-hydroxyaspartate (L-CCG II)] (Bittner et al. 2011; Pellerin and Magistretti 1994; Porras et al. 2008; Sokoloff et al. 1996). Furthermore, agonists for ionotropic glutamate receptors [N-methyl-D-aspartate (NMDA), DL-α-amino-3-hydroxy-5-methyl-4-isoxazolepropionic acid (AMPA), quisqualate (non-NMDA)] and metabotropic glutamate receptors [(±)-1-aminocyclopentane-trans-1,3-dicarboxylic acid (*trans*-ACPD)] do not mimic the effect; and the antagonist of glutamate receptors [D(−)-2-amino-5-

phosphonopentanoic acid (APV) for NMDA, 6-cyano-7-nitroquinoxaline-2,3-dione (CNQX) for non-NMD, and L-(+)-2-amino-3-phosphonopropionic acid (L-AP3) for metabotropic 1)] do not interfere with the induction of glucose uptake (Pellerin and Magistretti 1994; Sokoloff et al. 1996). These results clearly show that glutamate is a signal that couples neuronal activity to glucose utilization though activation of the EAAT.

Glutamate-induced increase on glycolytic rate is dependent on Na^+ entry to the astrocytic cell. Nonetheless, early glycolysis stimulation is not observed with the ionophore gramicidin which induces a substantial increase on intracellular Na^+, only partially simulating the effect of EAAT activation (Bittner et al. 2011). A similar effect is observed when extracellular Na^+ is replaced in equimolar proportions with the non-permeant cation N-methyl-D-glucamine ($NMDG^+$) (Porras et al. 2008). This result highlights that there is an interrelationship between ion transport and metabolism in which the increment on extracellular K^+ is also important, because it allows activation of the Na^+/K^+-ATPase. In accordance, the short- and the long-term stimulatory effect of glutamate plus high K^+ on glycolysis is reversibly inhibited by ouabain, an inhibitor of the Na^+ pump. That means that during EAAT activation, the Na^+/K^+-ATPase plays a central role in the control of glycolysis in astrocytes (Bittner et al. 2011). The implicated molecular mechanism has not been elucidated; however, evidence suggests that there is a close relationship among the Na^+/K^+-ATPase and certain key enzymes of glycolysis. In fact, GLAST interacts with the $\alpha1$, $\alpha2$, and $\beta1$ subunits of the Na^+/K^+-ATPase and with proteins involved in energy metabolism such as the glycolytic enzymes glyceraldehyde-3-phosphate dehydrogenase and hexokinase (Robinson and Jackson 2016). Additionally, EAATs have been found co-localized with mitochondrial proteins (Bauer et al. 2012; Genda et al. 2011) and mitochondrial dehydrogenases and glycogen phosphorylase. Interestingly, these enzymes are stimulated by Ca^{+2} ions, concentration of which is also increased after EAAT activation (Denton 2009; Robinson and Jackson 2016). These results strongly suggest that signal of EAATs initiates glycolysis and oxidative phosphorylation, processes that will favor synthesis of energetic sources.

Remarkably, glutamate (IC_{50} of 5 μM) triggers the opposite response in neurons. While glutamate inhibits hexose uptake by 80 ± 9% in neurons, it stimulates 151 ± 21% in astrocytes, meaning that hexose transport rises 12-fold, but it is triggered mainly by astrocytes. Neurons required higher concentration than 100 μM of glutamate to gain the maximum inhibition of glucose uptake. In these cells inhibition of glucose transport is mediated by the AMPA receptor, and in contrast to what is observed in astrocytes, the EAAT blocker THA (200 μM) has no effect. Na^+ entry mimicked with veratridine effectively inhibits glucose transport, but membrane depolarization by itself does not induce the effect (Porras et al. 2004). Glutamate mediates glucose transport in brain cells by different mechanisms in which Na^+ is participating: in neurons AMPA receptors are activated, while in astrocytes the EAATs are key mediators of the response. These data are in concordance with the Astrocyte-to-Neuron Lactate Shuttle Hypothesis since neurotransmission favors glucose utilization in astrocyte in order to produce lactate and deliver it to neurons.

5 Regulation of GLUT Activity by Glutamate

In order to increase glucose metabolism in astrocytes, EAAT activation also stimulates GLUTs. In intact brain, exposure of astrocytes to 500 μM glutamate causes a sharp increase in the rate of hexose uptake via astrocytic GLUT. Since astrocyte in primary cultures expresses abundantly GLUT1 but not GLUT3, transport should be mediated by GLUT1 (Loaiza et al. 2003). The concentration that glutamate reaches in the synaptic cleft does not appear to be fundamental to stimulate the glucose transport in astrocytes because either low (5 μM) or high (500 μM) micromolar concentration of glutamate has the same effect (half-stimulation value of 1.8 ± 0.7 μM) (Porras et al. 2008). Glutamate modulates GLUT activity by rapid (seconds) and slow (minutes) mechanisms, but neither of them has been completely characterized (Porras et al. 2008; Loaiza et al. 2003; Mendez-Flores et al. 2016). In both cases, glutamate alters the V_{max} of GLUT. A mechanism activated in the range of a few seconds (9 ± 3), effect that is withdrawn by inhibitors of GLUT [the exofacial 4,6-ethylidine-D-glucose (ETDG) and the endofacial cytochalasin B] (Loaiza et al. 2003) or a long-lasting signaling event, in which an intracellular trafficking could be involved (Mendez-Flores et al. 2016) are possible options.

GLUT activation depends on a Ca^{2+} signal that induces a posttranslational modification. Astrocytic Ca^{2+} signals mediated by glutamate could be dependent on metabotropic glutamate receptor activation and on the Na^{+}-Ca^{2+} exchanger pushed in the reverse mode by the Na^{+} load after EAAT activation (Rojas et al. 2007). Evidence suggests that the Na^{+}-Ca^{2+} exchanger is involved because agonists of glutamate receptors do not induce the effect, but it is blocked by the EAAT inhibitors. In different non-cerebral cell types, increase of cytosolic Ca^{2+} (e.g., ionomycin and cytokines) or mobilization of intracellular Ca^{2+} (e.g., thapsigargin, depleting intracellular stores) upsurges catalytic efficiency of GLUT1. Chelation of cytosolic Ca^{2+} [e.g., 1, 2-bis(2-aminophenoxy) ethane-N,N,N′,N′-tetraacetic acid (BAPTA)] or depletion extracellular Ca^{2+} [e.g., 2-aminoethoxydiphenyl-borato (2-APB)] also decreases hexose uptake (Quintanilla et al. 2000; Maraldi et al. 2006). In agreement, experiments performed in astrocytes proved that Na^{+} gain is involve in stimulation of glucose transport; however, induction of cation entry after the ionophore gramicidin is not effective because an increase in cytosolic Ca^{2+} is also required. An effective stimulation of GLUT1 arises only when an increase on Na^{+} and Ca^{2+} coincide, showing that signaling depends on both (Porras et al. 2008). In primary cultures of Bergmann glial cells, L-glutamate or D-aspartate (EC_{50} = 9 and 11.7 μM, respectively) treatment induces a 30% increment on glucose uptake that depends on GLAST activity. Equimolar substitution of Na^{+} or removal of Ca^{2+} in the buffer and the use of the glutamate blocker TBOA prevent the glucose uptake induced by glutamate (Mendez-Flores et al. 2016).

GLUT4 is the most studied glucose transporter which is expressed abundantly in muscle cells. The increase on cytosolic Ca^{2+} induces GLUT4 exocytosis in muscle cells by activation of Ca^{2+}-calmodulin-dependent protein kinase II and AMP-activated protein kinase (AMPK), while novel protein kinase C (PKC) activation reduces endocytosis of GLUT4 (Li et al. 2014). Therefore, it seems likely that kinases could be involved in the regulation of intracellular GLUT1 traffic in astrocytes. In Bergmann glial cells, the increment on glucose entry to the cell is couple to a movement of GLUT to the plasma membrane. This glucose uptake is associated to a two-fold augmentation in GLUT V_{max} 30 min after glutamate stimulus. Additionally, Ca^{2+} influx induces PKC activation with the consequent translocation of GLUT to the cell membrane (Mendez-Flores et al. 2016). Consistent with this study, stimulation for 30 min with the PKC activator 2-O-tetradecanoylphorbol-13-acetate (TPA) induces GLUT1 phosphorylation in serine 226, enriches its localization at the cell membrane, and rises glucose transport mostly by increasing the Vmax (Lee et al. 2015). Interestingly, GLT-1 co-immunoprecipitates with PKCα which responds to the increase in Ca^{2+} induced by glutamate (González et al. 2005); it is probably that PKC might be recruited to multimeric complexes that include GLUT1 and GLT-1. Other possible activator of GLUT1 is the Ca^{2+}-calmodulin-dependent protein kinase II that in human hematopoietic cells modulated GLUT1 intracellular trafficking (Maraldi et al. 2006).

On the other hand, it is interesting that regulation of GLUT trafficking induced by transient glutamate excitation (10 min/100 μM) has also been reported in cerebellar granule neurons. After 30 min of excitation with glutamate, GLUT3 increases in the cell surface for up to 4 h and its translocation dependents on activation of the AMPK (Weisová et al. 2009). This enzyme is an energy sensor that activates catabolic processes such as glucose uptake when energy supply is required. Importantly, other GLUTs are known to be modulated in this way (Barnes et al. 2002); in skeletal muscle AMPK causes GLUT4 translocation (Li et al. 2014). However, AMPK activation by AICAR does not affect the glucose uptake in primary cultures of cortical astrocytes from mouse brain; although it participates in other processes involved in glutamate metabolism (Voss et al. 2015) (Fig. 1).

6 EAAT Function During Energy Deficiency

The regular glutamate concentration in the synaptic cleft is about 25 nM and reaches concentration between 200 and 1000 μM following neuronal activity. Under normal conditions glutamate is rapidly cleared from extracellular space and is not present in the cleft at concentration >1000 μM for more than 100–200 μs and is not greater than 50 μM for more than 5 ms (Clements et al. 1992). However, pathological conditions that attain energy deprivation bring on neurotoxic glutamate levels in the range of 10–100 μM during minutes as a result of two events (Lipton 1999):1) Enhanced vesicular release, process that contributes approximately with half of the extracellular glutamate concentration (Jabaudon et al. 2000). Nevertheless, the proportion of glutamate released from vesicles will decrease gradually because Ca^{2+}-dependent release requires ATP, and therefore, it is blocked under inhibition of oxidative phosphorylation and glycolysis (Sanchez-Prieto et al. 1987), and 2) Electrochemical gradients that drive the glutamate uptake (i.e., Na^+/K^+-ATPase) are altered; as a consequence, there is a failure on EAATs activity that induces a rapidly (within minutes) net reduction in glutamate uptake. This event involves decreased uptake and induces EAAT function in reverse, effect that can be associated to a rise in glutamate extracellular concentration (Jabaudon et al. 2000).

The situation described about is observed in different anomalous situations (e.g., ischemia and hypoglycemia) in which a reduction on cerebral glucose is observed. Reduction on glucose levels in the medium decreases glutamate uptake but also GLUT activity (Quincozes-Santos et al. 2017). However, the uptake of glutamate following glucose withdrawal during 20 min is not altered unless the glycogen degradation is inhibited (Sickmann et al. 2009). In fact, intracellular Na^+ and ATP levels are only marginally affected in astrocytes deprived of glucose for 30 min (Pauwels et al. 1985; Rose et al. 1998). In this situation, brain glycogen stored in astrocytes is used as emergency energy reserve and is recognized as a dynamic participant in glutamatergic neurotransmission affecting both astrocytic (e.g., Na^+/K^+-ATPase and EAAT) and neuronal (e.g., vesicular release) function. Interestingly, mild-to-moderate hypoglycemia is a weak activator of glycogen utilization because glucose is still present in blood. Only prolonged and severe energy failure speeds up glycogen mobilization (Dienel and Cruz 2015). Additionally, the activity of transporters can be restored partially after glucose return to basal levels. Glucose delivery through a dialysis probe reduces the rise in extracellular glutamate to eight times baseline after 40 min, meaning that the transporter will be working even under hypoxia conditions while there is available glucose (Swanson et al. 1994).

Constant stimulation with a high concentration of glutamate (e.g., 1 mM), which could emulate firing, also induces glycogen accumulation reaching maximal levels after 4 h. Lower concentration of glutamate does not induce the effect, and the agonist of the NMDA receptor does not alter astrocyte glycogen content, suggesting that the process is mediated by activation of the EAAT. In agreement, inhibition of the EAAT with THPA or replacement of Na^+ in the buffer with choline prevents the effect (Swanson et al. 1990). These results add an evidence of the link existing between EAAT and the regulation of glucose utilization, pointing out that astrocytes perform an important role to favor suitable synaptic functioning and to prevent impairment of neuronal function.

On the other hand, astrocytes which express merely GLUT1 also has the capacity to express GLUT3. GLUT3 has higher affinity for glucose than GLUT1; therefore, its expression might confer astrocytes a greater capacity to uptake glucose from the extracellular space even at low glucose concentration. Interestingly, expression of both transporters is induced under hypoxic, hypoglycemic, and ischemic conditions in neurons and astrocytes (Behrooz and Ismail-Beigi 1999; Bruckner et al. 1999; Cidad et al. 2001; Iwabuchi and Kawahara 2011; Iwabuchi et al. 2014). For example, incubation of primary cultures of astrocytes with glucose-free DMEM or with DMEM under a hypoxic atmosphere for 4 h enhances expression of GLUT1 and also GLUT3 mRNA levels (Cidad et al. 2001). Also, exposition to oxygen and glucose deprivation induces GLUT3. This change in the expression patron of GLUTs allowed astrocytes to transport glucose more efficiently when they recovered from oxygen and glucose deprivation (Iwabuchi and Kawahara 2011; Iwabuchi et al.

2014). Induction of GLUT3 in astrocytes is important given that this isoform is rarely detected under control conditions. Therefore, it is possible that neuronal activity (i.e., glutamate released to the synaptic cleft) will induce GLUT3 mRNA expression in astrocytes under conditions of energy failure. Given that EAATs induce a reduction on energetic substrates, AMPK could be involved on GLUT3 expression (Iwabuchi et al. 2014). However, regulation of GLUTs in astrocytes is a mechanism that unfortunately has been barely studied.

7 Concluding Remarks

Neuronal activity on glutamatergic synapses is closely dependent on astrocytic function; however, little advances have been made over the past few years in our understanding of the molecular mechanisms that coordinate the metabolic adaptations of glial cells to neurotransmission. Evidence suggests that the well-known glycolytic metabolism in astrocytes, which is not observed in neurons, is improved by the activity of the EAAT. Thus, the EAAT acts as a sensor of the glutamatergic activity and through the Na^+/K^+-ATPase and the Na^+-Ca^{2+} exchanger triggers the signaling pathway needed to increase the glucose metabolism in the astrocyte. An important point of regulation is the glucose uptake mediated by the GLUTs (essentially GLUT1). Properties of GLUTs are altered by posttranslational modifications affecting either their activity or traffic to cell membrane, but not exclude the occurrence of other mechanisms such as transcriptional regulation that could be modified under conditions of metabolic stress. However, the information about the topic is scarce; therefore, it seems relevant to undertake the study of the process.

Acknowledgment This work was supported by Basic Science SEP-CONACYT under grant number 182266.

Germán Fernando Gutiérrez Aguilar and Iván Alquisiras-Burgos are doctoral students from Programa de Doctorado en Ciencias Biomédicas, Universidad Nacional Autónoma de México (U.N.A.M.) and receiver scholarship numbers 365585 and 413994 from CONACYT.

Conflict of Interest The author declares no conflicts of interest.

References

Arriza JL, Eliasof S, Kavanaugh MP, Amara SG. Excitatory amino acid transporter 5, a retinal glutamate transporter coupled to a chloride conductance. Neurobiology. 1997;94:4155–60.

Barnes K, Ingram JC, Porras OH, Barros LF, Hudson ER, Fryer LGD, et al. Activation of GLUT1 by metabolic and osmotic stress: potential involvement of AMP-activated protein kinase (AMPK). J Cell Sci. 2002;115(11):2433–42.

Bauer DE, Jackson JG, Genda EN, Montoya MM, Yudkoff M, Robinson MB. The glutamate transporter, GLAST, participates in a macromolecular complex that supports glutamate metabolism. Neurochem Int. 2012;61(4):566–74.

Behrooz A, Ismail-Beigi F. Stimulation of glucose transport by hypoxia: signals and mechanisms. News Physiol Sci. 1999;14:105–10.

Bernardinelli Y, Magistretti PJ, Chatton JY. Astrocytes generate Na+-mediated metabolic waves. Proc Natl Acad Sci U S A. 2004;101(41):14937–42.

Bittner CX, Valdebenito R, Ruminot I, Loaiza A, Larenas V, Sotelo-Hitschfeld T, et al. Fast and reversible stimulation of astrocytic glycolysis by K+ and a delayed and persistent effect of glutamate. J Neurosci [Internet]. 2011;31(12):4709–13. Available from: http://www.jneurosci.org/cgi/doi/10.1523/JNEUROSCI.5311-10.2011

Bruckner BA, Ammini CV, Otal MP, Raizada MK, Stacpoole PW. Regulation of brain glucose transporters by glucose and oxygen deprivation. Metabolism [Internet]. 1999;48(4):422–31. Available from: http://www.ncbi.nlm.nih.gov/pubmed/10206432

Chatton J-Y, Magistretti PJ, Barros LF. Sodium signaling and astrocyte energy metabolism. Glia. 2016;64:1667–76.

Chuquet J, Quilichini P, Nimchinsky EA, Buzsáki G. Predominant enhancement of glucose uptake in astrocytes versus neurons during activation of the somatosensory cortex. J Neurosci [Internet]. 2010;30(45):15298–303. doi: 10.1523/JNEUROSCI.0762-10.2010.

Cidad P, Garcia-Nogales P, Almeida A, Bolaños JP. Expression of glucose transporter GLUT3 by endotoxin in cultured rat astrocytes: the role of nitric oxide. J Neurochem. 2001;79(79):17–24.

Clements JD, Lester RA, Tong G, Jahr CE, Westbrook GL. The time course of glutamate in the synaptic cleft. Science. 1992;258(5087):1498–501.

Danbolt NC, Storm-Mathisen J, Kanner BI. An [Na+ + K+] coupled l-glutamate transporter purified from rat brain is located in glial cell processes. Neuroscience. 1992;51(2):295–310.

Deng D, Yan N. GLUT, SGLT, and SWEET: structural and mechanistic investigations of the glucose transporters. Protein Sci. 2016;25(3):546–58.

Denton RM. Regulation of mitochondrial dehydrogenases by calcium ions. Biochim Biophys Acta Bioenerg. 2009;1787(11):1309–16.

Dienel GA, Cruz NF. Contributions of glycogen to astrocytic energetics during brain activation. Metab Brain Dis. 2015;30(1):281–98.

Fairman WA, Vandenberg RJ, Arriza JL, Kavanaugh MPAS. An excitatory amino-acid transporter with properties of a ligand-gated chloride channel. Nature. 1995;375(6532):599–603.

Fernández-Moncada I, Barros LF. Non-preferential fuelling of the Na. Biochem J. 2014;460:353–61.

Focke PJ, Wang X, Larsson HP. Neurotransmitter transporters: structure meets function. Structure. 2013;21(5):694–705.

Furuta A, Martin L, Lin C-L, Dykes-Hoberg M, Rothstein JD. Cellular and synaptic localization of the neuronal glutamate transporters excitatory amino acid transporter 3 and 4. Neuroscience [Internet]. 1997;81(4):1031–42. Available from: http://www.sciencedirect.com/science/article/pii/S0306452297002522\nhttp://linkinghub.elsevier.com/retrieve/pii/S0306452297002522

Genda EN, Jackson JG, Sheldon AL, Locke SF, Greco TM, O'Donnell JC, et al. Co-compartmentalization of the astroglial glutamate transporter, GLT-1, with glycolytic enzymes and mitochondria. J Neurosci [Internet]. 2011;31(50):18275–88. Available from: http://www.pubmedcentral.nih.gov/articlerender.fcgi?artid=3259858&tool=pmcentrez&rendertype=abstract

González MI, Susarla BTS, Robinson MB. Evidence that protein kinase Cα interacts with and regulates the glial glutamate transporter GLT-1. J Neurochem. 2005;94(5):1180–8.

Hu X, Yuan Y, Wang D, Su Z. Heterogeneous astrocytes: active players in CNS. Brain Res Bull [Internet]. 2016;125:1–18. doi:10.1016/j.brainresbull.2016.03.017.

Iwabuchi S, Kawahara K. Inducible astrocytic glucose transporter-3 contributes to the enhanced storage of intracellular glycogen during reperfusion after ischemia. Neurochem Int. 2011;59(2):319–25.

Iwabuchi S, Kawahara K, Harata NC. Effects of pharmacological inhibition of AMP-activated protein kinase on GLUT3 expression and the development of ischemic tolerance in astrocytes. Neurosci Res. 2014;84:68–71.

Jabaudon D, Scanziani M, Gähwiler BH, Gerber U. Acute decrease in net glutamate uptake during energy deprivation. Proc Natl Acad Sci U S A [Internet]. 2000;97(10):5610–5. Available from: http://www.pubmedcentral.nih.gov/articlerender.fcgi?artid=25876&tool=pmcentrez&rendertype=abstract

Jakoby P, Schmidt E, Ruminot I, Gutiérrez R, Barros LF, Deitmer JW. Higher transport and metabolism of glucose in astrocytes compared with neurons: a multiphoton study of hippocampal and cerebellar tissue slices. Cereb Cortex. 2014;24(1):222–31.

Kanai Y, Hediger MA. Primary structure and functional characterization of a high-affinity glutamate transporter. Nature [Internet]. 1992;360(6403):467–71. Available from: http://www.ncbi.nlm.nih.gov/pubmed/1280334

Lee EE, Ma J, Sacharidou A, Mi W, Salato VK, Nguyen N, et al. A protein kinase C phosphorylation motif in GLUT1 affects glucose transport and is mutated in GLUT1 deficiency syndrome. Mol Cell. 2015;58(5):845–53.

Li Q, Zhu X, Ishikura S, Zhang D, Gao J, Sun Y, et al. Ca2+ signals promote GLUT4 exocytosis and reduce its endocytosis in muscle cells. Am J Physiol Endocrinol Metab. 2014;307:209–24.

Lipton P. Ischemic cell death in brain neurons. Physiol Rev. 1999;79(4):1431–568.

Loaiza A, Porras OH, Barros LF. Glutamate triggers rapid glucose transport stimulation in astrocytes as evidenced by real-time confocal microscopy. J Neurosci. 2003;23(19):7337–42.

Mächler P, Wyss MT, Elsayed M, Stobart J, Gutierrez R, Von Faber-Castell A, et al. In vivo evidence for a lactate gradient from astrocytes to neurons. Cell Metab. 2016;23(1):94–102.

Maher F, Davies-Hill TM, Lysko PG, Henneberry RC, Simpson IA. Expression of two GLUTs, GLUT1 and GLUT3, in cultured cerebellar neurons: evidence for neuron-specific expression of GLUT3. Mol Cell Neurosci [Internet]. 1991;2(4):351–60. Available from: http://www.ncbi.nlm.nih.gov/pubmed/19912819

Maher F, Vannucci S, Takeda J, Simpson IA. Expression of mouse-GLUT3 and human-GLUT3 glucose transporter proteins in brain. Biochem Biophys Res Commun. 1992;182(2):703–11.

Maher F, Vannucci SJ, Simpson IA. Glucose transporter proteins in brain. FASEB J. 1994;8(13):1003–11.

Maraldi T, Rugolo M, Fiorentini D, Landi L, Hakim G. Glucose transport activation in human hematopoietic cells M07e is modulated by cytosolic calcium and calmodulin. Cell Calcium. 2006;40(4):373–81.

McKenna MC. Glutamate pays its own way in astrocytes. Front Endocrinol (Lausanne). 2013;4:191.

Mendez-Flores OG, Hernández-Kelly LC, Suárez-Pozos E, Najimi M, Ortega A. Coupling of glutamate and glucose uptake in cultured Bergmann glial cells. Neurochem Int. 2016;98:72–81.

Minelli A, Barbaresi P, Reimer RJ, Edwards RH, Conti F. The glial glutamate transporter GLT-1 is localized both in the vicinity of and at distance from axon terminals in the rat cerebral cortex. Neuroscience. 2001;108(1):51–9.

Pauwels PJ, Opperdoes FR, Trouet A. Effects of antimycin, glucose deprivation, and serum on cultures of neurons, astrocytes, and neuroblastoma cells. J Neurochem [Internet]. 1985;44(1):143–8. Available from: http://www.ncbi.nlm.nih.gov/pubmed/3964824

Pellerin L, Magistretti PJ. Glutamate uptake into astrocytes stimulates aerobic glycolysis: a mechanism coupling neuronal activity to glucose utilization. Proc Natl Acad Sci U S A [Internet]. 1994;91(22):10625–9. Available from: http://www.ncbi.nlm.nih.gov/pubmed/7938003\nhttp://www.pubmedcentral.nih.gov/articlerender.fcgi?artid=PMC45074

Porras OH, Loaiza A, Barros LF. Glutamate mediates acute glucose transport inhibition in hippocampal neurons. J Neurosci. 2004;24(43):9669–73.

Porras OH, Ruminot I, Loaiza ABL. Na(+)-Ca(2+) cosignaling in the stimulation of the glucose transporter GLUT1 in cultured astrocytes. Glia. 2008;56(1):59–68.

Quincozes-Santos A, Bobermin LD, de Assis AM, Gonçalves C-A, Souza DO. Fluctuations in glucose levels induce glial toxicity with glutamatergic, oxidative and inflammatory implications. Biochim Biophys Acta – Mol Basis Dis [Internet]. 2017;1863(1):1–14. Available from: http://linkinghub.elsevier.com/retrieve/pii/S092544391630237X

Quintanilla RA, Porras OH, Castro J, Barros LF, Porras OH, Castro J, Barros LF. Cytosolic [Ca(2+)] modulates basal GLUT1 activity and plays a permissive role in its activation by metabolic stress and insulin in rat epithelial cells. Cell Calcium. 2000;28(2):97–106.

Robinson MB, Jackson JG. Astroglial glutamate transporters coordinate excitatory signaling and brain energetics. Neurochem Int [Internet]. 2016;98:56–71. doi:10.1016/j.neuint.2016.03.014.
Rojas H, Colina C, Ramos M, Benaim G, Jaffe EH, Caputo C, et al. Na+ entry via glutamate transporter activates the reverse Na+/Ca2+ exchange and triggers Cai2+-induced Ca2+ release in rat cerebellar Type-1 astrocytes. J Neurochem. 2007;100(5):1188–202.
Rose CR, Waxman SG, Ransom BR. Effects of glucose deprivation, chemical hypoxia, and simulated ischemia on Na+ homeostasis in rat spinal cord astrocytes. J Neurosci. 1998;18(10):3554–62.
Rothstein JD, Martin L, Levey AI, Dykes-Hoberg M, Jin L, Wu D, Nash NKR. Localization of neuronal and glial glutamate transporters. Nature. 1994;13(3):713–25.
Sanchez-Prieto J, Sihra TS, Nicholls DG. Characterization of the exocytotic release of glutamate from guinea-pig cerebral cortical synaptosomes. J Neurochem [Internet]. 1987;49(1):58–64. Available from: http://www.ncbi.nlm.nih.gov/entrez/query.fcgi?cmd=Retrieve&db=PubMed&dopt=Citation&list_uids=2884280
Sickmann HM, Walls AB, Schousboe A, Bouman SD, Waagepetersen HS. Functional significance of brain glycogen in sustaining glutamatergic neurotransmission. J Neurochem. 2009;109(Suppl 1):80–6.
Simpson I, Dwyer D, Malide D, Moley KH, Travis A, Vannucci SJ. The facilitative glucose transporter GLUT3: 20 years of distinction. Am J Physiol Endocrinol Metab. 2008;295(2):E242–53.
Sokoloff L, Takahashi S, Gotoh J, Driscoll BF, Law MJ. Contribution of astroglia to functionally activated energy metabolism. Dev Neurosci [Internet]. 1996;18(5-6):344–52. Available from: http://www.ncbi.nlm.nih.gov/entrez/query.fcgi?cmd=Retrieve&db=PubMed&dopt=Citation&list_uids=8940605
Storck T, Schulte S, Hofmann K, Stoffel W. Structure, expression, and functional analysis of a Na(+)-dependent glutamate/aspartate transporter from rat brain. Proc Natl Acad Sci U S A. 1992;89(22):10955–9.
Swanson RA, Yu ACH, Chan PH, Sharp FR. Glutamate increases glycogen content and reduces glucose utilization in primary astrocyte culture. J Neurochem [Internet]. 1990;54(2):490–6. Available from: http://www.scopus.com/inward/record.url?eid=2-s2.0-0025093473&partnerID=40&md5=c4dcddb21078b0dc0b82765937b1e6b8
Swanson RA, Chen J, Graham SH. Glucose can fuel glutamate uptake in ischemic brain. J Cereb Blood Flow Metab. 1994;14(1):1–6.
Swanson RA, Liu J, Miller JW, Rothstein JD, Farrell K, Stein BA, et al. Neuronal regulation of glutamate transporter subtype expression in astrocytes. J Neurosci. 1997;17(3):932–40.
Verkhratsky A, Nedergaard M. The homeostatic astroglia emerges from evolutionary specialization of neural cells. Philos Trans R Soc B Biol Sci [Internet]. 2016;371(1700):20150428. Available from: http://rstb.royalsocietypublishing.org/lookup/doi/10.1098/rstb.2015.0428
Voss CM, Pajęcka K, Stridh MH, Nissen JD, Schousboe A, Waagepetersen HS. AMPK activation affects glutamate metabolism in astrocytes. Neurochem Res. 2015;40(12):2431–42.
Weisová P, Concannon CG, Devocelle M, Prehn JHM, Ward MW. Regulation of glucose transporter 3 surface expression by the AMP-activated protein kinase mediates tolerance to glutamate excitation in neurons. J Neurosci [Internet]. 2009 Mar 4 [cited 2012 Mar 14];29(9):2997–3008. Available from: http://www.ncbi.nlm.nih.gov/pubmed/19261894
Wyss MT, Jolivet R, Buck A, Magistretti PJ, Weber B. In vivo evidence for lactate as a neuronal energy source. J Neurosci. 2011;31(20):7477–85.

Astrocytic GABA Transporters: Pharmacological Properties and Targets for Antiepileptic Drugs

Arne Schousboe, Petrine Wellendorph, Bente Frølund, Rasmus P. Clausen, and Povl Krogsgaard-Larsen

Abstract Inactivation of GABA-mediated neurotransmission is achieved by high-affinity transporters located at both GABAergic neurons and the surrounding astrocytes. Early studies of the pharmacological properties of neuronal and glial GABA transporters suggested that different types of transporters might be expressed in the two cell types, and such a scenario was confirmed by the cloning of four distinctly different GABA transporters from a number of different species. These GABA-transport entities have been extensively characterized using a large number of GABA analogues of restricted conformation, and several of these compounds have been shown to exhibit pronounced anticonvulsant activity in a variety of animal seizure models. As proof of concept of the validity of this drug development approach, one GABA-transport inhibitor, tiagabine, has been developed as a clinically active antiepileptic drug. This review provides a detailed account of efforts to design new subtype-selective GABA-transport inhibitors aiming at identifying novel antiepileptic drug candidates.

Keywords Astrocytes • Neurons • GAT • BGT • GABAergic transmission • Epilepsy

1 Introduction

Subsequent to the seminal discovery that GABA is present in the brain and is synthesized by a specific enzyme, glutamate decarboxylase (Roberts and Frankel 1950), it has been firmly established that this amino acid fulfills the criteria for an inhibitory neurotransmitter (Krnjevic and Schwartz 1967; Roberts 1971; Curtis and Johnston 1974), and it may serve this purpose inasmuch as 70% of the synapses in the CNS (Durkin et al. 1995). One of the criteria for a neurotransmitter is the

A. Schousboe (✉) • P. Wellendorph • B. Frølund • R.P. Clausen • P. Krogsgaard-Larsen
Department of Drug Design and Pharmacology, Faculty of Health and Medical Sciences, University of Copenhagen, Universitetsparken 2, 2100 Copenhagen, Denmark
e-mail: arne.schousboe@sund.ku.dk

A. Ortega, A. Schousboe (eds.), *Glial Amino Acid Transporters*,
Advances in Neurobiology 16, DOI 10.1007/978-3-319-55769-4_14

presence of an efficient mechanism to inactivate its action at the receptors, and it was demonstrated by Elliott and van Gelder (1958) that brain slices are capable of concentrating GABA from the incubation medium. Later it was reported that high-affinity transport systems for GABA were present in both neurons and astrocytes (Iversen and Neal 1968), and this was confirmed using [^{3}H]GABA and autoradiography to identify the cellular elements responsible for GABA uptake (Hösli and Hösli 1976, 1978). Comparing the transport capacity for GABA in different brain preparations such as crude synaptosomes, bulk-prepared glial cells, and cultured neurons and astrocytes, it can be concluded that neuronal transport of GABA is more efficient than that into surrounding astrocytes (Schousboe 1981; Hertz and Schousboe 1987), and therefore it is likely that a major fraction of GABA released as neurotransmitter is transported back into the presynaptic GABAergic nerve endings to be reutilized as transmitter (Gram et al. 1988).

2 Glial GABA Transport: A Historical Perspective

As pointed out above, the general notion about the importance of neuronal GABA transport in relation to the function of GABAergic neurotransmission has detracted interest from glial GABA transport. However, the demonstration of high-affinity GABA uptake in bulk-prepared glial cells (Henn and Hamberger 1971) stimulated the interest in investigating the functional role of glial cells in GABA homeostasis (Iversen and Kelly 1975). Studies of GABA uptake in cultured glioma cells (C-6 cells) had found very low maximal velocities of the high-affinity transport (Hutchison et al. 1974; Schrier and Thompson 1974) indicating that such uptake might not be of functional importance. A subsequent study of GABA uptake in primary cultures of astrocytes from cerebral cortex of newborn mice, however, convincingly demonstrated that glial GABA uptake could be of functional importance since the uptake capacity (V_{max}) was of a considerable magnitude compared to that of the low V_{max} values reported for the C-6 cells (Schousboe et al. 1977). Almost at the same time, GABA analogues of restricted conformation had become available as pharmacological tools (see below), and a series of studies were initiated with the purpose of establishing a pharmacological avenue by which the function of astroglial GABA transport could be established (Krogsgaard-Larsen and Johnston 1975; Schousboe et al. 1978, 1979, 1981; Krogsgaard-Larsen 1980, 1981; Larsson et al. 1980, 1981, 1988; Larsson and Schousboe 1981). These aspects will be discussed in further detail below.

3 Cloning and Cellular Localization of GABA Transporters

Whereas a series of pharmacological studies had led to the notion that a distinction could be made between neuronal and glial GABA-transport systems (Schon and Kelly 1975; Iversen and Kelly 1975; Schousboe et al. 1981; Larsson et al. 1981; 1983a, b;

1985), the notion of a diversity of GABA transport could not be firmly established until the advent of molecular cloning of altogether four distinctly different transporters for GABA (Guastella et al. 1990; Nelson et al. 1990; Borden et al. 1992, 1994, 1995; Yamauchi et al. 1992; Liu et al. 1992, 1993; Lopez-Corcuera et al. 1992; Christiansen et al. 2007). As these different authors have used different nomenclatures for the cloned transporters from different species, the nomenclature has been summarized in Table 1. It should be noted that transporters cloned from rat and the human brain are written with a hyphen (e.g., GAT-1), while those cloned from mouse brain are not hyphenized (e.g., GAT1). One of the cloned transporters exhibits affinity not only for GABA but also for the osmolyte betaine and thus is referred to as the betaine-GABA transporter (BGT-1 for the rat and human protein and GAT2/BGT1 for the mouse variant). In the subsequent text, we will use the IUPHAR nomenclature (Table 1).

The cloning of the different subtypes of GABA transporters has provided the basis for detailed studies of the cellular and subcellular localization of the transporters to be performed (for overview see Fig. 1). It is clear that among the different transporters, GAT1 is the most abundant being expressed in GABAergic neurons in essentially all brain areas (Durkin et al. 1995; Borden 1996). It is predominantly expressed in axonal elements of GABAergic neurons and is believed to be responsible for the presynaptic transport of GABA allowing recycling of the transmitter (Radian et al. 1990; Pietrini et al. 1994; Borden 1996; Conti et al. 1998, 2004). Additionally, the GAT1 subtype is found in glial cells associated with GABAergic structures, and hence, it is likely to be responsible for glial GABA transport as well (Radian et al. 1990; Conti et al. 1998, 2004). It appears, however, to be expressed to a lower extent than that found for neuronal structures, albeit its expression level in glia may have been underestimated (Borden 1996). The higher expression level in neurons is clearly in agreement with the conclusion from studies at the cellular level of the kinetics of GABA transport in neurons and astrocytes mentioned above, i.e., a much higher capacity for transport in neurons compared to that in glial cells (Hertz and Schousboe 1987).

The cellular expression of GAT2 has also been investigated, and there may be some disagreements regarding the expression of GAT2. Liu et al. (1993) reported that GAT2 is only found in the neonatal brain and leptomeninges, while Conti et al. (1999, 2004) have reported its expression in both neurons and astrocytes in the mature brain, and it has been suggested that this GABA transporter may be involved in regulation of extrasynaptic GABA levels. Interestingly, this functional characteristic has also recently been suggested for BGT1 (Madsen et al. 2011), the expression level of which has been somewhat controversial. It was originally reported to

Table 1 Nomenclature of GABA transporters with IUPHAR nomenclature highlighted

Gene name	SLC6A1	SLC6A12	SLC6A13	SLC6A11
Human	GAT-1	BGT-1	GAT-2	GAT-3
Mouse	GAT1	GAT2	GAT3	GAT4
Rat	GAT-1	BGT-1	GAT-2	GAT-3
IUPHAR	**GAT1**	**BGT1**	**GAT2**	**GAT3**

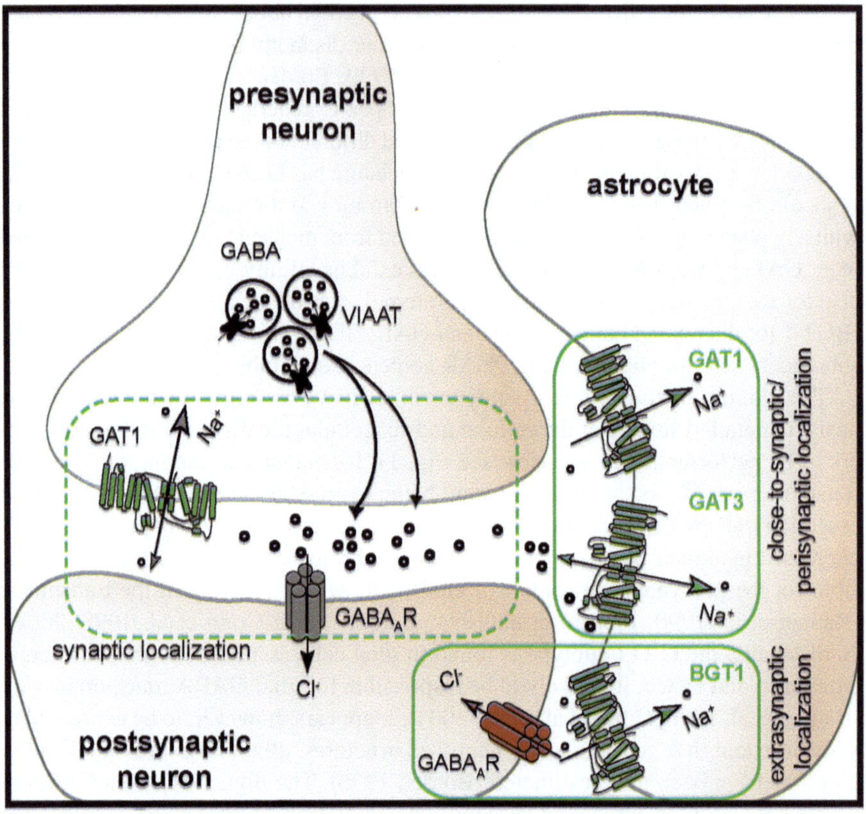

Fig. 1 Schematic representation of a typical GABAergic synapse with cellular localization of GAT1, BGT1, and GAT3. For further details, see main text

be abundantly expressed in extrasynaptic regions (Zhu and Ong 2004a, b), but more recently Zhou et al. (2012a) have claimed that its expression in the brain is very low. Of note, BGT1 and GAT2 are the only GAT subtypes found outside the CNS with BGT1 being highly expressed in the liver and to moderate levels in the kidney (Zhou et al. 2012b) and GAT2 being highly expressed in the kidney (Christiansen et al. 2007). The expression of GAT3 appears to be at a lower level than that for GAT1, and it is primarily associated with glial elements (Durkin et al. 1995; Minelli et al. 1996) suggesting that this transporter is primarily involved in glial GABA transport. It thus appears that GAT1 and GAT3 may be the two GABA transporters mainly responsible for the inactivation of GABA as neurotransmitter albeit BGT1 may also play an important role. These aspects will be elaborated upon in the sections devoted to a pharmacological characterization of the different subtypes of GABA transporters.

In addition to the cellular localization of the different subtypes of GABA transporters summarized above, it may be noted that these subtypes appear to be unevenly distributed in different membrane sub-domains. Thus, GAT1 and BGT1 have been shown to be associated with non-raft membrane domains, whereas GAT3 was found in raft domains (Madsen et al. 2015). Moreover, using FRET-based technology and GAT fuses with Cerulean and Venus (GPF proteins), it was demonstrated that these GAT subtypes can form either homo- or heterodimers (Madsen et al. 2015), a result also reported by others (Scholze et al. 2002). The fact that GAT1 and GAT3 are associated primarily with different cell types as discussed above and that they are expressed in different membrane micro-domains makes it unlikely that these two transporters may be able to form heterodimers in situ. Even GAT1 and BGT1, being associated with the same membrane micro-domain, would most likely be expressed in different cell types and, hence, would not likely form heterodimers. Whether or not such dimerization plays any functional role in the brain remains to be elucidated.

4 Pharmacological Tools to Distinguish Between GABA Transporters and Receptors

GABA is a highly flexible molecule with a high degree of mobility at the three carbon-carbon bonds (Krogsgaard-Larsen et al. 1977). It is therefore to be expected that different conformations of the molecule are associated with its binding as substrate/ligand at the proteins with which it interacts, i.e., GABA transaminase, GABA receptors, and GABA transporters as also discussed recently by Johnston (2016). Based on the molecular structure of the naturally occurring compound muscimol, a series of GABA analogues were prepared including guvacine, isoguvacine, and THIP (4,5,6,7-tetrahydroisoxazolo[5,4-*c*]pyridine-3-ol), and it turned out that while the two latter GABA analogues were found to specifically interact with GABA receptors, guvacine selectively was found to interact with GABA transport in analogy with nipecotic acid (Krogsgaard-Larsen and Johnston 1975; Krogsgaard-Larsen et al. 1977). These important observations laid the foundation for development of an enormous series of GABA analogues (see Fig. 2) acting specifically on the receptors and transporters, respectively (for reviews, see Krogsgaard-Larsen et al. (2006) and Clausen et al. (2006a)).

5 Pharmacology of Glial and Neuronal GABA Transporters

The GABA analogue nipecotic acid which was first described as a potent GABA-transport inhibitor with no affinity for the receptors (Krogsgaard-Larsen and Johnston 1975) was subsequently used to characterize the glial and neuronal transporters (Larsson and Schousboe 1981; Larsson et al. 1983a, 1986a). Based on these observations, it could be concluded that nipecotic acid and GABA act in the same way regarding the transporters in neurons and astrocytes. Interestingly, it was additionally observed that the *cis*-4-hydroxy analogue of nipecotic acid could distinguish between

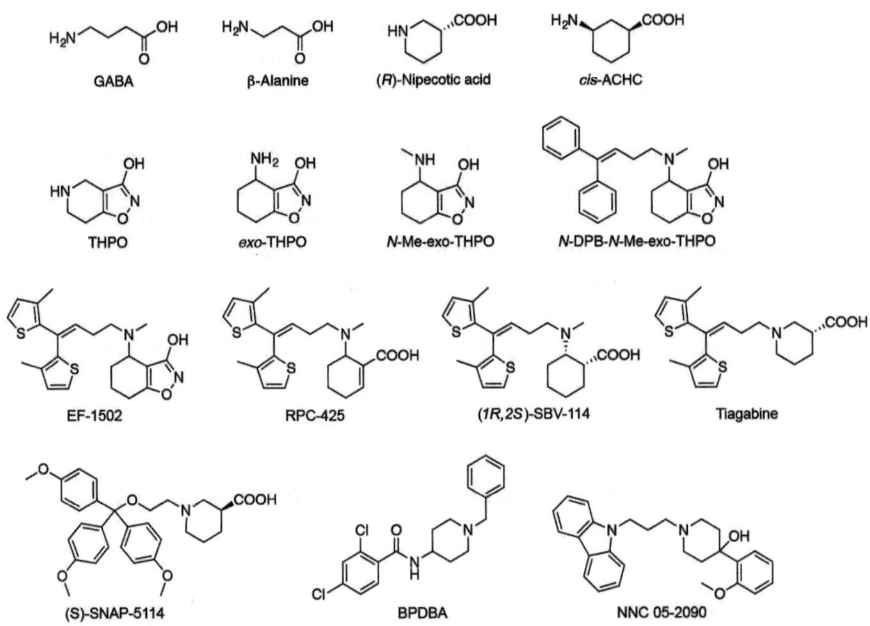

Fig. 2 Chemical structures of representative-selective GAT inhibitors. Please see Table 2 for comparable inhibitory (IC_{50}) values

the neuronal and glia GABA transporters acting competitively with GABA in neurons but noncompetitively in astrocytes (Larsson et al. 1985). This result indicated that neuronal and glial GABA transporters exhibit distinctly different pharmacological characteristics, a notion which was subsequently confirmed by the findings that other GABA analogues of restricted conformation such as THPO (4,5,6,7-tetrahydroisoxazolo[4,5-*c*]pyridin-3-ol), exo-THPO (*RS*)-3-hydroxy-4-amino-4,5,6,7-tetrahydro-1,2-benzisoxazole), and its *N*-methylated derivative, *N*-Me-*exo*-THPO, act as selective inhibitors of astrocytic GABA uptake, i.e., their inhibitory effect on neuronal GABA transport was less pronounced (Schousboe et al. 1981; Clausen et al. 2006b) and Table 2.

In the context of the development of inhibitors selective for astrocytic GABA transport, it may be mentioned that an unexpected result was reported concerning the inhibitory action of *cis*-3-aminocyclohexane carboxylic acid (ACHC) on neuronal and astrocytic GABA transport (Larsson et al. 1983b). ACHC had been described as a selective inhibitor of neuronal GABA uptake (Bowery et al. 1976), but, nevertheless, it was found to be not only inhibitory on astrocytic GABA uptake but also to be transported into astrocytes in a manner exhibiting competitive inhibition by GABA. It was concluded that both astrocytes and neurons express a GABA-like transport system which can also transport ACHC. The exact nature of such a transporter remains, however, to be elucidated.

In an attempt to facilitate blood-brain-barrier penetration of the two GABA analogues, nipecotic acid and guvacine, acting specifically as inhibitors of GABA transport (see above), lipophilic derivatives were synthesized by adding covalently a diphenylbu-

Table 2 Inhibitory activities of GABA analogues and other small molecules on GABA uptake in astrocytes (glia), neurons, and cloned GABA transporters from mice

Compound	Glia	Neuron	GAT1	BGT1	GAT2	GAT3
	IC_{50} (μM)					
(*R*)-Nipecotic acid	16	12	24	>1000	113	159
ACHC	700	200	132	>1000	>1000	>10,000
β-Alanine	843[a]	1666[a]	2920	1100	66	110
Tiagabine	0.18	0.36	0.8	300	>300	800
THPO	262	501	1300	3000	800	5000
Exo-THPO	250	780	1000	3000	>3000	>3000
N-Me-exo-THPO	48	405	450	>3000	>3000	>3000
N-DPB-*N*-Me-exo-THPO	2	5	2	200	>100	>100
(*RS*)-EF-1502	2	2	7	26	>300	>300
(*R*)-EF-1502	0.65	1.5	4	22	>150	>150
(*R*)-EF-1502	>100	>100	120	34	>150	>150
(*S*)-SNAP-5114	–	–	388	140	21	5
NNC 05–2090	–	–	19	1.4	41	15
RPC-425	–	–	307	45	>300	286
(*1R*,*2S*)-SBV-114	–	–	562	21.6	941	289
BPDBA	–	–	>100	39	>200	>100

[a]Indicates a K_i value. For references, see main text

Chemical names:

ACHC: *cis*-3-aminocyclohexane carboxylic acid

Tiagabine: (*R*)-*N*-(4,4-bis(3-methylthiophen-2-yl)but-3-en-1-yl)piperidine-3-carboxylic acid

THPO: 4,5,6,7-tetrahydroisoxazolo[4,5-*c*]pyridin-3-ol

Exo-THPO: (*RS*)-3-hydroxy-4-amino-4,5,6,7-tetrahydro-1,2-benzisoxazole

N-Me-exo-THPO: *N*-methyl-3-hydroxy-4-amino-4,5,6,7-tetrahydro-1,2-benzisoxazole

(*RS*)-EF-1502: (*RS*)-4-[*N*-[1,1bis(3-methyl-2-thienyl)but-1-en-4-yl]-*N*-methylamino]-4,5,6,7-tetrahydrobenzo[*d*]isoxazol-3-ol

N-DPB-*N*-Me-exo-THPO: *N*-diphenylbutenyl-*N*-methyl-3-hydroxy-4-amino-4,5,6,7-tetrahydro-1,2-benzisoxazole

(*S*)-SNAP-5114: (*S*)-*N*-(2-(tris(4-methoxyphenyl)methoxy)ethyl)piperidine-3-carboxylic acid

NNC 05-2090: *N*-(3-(9H-carbazol-9-yl)propyl)-4-(2-methoxyphenyl)piperidin-4-ol

RPC-425: 6-((4,4-bis(3-methylthiophen-2-yl)but-3-en-1-yl)(methyl)amino)cyclohex-1-ene-1-carboxylic acid

SBV-114: (1*R*,2*S*)-2-((4,4-bis(3-methylthiophen-2-yl)but-3-en-1-yl)(methyl)amino)cyclohexane-1-carboxylic acid

BPDBA: *N*-(1-benzyl-4-piperidinyl)-2,4-dichlorobenzamide

tenyl (DPB) residue to the parent compounds, e.g., *N*-DPB-*N*-Me-exo-THPO (Yunger et al. 1984; Ali et al. 1985). These lipophilic GABA analogues were far more potent as inhibitors of GABA transport than the parent compounds, but in contrast to nipecotic acid and guvacine, they did not act as substrates for the transporters in neurons and astrocytes (Larsson et al. 1988). These lipophilic GABA analogues did, however, turn out to be instrumental in subsequent attempts to develop GABA-transport inhibitors that might act as anticonvulsant drugs (see discussion below).

The advent of the cloning of four different GABA transporters (see above) stimulated the interest in designing GABA analogues of restricted conformation that would be able to selectively inhibit either one of these transporters (see, Sarup et al. (2003) and Clausen et al. (2005)). A very large number of such compounds have been synthesized (for overviews see Høg et al. (2006), Madsen et al. (2007, 2008, 2010), Schousboe et al. (2011a, b, 2014), and Wellendorph et al. (2017)). A number of the most pertinent compounds have been summarized in Table 2 with chemical structures shown in Fig. 2. It is seen that among the GABA analogues which are substrates for the transporter in question, nipecotic acid is selective for GAT1, (*R*)-nipecotic acid being the active stereoisomer (Krogsgaard-Larsen 1980), whereas β-alanine is selective for GAT2 and GAT3. Interestingly, β-alanine had many years earlier been described as a glial selective GABA-transporter inhibitor (Schon and Kelly 1975). However, since β-alanine is also an inhibitor of astrocytic taurine transport and is transported by this carrier (Schousboe et al. 1976; Larsson et al. 1986b), the specificity of β-alanine as a glial GABA-transport inhibitor has been questioned (Larsson et al. 1986b). The majority of GABA taken up by astrocytes is transported by GAT1 as it is almost completely inhibited by tiagabine (Borden 1996; A. Schousboe, unpublished), which is GAT1 selective (Madsen et al. 2007).

The β-amino acid backbone of β-alanine has, however, recently been used to synthesize a number of lipophilic GABA analogues. In contrast to the first GABA analogue to be identified as a potent inhibitor of BGT1, namely, EF-1502 ((*RS*)-4-[*N*-[1,1bis(3-methyl-2-thienyl)but-1-en-4-yl]-*N*-methylamino]-4,5,6,7-tetrahydrobenzo[*d*]isoxazol-3-ol) that additionally potently inhibited GAT1 (Clausen et al. 2005), these new lipophilic β-amino acid-based GABA analogues such as RPC-425 and SBV-114 (Fig. 2 and Table 2) were found to be selective inhibitors of the BGT1 GABA-transporter subtype (Vogensen et al. 2013, 2015). Also, non-amino acid-based inhibitors with BGT1 selectivity, such as NNC 05-2090, have been identified. Additionally, a nonrational screening of a compound library of over 3000 diverse compounds led to discovery of another selective inhibitor, BPDBA (*N*-(1-benzyl-4-piperidinyl)-2,4-dichlorobenzamide), of the human and the mouse BGT1 with a noncompetitive pharmacological profile (Kragholm et al. 2013). As will be discussed below, these discoveries may be interesting in light of the finding that inhibition of GAT2/BGT1 provides protection against sound and chemically induced convulsions in mice (White et al. 2005).

6 Astrocytic GABA Transporters and Development of Anticonvulsant Drugs

Epilepsy is a neurological disorder involving spontaneous convulsive and nonconvulsive seizures, and only about half of the patients can be adequately treated with the antiepileptic drugs currently available (White 1999). Epilepsy reflects an imbalance between excitatory and inhibitory neurotransmission (Dalby and Mody 2001), but most of the clinically active drugs have mechanisms of action that do not directly involve the GABAergic neurotransmission (White 1999; Schousboe and White 2009). Actually, only two currently available, clinically active antiepileptic drugs, vigabatrin and tiagabine, selectively target GABAergic entities, the former specifically inhibiting GABA transaminase and the latter inhibiting the GABA transporter 1 (for review, see Rowley et al. (2012)). This provides proof of concept that GABA transporters may be of interest for the development of a new class of GABA-transport inhibitors with efficacy as anticonvulsant agents. The finding that there is in general a better correlation between anticonvulsant activity and inhibition of glial GABA transport than between anticonvulsant activity and inhibitory action on neuronal GABA uptake led to the conclusion that inhibitors selectively targeting the glial GABA transport are particularly effective in protecting against audiogenic seizures in the Frings mouse model (White et al. 2002). This finding confirmed the hypothesis that inhibition of glial GABA transport would be beneficial in seizure management (Schousboe et al. 1983).

The advent of the synthesis of the dithienylbutenyl derivative of N-Me-exo-THPO, EF-1502, acting as a potent inhibitor of both GAT1 and BGT1 (Clausen et al. 2005), prompted an investigation of the ability of this GABA analogue to protect against seizures induced either by sound or by chemical agents (White et al. 2005). EF-1502 was established as a highly efficacious anticonvulsant both in the Frings mouse audiogenic seizure model and in pentylenetetrazole-induced seizures, and, interestingly, it did not exhibit any adverse effects contrary to tiagabine. Using the isobologram procedure (for details, see Schousboe et al. (2011b)), it could be demonstrated that tiagabine and EF-1502 acted synergistically as anticonvulsants (White et al. 2005). A similar result was obtained with a combination of EF-1502 and the GAT4-selective inhibitor SNAP-5114 (Madsen et al. 2009) albeit this was only statistically significant at a combinatorial dose of 3:1 (tiagabine: SNAP-5114). This finding that tiagabine, a specific GAT1 inhibitor, in combination with the BGT1 and GAT3 inhibitors acted synergistically led to the notion that extrasynaptic GABA transporters exemplified by BGT1 and GAT3 (see above) may play an important role in the anticonvulsant action of GABA-transport inhibitors (Madsen et al. 2009; Eulenburg and Gomeza 2010). This was further investigated taking advantage of the previous demonstration that the $GABA_A$-specific agonist Gaboxadol (Krogsgaard-Larsen et al. 2006) acts primarily on the extrasynaptic GABA receptors (Stórustovu and Ebert 2006), and it could be shown that the anticonvulsant action of Gaboxadol (THIP) was reversed by EF-1502 but not by tiagabine. This finding can best be explained by assuming that the anticonvulsant action of a GABA-transport inhibitor is most pronounced by an inhibitory action on extrasynaptically located transporters (Madsen et al. 2011). It is, however, somewhat enigmatic that deletion of the BGT1 gene had no effect on the anticonvulsant activity of EF-1502 (Lehre et al. 2011), and this needs further clarification.

7 Concluding Remarks

From early functional studies on glial preparations and cultured astrocytes to more recent recombinant studies of cloned GABA transporters, glial GABA transporters have emerged as key mediators in controlling extracellular GABA levels and hence neuronal inhibition. Given the metabolic degradation of GABA upon uptake into glial cells, inhibiting glial uptake over neuronal uptake may be advantageous in terms of increasing synaptic GABA levels (Schousboe et al. 1983). In the field of epilepsy, glial GABA transporters are already rendered relevant drug targets based on the anticonvulsant effects of inhibitors of GAT3 and BGT1/GAT1 (Clausen et al. 2005; Madsen et al. 2011). This effect may relate to the peri- or extrasynaptic localization of GAT3 and BGT1 and hence an overall effect on increasing tonic inhibition in the brain. Along these lines, GAT3 is also emerging as a drug target in relation to functional recovery after stroke (Clarkson et al. 2010). The pathophysiology and therapeutic potential of modulating GABAergic inhibition through GATs will be further elaborated in Chapter Glial GABA Transporters as Modulators of Inhibitory Signalling in Epilepsy and Stroke.

To advance the field further, there is a continuous need for better and more subtype-selective ligands that are at the same time able to permeate the blood-brain barrier. This requires meticulous medicinal chemistry efforts, which will be discussed in Chapter Development of Non-GAT1-Selective Inhibitors: Challenges and Achievements.

Conflict of Interest The author declares no conflicts of interest.

References

Ali FE, Bondinell WE, Dandridge PA, Frazee JS, Garvey E, Girard GR, Kaiser C, Ku TW, Lafferty JJ, Moonsammy GI. Orally active and potent inhibitors of gamma-aminobutyric acid uptake. J Med Chem. 1985;28:653–60.

Borden LA. GABA transporter heterogeneity: pharmacology and cellular localization. Neurochem Int. 1996;29:335–56.

Borden LA, Smith KE, Hartig PR, Branchek TA, Weinshank RL. Molecular heterogeneity of the gamma-aminobutyric acid (GABA) transport system. Cloning of two novel high affinity GABA transporters form rat brain. J Biol Chem. 1992;267:21098–104.

Borden LA, Dhar TG, Smith KE, Branchek TA, Gluchowski C, Weinshank RL. Cloning of the human homologue of the GABA transporter GAT-3 and identification of a novel inhibitor with selectivity for this site. Recept Channels. 1994;2:207–13.

Borden LA, Smith KE, Gustafson EL, Branchek TA, Weinshank RL. Cloning and expression of a betaine/GABA transporter from human brain. J Neurochem. 1995;64:977–84.

Bowery NG, Jones GP, Neal MJ. Selective inhibition of neuronal GABA uptake by *cis*-1,3-aminocyclohexane carboxylic acid. Nature. 1976;264:281–4.

Christiansen B, Meinild A-K, Jensen AA, Bräuner-Osborne H. Cloning and characterization of a functional human γ-aminobutyric acid, GABA transporter, human GAT2. J Biol Chem. 2007;282:19331–41.

Clarkson A, Huang BS, MacIsaac SA, Mody I, Carmichael T. Reducing excessive GABA-mediated tonic inhibition promotes functional recovery after stroke. Nature. 2010;468:305–9.

Clausen RP, Moltzen E, Perregaard J, Lenz SM, Sanchez C, Falch E, Frølund B, Bolvig T, Sarup A, Larsson OM, Schousboe A, Krogsgaard-Larsen P. Selective inhibitors of GABA uptake: synthesis and molecular pharmacology of 4-*N*-methylamino-4,5,6,7-tetrahydrobenzo[*d*]isoxazol-3-ol analogues. Bioorg Med Chem. 2005;13:895–908.

Clausen RP, Madsen K, Larsson OM, Frølund B, Krogsgaard-Larsen P, Schousboe A. Structure-activity relationship and pharmacology of gamma-aminobutyric acid (GABA) transport inhibitors. Adv Pharmacol. 2006a;54:265–84.

Clausen RP, Frølund B, Larsson OM, Schousboe A, Krogsgaard-Larsen P, White HS. A novel selective y-aminobutyric acid transport inhibitor demonstrates a functional role for GABA transporter subtype GAT2/BGT-1 in the CNS. Neurochem Int. 2006b;48:637–42.

Conti F, Melone M, De Biasi S, Minelli A, Brecha NC, Ducati A. Neuronal and glial localization of GAT-1, a high-affinity gamma-aminobutyric acid plasma membrane transporter, in human cerebral cortex: with a note on its distribution in monkey cortex. J Comp Neurol. 1998;396:51–63.

Conti F, Zuccarello LV, Barberesi P, Minelli A, Brecha NC, Melone M. Neuronal, glial, and epithelial location of gamma-aminobutyric acid transporter 2, a high-affinity gamma-aminobutyric acid plasma membrane transporter, in the cerebral cortex and neighboring structures. J Comp Neurol. 1999;409:482–94.

Conti F, Minelli A, Melone M. GABA transporters in the mammalian cerebral cortex: localization, development and pathological implications. Brain Res Rev. 2004;45:196–212.

Curtis DA, Johnston GA. Amino acid transmitters in the mammalian central nervous system. Ergeb Physiol. 1974;69:97–188.

Dalby NO, Mody I. The process of epileptogenesis: a pathophysiological approach. Curr Opin Neurol. 2001;14:187–92.

Durkin MM, Smith KE, Borden LA, Weinshank RL, Branchek TA, Gustafson EL. Localization of messenger RNAs encoding three GABA transporters in rat brain: an in situ hybridization study. Brain Res Mol Brain Res. 1995;33:7–21.

Elliott KA, van Gelder NM. Occlusion and metabolism of gamma-aminobutyric acid by brain tissue. J Neurochem. 1958;3:28–40.

Eulenburg V, Gomeza J. Neurotransmitter transporters expressed in glial cells as regulators of synapse function. Brain Res Rev. 2010;63:103–12.

Gram L, Larsson OM, Johnsen AH, Schousboe A. Effects of valproate, vigabatrin and aminooxyacetic acid on release of endogenous and exogenous GABA from cultured neurons. Epilepsy Res. 1988;2:87–95.

Guastella J, Nelson N, Nelson H, Czyzyk L, Keynan S, Miedel MC, Davidson M, Lester HA, Kanner BI. Cloning and expression of a rat brain GABA transporter. Science. 1990;249:1303–6.

Henn FA, Hamberger A. Glial cell function: uptake of transmitter substances. Proc Natl Acad Sci U S A. 1971;68:2686–90.

Hertz L, Schousboe A. Primary cultures of GABAergic and glutamatergic neurons as model systems to study neurotransmitter function. I. Differentiated cells. In: Vernadakis A, Privat A, Lauder JM, Timiras PS, Giacobini E, editors. Model systems of development and aging of the nervous system. Boston: Martinus Nijhoff; 1987. p. 19–31.

Høg S, Greenwood JR, Madsen KB, Larsson OM, Frølund B, Schousboe A, Krogsgaard-Larsen P, Clausen RP. Structure-activity relationships of selective GABA uptake inhibitors. Curr Top Med Chem. 2006;6:1861–82.

Hösli E, Hösli L. Autoradiographic studies on the uptake of [^{3}H]noradrenaline and [^{3}H]GABA in cultured rat cerebellum. Exp Brain Res. 1976;26:319–24.

Hösli E, Hösli L. Autoradiographic localization of the uptake of [^{3}H]GABA and L-[^{3}H]glutamic acid in neurons and glial cells of cultured dorsal root ganglia. Neurosci Lett. 1978;7:173–6.

Hutchison HT, Werrbach K, Vance C, Haber B. Uptake of neurotransmitters by clonal cell lines of astrocytoma and neuroblastoma in cell culture. Brain Res. 1974;66:265–74.

Iversen LL, Kelly JS. Uptake and metabolism of y-aminobutyric acid by neurons and glial cells. Biochem Pharmacol. 1975;24:933–8.

Iversen LL, Neal MJ. The uptake of [^{3}H]GABA by slices of rat cerebral cortex. J Neurochem. 1968;15:1141–9.

Johnston GAR. Unsaturated analogues of the neurotransmitter GABA: *trans*-4-aminocrotonic, *cis*-4-aminocrotonic and 4-aminotetrolic acids. Neurochem Res. 2016;41:476–80.

Kragholm B, Kvist T, Madsen KK, Jørgensen L, Vogensen SB, Schousboe A, Clausen RP, Jensen AA, Bräuner-Osborne H. Discovery of a subtype selective inhibitor of the human betaine/GABA transporter 1 (BGT-1) with a non-competitive pharmacological profile. Biochem Pharmacol. 2013;86:521–8.

Krnjevic K, Schwartz S. The action of gamma-aminobutyric acid on cortical neurons. Exp Brain Res. 1967;3:320–36.

Krogsgaard-Larsen P. Inhibitors of the GABA uptake system. Mol Cell Biochem. 1980;31:105–21.

Krogsgaard-Larsen P. Gamma-aminobutyric acid agonists, antagonists, and uptake inhibitors. J Med Chem. 1981;24:1377–83.

Krogsgaard-Larsen P, Johnston GA. Inhibition of GABA uptake in rat brain slices by nipecotic acid, various isoxazoles and related compounds. J Neurochem. 1975;25:797–802.

Krogsgaard-Larsen P, Johnston GAR, Lodge D, Curtis DR. A new class of GABA agonist. Nature. 1977;268:53–5.

Krogsgaard-Larsen P, Frølund B, Liljefors T. GABA(A) agonists and partial agonists: THIP (Gaboxadol) as a non-opioid analgesic and a novel type of hypnotic. Adv Pharmacol. 2006;54:53–71.

Larsson OM, Schousboe A. Comparison between (*RS*)-nipecotic acid and GABA transport in cultured astrocytes: coupling with two sodium ions. Neurochem Res. 1981;6:257–66.

Larsson OM, Krogsgaard-Larsen P, Schousboe A. High-affinity uptake of (*RS*)-nipecotic acid in astrocytes cultured from mouse brain. Comparison with GABA transport. J Neurochem. 1980;34:970–7.

Larsson OM, Thorbek P, Krogsgaard-Larsen P, Schousboe A. Effect of homo-β-proline and other heterocyclic GABA analogs on GABA uptake in neurons and astroglial cells and on GABA receptor binding. J Neurochem. 1981;37:1509–16.

Larsson OM, Drejer J, Hertz L, Schousboe A. Ion dependency of uptake and release of GABA and (*RS*)-nipecotic acid studied in cultured mouse brain cortex neurons. J Neurosci Res. 1983a;9:291–302.

Larsson OM, Johnston GAR, Schousboe A. Differences in uptake kinetics of *cis*-3-aminocyclohexane carboxylic acid into neurons and astrocytes in primary cultures. Brain Res. 1983b;260:279–85.

Larsson OM, Krogsgaard-Larsen P, Schousboe A. Characterization of uptake of GABA and its analogues nipecotic acid and *cis*-4-OH-nipecotic acid in cultured neurons and astrocytes. Neurochem Int. 1985;5:853–60.

Larsson OM, Hertz L, Schousboe A. Uptake of GABA and nipecotic acid in astrocytes and neurons in primary cultures: changes in the sodium coupling ratio during differentiation. J Neurosci Res. 1986a;16:699–708.

Larsson OM, Griffiths R, Allen IC, Schousboe A. Mutual inhibition kinetic analysis of γ-aminobutyric acid, taurine, and β-alanine high-affinity transport into neurons and astrocytes: evidence for similarity between the taurine and β-alanine carriers in both cell types. J Neurochem.. 1986b;47:426–432.

Larsson OM, Falch E, Krogsgaard-Larsen P, Schousboe A. Kinetic characterization of inhibition of gamma-aminobutyric acid uptake into cultured neurons and astrocytes by 4,4-diphenyl-3-butenyl derivatives of nipecotic acid and guvacine. J Neurochem. 1988;50:818–23.

Lehre AC, Rowley NM, Zhou Y, Holmseth S, Guo C, Holen T, Hua R, Laake P, Olofsson AM, Poblete-Naredo I, Rusakov DA, Madsen KK, Clausen RP, Schousboe A, White HS, Danbolt NC. Deletion of the betaine-GABA transporter (BGT1;slc6a12) gene does not affect seizure thresholds of adult mice. Epilepsy Res. 2011;95:70–81.

Liu QR, Mandiyan S, Nelson H, Nelson N. A family of genes encoding neurotransmitter transporters. Proc Natl Acad Sci U S A. 1992;89:6639–43.

Liu QR, Lopez-Corcuera B, Mandiyan S, Nelson H, Nelson N. Molecular characterization of four pharmacologically distinct gamma-aminobutyric acid transporters in mouse brain. J Biol Chem. 1993;268:2106–12.

Lopez-Corcuera B, Liu QR, Mandiyan S, Nelson H, Nelson N. Expression of a mouse brain cDNA encoding novel y-aminobutyric acid transporter. J Biol Chem. 1992;267:17491–3.

Madsen K, White HS, Clausen RP, Frølund B, Larsson OM, Krogsgaard-Larsen P, Schousboe A. Functional and pharmacological aspects of GABA transporters. In: Reith M, editor. Neural membranes and transport: handbook of neurochemistry and molecular neurobiology. Third ed. NY: Springer; 2007. p. 285–303.

Madsen KK, Larsson OM, Schousboe A. Regulation of excitation by GABA neurotransmission: focus on metabolism and transport. Results Probl Cell Differ. 2008;44:201–21.

Madsen KK, Clausen RP, Larsson OM, Krogsgaard-Larsen P, Schousboe A, White HS. Synaptic and extrasynaptic GABA transporters as targets for anti-epileptic drugs. J Neurochem. 2009;109(Suppl 1):139–44.

Madsen KK, White HS, Schousboe A. Neuronal and non-neuronal GABA transporters as targets for antiepileptic drugs. Pharmacol Therap. 2010;125:394–401.

Madsen KK, Ebert B, Clausen RP, Krogsgaard-Larsen P, Schousboe A, White HS. Selective GABA transporter inhibitors tiagabine and EF1502 exhibit mechanistic differences in their ability to modulate the ataxia and anticonvulsant action of the extrasynaptic GABA(A) receptor agonist gaboxadol. J Pharmacol Exp Ther. 2011;338:214–9.

Madsen KK, Hansen GH, Danielsen EM, Schousboe A. The subcellular localization of GABA transporters and its implication for seizure management. Neurochem Res. 2015;40:410–9.

Minelli A, De Biasi S, Brecha NC, Zuccarello LV, Conti E. GAT-3, a high-affinity GABA plasma membrane transporter, is localized to astrocytic processes, and it is not confined to the vicinity of the GABAergic synapses in the cerebral cortex. J Neurosci. 1996;16:6255–64.

Nelson H, Mandiyan S, Nelson N. Cloning of the human brain GABA transporter. FEBS Lett. 1990;269:181–4.

Pietrini G, Suh YJ, Edelmann L, Rudnick G, Caplan MJ. The axonal gamma-aminobutyric acid transporter GAT-1 is sorted to the apical membranes of polarized epithelial cells. J Biol Chem. 1994;269:4668–74.

Radian R, Ottersen OP, Storm-Mathisen J, Castel M, Kanner BI. Immunocytochemical localization of the GABA transporter in rat brain. J Neurosci. 1990;10:1319–30.

Roberts E. The GABA system in brain development. In: Paoletti R, Davison AN, editors. Chemistry and brain development. New York: Plenum Press; 1971. p. 207–14.

Roberts E, Frankel S. Gamma-aminobutyric acid in brain: its formation from glutamic acid. J Biol Chem. 1950;187:55–63.

Rowley NM, Madsen KK, Schousboe A, White HS. Glutamate and GABA synthesis, release, transport and metabolism as targets for seizure control. Neurochem Int. 2012;61:546–58.

Sarup A, Larsson OM, Bolvig T, Frølund B, Krogsgaard-Larsen P, Schousboe A. Effects of 3-hydroxy-4-amino-4,5,6,7-tetrahydro-1,2-benzisoxazol (*exo*-THPO) and its N-substituted analogs on GABA transport in cultured neurons and astrocytes and by the four cloned mouse GABA transporters. Neurochem Int. 2003;43:445–51.

Scholze P, Freissmuth M, Sitte HH. Mutations within an intramembrane leucine heptad repeat disrupt oligomer formation of the rat GABA transporter 1. J Biol Chem. 2002;277:43682–90.

Schon F, Kelly JS. Selective uptake of β-[^{3}H]alanine by glia: association with the glial uptake system for GABA. Brain Res. 1975;86:243–57.

Schousboe A. Transport and metabolism of glutamate and GABA in neurons and glial cells. Int Rev Neurobiol. 1981;22:1–45.

Schousboe A, White HS. Glial modulation of excitability via glutamate and GABA transporters. In: Schwartzkroin PA, editor. Encyclopedia of basic epilepsy research. Oxford: Academic Press; 2009. p. 397–401.

Schousboe A, Fosmark H, Svenneby G. Taurine uptake in astrocytes cultured from dissociated mouse brain hemispheres. Brain Res. 1976;116:158–64.

Schousboe A, Hertz L, Svenneby G. Uptake and metabolism of GABA in astrocytes cultured from dissociated mouse brain hemispheres. Neurochem Res. 1977;2:217–29.

Schousboe A, Krogsgaard-Larsen P, Svenneby G, Hertz L. Inhibition of high affinity, net uptake of GABA into cultured astrocytes by β-proline, nipecotic acid and other compounds. Brain Res. 1978;153:623–6.

Schousboe A, Thorbek P, Hertz L, Krogsgaard-Larsen P. Effects of GABA analogues of restricted conformation on GABA transport in astrocytes and brain cortex slices and on GABA receptor binding. J Neurochem. 1979;33:181–9.

Schousboe A, Larsson OM, Hertz L, Krogsgaard-Larsen P. Heterocyclic GABA analogues as new selective inhibitors of astroglial GABA transport. Drug Dev Res. 1981;1:115–27.

Schousboe A, Larsson OM, Wood JD, Krogsgaard-Larsen P. Transport and metabolism of γ-aminobutyric acid in neurons and glia: implications for epilepsy. Epilepsia. 1983;24:531–8.

Schousboe A, Madsen KK, White SH. GABA transport inhibitors and seizure protection: the past and the future. Future Med Chem. 2011a;3:183–7.

Schousboe A, Madsen KK, White SH. Neurotransmitter transporters and anticonvulsant drug development. NeuroMethods. 2011b;56:431–46.

Schousboe A, Madsen KK, Barker-Haliski M, White HS. The GABA synapse as a target for antiepileptic drugs: a historical overview focused on GABA transporters. Neurochem Res. 2014;39:1980–7.

Schrier BH, Thompson EJ. On the role of the glial cells in the mammalian nervous system. J Biol Chem. 1974;249:1769–80.

Stórustovu SI, Ebert B. Pharmacological characterization of agonists at δ-containing $GABA_A$ receptors: functional selectivity of receptors is dependent on the absence of y_2. J Pharmacol Exp Ther. 2006;316:1351–9.

Vogensen SB, Jørgensen L, Madsen KK, Borkar N, Wellendorph P, Skovgaard-Petersen J, Schousboe A, White HS, Krogsgaard-Larsen P, Clausen RP. Selective mGAT2 (BGT-1) GABA uptake inhibitors: design, synthesis, and pharmacological characterization. J Med Chem. 2013;56:2160–4.

Vogensen SB, Jørgensen L, Madsen KK, Jurik A, Borkar N, Rosatelli E, Nielsen B, Ecker G, Schousboe A, Clausen RP. Structure-activity relationship of selective GABA uptake inhibitors. Bioorg Med Chem. 2015;23:2480–8.

Wellendorph P, Jacobsen J, Skovgaard-Petersen J, Jurik A, Vogensen SB, Ecker G, Schousboe A, Krogsgaard-Larsen P, Clausen RP. γ-Aminobutyric acid and glycine neurotransmitter transporters. In: Ecker G, Clausen RP, Sitte H, editors. Methods and principles in medicinal chemistry – transporters as drug targets. Weinheim: Wiley-VCH Verlag GmBH & Co. KGaA; 2017. doi:10.1002/9783527679430.ch4.

White HS. Comparative anticonvulsant and mechanistic profile of the established and newer antiepileptic drugs. Epilepsia. 1999;40(Suppl 5):S2–S10.

White HS, Sarup A, Bolvig T, Kristensen AS, Petersen G, Nelson N, Pickering DS, Larsson OM, Frølund B, Krogsgaard-Larsen P, Schousboe A. Correlation between anticonvulsant activity and inhibitory action on glial γ-aminobutyric acid uptake of the highly selective mouse γ-aminobutyric acid transporter 1 inhibitor 3-hydroxy-4-amino-4,5,6,7-tetrahydro-1,2-benzisoxazole and its *N*-alkylated analogs. J Pharmacol Exp Ther. 2002;202:636–47.

White HS, Watson WP, Hansen SL, Slough S, Perregaard J, Sarup A, Bolvig T, Petersen G, Larsson OM, Clausen RP, Frølund B, Falck E, Krogsgaard-Larsen P, Schousboe A. First demonstration of a functional role for central nervous system betaine/γ-aminobutyric acid transporter (mGAT2) based on synergistic anticonvulsant action among inhibitors of mGAT1 and mGAT2. J Pharmacol Exp Ther. 2005;312:866–74.

Yamauchi A, Uchida S, Kwon HM, Preston AS, Robey RB, Garcia-Perz A, Burg MG, Handler JS. Cloning of the Na^+- and Cl^--dependent betaine transporter that is regulated by hypertonicity. J Biol Chem. 1992;267:649–52.

Yunger LM, Fowler PJ, Zarevics P, Settler PE. Novel inhibitors of gamma-aminobutyric acid (GABA) uptake: anticonvulsant actions in rats and mice. J Pharmacol Exp Ther. 1984;288:109–15.

Zhou Y, Holmseth S, Guo C, Hassel B, Höfner G, Wanner KT, Danbolt NC. Deletion of the γ-aminobutyric acid transporter 2 (GAT2 and SLC6A13) gene in mice leads to changes in liver and brain taurine contents. J Biol Chem. 2012a;287:35733–46.

Zhou Y, Holmseth S, Hua R, Lehre AC, Olofsson AM, Poblete-Naredo I, Kempson SA, Danbolt NC. The betaine-GABA transporter (BGT1, slc6a12) is predominantly expressed in the liver and at lower levels in the kidneys and at the brain surface. Am J Physiol Ren Physiol. 2012b;302:F316–28.

Zhu XM, Ong WY. A light and electron microscopic study of betaine/GABA transporter distribution in the monkey cerebral neocortex and hippocampus. J Neurocytol. 2004a;33:233–40.

Zhu XM, Ong WY. Changes in the GABA transporters in the rat hippocampus after kainate-induced neuronal injury: decrease in GAT-1 and GAT-3 but upregulation of betaine/GABA transporter BGT-1. J Neurosci Res. 2004b;77:402–9.

Glutamate Transporters in the Blood-Brain Barrier

Hans Christian Cederberg Helms, Carsten Uhd Nielsen, Helle Sønderby Waagepetersen, and Birger Brodin

Abstract The amino acid L-glutamate serves a number of roles in the central nervous system, being an excitatory neurotransmitter, metabolite, and building block in protein synthesis. During pathophysiological events, where L-glutamate homeostasis cannot be maintained, the increased brain interstitial fluid concentration of L-glutamate causes excitotoxicity. A tight control of the brain interstitial fluid L-glutamate levels is therefore imperative, in order to maintain optimal neurotransmission and to avoid such excitotoxicity. The blood-brain barrier, i.e., the endothelial lining of the brain capillaries, regulates the exchange of nutrients, gases, and metabolic waste products between plasma and brain interstitial fluid. It has been suggested that brain capillary endothelial cells could play an important role in L-glutamate homeostasis by mediating brain-to-blood L-glutamate efflux. Both in vitro and in vivo studies have demonstrated blood-to-brain transport of L-glutamate, at least during pathological events. A number of studies have shown that brain endothelial cells express excitatory amino acid transporters, which may account for abluminal concentrative uptake of L-glutamate into the capillary endothelial cells. The mechanisms underlying transendothelial L-glutamate transport are however still not well understood. The present chapter summarizes the current knowledge on blood-brain barrier L-glutamate transporters and the suggested pathways for the brain-to-blood L-glutamate efflux.

Keywords Excitotoxicity • EAAT • Neurovascular unit • Brain glutamate efflux • Glutamate metabolism

---

H.C.C. Helms • B. Brodin (✉)
Department of Pharmacy, The Faculty of Health and Medical Sciences,
University of Copenhagen, Universitetsparken 2, DK-2100 Copenhagen, Denmark
e-mail: birger.brodin@sund.ku.dk

C.U. Nielsen
Department of Physics, Chemistry and Pharmacy, University of Southern Denmark,
Campusvej 55, DK-5230 Odense, Denmark

H.S. Waagepetersen
Department of Drug Design and Pharmacology, The Faculty of Health and Medical Sciences,
University of Copenhagen, Universitetsparken 2, DK-2100 Copenhagen, Denmark

A. Ortega, A. Schousboe (eds.), *Glial Amino Acid Transporters*,
Advances in Neurobiology 16, DOI 10.1007/978-3-319-55769-4_15

1 Introduction

L-glutamate is one of the most common amino acids, a building block in protein synthesis and an important metabolite in intermediary metabolism linking carbohydrate and amino acid metabolism. L-glutamate is also considered to be the most abundant excitatory neurotransmitter in the vertebrate brain (Zhou and Danbolt 2014). L-glutamate is stored in vesicles in the synapses of glutamatergic neurons and released into the synaptic cleft upon nerve stimulation. The transmitter is rapidly removed from the synaptic cleft and the brain interstitial fluid (ISF) by the concerted actions of glutamate transporters in neurons and astrocytes. Prolonged elevated concentrations of L-glutamate in the ISF are highly cytotoxic since continued and excessive stimulation of glutamate receptors will cause abnormal high intracellular calcium levels in neurons, followed by activation of a number of enzyme cascades and, eventually, cell death. This phenomenon has been termed excitotoxicity (Nakanishi 1992; Rothman 1984) and occurs during a number of pathophysiological conditions (Gillessen et al. 2002; Leibowitz et al. 2012).

The blood-brain barrier, i.e., the endothelium of the brain capillaries, acts as a gatekeeper between the blood and the brain ISF. One major role of the brain endothelium is to separate the blood plasma constituents and the ISF, since the blood concentration of L-glutamate is 10–100-fold higher than in the brain, and free passage across the barrier would cause excitotoxicity. A number of studies have indicated that the brain endothelium might also play an active role in the regulation of brain L-glutamate concentrations beyond just being a static barrier to the peripheral circulation. Brain endothelial cells express excitatory amino acid transporters (EAAT's) and have been proposed to efflux L-glutamate via the concerted actions of abluminal (brain-facing) and luminal (blood-facing) transport proteins (Boyko et al. 2014; Roberts et al. 2014; Teichberg et al. 2009; Campos et al. 2011a; Cohen-Kashi-Malina et al. 2012; Helms et al. 2012; Lyck et al. 2009).

Previous reviews of the field have focused on the role of possible blood-to-brain transport of L-glutamate originating from dietary monosodium glutamate (Hawkins 2009), on EAATs (Danbolt et al. 2016) and blood L-glutamate scavenging (Teichberg et al. 2009), on possible therapeutic applications of blood glutamate scavenging (Castillo et al. 2016) and on potential clinical relevance of BBB-glutamate efflux (Cederberg et al. 2014). The aim of the present chapter is to supply an overview of L-glutamate transport proteins in the brain capillary endothelium, as well as their role in uptake and transendothelial transport. Possible metabolism of glutamate in the endothelial cells and astrocytes will be discussed. Finally, we will address the proposed concept of "glutamate scavenging" and point toward areas which in our point of view warrants further investigations.

2 Early Studies on Glutamate Transport Across Brain Endothelium

There is a large difference between L-glutamate concentrations in blood and in brain tissue interstitial fluid. Free plasma concentrations have been estimated to be in the range of 30–90 μM (Hagenfeldt and Arvidsson 1980; Graham et al. 1995), whereas erythrocyte concentrations are in the range of 280 μM (Hagenfeldt and Arvidsson 1980) to 450 μM (Divino Filho et al. 1997), thus giving whole blood concentrations of L-glutamate at approximately 140 μM taking the volume fractions into consideration (Hagenfeldt and Arvidsson 1980).

Brain ISF L-glutamate concentrations are in the lower μmolar range. Sampling and analysis of brain ISF from the resting undisturbed brain is not trivial, but microdialysis studies have indicated values of 0.1–3 μM (Persson and Hillered 1992). Estimations of L-glutamate concentrations in the cerebrospinal fluid yield values in a similar range, i.e., 0.5–3 μM (Iijima et al. 1978; Spink et al. 1986).

The large concentration difference of L-glutamate across the blood-brain barrier could indicate that the brain endothelial cells have a very low permeability for L-glutamate, an active transport of L-glutamate from brain ISF to blood, a fast metabolism of L-glutamate, or a combination of these, along with a fast metabolism/clearance of L-glutamate in the ISF by astrocytes and neurons (Schousboe et al. 2014; Waagepetersen et al. 2005).

The concentrations of L-glutamate in brain endothelial cells have been estimated to be approximately 750 nmol g^{-1} (Cardelli-Cangiano et al. 1981), equivalent to approximately 785 μM (assuming an endothelial cell density of 1.048 g ml^{-1} (Pries et al. 2000)). Considering that the free L-glutamate concentration is ~60 μM in the blood, the concentration gradient will favor transport from the cell cytosol to the blood and not favor uptake from blood to cell. Early studies on blood-to-brain transport of tracer amounts of radiolabeled glutamate in rats showed a low brain uptake (Oldendorf 1971; Sershen and Lajtha 1976), as measured by the brain uptake index (BUI) method. The uptake of radiolabeled tracer was shown to be inhibited by excess non-labeled L-glutamate, indicating that the transport from blood to brain took place via a saturable carrier-type mechanism (Oldendorf and Szabo 1976). These estimates were, however, based on unidirectional influx values.

Despite the cell-to-blood concentration gradient of L-glutamate, there would still exist a unidirectional isotope flux in the blood-to-cell direction, as observed by Oldendorf and Szabo, but not a net uptake (Oldendorf and Szabo 1976). In summary, the initial studies indicated the presence of a carrier mechanism for L-glutamate in the luminal membrane of the brain endothelium but also a large gradient for L-glutamate transport from plasma to cell, indicating either that the observed fluxes only were indicative of unidirectional influx (plasma to cell) in a situation where a net efflux was occurring or that the thermodynamically "uphill" transport of L-glutamate from plasma to cell was coupled to a dissipation of a concentration gradient (either a co- or counter transport) or a pump-type transporter (driven by hydrolysis of ATP). This issue was addressed by Drewes and colleagues, who showed a net efflux of L-glutamate from the brain to the blood in studies on perfused dog brains (Drewes et al. 1977). In brief, the author's measured concentrations of amino acids in the arterial and venous perfusates using chromatography techniques, and net movements of amino acids, were calculated from the concentration differences in arterial and venous concentrations. They observed a net efflux of L-glutamate from the brain in the order of ~1 μmol 100 g brain^{-1} min^{-1} under basal conditions (Drewes et al. 1977). On the basis of these data, Pardridge suggested that an active L-glutamate brain-to-blood efflux system was present and functional under resting conditions in the brain capillary endothelium (Pardridge 1979). Hutchison and colleagues investigated uptake kinetics of L-glutamate into isolated rat brain capillaries (Hutchison et al. 1985). They demonstrated that a high-affinity (K_M of approximately 2 μM), temperature-dependent, and ouabain-sensitive L-glutamate uptake system was present in the capillaries, which could account for the previously observed brain efflux. This was further supported by Hosoya et al., who performed intracerebral microinjections of radioactive L-glutamate, L-aspartate, and D-aspartate in rats (Hosoya et al. 1999). They demonstrated rapid clearance from the brain of L-glutamate and L-aspartate, which correlated with the appearance of the two isotopes in jugular vein samples, whereas D-aspartate stayed in the brain compartment. Hosoya et al. used thin layer chromatography to verify that at least the main part of the appearing radioactivity in the blood originated from intact L-aspartate/L-glutamate (approximately 70% and 84%, respectively) (Hosoya et al. 1999). Initially, this efflux was not believed to be associated with EAAT transporters, partly because of lack of D-aspartate efflux and partly because immunolabeling studies had not shown EAAT1 and EAAT2 expression in rat brain endothelial cells (Hosoya et al. 1999; Chaudhry et al. 1995; Lehre et al. 1995). However, in the recent decades, evidence has accumulated that EAATs are present in brain capillary endothelial cells and may take part in the brain L-glutamate efflux (see sections below).

3 Expression of Glutamate Transporters in the Brain Capillary Endothelium

3.1 Transporters of Glutamate

Glutamate transporters are transmembrane proteins belonging to the solute carrier (SLC) family 1, SLC 7, SLC 17, or SLC25 (see Table 1) (Fotiadis et al. 2013; Kanai et al. 2013; Palmieri 2013; Reimer 2013; Hegedus and Taale 2013). The SLC1 family consists of seven members, with five members being excitatory amino acid transporters (EAATs) transporting glutamate and aspartate and two members being alanine, serine, and cystine (ASC) transporters transporting alanine, serine, cystine, and threonine (see Table 1 (Kanai et al. 2013; Hegedus and Taale 2013)). The K_m values described for L-glutamate transport by EAAT1 are in the range of 7–20 μM;

Table 1 Transporters of glutamate

Gene	Protein	Main cell types in the CNS	Cellular localization	References
SLC1A1	EAAT3 (EAAC1)	Neurons and substantia nigra, red nucleus, hippocampus, cerebral cortical layers	Plasma membrane of postsynaptic neurons	Shashidharan et al. (1994), Kanai and Hediger (1992), Kanai et al. (1994), Rothstein et al. (1994)
SLC1A2	EAAT2 (GLT-1)	Astrocytes Glia cells	Plasma membrane	Pines et al. (1992), Danbolt et al. (1992)
SLC1A3	EAAT1 (GLAST)	Glia cells Astrocytes	Plasma membrane	Storck et al. (1992), Danbolt et al. (1994)
SLC1A6	EAAT4	Purkinje cell somas and dendrites	Plasma membrane	Fairman et al. (1995), Lin et al. (1998)
SLC1A7	EAAT5	Retina	Plasma membrane	Arriza et al. (1997)
SLC7A11	xCT (associate with 4F2hc to form system x_c^-)	Brain (neurons, astrocytes, and glial cells) and spinal cord	Plasma membrane	Kim et al. (2001), Jackman et al. (2010), Fogal et al. (2007)
SLC7A13	AGT-1 (associate with rBAT as heavy chain)	Not expressed	Basolateral plasma membrane in proximal straight tubules and distal convoluted tubules (Matsuo et al. 2002) but apical in another study (Nagamori et al. 2016)	Blondeau (2002), Matsuo et al. (2002)
SLC17A6	VGLUT2	Small: (0–500 μm²) Medium: (500–1300 μm²) DRG	Vesicular membrane	Malet et al. (2013), Aihara et al. (2000)
SLC17A7	VGLUT1	Medium: (500–1300 μm²) Large: (1300–2900 μm²) DRG	Vesicular membrane	Malet et al. (2013), Ni et al. (1994)
SLC17A8	VGLUT3	Small: (0–500 μm²) Medium: (500–1300 μm²) DRG	Vesicular membrane	Malet et al. (2013), Takamori et al. (2002)
SLC25A12	AGC1 (aralar)	Neurons in the brain stem and spinal cord	Mitochondrion inner membrane	del Arco and Satrustegui (1998), Ramos et al. (2003)

(continued)

Table 1 (continued)

Gene	Protein	Main cell types in the CNS	Cellular localization	References
SLC25A13	AGC2 (citrin)	Limited to neuronal cluster	Mitochondrion inner membrane	Kobayashi et al. (1999), Del Arco et al. (2000), Contreras et al. (2010)
SLC25A18	GC2	Expressed in brain panels	Mitochondrion inner membrane	Fiermonte et al. (2002)
SLC25A22	GC1	Expressed in brain panels, and in developing human nervous system, notably in the hippocampus, cortex, and brainstem	Mitochondrion inner membrane	Fiermonte et al. (2002), Molinari et al. (2005)

DRG dorsal root ganglia

for EAAT2 and EAAT3, they are 12–18 μM and 8–30 μM, respectively. EAAT4 has the highest affinity reported for glutamate transport via EAATs with K_m values in the range of 0.6–3.3 μM, while EAAT5 has the lowest affinity with K_m values in the range of 61–63 μM (for references see (Danbolt et al. 2016)). The uptake of L-glutamate via EAATs is indirectly dependent on the energy status of the cell since the translocation cycle includes cotransport of one molecule of L-glutamate with three sodium ions and one hydrogen ion and an exchange with one potassium ion (Levy et al. 1998; Zerangue and Kavanaugh 1996). The transport activity and direction is thus coupled to the sodium-potassium ATPase (Rose et al. 2009). SLC7 is the family of cationic amino acid transporters and glycoprotein-associated proteins. SLC7A11 is the light chain that coupled with the heavy chain 4F2hc forms system x_c^- that is a cystine/L-glutamate exchanger and SLC7A13 which associate with rBAT as heavy chain forming an L-aspartate/L-glutamate exchanger, apparently not expressed in the brain (SLC tables). The SLC 17 family contains vesicular glutamate transporters which are involved in glutamate accumulation in membrane vesicles in presynaptic neurons. These transporters are electrogenic chloride-dependent glutamate transporters. The SLC25 family contains mitochondrial glutamate transporters. SLC25A12 and SLC25A13 are calcium- and proton-dependent exchangers of cytoplasmic glutamate with mitochondrial aspartate which takes place across the inner mitochondrial membrane. SLC25A18 and SLC25A22 are proton-coupled glutamate transporters involved in glutamate transport across the inner mitochondrial membrane. In the context of glutamate transport across the brain endothelium, EAAT transporters are relevant to consider (Table 2), acknowledging that both VGLUT and EAATs are important for overall glutamate levels in the healthy brain and in disease.

Table 2 Overview of EAAT subtype expression patterns in capillaries and capillary-derived cell cultures from different species

		EAAT1	EAAT2	EAAT3	References
Capillaries or intact tissue preparations	mRNA	B,H, M	B, M	M	Helms et al. (2012), Lyck et al. (2009), Daneman et al. (2010), Guo et al. (2012), O'Kane et al. (1999), Shawahna et al. (2011)
	Protein	B, H, M, Ma	M	M	Roberts et al. (2014), Helms et al. (2012), Chun et al. (2011), Uchida et al. (2011), Lecointre et al. (2014), O'Kane et al. (1999), Hoshi et al. (2013)
Cell culture	mRNA	B		B, M	Helms et al. (2012), Lyck et al. (2009)
	Protein	B, M, P, R	M, P	B, M, P, R	Campos et al. (2011a), Cohen-Kashi-Malina et al. (2012), Helms et al. (2012), Lecointre et al. (2014)

B bovine, *H* human, *M* mouse, *P* porcine, *R* rat, *Ma* marmoset

3.2 Expression of EAATs in Intact or Isolated Brain Capillaries

Chun and co-workers performed a proteomics study on freshly isolated mouse brain microvessel membranes, and, using multidimensional protein identification technology, they identified 1143 proteins of which 101 were membrane transporters of mainly the SLC family (Chun et al. 2011). The gene products of plasma membrane glutamate transporters Slc1a2 and Slc1a3 were among the most abundant transporters, but also mitochondrial glutamate transporters from Slc25a22 and Slc25a18 genes were found (Chun et al. 2011). Gene products of Slc17a7 and Slc1a1 were also identified with spectral counts above 5 (Chun et al. 2011). In contrast, Ushida and co-worker found 20 membrane transporters in human brain microvessels and found only EAAT1 and not EAAT3 protein amounts above the detection limit (Uchida et al. 2011). In brain capillaries isolated from marmoset brain, EAAT1 protein was found highly expressed (Hoshi et al. 2013). Moreover, a recent electron microscopy study by Roberts et al. showed a clear EAAT1 expression in brain capillary endothelial cells from human postmortem cortex samples (Roberts et al. 2014). Brain capillaries from mice have been shown to have a high expression of *EAAT3 mRNA* (Lyck et al. 2009; Daneman et al. 2010; Guo et al. 2012) as well as protein expression of EAAT1, EAAT2 and EAAT3 (although mainly subtypes 2 and 3) (Lecointre et al. 2014). Freshly isolated bovine brain capillaries express *EAAT1*, *EAAT2*, and *EAAT3 mRNA*, whereas only EAAT1 has been detected at the protein level (Helms et al. 2012; O'Kane et al. 1999). Large quantities of *EAAT1 mRNA* (Shawahna et al. 2011) were found in human brain capillaries. These studies indicate that EAATs are present in endothelial cells, even though a study on rat capillaries revealed little or no EAAT staining on the abluminal surface of brain capillaries (Chaudry et al. 1995). However the subtype specific expression pattern may vary between species. It has to be kept in mind that contamination by glial tissue or remnants of astrocyte end feet may be a general concern as protein fragments from these tissues will greatly affect conclusions made.

3.3 Expression and Functional Activity of Excitatory Amino Acid Transporters in Cultured Endothelial Cells and Vesicle Preparations

Abluminal sodium-dependent saturable uptake was initially demonstrated in membrane vesicles from bovine brain endothelial cells (Lee et al. 1998). The uptake was attributed to EAATs through protein expression and uptake studies. The studies demonstrated expression of EAAT1, EAAT2, and EAAT3 in abluminal membrane vesicles as well as high-affinity uptake matching that of EAATs in other cell types (O'Kane et al. 1999). In cell cultures, EAAT3 seemed to be the main transporter in mouse brain endothelial cells (Lyck et al. 2009; Lecointre et al. 2014), whereas EAAT1 displayed the highest expression level in bovine endothelial cells co-cultured with rat astrocytes (Helms et al. 2012). Helms et al. showed that EAAT1 is the dominant abluminal L-glutamate transporter in this model (Helms et al. 2016). EAAT1, EAAT2, and EAAT3 were all present in cultured endothelial cells from rat and porcine brains although EAAT3 seemed to be expressed at a higher level than the other subtypes (Campos et al. 2011a; Cohen-Kashi-Malina et al. 2012).

3.4 The Luminal Glutamate Transport System

Early in vivo studies indicate the presence of a facilitative L-glutamate transporter at the luminal membrane of the endothelial cells. Uptake of L-glutamate from the blood circulation and into the brain was shown in rats after a carotid bolus injection of radiolabeled L-glutamate (Oldendorf and Szabo 1976). The putative transporter was named x_G^- (Christensen 1984) and was shown to be independent of sodium and inhibited by L-glutamate and L-aspartate (Oldendorf and Szabo 1976; Lee et al. 1998; Benrabh and Lefauconnier 1996). The transporter has not been cloned, and it is possible that an already known amino acid transporter facilitates the passage of the luminal membrane. The glutamate-cystine exchanger (x_c) could be a candidate to a transporter which could efflux intracellular L-glutamate from the endothelial cells to the blood in exchange for cystine uptake. However, previous studies have shown no effects of cystine on luminal endothelial L-glutamate uptake (O'Kane et al. 1999; Benrabh and Lefauconnier 1996). In the mouse brain endothelial cell line, MBEC4A, *xCT*, and *4F2hc mRNA* are expressed, and carrier-mediated cystine and glutamate uptake has been measured, although it may be difficult to distinguish luminal and abluminal uptake in cells cultured on the bottom of cell culture wells (Hosoya et al. 2002). In rat brain slices, a low-affinity L-glutamate transporter has been described (Balcar and Johnston 1972; Benjamin and Quastel 1976). This transporter of L-glutamate was inhibited by L-glutamate, L-aspartate, and L-homocysteate (Cox et al. 1977), which matches the inhibition pattern observed for the x_G^- transporter (Benrabh and Lefauconnier 1996). The low-affinity L-glutamate uptake has a K_M of 1–2 mM (Balcar and Johnston 1972; Benjamin and Quastel 1976). This matches the K_M value for L-glutamate in luminal membrane vesicles from bovine brain endothelium, as determined by Lee et al. (1998). Helms et al. found kinetic and inhibition patterns matching the presumed low-affinity glutamate transporter by measuring luminal glutamate uptake in cultured bovine brain endothelial cells (Helms et al. 2016). An experimental limitation to the study of exchanges may be to have appropriate intracellular glutamate levels in intact cell cultures in vitro and to functionally show the exchange mechanism via cystine uptake.

4 Glutamate Metabolism in Brain Endothelial Cells

Efficient uptake of glutamate into astrocytes is followed by synthesis of glutamine catalyzed by the cytosolic enzyme glutamine synthetase. Glutamine is excreted from the astrocytes and known to be taken up by neurons to be reused as precursor for neurotransmitter glutamate as part of the glutamate-glutamine cycle. In addition, L-glutamate is extensively oxidatively metabolized in astrocytes, initially to α-ketoglutarate either via an oxidative deamination catalyzed by glutamate dehydrogenase or an aminotransferase of which aspartate aminotransferase (AAT), alanine aminotransferase (ALAT), and branched chain aminotransferase (BCAT) have the highest activities in the brain (Danbolt 2001). The observed efflux across the blood-brain barrier obtained using radiolabeled L-glutamate could therefore in theory be explained by uptake and metabolism in astrocytes and transport of labeled metabolites across the endothelial cells. Two studies tested the effect of glutamine synthetase inhibition using methionine sulfoximine in endothelial/astrocyte co-cultures and found no effects on the transcellular L-glutamate transport (Cohen-Kashi-Malina et al. 2012; Helms et al. 2012). However, conversion to α-ketoglutarate and further oxidation of the radiolabeled L-glutamate may still result in radiolabeled metabolites, which would influence estimates of the transendothelial transport of L-glutamate, using radiolabel. Astrocytes are known to release lactate as product of glutamate oxidation, via malic enzyme catalyzed oxidation of malate to pyruvate and subsequently operation of lactate dehydrogenase (Olsen and Sonnewald 2015). Another candidate is the TCA cycle intermediate citrate, which is also released from astrocytes (Westergaard et al. 1994). In agreement with this, citrate and lactate formed from exogenously applied glutamate was found in the abluminal medium (where astrocytes are present) of a noncontact co-culture of astrocytes and endothelial cells (Helms et al. 2016).

Alternatively, glutamate may be metabolized in the endothelial cells following EAAT-mediated uptake. Although endothelial cells in general are known to rely mostly on aerobic glycolytic energy production, Oldendorf et al. demonstrated a high density of mitochondria in brain endothelial cells (Oldendorf et al. 1977). Interestingly, mitochondria and EAAT1 and EAAT2 co-localize in astrocytes, supporting rapid glutamate oxidation and fueling of the energy requiring uptake by glutamate degradation (Pajecka et al. 2015; Robinson and Jackson 2016). It has been suggested that endothelial cells may oxidize L-glutamate to fuel the ATP-binding cassette (ABC) drug efflux transporters (Mann et al. 2003). It was recently shown, using a noncontact co-culture of astrocytes and endothelial cells exposed to ^{13}C glutamate in the abluminal medium, that glutamate is indeed oxidatively metabolized in the endothelial cells (Helms et al. 2016). This was evident from the ^{13}C labeling of glutamate and the TCA cycle intermediates, α-ketoglutarate, succinate, and malate observed in the endothelial compartment. A complete oxidative degradation of glutamate requires an initial oxidation of malate or oxaloacetate to pyruvate either via malic enzyme activity or the concerted action of phosphoenolpyruvate carboxykinase and pyruvate kinase, respectively. Pyruvate is next either reduced to lactate or decarboxylated to acetyl CoA. Lactate is released, whereas acetyl CoA is further processed to CO_2 in the TCA cycle. The study demonstrated formation of pyruvate and further metabolism to lactate and acetyl CoA in the endothelial cells. The relative extent to which glutamate was metabolized to acetyl CoA and further metabolized in the TCA cycle was higher in the endothelial cells compared to the astrocytes. This may at least partly be explained by a low oxidative glucose metabolism in endothelial cells, and it supports an important role of glutamate as an energy substrate for the endothelial cells. The complete oxidation of glutamate is in accordance with the expression of glutamate dehydrogenase in endothelial cells (Zhang et al. 2014). The operation of glutamate dehydrogenase provides a net entrance of α-ketoglutarate to the TCA cycle, in contrast to aminotransferases that involves another keto acid. Citrate, which is formed in the TCA cycle from the condensation of oxaloacetate and acetyl CoA, was released into the luminal medium and is together with lactate products of glutamate oxidation. In addition, a non-negligible amount of aspartate was found in the luminal medium most likely originating from endothelial metabolism. Net formation of aspartate is formed from glutamate via aspartate aminotransferase in combination with several oxidative steps in the TCA cycle converting α-ketoglutarate to oxaloacetate, a process known as "the truncated TCA cycle" (Skytt et al. 2012). The operation of "the truncated TCA cycle" is also in line with a low oxidative glycolytic rate and availability of acetyl CoA in endothelial cells. Thus, metabolism of glutamate in endothelial cells should be taken into account using radiolabeled glutamate to investigate transcellular processes.

As mentioned above, intact labeled L-glutamate was recovered in the blood after intracerebral injections in rats (Hosoya et al. 1999), indicating that some L-glutamate is transported intact through the cells. Alternatively, paracellular leakage of radiolabeled L-glutamate following the injection may add to the explanation. However, the true balance between metabolism and transport of intact L-glutamate has not yet been fully established and warrants further investigations. The possible transport mechanisms are summarized in Fig. 1.

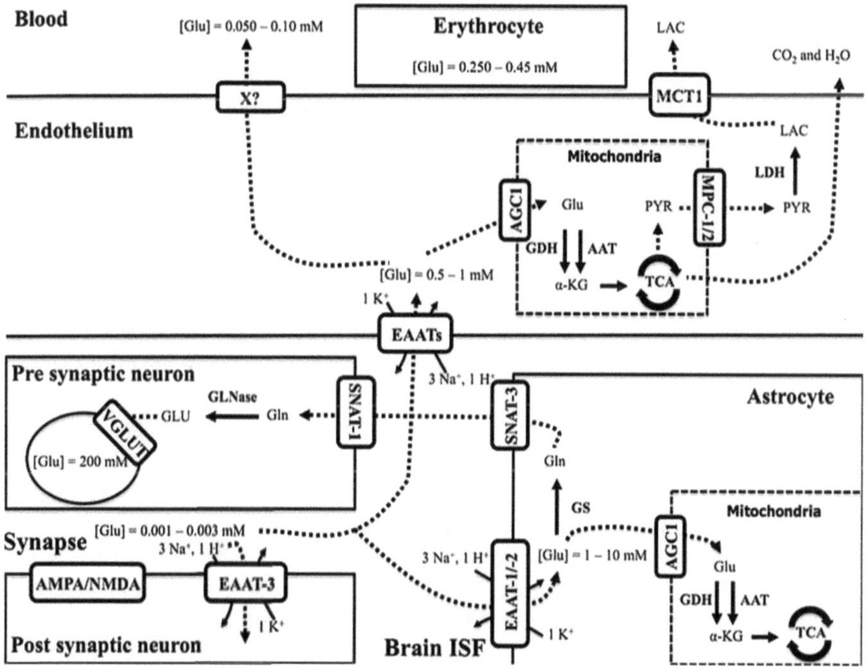

Fig. 1 Glutamate transport and metabolism pathways in the neurovascular unit and how metabolites may appear in the blood after transcellular transport. *α-KG* α-ketoglutarate, *AT* amino acid aminotransferase, *AGC1* mitochondrial aspartate/glutamate carrier-1, *EAAT* excitatory amino acid transporter, *GDH* glutamate dehydrogenase, *GLN* glutamine, *GLNase* glutaminase, *GLU* L-glutamate, *ISF* interstitial fluid, *LAC* lactate, *LDH* lactate dehydrogenase, *MCT-1* monocarboxylic acid transporter-1, *MPC1/2* mitochondrial pyruvate carrier-1/2, *Pyr* pyruvate, *SNAT* small neutral amino acid transporter, *TCA* tricarboxylic acid cycle, *VGLUT* vesicular glutamate transporter, *X?* unknown luminal glutamate transporter (figure modified from (Cederberg et al. 2014) with permission)

5 Blood Glutamate Scavenging: Treatment Paradigm, Clinical Relevance, and Possible Involvement of Endothelial Excitatory Amino Acid Transporters

Based on the observations that L-glutamate can be transported from brain ISF to the blood in animal studies (Drewes et al. 1977; Hosoya et al. 1999) combined with observations of expression of EAATs in brain endothelial cells (O'Kane et al. 1999), it has been proposed that the blood-brain barrier can mediate clearance of L-glutamate and that this might be a mechanism to minimize the excitotoxic damage which may occur under pathophysiological situations. However, the blood glutamate levels of approximately 140 μM (Hagenfeldt and Arvidsson 1980; Graham et al. 1995) limits the blood-brain-barrier-mediated L-glutamate clearance by providing an unfavorable concentration gradient. The combination of endothelial EAAT expression and the relatively high blood glutamate levels leads to the concept of blood glutamate scavenging, where active lowering of blood glutamate concentrations is hypothesized to increase brain-to-blood transport of L-glutamate because of a more favorable concentration gradient from endothelium to blood (Gottlieb et al. 2003). This was initially demonstrated in rats, where lowering blood glutamate by intravenous injections of oxaloacetate and pyruvate caused lowering of glutamate levels in the cerebrospinal fluid and increased presence of ^{3}H-labeled glutamate in the blood after intracerebroventricular injection (Gottlieb et al. 2003). This experiment demonstrated transport of glutamate from the cerebrospinal fluid to the blood, which is not directly comparable to transport across the blood-brain barrier, but the experiment inspired other investigations of similar concept regarding blood-brain-barrier-mediated efflux.

These experiments have shown beneficial effects of blood glutamate scavenging in rats after middle cerebral artery occlusion, closed head injury, subarachnoid hemorrhage, traumatic brain injury, paraoxon intoxication, and amyloid-beta toxicity (Campos et al. 2011a; Zlotnik et al. 2012; Ruban et al. 2014; Boyko et al. 2012; Perez-Mato et al. 2014; Zlotnik et al. 2009; Zhang et al. 2016) (for review see (Castillo et al. 2016)). Furthermore, blood levels of the endogenous enzyme aspartate aminotransferase have been shown to correlate positively with lower infarct volume and decreased early neurological deterioration (END) after ischemic stroke in humans (Campos et al. 2011b). Scavenging of L-glutamate from the blood circulation seems to be an effective treatment in animal models, and may hold clinical potential, but the exact clearance pathways from the brain to the blood are not yet clearly elucidated. The general assumption is that glutamate is taken up into the brain endothelial cells via EAATs and subsequently transported across the luminal membrane to the blood, when the concentration gradient (cell-to-blood) favors cellular efflux. Breakdown of glutamate in the blood thus creates a more favorable concentration gradient for the glutamate to exit the endothelial cells and thus accelerates the overall clearance from the brain ISF (Castillo et al. 2016). However, an alternative explanation could be that glutamate diffuses across the paracellular space between the brain endothelial cells, which is at least partly opened during stroke, traumatic brain injury, and subarachnoid hemorrhage (Chodobski et al. 2011; Fredriksson et al. 1987; Krueger et al. 2013; Doczi et al. 1986). Diffusion across the paracellular space would also be dependent on the brain-to-blood glutamate concentration gradient, and breakdown of glutamate in the blood would have a similar beneficial effect in this mechanism. Interestingly, a recent study on the effect of blood glutamate scavenging in amyloid-beta toxicity showed that blood glutamate scavenging ameliorates the effects on long-term potentiation of both amyloid-beta and of the EAAT inhibitor, TBOA (Zhang et al. 2016). TBOA has previously been shown to cause neuronal damage by inhibiting glutamate uptake into astrocytes (Montiel et al. 2005), but the fact that this could be ameliorated by blood glutamate scavenging indicates that the brain-to-blood efflux is not dependent on endothelial EAATs exclusively, since these would be inhibited by the TBOA in the ISF.

6 Summary and Perspectives

Since the original formulation of the "brain glutamate efflux hypothesis" (Pardridge 1979), a number of studies have demonstrated that excitatory amino acid transporters are present in brain endothelial cells both in vitro and in vivo, although the subtype expression pattern varies among species. EAAT-mediated L-glutamate uptake into the endothelial cells constitutes the first step of the brain L-glutamate efflux; however, the following steps remain unclear. Saturable, low-affinity luminal uptake has been demonstrated, indicating the presence of an L-glutamate transporter in the luminal membrane of endothelial cells. Such a transporter could also in theory account for L-glutamate transport from endothelial cell interior to blood, given an L-glutamate concentration gradient from cell to blood which would favor this direction of transport. However, it has also recently been demonstrated that some metabolism of L-glutamate occurs in the endothelial cells (and/or the astrocytes), followed by efflux of the metabolism products to the blood. Experimental evidence suggests that the glutamate-glutamine cycle does not play a role in the brain efflux of L-glutamate. However, metabolism to α-ketoglutarate via aminotransferases or glutamate dehydrogenase is likely to take place both in endothelial cells and astrocytes. The relative contributions of transport and metabolism to the apparent L-glutamate efflux have not been fully elucidated and warrant further studies.

A number of in vivo studies have indicated that scavenging of blood glutamate could improve the outcome after stroke, subarachnoid hemorrhage, traumatic brain injury, and paraoxon intoxication. The mechanism suggested to be responsible for this phenomenon is transporter-mediated brain glutamate efflux, i.e., when blood glutamate concentrations are kept low, the concentration gradient for transporter-mediated brain-to-blood efflux is kept favorable for efflux, but as yet no firm in vivo evidence exists to support this point of view. A direction for future studies would be to focus on verifying the presence of EAATs on brain endothelium in vivo, determine EAAT subtype expression profile, identify the possible transporter responsible for the transport of L-glutamate across the luminal membrane, and investigate to what extent metabolism and transporter-mediated transport of L-glutamate contributes to the apparent L-glutamate efflux. This could lead to a clarification of the mechanisms behind the apparently beneficial effects of blood L-glutamate scavenging.

Conflict of Interest The author declares no conflicts of interest.

References

Aihara Y, Mashima H, Onda H, Hisano S, Kasuya H, Hori T, et al. Molecular cloning of a novel brain-type Na(+)-dependent inorganic phosphate cotransporter. J Neurochem. 2000;74(6):2622–5.

Arriza JL, Eliasof S, Kavanaugh MP, Amara SG. Excitatory amino acid transporter 5, a retinal glutamate transporter coupled to a chloride conductance. Proc Natl Acad Sci U S A. 1997;94(8):4155–60.

Balcar VJ, Johnston GA. Glutamate uptake by brain slices and its relation to the depolarization of neurones by acidic amino acids. J Neurobiol. 1972;3(4):295–301.

Benjamin AM, Quastel JH. Cerebral uptakes and exchange diffusion in vitro of L- and D-glutamates. J Neurochem. 1976;26(3):431–41.
Benrabh H, Lefauconnier JM. Glutamate is transported across the rat blood-brain barrier by a sodium-independent system. Neurosci Lett. 1996;210(1):9–12.
Blondeau JP. Homologues of amino acid permeases: cloning and tissue expression of XAT1 and XAT2. Gene. 2002;286(2):241–8.
Boyko M, Melamed I, Gruenbaum BF, Gruenbaum SE, Ohayon S, Leibowitz A, et al. The effect of blood glutamate scavengers oxaloacetate and pyruvate on neurological outcome in a rat model of subarachnoid hemorrhage. Neurotherapeutics. 2012;9(3):649–57.
Boyko M, Gruenbaum SE, Gruenbaum BF, Shapira Y, Zlotnik A. Brain to blood glutamate scavenging as a novel therapeutic modality: a review. J Neural Transm. 2014;121(8):971–9.
Campos F, Sobrino T, Ramos-Cabrer P, Argibay B, Agulla J, Perez-Mato M, et al. Neuroprotection by glutamate oxaloacetate transaminase in ischemic stroke: an experimental study. J Cereb Blood Flow Metab. 2011a;31(6):1378–86.
Campos F, Rodriguez-Yanez M, Castellanos M, Arias S, Perez-Mato M, Sobrino T, et al. Blood levels of glutamate oxaloacetate transaminase are more strongly associated with good outcome in acute ischaemic stroke than glutamate pyruvate transaminase levels. Clin Sci (Lond). 2011b;121(1):11–7.
Cardelli-Cangiano P, Cangiano C, James JH, Jeppsson B, Brenner W, Fischer JE. Uptake of amino acids by brain microvessels isolated from rats after portacaval anastomosis. J Neurochem. 1981;36(2):627–32.
Castillo J, Loza MI, Mirelman D, Brea J, Blanco M, Sobrino T, et al. A novel mechanism of neuroprotection: blood glutamate grabber. J Cereb Blood Flow Metab. 2016;36(2):292–301.
Cederberg HH, Uhd NC, Brodin B. Glutamate efflux at the blood-brain barrier: cellular mechanisms and potential clinical relevance. Arch Med Res. 2014;45(8):639–45.
Chaudhry FA, Lehre KP, van Lookeren Campagne M, Ottersen OP, Danbolt NC, Storm-Mathisen J. Glutamate transporters in glial plasma membranes: highly differentiated localizations revealed by quantitative ultrastructural immunocytochemistry. Neuron. 1995;15(3):711–20.
Chodobski A, Zink BJ, Szmydynger-Chodobska J. Blood-brain barrier pathophysiology in traumatic brain injury. Transl Stroke Res. 2011;2(4):492–516.
Christensen HN. Organic ion transport during seven decades. The amino acids. Biochim Biophys Acta. 1984;779(3):255–69.
Chun HB, Scott M, Niessen S, Hoover H, Baird A, Yates J 3rd, et al. The proteome of mouse brain microvessel membranes and basal lamina. J Cereb Blood Flow Metab. 2011;31(12):2267–81.
Cohen-Kashi-Malina K, Cooper I, Teichberg VI. Mechanisms of glutamate efflux at the blood-brain barrier: involvement of glial cells. J Cereb Blood Flow Metab. 2012;32(1):177–89.
Contreras L, Urbieta A, Kobayashi K, Saheki T, Satrustegui J. Low levels of citrin (SLC25A13) expression in adult mouse brain restricted to neuronal clusters. J Neurosci Res. 2010;88(5):1009–16.
Cox DW, Headley MH, Watkins JC. Actions of L- and D-homocysteate in rat CNS: a correlation between low-affinity uptake and the time courses of excitation by microelectrophoretically applied L-glutamate analogues. J Neurochem. 1977;29(3):579–88.
Danbolt NC. Glutamate uptake. Prog Neurobiol. 2001;65(1):1–105.
Danbolt NC, Storm-Mathisen J, Kanner BI. An [Na+ + K+]coupled L-glutamate transporter purified from rat brain is located in glial cell processes. Neuroscience. 1992;51(2):295–310.
Danbolt NC, Storm-Mathisen J, Ottersen OP. Sodium/potassium-coupled glutamate transporters, a "new" family of eukaryotic proteins: do they have "new" physiological roles and could they be new targets for pharmacological intervention? Prog Brain Res. 1994;100:53–60.
Danbolt NC, Furness DN, Zhou Y. Neuronal vs glial glutamate uptake: resolving the conundrum. Neurochem Int. 2016;98:29–45.
Daneman R, Zhou L, Agalliu D, Cahoy JD, Kaushal A, Barres BA. The mouse blood-brain barrier transcriptome: a new resource for understanding the development and function of brain endothelial cells. PLoS One. 2010;5(10):e13741.
del Arco A, Satrustegui J. Molecular cloning of Aralar, a new member of the mitochondrial carrier superfamily that binds calcium and is present in human muscle and brain. J Biol Chem. 1998;273(36):23327–34.

Del Arco A, Agudo M, Satrustegui J. Characterization of a second member of the subfamily of calcium-binding mitochondrial carriers expressed in human non-excitable tissues. Biochem J. 2000;345(Pt 3):725–32.
Divino Filho JC, Barany P, Stehle P, Furst P, Bergstrom J. Free amino-acid levels simultaneously collected in plasma, muscle, and erythrocytes of uraemic patients. Nephrol Dial Transplant. 1997;12(11):2339–48.
Doczi T, Joo F, Adam G, Bozoky B, Szerdahelyi P. Blood-brain barrier damage during the acute stage of subarachnoid hemorrhage, as exemplified by a new animal model. Neurosurgery. 1986;18(6):733–9.
Drewes LR, Conway WP, Gilboe DD. Net amino acid transport between plasma and erythrocytes and perfused dog brain. Am J Phys. 1977;233(4):E320–5.
Fairman WA, Vandenberg RJ, Arriza JL, Kavanaugh MP, Amara SG. An excitatory amino-acid transporter with properties of a ligand-gated chloride channel. Nature. 1995;375(6532):599–603.
Fiermonte G, Palmieri L, Todisco S, Agrimi G, Palmieri F, Walker JE. Identification of the mitochondrial glutamate transporter. Bacterial expression, reconstitution, functional characterization, and tissue distribution of two human isoforms. J Biol Chem. 2002;277(22):19289–94.
Fogal B, Li J, Lobner D, McCullough LD, Hewett SJ. System x(c)- activity and astrocytes are necessary for interleukin-1 beta-mediated hypoxic neuronal injury. J Neurosci. 2007;27(38):10094–105.
Fotiadis D, Kanai Y, Palacin M. The SLC3 and SLC7 families of amino acid transporters. Mol Asp Med. 2013;34(2–3):139–58.
Fredriksson K, Kalimo H, Westergren I, Kahrstrom J, Johansson BB. Blood-brain barrier leakage and brain edema in stroke-prone spontaneously hypertensive rats. Effect of chronic sympathectomy and low protein/high salt diet. Acta Neuropathol. 1987;74(3):259–68.
Gillessen T, Budd SL, Lipton SA. Excitatory amino acid neurotoxicity. Adv Exp Med Biol. 2002;513:3–40.
Gottlieb M, Wang Y, Teichberg VI. Blood-mediated scavenging of cerebrospinal fluid glutamate. J Neurochem. 2003;87(1):119–26.
Graham TE, Turcotte LP, Kiens B, Richter EA. Training and muscle ammonia and amino acid metabolism in humans during prolonged exercise. J Appl Physiol (1985). 1995;78(2):725–35.
Guo S, Zhou Y, Xing C, Lok J, Som AT, Ning M, et al. The vasculome of the mouse brain. PLoS One. 2012;7(12):e52665.
Hagenfeldt L, Arvidsson A. The distribution of amino acids between plasma and erythrocytes. Clin Chim Acta. 1980;100(2):133–41.
Hawkins RA. The blood-brain barrier and glutamate. Am J Clin Nutr. 2009;90(3):867S–74S.
Hegedus T, Taale M. SLC tables: bioparadigms.org. 2013 [cited 2016 03-10-2016]. Available from: http://slc.bioparadigms.org/.
Helms HC, Madelung R, Waagepetersen HS, Nielsen CU, Brodin B. In vitro evidence for the brain glutamate efflux hypothesis: brain endothelial cells cocultured with astrocytes display a polarized brain-to-blood transport of glutamate. Glia. 2012;60(6):882–93.
Helms HC, Aldana BI, Groth S, Jensen MM, Waagepetersen HS, Nielsen CU, et al. Characterization of the L-glutamate clearance pathways across the blood-brain barrier and the effect of astrocytes in an in vitro blood-brain barrier model. J Cereb Blood Flow Metab. 2017; doi:10.1177/271678X17690760.
Hoshi Y, Uchida Y, Tachikawa M, Inoue T, Ohtsuki S, Terasaki T. Quantitative atlas of blood-brain barrier transporters, receptors, and tight junction proteins in rats and common marmoset. J Pharm Sci. 2013;102(9):3343–55.
Hosoya K, Sugawara M, Asaba H, Terasaki T. Blood-brain barrier produces significant efflux of L-aspartic acid but not D-aspartic acid: in vivo evidence using the brain efflux index method. J Neurochem. 1999;73(3):1206–11.
Hosoya K, Tomi M, Ohtsuki S, Takanaga H, Saeki S, Kanai Y, et al. Enhancement of L-cystine transport activity and its relation to xCT gene induction at the blood-brain barrier by diethyl maleate treatment. J Pharmacol Exp Ther. 2002;302(1):225–31.
Hutchison HT, Eisenberg HM, Haber B. High-affinity transport of glutamate in rat brain microvessels. Exp Neurol. 1985;87(2):260–9.

Iijima K, Takase S, Tsumuraya K, Endo M, Itahara K. Changes in free amino acids of cerebrospinal fluid and plasma in various neurological diseases. Tohoku J Exp Med. 1978;126(2):133–50.
Jackman NA, Uliasz TF, Hewett JA, Hewett SJ. Regulation of system x(c)(-)activity and expression in astrocytes by interleukin-1beta: implications for hypoxic neuronal injury. Glia. 2010;58(15):1806–15.
Kanai Y, Hediger MA. Primary structure and functional characterization of a high-affinity glutamate transporter. Nature. 1992;360(6403):467–71.
Kanai Y, Stelzner M, Nussberger S, Khawaja S, Hebert SC, Smith CP, et al. The neuronal and epithelial human high affinity glutamate transporter. Insights into structure and mechanism of transport. J Biol Chem. 1994;269(32):20599–606.
Kanai Y, Clemencon B, Simonin A, Leuenberger M, Lochner M, Weisstanner M, et al. The SLC1 high-affinity glutamate and neutral amino acid transporter family. Mol Asp Med. 2013;34(2–3):108–20.
Kim JY, Kanai Y, Chairoungdua A, Cha SH, Matsuo H, Kim DK, et al. Human cystine/glutamate transporter: cDNA cloning and upregulation by oxidative stress in glioma cells. Biochim Biophys Acta. 2001;1512(2):335–44.
Kobayashi K, Sinasac DS, Iijima M, Boright AP, Begum L, Lee JR, et al. The gene mutated in adult-onset type II citrullinaemia encodes a putative mitochondrial carrier protein. Nat Genet. 1999;22(2):159–63.
Krueger M, Hartig W, Reichenbach A, Bechmann I, Michalski D. Blood-brain barrier breakdown after embolic stroke in rats occurs without ultrastructural evidence for disrupting tight junctions. PLoS One. 2013;8(2):e56419.
Lecointre M, Hauchecorne M, Chaussivert A, Marret S, Leroux P, Jegou S, et al. The efficiency of glutamate uptake differs between neonatal and adult cortical microvascular endothelial cells. J Cereb Blood Flow Metab. 2014;34(5):764–7.
Lee WJ, Hawkins RA, Vina JR, Peterson DR. Glutamine transport by the blood-brain barrier: a possible mechanism for nitrogen removal. Am J Phys. 1998;274(4 Pt 1):C1101–7.
Lehre KP, Levy LM, Ottersen OP, Storm-Mathisen J, Danbolt NC. Differential expression of two glial glutamate transporters in the rat brain: quantitative and immunocytochemical observations. J Neurosci. 1995;15(3 Pt 1):1835–53.
Leibowitz A, Boyko M, Shapira Y, Zlotnik A. Blood glutamate scavenging: insight into neuroprotection. Int J Mol Sci. 2012;13(8):10041–66.
Levy LM, Warr O, Attwell D. Stoichiometry of the glial glutamate transporter GLT-1 expressed inducibly in a Chinese hamster ovary cell line selected for low endogenous Na+-dependent glutamate uptake. J Neurosci. 1998;18(23):9620–8.
Lin CL, Tzingounis AV, Jin L, Furuta A, Kavanaugh MP, Rothstein JD. Molecular cloning and expression of the rat EAAT4 glutamate transporter subtype. Brain Res Mol Brain Res. 1998;63(1):174–9.
Lyck R, Ruderisch N, Moll AG, Steiner O, Cohen CD, Engelhardt B, et al. Culture-induced changes in blood-brain barrier transcriptome: implications for amino-acid transporters in vivo. J Cereb Blood Flow Metab. 2009;29(9):1491–502.
Malet M, Vieytes CA, Lundgren KH, Seal RP, Tomasella E, Seroogy KB, et al. Transcript expression of vesicular glutamate transporters in lumbar dorsal root ganglia and the spinal cord of mice – effects of peripheral axotomy or hindpaw inflammation. Neuroscience. 2013;248:95–111.
Mann GE, Yudilevich DL, Sobrevia L. Regulation of amino acid and glucose transporters in endothelial and smooth muscle cells. Physiol Rev. 2003;83(1):183–252.
Matsuo H, Kanai Y, Kim JY, Chairoungdua A, Kim DK, Inatomi J, et al. Identification of a novel Na+-independent acidic amino acid transporter with structural similarity to the member of a heterodimeric amino acid transporter family associated with unknown heavy chains. J Biol Chem. 2002;277(23):21017–26.
Molinari F, Raas-Rothschild A, Rio M, Fiermonte G, Encha-Razavi F, Palmieri L, et al. Impaired mitochondrial glutamate transport in autosomal recessive neonatal myoclonic epilepsy. Am J Hum Genet. 2005;76(2):334–9.
Montiel T, Camacho A, Estrada-Sanchez AM, Massieu L. Differential effects of the substrate inhibitor l-trans-pyrrolidine-2,4-dicarboxylate (PDC) and the non-substrate inhibitor DL-threo-

beta-benzyloxyaspartate (DL-TBOA) of glutamate transporters on neuronal damage and extracellular amino acid levels in rat brain in vivo. Neuroscience. 2005;133(3):667–78.
Nagamori S, Wiriyasermkul P, Guarch ME, Okuyama H, Nakagomi S, Tadagaki K, et al. Novel cystine transporter in renal proximal tubule identified as a missing partner of cystinuria-related plasma membrane protein rBAT/SLC3A1. Proc Natl Acad Sci U S A. 2016;113(3):775–80.
Nakanishi S. Molecular diversity of glutamate receptors and implications for brain function. Science. 1992;258(5082):597–603.
Ni B, Rosteck PR Jr, Nadi NS, Paul SM. Cloning and expression of a cDNA encoding a brain-specific Na(+)-dependent inorganic phosphate cotransporter. Proc Natl Acad Sci U S A. 1994;91(12):5607–11.
O'Kane RL, Martinez-Lopez I, DeJoseph MR, Vina JR, Hawkins RA. Na(+)-dependent glutamate transporters (EAAT1, EAAT2, and EAAT3) of the blood-brain barrier. A mechanism for glutamate removal. J Biol Chem. 1999;274(45):31891–5.
Oldendorf WH. Brain uptake of radiolabeled amino acids, amines, and hexoses after arterial injection. Am J Phys. 1971;221(6):1629–39.
Oldendorf WH, Szabo J. Amino acid assignment to one of three blood-brain barrier amino acid carriers. Am J Phys. 1976;230(1):94–8.
Oldendorf WH, Cornford ME, Brown WJ. The large apparent work capability of the blood-brain barrier: a study of the mitochondrial content of capillary endothelial cells in brain and other tissues of the rat. Ann Neurol. 1977;1(5):409–17.
Olsen GM, Sonnewald U. Glutamate: where does it come from and where does it go? Neurochem Int. 2015;88:47–52.
Pajecka K, Nissen JD, Stridh MH, Skytt DM, Schousboe A, Waagepetersen HS. Glucose replaces glutamate as energy substrate to fuel glutamate uptake in glutamate dehydrogenase-deficient astrocytes. J Neurosci Res. 2015;93(7):1093–100.
Palmieri F. The mitochondrial transporter family SLC25: identification, properties and physiopathology. Mol Asp Med. 2013;34(2–3):465–84.
Pardridge W. Regulation of amino acid availability to brain: selective control mechanisms for glutamate. In: Filer Jr L, editor. Glutamic acid: advances in biochemistry and physiology. New York: Raven Press; 1979. p. 125–36.
Perez-Mato M, Ramos-Cabrer P, Sobrino T, Blanco M, Ruban A, Mirelman D, et al. Human recombinant glutamate oxaloacetate transaminase 1 (GOT1) supplemented with oxaloacetate induces a protective effect after cerebral ischemia. Cell Death Dis. 2014;5:e992.
Persson L, Hillered L. Chemical monitoring of neurosurgical intensive care patients using intracerebral microdialysis. J Neurosurg. 1992;76(1):72–80.
Pines G, Danbolt NC, Bjoras M, Zhang Y, Bendahan A, Eide L, et al. Cloning and expression of a rat brain L-glutamate transporter. Nature. 1992;360(6403):464–7.
Pries AR, Secomb TW, Gaehtgens P. The endothelial surface layer. Pflugers Arch. 2000; 440(5):653–66.
Ramos M, del Arco A, Pardo B, Martinez-Serrano A, Martinez-Morales JR, Kobayashi K, et al. Developmental changes in the Ca^{2+}-regulated mitochondrial aspartate-glutamate carrier aralar1 in brain and prominent expression in the spinal cord. Brain Res Dev Brain Res. 2003;143(1):33–46.
Reimer RJ. SLC17: a functionally diverse family of organic anion transporters. Mol Asp Med. 2013;34(2–3):350–9.
Roberts RC, Roche JK, McCullumsmith RE. Localization of excitatory amino acid transporters EAAT1 and EAAT2 in human postmortem cortex: a light and electron microscopic study. Neuroscience. 2014;277C:522–40.
Robinson MB, Jackson JG. Astroglial glutamate transporters coordinate excitatory signaling and brain energetics. Neurochem Int. 2016;98:56–71.
Rose EM, Koo JC, Antflick JE, Ahmed SM, Angers S, Hampson DR. Glutamate transporter coupling to Na,K-ATPase. J Neurosci. 2009;29(25):8143–55.
Rothman S. Synaptic release of excitatory amino acid neurotransmitter mediates anoxic neuronal death. J Neurosci. 1984;4(7):1884–91.

Rothstein JD, Martin L, Levey AI, Dykes-Hoberg M, Jin L, Wu D, et al. Localization of neuronal and glial glutamate transporters. Neuron. 1994;13(3):713–25.

Ruban A, Mohar B, Jona G, Teichberg VI. Blood glutamate scavenging as a novel neuroprotective treatment for paraoxon intoxication. J Cereb Blood Flow Metab. 2014;34(2):221–7.

Schousboe A, Scafidi S, Bak LK, Waagepetersen HS, McKenna MC. Glutamate metabolism in the brain focusing on astrocytes. Adv Neurobiol. 2014;11:13–30.

Sershen H, Lajtha A. Capillary transport of amino acids in the developing brain. Exp Neurol. 1976;53(2):465–74.

Shashidharan P, Huntley GW, Meyer T, Morrison JH, Plaitakis A. Neuron-specific human glutamate transporter: molecular cloning, characterization and expression in human brain. Brain Res. 1994;662(1–2):245–50.

Shawahna R, Uchida Y, Decleves X, Ohtsuki S, Yousif S, Dauchy S, et al. Transcriptomic and quantitative proteomic analysis of transporters and drug metabolizing enzymes in freshly isolated human brain microvessels. Mol Pharm. 2011;8(4):1332–41.

Skytt DM, Klawonn AM, Stridh MH, Pajecka K, Patruss Y, Quintana-Cabrera R, et al. siRNA knock down of glutamate dehydrogenase in astrocytes affects glutamate metabolism leading to extensive accumulation of the neuroactive amino acids glutamate and aspartate. Neurochem Int. 2012;61(4):490–7.

Spink DC, Swann JW, Snead OC, Waniewski RA, Martin DL. Analysis of aspartate and glutamate in human cerebrospinal fluid by high-performance liquid chromatography with automated precolumn derivatization. Anal Biochem. 1986;158(1):79–86.

Storck T, Schulte S, Hofmann K, Stoffel W. Structure, expression, and functional analysis of a Na(+)-dependent glutamate/aspartate transporter from rat brain. Proc Natl Acad Sci U S A. 1992;89(22):10955–9.

Takamori S, Malherbe P, Broger C, Jahn R. Molecular cloning and functional characterization of human vesicular glutamate transporter 3. EMBO Rep. 2002;3(8):798–803.

Teichberg VI, Cohen-Kashi-Malina K, Cooper I, Zlotnik A. Homeostasis of glutamate in brain fluids: an accelerated brain-to-blood efflux of excess glutamate is produced by blood glutamate scavenging and offers protection from neuropathologies. Neuroscience. 2009;158(1):301–8.

Uchida Y, Ohtsuki S, Katsukura Y, Ikeda C, Suzuki T, Kamiie J, et al. Quantitative targeted absolute proteomics of human blood-brain barrier transporters and receptors. J Neurochem. 2011;117(2):333–45.

Waagepetersen HS, Qu H, Sonnewald U, Shimamoto K, Schousboe A. Role of glutamine and neuronal glutamate uptake in glutamate homeostasis and synthesis during vesicular release in cultured glutamatergic neurons. Neurochem Int. 2005;47(1–2):92–102.

Westergaard N, Sonnewald U, Unsgard G, Peng L, Hertz L, Schousboe A. Uptake, release, and metabolism of citrate in neurons and astrocytes in primary cultures. J Neurochem. 1994;62(5):1727–33.

Zerangue N, Kavanaugh MP. Flux coupling in a neuronal glutamate transporter. Nature. 1996;383(6601):634–7.

Zhang Y, Chen K, Sloan SA, Bennett ML, Scholze AR, O'Keeffe S, et al. An RNA-sequencing transcriptome and splicing database of glia, neurons, and vascular cells of the cerebral cortex. J Neurosci. 2014;34(36):11929–47.

Zhang D, Mably AJ, Walsh DM, Rowan MJ. Peripheral interventions enhancing brain glutamate homeostasis relieve amyloid beta- and TNFalpha- mediated synaptic plasticity disruption in the rat hippocampus. Cereb Cortex 2016:1–12. doi: 10.1093/cercor/bhw193.

Zhou Y, Danbolt NC. Glutamate as a neurotransmitter in the healthy brain. J Neural Transm. 2014;121(8):799–817.

Zlotnik A, Gruenbaum SE, Artru AA, Rozet I, Dubilet M, Tkachov S, et al. The neuroprotective effects of oxaloacetate in closed head injury in rats is mediated by its blood glutamate scavenging activity: evidence from the use of maleate. J Neurosurg Anesthesiol. 2009;21(3):235–41.

Zlotnik A, Sinelnikov I, Gruenbaum BF, Gruenbaum SE, Dubilet M, Dubilet E, et al. Effect of glutamate and blood glutamate scavengers oxaloacetate and pyruvate on neurological outcome and pathohistology of the hippocampus after traumatic brain injury in rats. Anesthesiology. 2012;116(1):73–83.

Development of Non-GAT1-Selective Inhibitors: Challenges and Achievements

Maria Damgaard, Anne Stæhr Haugaard, Stefanie Kickinger, Anas Al-Khawaja, Maria E.K. Lie, Gerhard F. Ecker, Rasmus Prætorius Clausen, and Bente Frølund

Abstract γ-Aminobutyric acid (GABA) neurotransmission is terminated by the GABA transporters (GATs) via uptake of GABA into neurons and surrounding glial cells. Four different transporters have been identified: GAT1, GAT2, GAT3, and the betaine/GABA transporter 1 (BGT1). The GAT1 subtype is the most explored transporter due to its high abundance in the brain and the existence of selective and potent GAT1 inhibitors. Consequently, less is known about the role and therapeutic potential of the non-GAT1 subtypes. Emerging pharmacological evidence suggests that some of these transporters pose interesting targets in several brain disorders. Pharmacological non-GAT1-selective tool compounds are important to further investigate the involvement of GATs in different pathological conditions. Extensive medicinal chemistry efforts have been put into the development of subtype-selective inhibitors, but truly selective and potent inhibitors of non-GAT1 subtypes are still limited. This review covers the advances within the medicinal chemistry area and the structural basis for obtaining non-GAT1-selective inhibitors.

Keywords GABA transporters • GAT • GAT1 • GAT3 • BGT1 • Non-GAT1-selective inhibitors

1 Introduction

γ-Aminobutyric acid (GABA), the major inhibitory neurotransmitter in the mammalian brain, plays a fundamental role in the overall balance between neuronal excitation and inhibition. Dysfunctions in the GABAergic neuronal system have been

Maria Damgaard and Anne Stæhr Haugaard contributed equally to this work.

M. Damgaard • A.S. Haugaard • A. Al-Khawaja • M.E.K. Lie • R.P. Clausen • B. Frølund (✉)
Department of Drug Design and Pharmacology, Faculty of Health and Medical Sciences, University of Copenhagen, Universitetsparken 2, 2100 Copenhagen, Denmark
e-mail: bfr@sund.ku.dk

S. Kickinger • G.F. Ecker
Department of Pharmaceutical Chemistry, University of Vienna, Althanstrasse 14, 1090 Vienna, Austria

A. Ortega, A. Schousboe (eds.), *Glial Amino Acid Transporters*, Advances in Neurobiology 16, DOI 10.1007/978-3-319-55769-4_16

implicated in the pathophysiology of a variety of neurological disorders, e.g., epilepsy and stroke (Treiman 2001; Clarkson et al. 2010).

GABA transporters (GATs) are transmembrane proteins essential for keeping GABAergic homeostasis by facilitating uptake of GABA into the neurons and the surrounding glial cells, terminating the GABAergic signaling. Inhibition of the GABA uptake by these transporters has been an approach taken to enhance GABAergic inhibitory neurotransmission and has been shown to be effective in the treatment of seizures in epileptic disorders (Kwan et al. 2001).

Four different GATs have been identified in the human and rodent central nervous system. The four human GAT subtypes are named GAT1, BGT1, GAT2, and GAT3 (Alexander et al. 2015) corresponding to the mouse mGAT1-mGAT4 (Liu et al. 1993). GAT1 is the most abundant subtype and has the pharmacological characteristics of the neuronal transport system; however, GAT1 has also been shown to be present in astrocytes (Borden 1996; Conti et al. 1998). GAT3 is also found at high levels in the CNS, although at much lower levels than GAT1, and is localized predominantly to astrocytes (Conti et al. 2004; Melone et al. 2005). BGT1 is expressed at much lower levels in the brain and has primarily been assigned to the astrocytes (Borden et al. 1995). GAT2 is expressed at an even lower level and in very defined cell regions (Conti et al. 1999) (see Lie et al. 2017; Schousboe et al. 2017).

Extensive medicinal chemistry efforts have been put into the development of potent and specific GAT inhibitors. Over the years, the majority of the developed GAT inhibitors have been targeting the GAT1 subtype with high selectivity and potency. The non-GAT1 subtypes are today not targeted by inhibitors with similar selectivity and potency, although some non-GAT1-selective compounds have emerged (Damgaard et al. 2015; Dhar et al. 1994; Kragholm et al. 2013; Vogensen et al. 2013) A major challenge in developing selective GAT inhibitors lies in the high sequence similarity of the four GAT subtypes. The GAT2 and GAT3 subtypes display an amino acid identity with one another of 67%, and GAT2 shares an amino acid identity of 68% with the BGT1 subtype and 65% with GAT3. GAT1 diverges more from the other subtypes as it shares an amino acid identity of ≈52% with GAT3 and GAT2 (Borden 1996).

Traditionally, the GABA uptake inhibitors have been divided into two main categories based on differences in chemical structure and pharmacological profiles: the small competitive substrate-related analogues and the corresponding analogues containing lipophilic side chains. Due to screening of large compound libraries, a few compounds with a non-substrate-related structural scaffold have recently emerged (Damgaard et al. 2015; Kragholm et al. 2013).

The aim of this chapter is to present and discuss the advances within the medicinal chemistry and the structural basis for obtaining non-GAT1-selective inhibitors.

2 Assessing GAT Subtype Selectivity

The methods to assay the potency of GAT inhibitors have evolved over the past decades. The majority of compounds in the literature have been characterized in vitro using synaptosomes (isolated synaptic terminals from neurons), brain slices and cultured neurons and astrocytes (Fjalland 1978). These assays typically utilize [^{3}H]-GABA as the radiolabeled substrate. Brain slices and synaptosomal preparations are believed to represent primarily neuronal uptake, although synaptosomes contain a marker for astrocytes (glial fibrillary acidic protein) (Bock and Hamberger 1976). Cortical astrocytes are cultured to provide astroglial cells with practically no remaining neurons (Hertz et al. 1989b). Likewise, cortical neurons are prepared to provide GABAergic neurons with no significant astrocytic contamination (Hertz et al. 1989a). The latter two methods are usually applied in characterization of the inhibitors as inhibitors of the glial or neuronal GABAergic uptake systems, respectively (Krogsgaard-Larsen 1980; Krogsgaard-Larsen et al. 1987) (see also Schousboe et al. 2017).

With the cloning of the four GATs, it became possible to recombinantly express the transporters in different cell systems and test the inhibitors for selectivity among the four GATs (Kvist et al. 2009). With this approach it became evident that many of the previously identified potent uptake inhibitors were GAT1 selective, including tiagabine, a clinically active antiepileptic drug (Borden 1996). The majority of these compounds was originally tested in synaptosome preparations and therefore may have been probed mainly toward GAT1 activity, due to the high abundance of GAT1 compared to the non-GAT1 subtypes in synaptosomes. Thus, the optimization of inhibitors would have been directed toward GAT1 selectivity and may explain the high abundance of GAT1-selective compounds compared to non-GAT1-selective compounds.

Although uptake assays using radiolabeled substrates represent the most commonly used assay technique to identify new GAT inhibitors, an LC-MS-/MS-based method has been reported applying deuterium-labeled [$^{2}H_{6}$]GABA as a substitute for [^{3}H]-GABA (Schmitt et al. 2015).

3 Development of Non-GAT1 Subtype-Selective Inhibitors

3.1 *Small Amino Acid-Based Inhibitors*

Throughout the years, a large number of GABA and β-alanine analogues has been studied as GAT inhibitors. In general, the GABA and β-alanine analogues display competitive inhibition and are often substrates for the transporters (Schousboe et al. 1983). However, few of the compounds display high selectivity as they are often recognized by several GAT subtypes as shown in Table 1.

GABA itself is recognized by all four GAT subtypes and is transported in the low-mid micromolar range, although with a reduced potency at BGT1 (Bolvig et al. 1999). The neutral, zwitterionic compound betaine is exclusively transported by BGT1 (Borden et al. 1992), and the short-chain amino acid, β-alanine, is selectively transported by GAT2 and GAT3 (IC_{50} 42 μM and 36 μM, respectively) (Kvist et al. 2009) and shows a twofold selectivity for astrocytic GAT compared to neuronal GAT (K_i 843 vs. 1666 μM, respectively) (Clausen et al. 2006; Madsen et al. 2010). The hydroxylated analogue of β-alanine, isoserine, also displays a clear preference

Table 1 Pharmacological data for small amino acid-based GAT inhibitors

Compound	Species	GAT1 (mGAT1)	BGT1 (mGAT2)	GAT2 (mGAT3)	GAT3 (mGAT4)	Ref.
		IC_{50} (μM)				
GABA	Human	10	26	11	10	Kvist et al. (2009)
Betaine	Human	>10,000	590	>3000	>10,000	Kvist et al. (2009)
β-Alanine	Human	5000	660	42	36	Kvist et al. (2009)
Isoserine	Mouse	2500	230	4.3	5.9	Kragler et al. (2005)
THPO	Human	1300	2100	1500	2200	Kvist et al. (2009)
Guvacine	Human	15	2300	1000	94	Kvist et al. (2009)
Nipecotic acid	Human	19	3000	530	94	Kvist et al. (2009)
exo-THPO	Mouse	1000	3000	>3000	>3000	White et al. (2002)
N-methyl-*exo*-THPO	Mouse	450	>3000	>3000	>3000	White et al. (2002)
GPA	Human	460	96	17	38	Al-Khawaja et al. (2014)
GBA	Human	>1000	360	55	150	Al-Khawaja et al. (2014)
ATPCA	Human	>1000	2.5	41	99	Al-Khawaja et al. (2014)
1	Human	>100	9.8	>100	>100	Al-Khawaja et al. (2014)
2	Human	>100	5.48	36.9	13.9	Nakada et al. (2013)
3	Human	>100	0.59	>100	76.3	Kobayashi et al. (2014)

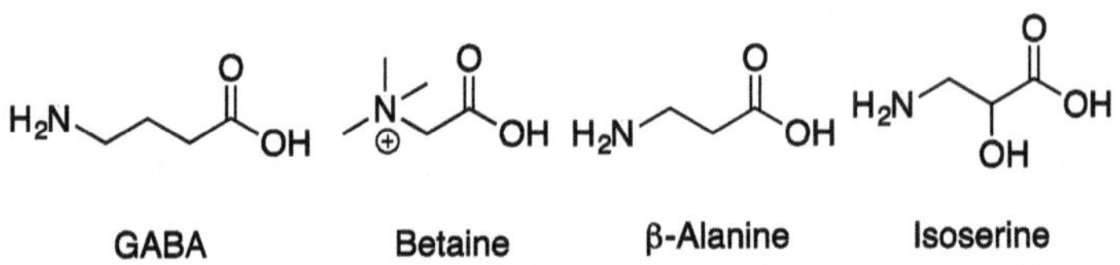

Fig. 1 Chemical structures of amino acid-based inhibitors. GABA, betaine, β-alanine, and isoserine

for the GAT2 and GAT3 subtypes with potencies in the low micromolar range (IC_{50} 4.3 μM and 5.9 μM, respectively) (Fig. 1) (Kragler et al. 2005).

The initial characterization of GABA transport was to a large extent enabled by the discovery of the naturally occurring compound muscimol (Fig. 2) originating from the fly agaric mushroom (Madsen et al. 2010). Muscimol is a GABA analogue, as the 3-hydroxyisoxazole group functions as a bioisostere for the carboxylic acid group. Muscimol was recognized as a potent $GABA_A$ receptor agonist and a substrate for all GABA transporter subtypes in rat brain slices (Krogsgaard-Larsen 1980; Krogsgaard-Larsen et al. 1981). Conformational restriction of the basic side chain of muscimol by incorporation into a ring system led to the rigid analogue 4,5,6,7-tetrahydroisoxazolo[5,4-

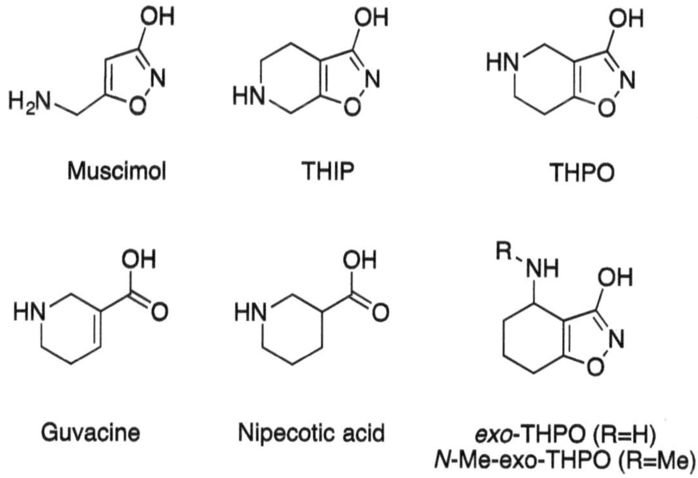

Fig. 2 Chemical structures of conformationally restricted GABA uptake inhibitors: muscimol, THPO, guvacine, nipecotic acid, *exo*-THPO, *N*-Me-*exo*-THPO, and THIP (GABA uptake inactive)

c]pyridin-3-ol (THIP) (Krogsgaard-Larsen and Johnston 1975; Krogsgaard-Larsen 1980; Schousboe et al. 1981). THIP is devoid of GABA transport inhibitory activity and has been characterized as a functionally selective and potent agonist at δ-containing extrasynaptic $GABA_A$ receptors (Brown et al. 2002). The closely related bicyclic β-alanine analogue, 4,5,6,7-tetrahydroisoxazolo[4,5-*c*]pyridin-3-ol (THPO), retained activity for GABA uptake and showed only weak GABA receptor binding (Krogsgaard-Larsen and Johnston 1975; Krogsgaard-Larsen 1980; Schousboe et al. 1981). Like β-alanine, THPO shows a twofold selectivity for astrocytic uptake over neuronal uptake (K_i 262 μM vs. 501 μM) (Madsen et al. 2010). However, substitution of the 3-hydroxyisoxazole group of THPO for a carboxyl group leads to guvacine and nipecotic acid that are much more potent inhibitors but at the expense of selectivity toward astrocytic uptake (Fig. 2) (Schousboe et al. 1981; Madsen et al. 2010). *N*-methylation of nipecotic acid impaired the activity at both neuronal and astrocytic uptake (Schousboe et al. 1979).

Selectivity among GAT subtypes has later been studied showing that guvacine and nipecotic acid display GAT1 selectivity with IC_{50} values of 15 and 19 μM, respectively (Table 1) (Kvist et al. 2009). Nipecotic acid and guvacine have subsequently served as valuable lead structures in the development of potent lipophilic GAT inhibitors.

Prompted by the preference of THPO for astrocytic uptake, new ligands were developed by modifying the THPO structure. Moving the amino group in THPO to an *exo*-cyclic position resulted in 4-amino-4,5,6,7-tetrahydrobenzo[*d*]isoxazol-3-ol (*exo*-THPO), which displays increased selectivity for astrocytic uptake (Falch et al. 1999). Further modifications of the *exo*-THPO structure led to a series of *N*-alkylated derivatives in which the mono-methylated compound, *N*-methyl-*exo*-THPO, was the most selective and potent glial uptake inhibitor (Fig. 2) (Falch et al. 1999).

Based on the low potency toward GAT1, β-alanine seems to be a promising lead structure for the development of non-GAT1-selective inhibitors. Applying this approach, the guanidine analogues 3-guanidinopropionic acid (GPA) and 4-guanidinobutyric acid (GBA) were found to be distinctly more potent at GAT2/GAT3 than at GAT1/BGT1 (Kragler et al. 2005; Al-Khawaja et al. 2014). Incorporating the guanidine or amidine group as a bioisosteric replacement for the amino group in the structure of nipecotic acid resulted in a series of cyclic 2-amino-tetrahydropyridine/pyrimidine analogues recently identified as substrate-inhibitors with pronounced selectivity for BGT1 over the three other GAT subtypes (Fig. 3) (Al-Khawaja et al. 2014). Within the series, 2-amino-1,4,5,6-tetrahydropyrimidine-5-carboxylic acid (ATPCA) and the corresponding amidine analogue (6-amino-2,3,4,5-tetrahydropyridine-3-carboxylic acid, **1**) displayed the highest potencies toward BGT1 (IC_{50} of 2.5 μM and 9.8 μM, respectively). GAT1 homology model studies suggest that the amino group of GABA has an interaction with tyrosine (Tyr60). In the other subtypes, this residue is a glutamate (Glu52) offering opportunity for a strong bidentate interaction with the guanidinium group in ATCPA, which could be the basis for the observed selectivity toward BGT1 over GAT1 (Al-Khawaja et al. 2014), whereas the lower activity at GAT2 and GAT3 must be attributed to other residues. ATPCA was furthermore identified as a substrate for BGT1 (Al-Khawaja et al. 2014) and as a $GABA_A$ receptor agonist (Petersen et al. 2014).

Conformationally restricting GABA into a *trans* cyclopropane structure resulted in the inhibitor 2-((1*S*,2*R*)-2-aminocyclopropyl)acetic acid (**2**), selective for the GAT3 and BGT1 subtypes (IC_{50} of 13.9 μM and 5.48 μM, respectively) (Fig. 3) (Nakada et al. 2013). This small-molecule GABA analogue also exhibited affinity toward $GABA_A$ receptors (K_i of 0.34 μM). By further restricting **2** into a bicyclic structure, (1*S*,2*S*,5*R*)-5-aminobicyclo[3.1.0]hexane-2-carboxylic acid (**3**) led to a highly potent and selective inhibitor of the BGT1 subtype transporter (IC_{50} 0.59 μM) (Fig. 3) (Kobayashi et al. 2014). The preference for the *syn* versus the *anti* conformation of **3** in BGT1 binding provides information of the preferred orientation of the carboxylic acid group for future development of compounds targeting the BGT1 subtype.

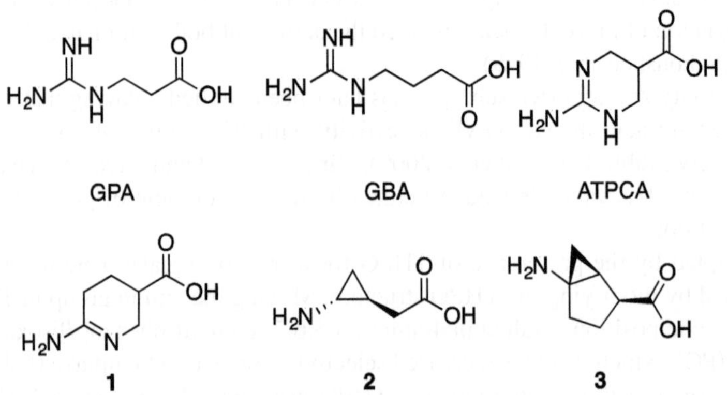

Fig. 3 Chemical structures of substrate-related GABA uptake inhibitors. 3-Guanidinopropionic acid (GPA), 4-guanidinobutyric acid (GBA), ATPCA, and compounds **1**, **2**, and **3**

Considering the zwitterionic nature at physiological pH of the small amino acid-based uptake inhibitors, which often is associated with restricted passive brain penetration, the therapeutic potential of these compounds might be limited. However, the substrate-related analogues have not only served as important pharmacological tools, but also as lead scaffolds for selective GABA uptake inhibitors.

3.2 *Lipophilic Aromatic Amino Acid-Based Non-GAT1 Inhibitors*

In the mid-1980s, it was found that *N*-alkylation of the amino group of small GABA uptake inhibitors with a bulky lipophilic tail increases the potency and enables compounds to cross the blood-brain barrier (Ali et al. 1985). One of the first examples of this was 4,4-diphenyl-3-butenyl (DPB)-nipecotic acid (Fig. 4) (Ali et al. 1985). This strategy also provided compounds that were not substrates for the GATs. Tiagabine is the most well known of these lipophilic GAT inhibitors (Braestrup et al. 1990) and is now a registered drug as add-on therapy in the treatment of epilepsy (Kalviainen 2001).

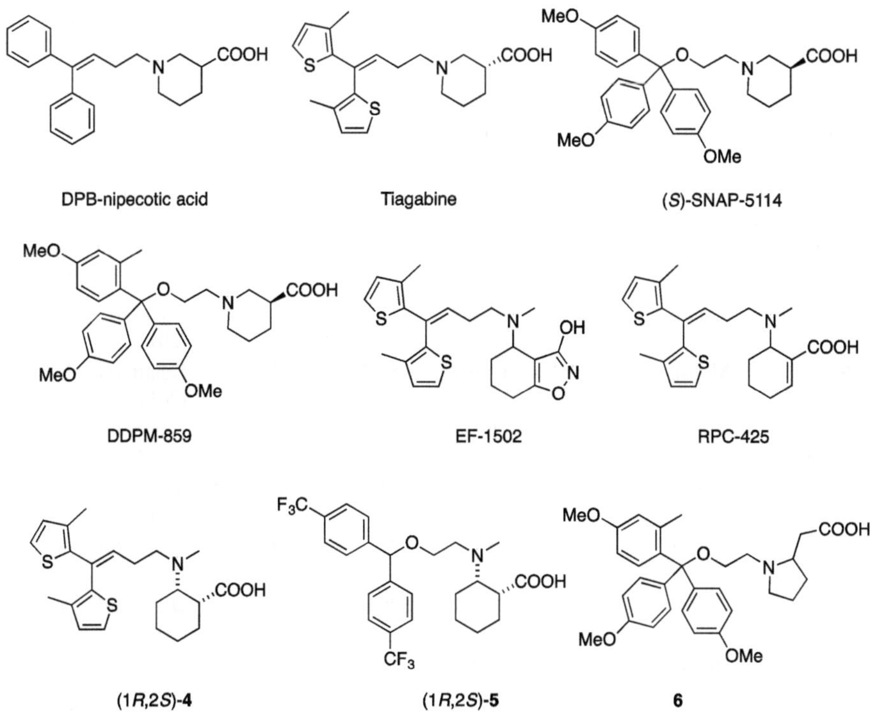

Fig. 4 Chemical structures of DPB-nipecotic acid, tiagabine, (*S*)-SNAP5114, DDPM-859, EF1502, RPC-425, (1*R*, 2*S*)-**4**, (1*R*, 2*S*)-**5,** and (1*R*, 2*S*)**6**

With the cloning of the four GAT subtypes, it was demonstrated that most of these lipophilic compounds, including tiagabine and DPB-nipecotic acid, have a clear preference for GAT1 (Table 2) (White et al. 2005; Petrera et al. 2016). An exception from this is the moderately selective inhibitor of GAT3, (*S*)-SNAP-5114 (Fig. 4) (Dhar et al. 1994). (*S*)-SNAP-5114 has a low micromolar affinity toward GAT3

Table 2 Pharmacological data for lipophilic non-GAT1 inhibitors

Compound	Species	GAT1 (mGAT1)	BGT1 (mGAT2)	GAT2 (mGAT3)	GAT3 (mGAT4)	Ref.
		IC_{50} (μM)				
DPB-nipecotic acid	Mouse	0.69	372	195	275	Petrera et al. (2016)
Tiagabine	Mouse	0.8	300	>300	800	White et al. (2005)
(*S*)-SNAP-5114	Human/rat	388	140	21	5	Dhar et al. (1994)
DDPM-859	Mouse	65	76	14	1.7	Pabel et al. (2012)
EF-1502	Mouse	7	26	>300	>300	Clausen et al. (2005)
(*S*)-EF1502	Mouse	120	34	>150	>150	Clausen et al. (2005)
(*R*)-EF1502	Mouse	4	22	>150	>150	Clausen et al. (2005)
RPC-425	Mouse	307	45	>300	286	Vogensen et al. (2013)
(1*R*,2*S*)-**4**	Mouse	562	21.6	941	289	Vogensen et al. (2015)
(1*R*,2*S*)-**5**	Mouse	631	90.0	392	303	Vogensen et al. (2015)
NNC 05-2090	Mouse	19[a]	1.4[a]	41[a]	15[a]	Thomsen et al. (1997)
(*R*)-**6**	Bovine	67.8	–	–	3.10	Fülep et al. (2006)
(*S*)-**6**	Bovine	35.4	–	–	28.7	Fülep et al. (2006)
7a	Mouse	72.4%[b]	67.1%[b]	45	52.3%[b]	Schaffert et al. (2011)
7b	Mouse	73.9%[b]	61.6%[b]	51.7%[b]	63	Schaffert et al. (2011)
8a (BPDBA)	Human	NE	20[h]	>100	NE	Kragholm et al. (2013)
8b	Human	NE	42	63	NE	Kragholm et al. (2013)
8c	Human	NE	15	23	NE	Kragholm et al. (2013)
9	Human	>1000	749	203	6	Damgaard et al. (2015)
10	Human	>1000	>1000	>1000	29	Damgaard et al. (2015)

NE: no or negligible effect at 100 μM
[a]K_i (μM)
[b]Percent inhibition at 100 μM

(IC_{50} 5 μM), whereas the potencies of (*S*)-SNAP-5114 on GAT1, GAT2, and BGT1 were shown to be 388 μM, 21 μM and 140 μM, respectively (Dhar et al. 1994). The use of (*S*)-SNAP-5114, however, is limited by low solubility and poor chemical stability (Pabel et al. 2012).

The selectivity of (*S*)-SNAP-5114 is very much dependent on the substitution pattern on the aromatic rings, and the presence of lipophilic substituents in the para-position of the three aromatic groups is necessary to retain selectivity (Dhar et al. 1994). Further development has led to the chemically more stable, but less selective analogue, DDPM-859 (Fig. 4), shown to be equipotent with (*S*)-SNAP-5114 (Pabel et al. 2012).

Replacing the nipecotic acid part of the lipophilic compounds with THPO, inspired by the selectivity shown for *N*-Me-exo-THPO, has led to the discovery of EF-1502, which shows preference for both GAT1 (IC_{50} 7 μM) and BGT1 (IC_{50} 26 μM) over GAT2 (IC_{50} > 300 μM) and GAT3 (IC_{50} > 300 μM) (Clausen et al. 2005). The selectivity of EF-1502 was found to be dependent on the stereochemistry of the compound. (*S*)-EF-1502 was found to be selective for BGT1, whereas (*R*)-EF-1502 was found to be a nonselective GAT1/BGT1 inhibitor (Clausen et al. 2005). Bioisosteric replacement of the 3-hydroxyisoxazole in EF-1502 for a carboxylic acid led to another BGT1-selective compound RPC-425 (Fig. 4) (Vogensen et al. 2013). RPC-425 is less potent than EF-1502, but more BGT1 selective (BGT1 IC_{50} 45 μM, GAT1 IC_{50} 307 μM) (Vogensen et al. 2013). Inhibitory activity and BGT1 selectivity were retained for the saturated and thereby conformational less restricted analogue of RPC-425, **4**, and the corresponding 2-(bis(4-(trifluoromethyl)phenyl)methoxy)ethyl analogue **5** (Vogensen et al. 2015). Interestingly, enhanced BGT1 selectivity of the *cis* forms compared to their *trans* counterparts was observed, suggesting a certain structural preference for determinants triggering substrate specificity in the binding pocket.

The stereochemistry has continuously proven to be of great importance for the selectivity among GAT subtypes. One example is the pyrrolidine-2-ylacetic acid-derived compounds reported by Fülep et al., (*R*)-pyrrolidine-2-acetic acid *N*-substituted with a 2-(tris(4-methoxyphenyl)methoxy)ethyl (**6**) was shown to be a potent GAT3-selective compound, with a GAT1/GAT3 selectivity ratio of 1:20 and an IC_{50} of 3.1 μM for GAT3. However, the selectivity between GAT2 and BGT1 was not described. The corresponding (*S*)-enantiomer is equipotent at GAT1 and GAT3 (IC_{50} 35.4 μM and 28.7 μM, respectively) (Fülep et al. 2006).

3.3 Non-amino Acid GAT Inhibitors

An amino acid structure element is not essential for GABA uptake inhibition. An example of this is 1-(3-(9H-carbazol-9-yl)-1-propyl)-4-(2-methoxyphenyl)-4-piperidinol (NNC 05-2090), shown to be a potent inhibitor of BGT1 (K_i 1.4 μM) displaying at least tenfold selectivity over the remaining GATs (Thomsen et al. 1997). As seen for other GAT inhibitors, the substitution pattern on the aromatic ring is important for the selectivity of the compound. Changing the position of the ortho-OCH_3 to a para-OCH_3 retains the potency of the compound, but greatly reduces the selectivity toward GAT2 and GAT3 (Thomsen et al. 1997). However, the selectivity of NNC 05-2090 could not be reproduced in a later study (Schaffert et al. 2011). Another class of compounds showing non-GAT1 selectivity is the 1,5-disubstituted tetrazole analogues that display diverse selectivity profiles among GAT subtypes depending on their substituents (Schaffert et al. 2011). As an example, the methyl-3,3-diphenylpropylamine substituted tetrazole (**7a**) shows GAT2 selectivity, whereas the corresponding diphenylbutylamine analogue (**7b**) was shown to be GAT3 selective (Table 2, Fig. 5) (Schaffert et al. 2011).

A more recent approach in the effort of obtaining non-GAT1-selective inhibitors is screening of compound libraries for compounds capable of inhibiting GABA transport. This offers the possibility to identify compounds binding to less conserved areas of the transporter proteins, hence, which may increase the likeliness of identifying selective compounds. Using this approach, the BGT1-selective compound BPDBA (**8a**) was identified (Kragholm et al. 2013). BPDBA displayed an IC_{50} value of 20 μM for BGT1 and no inhibitory activity for GAT1-3 at 100 μM (Table 2).

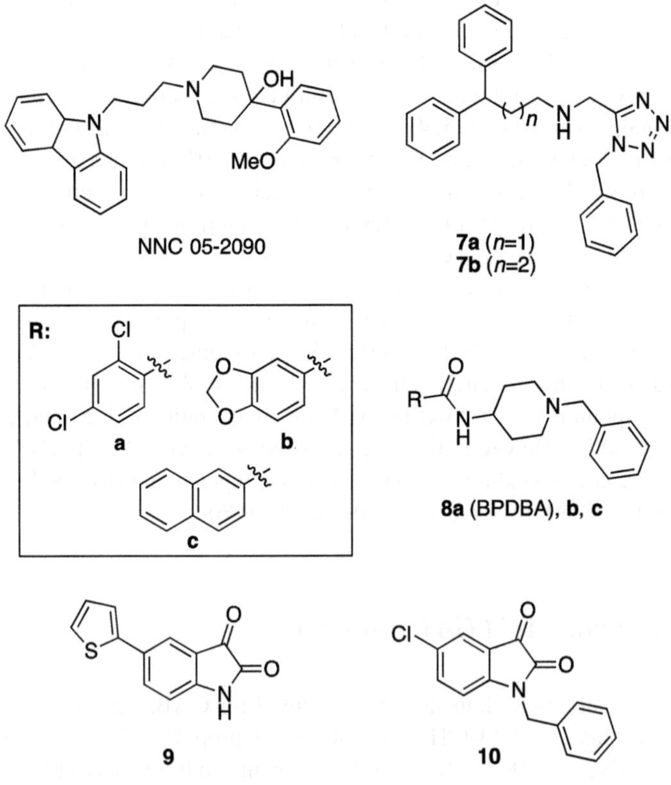

Fig. 5 Chemical structures of non-amino acid GAT inhibitors NNC 05-2090, **7a,b**, **8a** (BPDBA),**b,c**, and the isatin analogues **9** and **10**

A limited number of commercial analogues of BPDBA were tested in this study, of which two compounds, **8b** and **8c**, retained the inhibition of BGT1 (Fig. 5). However, the selectivity was reduced since these analogues also inhibited GAT2 (Kragholm et al. 2013), highlighting how subtle changes can alter the selectivity also for non-amino acid GAT inhibitors.

Another type of non-GAT1 inhibitors identified by library screening is the GAT3-selective isatin-derived compounds (Damgaard et al. 2015). Isatin compounds **9** and **10** were both shown to be potent GAT3 inhibitors (IC_{50} values of 6 μM and 29 μM) with a more than 30-fold selectivity for the other GAT subtypes (Table 2, Fig. 5) (Damgaard et al. 2015). Further characterization of compound **10** showed that the compound had a non-competitive mode of inhibition at GAT3. A molecular model of the binding site furthermore supported a unique binding site for this compound that matched the selectivity observed (Damgaard et al. 2015).

Many years of medicinal chemistry efforts have been put into the development of non-GAT1 inhibitors based on ligand design. More recently, structural knowledge of the transporters obtained from related crystal structures has given valuable information regarding the possibilities for obtaining selectivity among GATs.

4 Structural Basis for Design of Non-GAT1 Inhibitors

The determination of the first crystal structure of the prokaryotic Na^+/Cl^--dependent leucine transporter (LeuT) paved the way for studying the structure-function relationship of structural homologous such as the neurotransmitter-sodium symporter (NSS) family, including the GATs, the serotonin transporter, the dopamine transporter and the norepinephrine transporter (Kristensen et al. 2011; Yamashita et al. 2005). The X-ray structure comprises 12 transmembrane helices linked by multiple extra- and intracellular loops, where TM1-5 and TM6-10 show inherent pseudosymmetry as they are inversely repeated with respect to the membrane (Yamashita et al. 2005).

Although LeuT only displays about 20% sequence identity with the GATs, it has successfully functioned as a suitable template for homology modeling (Baglo et al. 2013; Beuming et al. 2006; Damgaard et al. 2015; Jurik et al. 2015; Schlessinger et al. 2012; Skovstrup et al. 2010; Vogensen et al. 2015; Wein et al. 2016; Wein and Wanner 2010). More recently, Penmatsa et al. revealed the X-ray structure of the *Drosophila* dopamine transporter (dDAT), and in 2016 the crystal structure of the human serotonin transporter (hSERT) was determined (Coleman et al. 2016; Penmatsa et al. 2013). These crystal structures function as new suitable templates for homology modeling, as they display higher sequence similarity with the GATs (dDAT about 45%, hSERT 42%) and contain fewer gaps in the alignment.

The combination of homology modeling and experimental testing successfully enabled the development of binding hypothesis for GAT substrates and inhibitors. Based on homology modeling, binding poses of four compounds were postulated: GABA, the GAT1 inhibitor tiagabine, the BGT1-selective inhibitor (1*S*, 2*R*)-**5**, and the isatin analogue **10**, a selective GAT3 inhibitor (Fig. 6). The carboxyl

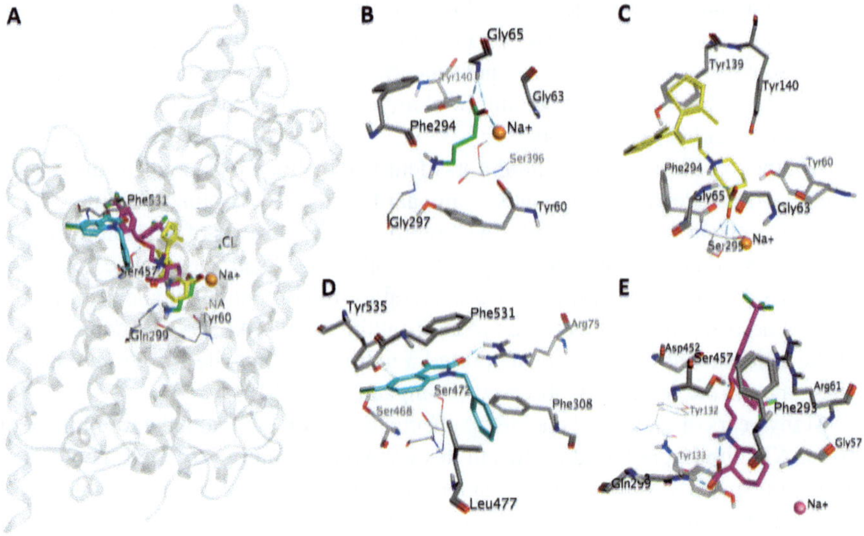

Fig. 6 Representative docking pose examples of GABA, tiagabine, **10** and (1*R*, 2*S*)-**5**. (**a**) Overview of prominent binding poses in the GABA transporter. (**b**) Docking pose of GABA (*green*) in the human GAT1 in the orthosteric pocket. GABA adopts an extended conformation where the carboxyl group interacts with Na1, Y140, and G65. (**c**) Docking pose of tiagabine (*yellow*) in the human GAT1 in the orthosteric pocket. The nipecotic acid moiety of tiagabine adopts an equatorial conformation where the carboxyl group coordinates Na1 and the positively charged nitrogen interacts with the backbone of F294. (**d**) Docking pose of compound **10** (*blue*) in the human GAT3 in a postulated new allosteric pocket. The ligand is stabilized by hydrophobic interactions with F530 (not shown), F531, and Y535. (**e**) Docking pose of (1*R*, 2*S*)-**5** (*pink*) in the mouse GAT2 in the orthosteric pocket. The carboxyl group can form a hydrogen bond with Q299, and the ether linker is coordinated by S457

groups of GABA (green) and tiagabine (yellow) coordinate the Na^+ ion and are positioned in a similar way in the orthosteric pocket (Fig. 6b, c). The carboxyl group of (1*R*, 2*S*)-**5** (pink) is orientated differently and can coordinate Gln299 (Fig. 6e). Compound **10** (blue) is structurally very different to the endogenous substrate GABA and shows a non-competitive binding mode. Therefore, it was suggested that compound **10** binds in a new potential binding site located between TM10 and TM11 in the extracellular vestibule above the orthosteric binding site (Fig. 6d) (Damgaard et al. 2015).

GABA adopts an extended conformation in the orthosteric binding site of GAT1 where the carboxyl group interacts with Na^+, the hydroxyl group of Tyr140, and the backbone of Gly65 (Fig. 6b). The position of the carboxyl group is identical with the position of the carboxyl group of leucine determined in the LeuT crystal structure (Skovstrup et al. 2010; Yamashita et al. 2005). The amino group of GABA can interact with tyrosine (Tyr60), which appears to be important for activity, since this residue is a glutamate (Glu52) in all non-GAT1 subtypes (Al-Khawaja et al. 2014; Skovstrup et al. 2010). Mutating Tyr60 in GAT1 to glutamate results in a complete loss of GABA uptake (Kanner 2003). Gly63, which is located closely to the carboxyl group of GABA and highly conserved in all GABA and glycine transporters, is also known to be an important trigger for selectivity (Kristensen et al. 2011). In monoamine transporters, this glycine is exchanged with asparagine, which facilitates the coordination of the Na^+ and therefore compensates for the missing carboxyl group in monoamines (Kristensen et al. 2011).

The acid moiety of tiagabine adopts an equatorial conformation in GAT1 (Fig. 6c). The carboxyl group coordinates Na^+ in a similar way as described for GABA binding, and the positively charged nitrogen interacts with the backbone of Phe294. The long bulky lipophilic side chain of tiagabine fits into the hydrophobic extracellular vestibule, which is part of the substrate entry pathway. Residues Phe294 and Tyr140, which are part of the extracellular lid of the transporter, need to change conformation to accommodate tiagabine. Consequently, tiagabine stabilizes the transporter in an outward-open conformation (Jurik et al. 2015).

The selective BGT1 inhibitor, (1*R*, 2*S*)-**5**, shows a possible exclusive interaction pattern (Fig. 6e). The carboxyl group of the β-amino acid moiety can maintain a hydrogen bond with Gln299 that might substitute for the missing coordination of Na^+. Gln299 is a BGT1 exclusive residue, being leucine in all other GATs. The ether linker of (1*R*, 2*S*)-**5** is coordinated by Ser457, which is the last amino acid of an interesting three-residue motif in the middle of an unwounded region of TM10, being SerAlaSer456 in GAT1, AlaSerSer457 in BGT1, and AlaAlaSer452/AlaAlaSer472 in GAT2/GAT3. This unwound region is unique for the GATs and the highly related taurine transporter (TauT). The last serine residue represents a one-amino acid insertion not seen in any other NSS. Today, there is no template available to model the local fold of this residue, and extensive modeling effort needs to be spent on the modeling of the three-residue motif as it causes a tighter substrate entry pathway, therefore possibly being an additional trigger for selectivity (Vogensen et al. 2015).

As previously mentioned, the isatin analogue **10** binds to a postulated new binding site that is located between TM10 and TM11 above the orthosteric pocket (Fig. 6a, d). Compound **10** is stabilized by hydrophobic interactions with Phe530 (not shown in Fig. 6d), Phe531, and Tyr535. Phe531 is unique for GAT3, being a serine in all other GATs and thus may act as a trigger for selectivity (Damgaard et al. 2015).

The discussed binding poses show that GAT subtype selectivity can be achieved. Although the binding pockets of all GATs are rather similar, this section could identify possible residues as triggers for substrate specificity: Tyr60 in GAT1 that corresponds to glutamate in all other GATs and Gln299 in BGT1 that corresponds to leucine in all other GATs (Al-Khawaja et al. 2014; Skovstrup et al. 2010; Vogensen et al. 2015). The aforementioned three-residue motif in TM10 might additionally contribute to substrate specificity by altering the substrate entry pathway (Vogensen et al. 2015). Furthermore, non-competitive inhibitors such as the isatin analogue **10** most probably bind to a different pocket, which makes it essential to experimentally confirm the actual binding site before performing in silico studies.

5 Conclusion and Perspectives

A major challenge in GAT drug discovery has for long been to obtain subtype-selective inhibitors. Despite extensive efforts from both academia and industry, the most promising selectivity ratios obtained have been for GAT1 over GAT2, BGT1 and GAT3, while the number of non-GAT1-selective inhibitors is limited. This could be due to the fundamental nature of the binding site with small differences between individual transporter subtypes and might also relate to the fact that pharmacological testing of most of the established and presumed inhibitors has been limited to one or few GAT subtypes.

To accelerate the process of understanding the molecular determinants of subtype selectivity in the GAT family, combining pieces of information from both structure and ligand-based sources is crucial. Lack of structural knowledge on the transporters has for long been the limiting factor for detailed insight into the molecular basis for binding and activation. However, related crystal structures have functioned as suitable templates for homology modeling of GATs, and recently new structures displaying higher sequence similarity with the GATs have been released, which probably will facilitate a more rational approach for future drug discovery within the field.

So far, structure-activity studies have served as a valuable basis for ligand development in the GAT area. The small amino acid inhibitors have contributed to differentiate between the GAT subtypes and served as templates for more potent lipophilic inhibitors. These lipophilic compounds have in general greatly improved potency and bioavailability of GAT inhibitors and have been important in the investigation of the pharmacological role and therapeutic potential of the transporters. However, since the substrate binding site is relatively conserved among the GAT subtypes, the use of GAT substrates as scaffolds in the development of inhibitors could be connected to a risk of generating non-selective ligands. This has been circumvented by screening compound libraries for non-GAT1-selective compounds allowing for new types of inhibitors. These compounds are mainly allosteric inhibitors exhibiting a large degree of structural diversity. This represents an opportunity for development of more subtype-selective compounds, as the lead structures no longer target the conserved substrate pocket.

Recently, highly selective non-GAT1 inhibitors, which include both substrate-based and allosteric inhibitors, were reported, further demonstrating that high selectivity can be achieved. However, the number of useful tool compounds for studying the non-GAT1 transporters is still limited, and they often lack either selectivity or potency. Therefore, selective and potent inhibitors of the non-GAT1 transporters are essential to further understand the physiological role of these transporters in various neurological disorders like epilepsy and stroke and the therapeutic potential of non-GAT1 inhibitors.

Acknowledgment SK and GE acknowledge funding provided by to the Austrian Science Fund (MolTag, grant # W1232).

Conflict of Interest The author declares no conflicts of interest.

References

Al-Khawaja A, Petersen JG, Damgaard M, Jensen MH, Vogensen SB, Lie ME, et al. Pharmacological identification of a guanidine-containing β-alanine analogue with low micromolar potency and selectivity for the betaine/GABA transporter 1 (BGT1). Neurochem Res. 2014;39(10):1988–96.

Alexander SP, Kelly E, Marrion N, Peters JA, Benson HE, Faccenda E, et al. The concise guide to pharmacology 2015/16: transporters. Br J Pharmacol. 2015;172(24):6110–202.

Ali FE, Bondinell WE, Dandridge PA, Frazee JS, Garvey E, Girard GR, et al. Orally active and potent inhibitors of gamma-aminobutyric acid uptake. J Med Chem. 1985;28(5):653–60.

Baglo Y, Gabrielsen M, Sylte I, Gederaas OA. Homology modeling of human gamma-butyric acid transporters and the binding of pro-drugs 5-aminolevulinic acid and methyl aminolevulinic acid used in photodynamic therapy. PLoS One. 2013;8(6):e65200.

Beuming T, Shi L, Javitch JA, Weinstein H. A comprehensive structure-based alignment of prokaryotic and eukaryotic neurotransmitter/Na+ symporters (NSS) aids in the use of the LeuT structure to probe NSS structure and function. Mol Pharmacol. 2006;70(5):1630–42.

Bock E, Hamberger A. Immunoelectrophoretic determination of brain-specific antigens in bulk-prepared neuronal and glial cells. Brain Res. 1976;112(2):329–35.

Bolvig T, Larsson OM, Pickering DS, Nelson N, Falch E, Krogsgaard-Larsen P, et al. Action of bicyclic isoxazole GABA analogues on GABA transporters and its relation to anticonvulsant activity. Eur J Pharmacol. 1999;375(1–3):367–74.

Borden LA. GABA transporter heterogeneity: pharmacology and cellular localization. Neurochem Int. 1996;29(4):335–56.

Borden LA, Smith KE, Hartig PR, Branchek TA, Weinshank RL. Molecular heterogeneity of the gamma-aminobutyric acid (GABA) transport system. Cloning of two novel high affinity GABA transporters from rat brain. J Biol Chem. 1992;267(29):21098–104.

Borden LA, Smith KE, Gustafson EL, Branchek TA, Weinshank RL. Cloning and expression of a betaine/Gaba transporter from human brain. J Neurochem. 1995;64(3):977–84.

Braestrup C, Nielsen EB, Sonnewald U, Knutsen LJ, Andersen KE, Jansen JA, et al. (R)-N-[4,4-bis(3-methyl-2-thienyl)but-3-en-1-yl]nipecotic acid binds with high affinity to the brain gamma-aminobutyric acid uptake carrier. J Neurochem. 1990;54(2):639–47.

Brown N, Kerby J, Bonnert TP, Whiting PJ, Wafford KA. Pharmacological characterization of a novel cell line expressing human alpha(4)beta(3)delta GABA(A) receptors. Br J Pharmacol. 2002;136(7):965–74.

Clarkson AN, Huang BS, Macisaac SE, Mody I, Carmichael ST. Reducing excessive GABA-mediated tonic inhibition promotes functional recovery after stroke. Nature. 2010;468(7321):305–9.

Clausen RP, Moltzen EK, Perregaard J, Lenz SM, Sanchez C, Falch E, et al. Selective inhibitors of GABA uptake: synthesis and molecular pharmacology of 4-N-methylamino-4,5,6,7-tetrahydrobenzo[d]isoxazol-3-ol analogues. Bioorg Med Chem. 2005;13(3):895–908.

Clausen RP, Madsen K, Larsson OM, Frolund B, Krogsgaard-Larsen P, Schousboe A. Structure-activity relationship and pharmacology of gamma-aminobutyric acid (GABA) transport inhibitors. Adv Pharmacol. 2006;54:265–84.

Coleman JA, Green EM, Gouaux E. X-ray structures and mechanism of the human serotonin transporter. Nature. 2016;532(7599):334–9.

Conti F, Melone M, De Biasi S, Minelli A, Brecha NC, Ducati A. Neuronal and glial localization of GAT-1, a high-affinity gamma-aminobutyric acid plasma membrane transporter, in human cerebral cortex: with a note on its distribution in monkey cortex. J Comp Neurol. 1998;396(1):51–63.

Conti F, Zuccarello LV, Barbaresi P, Minelli A, Brecha NC, Melone M. Neuronal, glial, and epithelial localization of gamma-aminobutyric acid transporter 2, a high-affinity gamma-aminobutyric acid plasma membrane transporter, in the cerebral cortex and neighboring structures. J Comp Neurol. 1999;409(3):482–94.

Conti F, Minelli A, Melone M. GABA transporters in the mammalian cerebral cortex: localization, development and pathological implications. Brain Res Rev. 2004;45(3):196–212.

Damgaard M, Al-Khawaja A, Vogensen SB, Jurik A, Sijm M, Lie MEK, et al. Identification of the first highly subtype-selective inhibitor of human GABA transporter GAT3. ACS Chem Neurosci. 2015;6(9):1591–9.

Dhar TGM, Borden LA, Tyagarajan S, Smith KE, Branchek TA, Weinshank RL, et al. Design, synthesis and evaluation of substituted triarylnipecotic acid-derivatives as Gaba uptake inhibitors – identification of a ligand with moderate affinity and selectivity for the cloned human Gaba transporter Gat-3. J Med Chem. 1994;37(15):2334–42.

Falch E, Perregaard J, FrLlund B, Søkilde B, Buur A, Hansen LM, et al. Selective inhibitors of glial GABA uptake: synthesis, absolute stereochemistry, and pharmacology of the enantiomers of 3-hydroxy-4-amino-4,5,6,7-tetrahydro-1,2-benzisoxazole (exo-THPO) and analogues. J Med Chem. 1999;42(26):5402–14.

Fjalland B. Inhibition by neuroleptics of uptake of 3H-GABA into rat brain synaptosomes. Acta Pharmacol Toxicol (Copenh). 1978;42(1):73–6.

Fülep GH, Hoesl CE, Hofner G, Wanner KT. New highly potent GABA uptake inhibitors selective for GAT-1 and GAT-3 derived from (R)- and (S)-proline and homologous pyrrolidine-2-alkanoic acids. Eur J Med Chem. 2006;41(7):809–24.

Hertz E, Yu ACH, Hertz L, Juurlink BHJ, Schousboe A. Preparation of primary cultures of mouse cortical neurons. In: Shahar A, de Vellis J, Vernadakis A, Haber B, editors. A dissection and tissue culture manual of the nervous system, vol. 40. New York: Alan R. Liss, Inc.; 1989a. p. 183–6.

Hertz L, Juurlink BHJ, Hertz E, Fosmark H, Schousboe A. Preparation of primary cultures of mouse (rat) astrocyte. In: Shahar A, de Vellis J, Vernadakis A, Haber B, editors. A dissection and tissue culture manual of the nervous system, vol. 21. New York: Alan R. Liss, Inc.; 1989b. p. 105–8.

Jurik A, Zdrazil B, Holy M, Stockner T, Sitte HH, Ecker GF. A binding mode hypothesis of tiagabine confirms liothyronine effect on gamma-aminobutyric acid transporter 1 (GAT1). J Med Chem. 2015;58(5):2149–58.

Kalviainen R. Long-term safety of tiagabine. Epilepsia. 2001;42(Suppl 3):46–8.

Kanner BI. Transmembrane domain I of the gamma-aminobutyric acid transporter GAT-1 plays a crucial role in the transition between cation leak and transport modes. J Biol Chem. 2003;278(6):3705–12.

Kobayashi T, Suemasa A, Igawa A, Ide S, Fukuda H, Abe H, et al. Conformationally restricted GABA with bicyclo[3.1.0]hexane backbone as the first highly selective BGT-1 inhibitor. ACS Med Chem Lett. 2014;5(8):889–93.

Kragholm B, Kvist T, Madsen KK, Jorgensen L, Vogensen SB, Schousboe A, et al. Discovery of a subtype selective inhibitor of the human betaine/GABA transporter 1 (BGT-1) with a non-competitive pharmacological profile. Biochem Pharmacol. 2013;86(4):521–8.

Kragler A, Hofner G, Wanner KT. Novel parent structures for inhibitors of the murine GABA transporters mGAT3 and mGAT4. Eur J Pharmacol. 2005;519(1–2):43–7.

Kristensen AS, Andersen J, Jorgensen TN, Sorensen L, Eriksen J, Loland CJ, et al. SLC6 neurotransmitter transporters: structure, function, and regulation. Pharmacol Rev. 2011;63(3):585–640.

Krogsgaard-Larsen P. Inhibitors of the GABA uptake systems. Mol Cell Biochem. 1980;31(2):105–21.

Krogsgaard-Larsen P, Johnston GA. Inhibition of GABA uptake in rat brain slices by nipecotic acid, various isoxazoles and related compounds. J Neurochem. 1975;25(6):797–802.

Krogsgaard-Larsen P, Brehm L, Schaumburg K. Muscimol, a psychoactive constituent of Amanita muscaria, as a medicinal chemical model structure. Acta Chem Scand B. 1981;35(5):311–24.

Krogsgaard-Larsen P, Falch E, Larsson OM, Schousboe A. GABA uptake inhibitors: relevance to antiepileptic drug research. Epilepsy Res. 1987;1(2):77–93.

Kvist T, Christiansen B, Jensen AA, Brauner-Osborne H. The four human gamma-aminobutyric acid (GABA) transporters: pharmacological characterization and validation of a highly efficient screening assay. Comb Chem High Throughput Screen. 2009;12(3):241–9.

Kwan P, Sills GJ, Brodie MJ. The mechanisms of action of commonly used antiepileptic drugs. Pharmacol Ther. 2001;90(1):21–34.

Lie MEK, Al-Khawaja A, Damgaard M, Haugaard AS, Schousboe A, Clarkson AN, Wellendorph P. Glial GABA transporters as modulators of inhibitory signalling in epilepsy and stroke. Adv Neurobiol. 2017; 16, in press.

Liu QR, Lopez-Corcuera B, Mandiyan S, Nelson H, Nelson N. Molecular characterization of four pharmacologically distinct gamma-aminobutyric acid transporters in mouse brain [corrected]. J Biol Chem. 1993;268(3):2106–12.

Madsen KK, White HS, Schousboe A. Neuronal and non-neuronal GABA transporters as targets for antiepileptic drugs. Pharmacol Ther. 2010;125(3):394–401.

Melone M, Barbaresi P, Fattorini G, Conti F. Neuronal localization of the GABA transporter GAT-3 in human cerebral cortex: a procedural artifact? J Chem Neuroanat. 2005;30(1):45–54.

Nakada K, Yoshikawa M, Ide S, Suemasa A, Kawamura S, Kobayashi T, et al. Cyclopropane-based conformational restriction of GABA by a stereochemical diversity-oriented strategy: identification of an efficient lead for potent inhibitors of GABA transports. Bioorg Med Chem. 2013;21(17):4938–50.

Pabel J, Faust M, Prehn C, Wörlein B, Allmendinger L, Höfner G, et al. Development of an (S)-1-{2-[Tris(4-methoxyphenyl)methoxy]ethyl}piperidine-3-carboxylic acid [(S)-SNAP-5114] carba analogue inhibitor for murine γ-aminobutyric acid transporter type 4. ChemMedChem. 2012;7(7):1245–55.

Penmatsa A, Wang KH, Gouaux E. X-ray structure of dopamine transporter elucidates antidepressant mechanism. Nature. 2013;503(7474):85–90.

Petersen JG, Sorensen T, Damgaard M, Nielsen B, Jensen AA, Balle T, et al. Synthesis and pharmacological evaluation of 6-aminonicotinic acid analogues as novel GABA(A) receptor agonists. Eur J Med Chem. 2014;84:404–16.

Petrera M, Wein T, Allmendinger L, Sindelar M, Pabel J, Hofner G, et al. Development of highly potent GAT1 inhibitors: synthesis of nipecotic acid derivatives by Suzuki-Miyaura cross-coupling reactions. ChemMedChem. 2016;11(5):519–38.

Schaffert ES, Hofner G, Wanner KT. Aminomethyltetrazoles as potential inhibitors of the gamma-aminobutyric acid transporters mGAT1-mGAT4: synthesis and biological evaluation. Bioorg Med Chem. 2011;19(21):6492–504.

Schlessinger A, Wittwer MB, Dahlin A, Khuri N, Bonomi M, Fan H, et al. High selectivity of the gamma-aminobutyric acid transporter 2 (GAT-2, SLC6A13) revealed by structure-based approach. J Biol Chem. 2012;287(45):37745–56.

Schmitt S, Hofner G, Wanner KT. Application of MS transport assays to the four human gamma-aminobutyric acid transporters. ChemMedChem. 2015;10(9):1498–510.

Schousboe A, Thorbek P, Hertz L, Krogsgaard-Larsen P. Effects of GABA analogues of restricted conformation on GABA transport in astrocytes and brain cortex slices and on GABA receptor binding. J Neurochem. 1979;33(1):181–9.

Schousboe A, Larsson OM, Hertz L, Krogsgaard-Larsen P. Heterocyclic GABA analogues as selective inhibitors of astroglial GABA uptake. Adv Biochem Psychopharmacol. 1981;29:135–41.

Schousboe A, Larsson OM, Wood JD, Krogsgaard-Larsen P. Transport and metabolism of gamma-aminobutyric acid in neurons and glia: implications for epilepsy. Epilepsia. 1983;24(5):531–8.

Schousboe A, Wellendorph P, Frølund B, Clausen RP, Krogsgaard-Larsen P. Astrocytic GABA-transporters: pharmacological properties and targets for antiepileptic drugs. Adv Nerobiol. 2017; 16, in press.

Skovstrup S, Taboureau O, Brauner-Osborne H, Jorgensen FS. Homology modelling of the GABA transporter and analysis of tiagabine binding. ChemMedChem. 2010;5(7):986–1000.

Thomsen C, Sorensen PO, Egebjerg J. 1-(3-(9H-carbazol-9-yl)-1-propyl)-4-(2-methoxyphenyl)-4-piperidinol, a novel subtype selective inhibitor of the mouse type II GABA-transporter. Br J Pharmacol. 1997;120(6):983–5.

Treiman DM. GABAergic mechanisms in epilepsy. Epilepsia. 2001;42(Suppl 3):8–12.

Vogensen SB, Jorgensen L, Madsen KK, Borkar N, Wellendorph P, Skovgaard-Petersen J, et al. Selective mGAT2 (BGT-1) GABA uptake inhibitors: design, synthesis, and pharmacological characterization. J Med Chem. 2013;56(5):2160–4.

Vogensen SB, Jorgensen L, Madsen KK, Jurik A, Borkar N, Rosatelli E, et al. Structure activity relationship of selective GABA uptake inhibitors. Bioorg Med Chem. 2015;23(10):2480–8.

Wein T, Wanner KT. Generation of a 3D model for human GABA transporter hGAT-1 using molecular modeling and investigation of the binding of GABA. J Mol Model. 2010;16(1):155–61.

Wein T, Petrera M, Allmendinger L, Hofner G, Pabel J, Wanner KT. Different binding modes of small and large binders of GAT1. ChemMedChem. 2016;11(5):509–18.

White HS, Sarup A, Bolvig T, Kristensen AS, Petersen G, Nelson N, et al. Correlation between anticonvulsant activity and inhibitory action on glial gamma-aminobutyric acid uptake of the highly selective mouse gamma-aminobutyric acid transporter 1 inhibitor 3-hydroxy-4-amino-4,5,6,7-tetrahydro-1,2-benzisoxazole and its N-alkylated analogs. J Pharmacol Exp Ther. 2002;302(2):636–44.

White HS, Watson WP, Hansen SL, Slough S, Perregaard J, Sarup A, et al. First demonstration of a functional role for central nervous system betaine/{gamma}-aminobutyric acid transporter (mGAT2) based on synergistic anticonvulsant action among inhibitors of mGAT1 and mGAT2. J Pharmacol Exp Ther. 2005;312(2):866–74.

Yamashita A, Singh SK, Kawate T, Jin Y, Gouaux E. Crystal structure of a bacterial homologue of Na+/Cl--dependent neurotransmitter transporters. Nature. 2005;437(7056):215–23.

Index

A. Ortega, A. Schousboe (eds.), *Glial Amino Acid Transporters*, Advances in Neurobiology 16, DOI 10.1007/978-3-319-55769-4

U

V

X

Y

Z

FSC
www.fsc.org
MIX
Papier aus verantwortungsvollen Quellen
Paper from responsible sources
FSC® C105338

If you have any concerns about our products,
you can contact us on
ProductSafety@springernature.com

In case Publisher is established outside the EU,
the EU authorized representative is:
Springer Nature Customer Service Center GmbH
Europaplatz 3, 69115 Heidelberg, Germany

Printed by Libri Plureos GmbH
in Hamburg, Germany